AF410809

Ecology and Diversity of Marine Decapods

Ecology and Diversity of Marine Decapods

Editors

Patricia Briones-Fourzán
Michel E. Hendrickx

MDPI • Basel • Beijing • Wuhan • Barcelona • Belgrade • Manchester • Tokyo • Cluj • Tianjin

Editors
Patricia Briones-Fourzán
Universidad Nacional Autónoma de México
México

Michel E. Hendrickx
Universidad Nacional Autónoma de México
México

Editorial Office
MDPI
St. Alban-Anlage 66
4052 Basel, Switzerland

This is a reprint of articles from the Special Issue published online in the open access journal *Diversity* (ISSN 1424-2818) (available at: https://www.mdpi.com/journal/diversity/special_issues/Marine_Decapods).

For citation purposes, cite each article independently as indicated on the article page online and as indicated below:

LastName, A.A.; LastName, B.B.; LastName, C.C. Article Title. *Journal Name* **Year**, *Volume Number*, Page Range.

ISBN 978-3-0365-6041-0 (Hbk)
ISBN 978-3-0365-6042-7 (PDF)

Cover image courtesy of Fernando Negrete Soto.

Contents

About the Editors

Patricia Briones-Fourzán

Patricia Briones-Fourzán is Senior Scientist at the Reef Systems Academic Unit (Institute of Marine Science and Limnology, UNAM) in Puerto Morelos, Mexico, where she founded the Crustacean Ecology Lab. She is a marine biologist specializing in marine crustacean biology, ecology, and behavior. She works with spiny lobsters, both from the eastern Pacific and Caribbean coasts, deep-sea crustaceans, and crustacean communities associated with coral reefs and seagrasses. She obtained her PhD from National Autonomous University of Mexico (UNAM) and has been visiting scientist in Australia, Spain, the USA, and Costa Rica. She has published 115 papers and book chapters and 20 outreach articles. She has supervised 30 graduate and postgraduate students, and lectures in postgraduate programs since 1994. She was head of the Reef Systems Academic Unit from 1995 to 2001 and has served in numerous committees inside and outside UNAM. She has been a member of the editorial team of several scientific journals. She is a member of Mexico's National System of Researchers since 1990 and is currently National Researcher Level III. She has maintained the highest level (D) in UNAM's Performance Program for Academics since 2005. She obtained an Honorary Mention in the National Oceanographic Award 1994 contest, and was awarded UNAM's "Sor Juana Inés de la Cruz Medal for Outstanding University Women" in 2005.

Michel E. Hendrickx

Michel E. Hendrickx is Senior Scientist at the Mazatlán Academic Unit (Instituto de Ciencias del Mar y Limnología (ICML), Universidad Nacional Autónoma de México (UNAM), in Mazatlán, Mexico. He specializes in marine sciences, with a focus on Marine Biology, and is currently studying the invertebrate fauna of western Mexico and the tropical eastern Pacific. In recent years, his particular interest has been in the deep-sea fauna, mostly including mollusks, crustaceans and echinoderms. Having graduated from the Free University of Brussels, Belgium, he is currently a Member Emeritus of the National System of Researchers (SNI, Mexico, 2017 to date), and has been a member of the Mexican Academy of Sciences since 1993. He is also a Scientific Collaborator (1989 to date) of the Royal Institute of Natural Sciences, Belgium. He was awarded a scientific recognition by the Brazilian Society of Carcinology in 2006 and an "Excellence in Research Award of The Crustacean Society" in 2016. He was also awarded the "State Award of Investigation" by the State of Sinaloa, Mexico, in 2014. A member of the DGAPA-PRIDE program at UNAM with the highest level since its foundation, he is the founder of the Laboratory of Benthic Invertebrates and of the Regional Collection of Marine Invertebrates. ME Hendrickx has published over 390 contributions, alone or in collaboration with colleagues from Mexico and 14 other countries, including the description of 4 new genera and 55 new species. He has presented 164 contributions in national and international meetings. Under his direction, 42 students have obtained their graduation thesis and he has been coordinating a local course of Methods in Oceanography (postgraduate level) since 1998. He ie head of the Marine Station in Mazatlán in 1995-2000; he has also acted in numerous committees and commissions within UNAM or in other institutions.

Editorial

Ecology and Diversity of Marine Decapod Crustaceans

Patricia Briones-Fourzán [1,*] and Michel E. Hendrickx [2]

[1] Unidad Académica de Sistemas Arrecifales, Instituto de Ciencias del Mar y Limnología, Universidad Nacional Autónoma de México, Puerto Morelos 77580, Quintana Roo, Mexico
[2] Unidad Académica Mazatlán, Instituto de Ciencias del Mar y Limnología, Universidad Nacional Autónoma de México, Mazatlán 82000, Sinaloa, Mexico; michel@ola.icmyl.unam.mx
* Correspondence: briones@cmarl.unam.mx

Citation: Briones-Fourzán, P.; Hendrickx, M.E. Ecology and Diversity of Marine Decapod Crustaceans. *Diversity* 2022, 14, 614. https://doi.org/10.3390/d14080614

Received: 21 July 2022
Accepted: 26 July 2022
Published: 29 July 2022

Publisher's Note: MDPI stays neutral with regard to jurisdictional claims in published maps and institutional affiliations.

Decapods are one of the most diverse crustacean orders, with around 17,500 extant species [1] of those morphotypes that are most easily identifiable as crustaceans, shrimps, crayfish, lobsters, hermit crabs, and crabs [2]. Although decapods include marine, freshwater and semiterrestrial species, the vast majority of species in this group are marine species [3]. Marine decapods occur in shallow and deep water, including the entire water column (pelagos), and the benthic ecosystems (benthos) over both hard and soft bottoms, including coral reefs, kelp forests, seagrass meadows, macroalgal beds, and hydrothermal vents, wherein they have important regulatory functions (e.g., [4–7]). The feeding habits in decapods include herbivores, detritivores, carnivores, and omnivores. They are consumed by a vast array of other higher-order consumers, thus constituting critical links in many food webs. Decapods have complex behaviors, many of which are mediated by chemical communication [8], and they are often involved in symbiotic relationships with members of many other phyla [9,10]. Many marine decapods sustain important fisheries [11], but their diversity and the ecological roles they play are far from being well understood [12]. Decapods exhibit a diversity of reproduction systems [13], and most species pass through a pelagic larval phase [3]. Therefore, in many species, population connectivity depends on pelagic larvae that develop as part of the meroplankton for weeks or even months, thus also playing a significant role in planktonic ecology.

The purpose of this Special Issue was to invite researchers from around the world to share part of their knowledge on any aspect of diversity and ecology of decapods, whether pelagic or benthic, from any type of marine environment. Manuscripts were received from the Americas, Europe, Asia, and Oceania, with contributing authors from numerous institutions in the USA, Mexico, Brazil, Chile, the Netherlands, Italy, Greece, Pakistan, China, Japan, and New Zealand. The 12 contributions published in this Special Issue address a variety of topics including integrative taxonomy and genetic diversity, DNA barcoding to match larvae to adults, predator-prey interaction, coral-crab symbiosis, Sargassum-shrimp symbiosis, population dynamics of pelagic shrimps, diversity and distribution of oceanic larvae, spatial distribution of crabs, biodiversity of lobsters, and ecology of cave decapods.

The taxonomy of decapods is challenging, but the increasing use of integrative taxonomy is providing new insight into phylogenetic relationships at different taxonomic levels [14,15]. For example, Nishikawa et al. [16] used molecular and morphological analyses to evaluate the variability of the hermit crab *Clibanarius antillensis*. They found no apparent population structure of this species despite its very broad distribution. The authors offer plausible explanations for their results, such as the dispersive potential of the species and the absence of barriers that could prevent gene flow. They also provide a redescription of the species.

Using publicly available mitochondrial and nuclear markers, Sultana et al. [17] evaluated the phylogenetic relationships and genetic diversity among species of the hermit crab genus *Pagurus*, which occurs in a wide variety of marine habitats throughout the world.

The authors established that "Provenzanoi" was the basal group of the genus. They also established several monophyletic species clusters corresponding to previously established morphology-based species groups and resolved the taxonomic position of a number of recently described species. Their study increases insight into the evolutionary relationships among the species within this genus.

Schnabel et al. [18] used a combination of morphological and molecular tools to review the New Zealand fauna of coral and sponge shrimps (Infraorder Stenopodidea). In their extensive and detailed work, these authors described three new species, reported another one in their region for the first time, synonymized two species, and reviewed the distribution range of another one. With this study, the species of Stenopodidea in the New Zealand region increased from three to seven.

As previously noted, most decapods have planktonic larvae, and the larvae of many species are morphologically very different from the adults, making their identification difficult. Varela and Bracken-Grisson [19] used DNA barcoding combined with morphological methods to match larval stages from the northern Gulf of Mexico and adjacent waters with their adult counterparts. They were able to identify 12 unknown larval and two juvenile species from the infraorder Caridea and the suborder Dendrobranchiata, providing taxonomic descriptions and illustrations as well as reviewing the state of knowledge for their respective families.

Phyllosomata, the pelagic larvae of Achelata lobsters, have a long larval duration encompassing multiple stages, but little is known about the mid- to late stages. Muñoz-de-Cote et al. [20] sampled and examined the diversity and distribution of mid- and late-stage phyllosomata in oceanic waters of the Mexican Caribbean during two seasons. They obtained thousands of larvae. Palinurids (five species) outnumbered scyllarids (three species), with *Panulirus argus* dominating over the other species. Overall low densities, lack of a clear spatial pattern, and overlap of the phyllosomata assemblage composition between seasons suggests extensive mixing of the organisms entrained in the strong Yucatan Current.

In addition to the larval stages of most decapods, some species are also pelagic in their adult phase. Sanvicente-Añorve et al. [21] studied the population ecology of two coexisting species of pelagic decapods: the luciferid shrimps *Belzebub faxoni* and *Lucifer typus* in the western Gulf of Mexico. They concluded that *B. faxoni* was far more abundant than *L. typus* and mainly occurred over the inner shelf, where food availability was higher, whereas *L. typus* was more abundant over the slope and oceanic waters, avoiding low salinity waters. The overall distribution pattern of both species could be the result of a long competition process causing partial resource partitioning. The authors also found differences in size and reproductive ecology between the two species.

At the population level, de los Ríos et al. [22] determined the spatial distribution patterns of the grapsoid crab *Cyclograpsus cinereus* at different sites along rocky shores of northern Chile, by counting individuals in random quadrats on the intertidal zone. They found that the negative binomial distribution was the best fit to the data in 15 out of 19 sampling events, and the positive binomial in the remaining four. Their results are consistent with previously reported models for the distribution of decapods on the rocky shores of central and southern Chile.

Most decapods are nocturnal and it is difficult to observe their behavior. Muller et al. [23] reported on a batwing coral crab, *Carpilius corallinus* preying on two individuals of the Christmas tree worm *Spirobranchus giganteus* in Bonaire and described the entire process in detail, even providing a short video. This study brought new light to the little-known predators of Christmas tree worms and their behavior, and on the diet of the Batwing coral crab.

Some decapods establish symbiotic relationships, and factors determining the maintenance of those relationships are often unknown. The effect of chemical cues on habitat choice by two species of *Sargassum*-associated shrimps, *Latreutes fucorum* and *Leander tenuicornis*, was tested by Frahm and Brooks [24]. Neither species showed a strong directional response to *Sargassum* cues, dimethylsulfoniopropionate (DMSP, a chemical excreted by the algae), or conspecific chemical cues, but DMSP cues did cause more shrimp to exhibit searching behavior. Also, males and females responded differently to each cue. The authors suggest that, in the absence of visual cues (previously found to be important for these shrimps), shrimps can detect chemical cues, which could affect both initiating and maintaining shrimp/algal symbiosis.

Canizales-Flores et al. [25] examined the time and conditions when coral symbiont *Trapezia* crabs recruit onto previously unrecruited fragments of *Pocillopora* coral attached to the substrate. They found a relationship between the space available (coral volume) and crab recruitment, given that an increase in substrate complexity is required to provide protection for the crabs and hence maintain the symbiosis. In contrast, abiotic conditions did not appear to influence recruitment. They also found that crabs can move among colonies, which is counter to the theory that, once recruited, these crabs become obligate residents on that specific colony.

Lobsters are important fishing resources wherever they occur. Cruz et al. [26] reviewed the biodiversity and distribution of lobsters in Brazil, with emphasis on fisheries aspects. They listed 24 species from five families: Palinuridae, Scyllaridae, Nephropidae, Enoplometopidae, and Polychelidae, ranging in maximum total size from 30 to 620 mm, with palinurids being the most important from the fisheries viewpoint. Based on available evidence of distribution, biodiversity, life cycle, connectivity, and abundance, these authors proposed a simplified theoretical scheme about the role of lobsters in the ecosystem and their interactions with species from other trophic levels.

Bianchi et al. [27] provide a review of the decapod fauna of Mediterranean marine caves on the basis of a dataset of 76 species from nine Infraorders recorded in 133 caves. The greatest number of species has been recorded in the northern Mediterranean. Most species were found in only a few marine caves, and the proportion of endemic species in caves is low. Decapod occurrence in caves is more correlated with the decrease in light intensity than with other factors that characterize the marine cave environment. These authors call attention to the dearth of knowledge on the population biology of cave decapods and on their ecological role.

The contributions constituting this Special Issue significantly increased knowledge concerning the biology, ecology, and diversity of shrimps, hermit crabs, true crabs, and lobsters. However, decapod populations throughout the world are threatened by climate change, habitat degradation and loss, invasive species, overfishing, and diseases (e.g., [28–30]). Consequently, there is much more to learn about decapods and we hope these contributions encourage other researchers, and especially students, to investigate these fascinating creatures and the roles they play in their ecosystems.

Acknowledgments: We wish to thank all of the authors who contributed to this special issue, as well as the reviewers who kindly shared their expertise and time to improve the quality of the contributions.

Conflicts of Interest: The authors declare no conflict of interest.

References

1. Davis, K.E.; de Grave, S.; Delmer, C.; Payne, A.R.D.; Mitchell, S.; Wills, M.A. Ecological transitions and the shape of the decapod tree of life. *Integr. Comp. Biol.* 2022; *in press*. [CrossRef] [PubMed]
2. Álvarez, F.; Villalobos, J.L.; Hendrickx, M.E.; Escobar-Briones, E.; Rodríguez-Almaraz, G.; Campos, E. Biodiversidad de crustáceos decápodos (Crustacea: Decapoda) en México. *Rev. Mex. Biodiv.* **2014**, *85* (Suppl. 1), S208–S219. [CrossRef]
3. Anger, K. *The Biology of Decapod Crustacean Larvae*; Balkema Publishers: Lisse, The Netherlands, 2001; Volume 14.
4. Briones-Fourzán, P.; Monroy-Velázquez, L.V.; Estrada-Olivo, J.; Lozano-Álvarez, E. Diversity of seagrass-associated decapod crustaceans in a tropical reef lagoon prior to large environmental changes: A baseline study. *Diversity* **2020**, *12*, 205. [CrossRef]
5. Pohle, G.; Iken, K.; Clarke, K.R.; Trott, T.; Konar, B.; Cruz-Motta, J.J.; Wong, M.; Benedetti-Cecchi, L.; Mead, A.; Miloslavich, P.; et al. Aspects of benthic decapod diversity and distribution from rocky nearshore habitat at geographically widely dispersed sites. *PLoS ONE* **2011**, *6*, e18606. [CrossRef] [PubMed]

6. Martin, J.W.; Haney, T.A. Decapod crustaceans from hydrothermal vents and cold seeps: A review through 2005. *Zool. J. Linn. Soc.* **2005**, *145*, 445–522. [CrossRef]

7. Hendrickx, M.E. Crustáceos decápodos (Arthropoda: Crustacea: Decapoda) de aguas profundas del Pacífico mexicano: Lista de especies y material recolectado durante el proyecto TALUD. In *Biodiversidad y Comunidades del Talud Continental del Pacífico Mexicano*; Zamorano, P., Hendrickx, M.E., Caso, M., Eds.; Secretaría de Medio Ambiente y Recursos Naturales e Instituto Nacional de Ecología: Ciudad de México, México, 2012; pp. 283–317.

8. Breithaupt, T.; Thiel, M. (Eds.) *Chemical Communication in Crustaceans*; Springer: Berlin/Heidelberg, Germany, 2011.

9. Briones-Fourzán, P.; Pérez-Ortiz, M.; Negrete-Soto, F.; Barradas-Ortiz, C.; Lozano-Álvarez, E. Ecological traits of Caribbean sea anemones and symbiotic crustaceans. *Mar. Ecol. Prog. Ser.* **2012**, *470*, 55–68. [CrossRef]

10. Salas-Moya, C.; Vargas-Castillo, R.; Alvarado, J.J.; Azofeifa-Solano, J.C.; Cortés, J. Decapod crustaceans associated with macroin-vertebrates in Pacific Costa Rica. *Mar. Biodiv. Rec.* **2021**, *14*, 6. [CrossRef]

11. Boenish, R.; Kritzer, J.P.; Kleisner, K.; Steneck, R.S.; Werner, K.M.; Zhu, W.; Schram, F.; Rader, D.; Cheung, W.; Ingles, J.; et al. The global rise of crustacean fisheries. *Front. Ecol. Environ.* **2022**, *20*, 102–110. [CrossRef]

12. Boudreau, S.A.; Worm, B. Ecological role of large benthic decapods in marine ecosystems: A review. *Mar. Ecol. Prog. Ser.* **2012**, *469*, 195–213. [CrossRef]

13. Baeza, J.A. Sexual systems in shrimps (Infraorder Caridea Dana, 1852), with special reference to the historical origin and adaptive value of protandric simultaneous hermaphroditism. In *Transitions between Sexual Systems*; Leonard, J.L., Ed.; Springer Nature: Cham, Switzerland, 2018; pp. 269–310.

14. de Grave, S.; Pentcheff, N.D.; Ahyong, S.T.; Chan, T.Y.; Crandall, K.A.; Dworschak, P.C.; Felder, D.L.; Feldmann, R.M.; Fransen, C.H.J.M.; Goulding, L.Y.D.; et al. A classification of living and fossil genera of decapod crustaceans. *Raffles Bull. Zool.* **2009**, (Suppl. 21), 1–109.

15. Wolfe, J.M.; Breinholt, J.W.; Crandall, K.A.; Lemmon, A.R.; Lemmon, E.M.; Timm, L.E.; Siddall, M.E.; Bracken-Grissom, H.D. A phylogenomic framework, evolutionary timeline and genomic resources for comparative studies of decapod crustaceans. *Proc. R. Sci. B* **2019**, *286*, 20190079. [CrossRef]

16. Nishikawa, K.S.; Negri, M.; Mantelatto, F.L. Unexpected absence of population structure and high genetic diversity of the Western Atlantic hermit crab *Clibanarius antillensis* Stimpson, 1859 (Decapoda: Diogenidae) based on mitochondrial markers and morphological data. *Diversity* **2021**, *13*, 56. [CrossRef]

17. Sultana, Z.; Babarinde, I.A.; Asakura, A. Diversity and molecular phylogeny of pagurid hermit crabs (Anomura: Paguridae: *Pagurus*). *Diversity* **2022**, *14*, 141. [CrossRef]

18. Schnabel, K.E.; Kou, Q.; Xu, P. Integrative taxonomy of New Zealand Stenopodidea (Crustacea: Decapoda) with new species and records for the region. *Diversity* **2021**, *13*, 343. [CrossRef]

19. Varela, C.; Bracken-Grissom, H. A mysterious world revealed: Larval-adult matching of deep-sea shrimps from the Gulf of Mexico. *Diversity* **2021**, *13*, 457. [CrossRef]

20. Muñoz de Cote-Hernández, R.; Briones-Fourzán, P.; Barradas-Ortiz, C.; Negrete-Soto, F.; Lozano-Álvarez, E. Diversity and distribution of mid- to late-stage phyllosomata of spiny and slipper lobsters (Decapoda: Achelata) in the Mexican Caribbean. *Diversity* **2021**, *13*, 485. [CrossRef]

21. Sanvicente-Añorve, L.; Hernández-González, J.; Lemus-Santana, E.; Hermoso-Salazar, M.; Violante-Huerta, M. Population structure and seasonal variability of two luciferid species (Decapoda: Sergestoidea) in the Western Gulf of Mexico. *Diversity* **2021**, *13*, 301. [CrossRef]

22. de los Ríos-Escalante, P.D.; Esse, C.; Retamal, M.A.; Zúñiga, O.; Fajardo, M.; Ghory, F. Spatial distribution of *Cyclograpsus cinereus* Dana 1851 on the rocky shores of Antofagasta (23°27′ S, Chile). *Diversity* **2022**, *14*, 418. [CrossRef]

23. Muller, E.; de Gier, W.; ten Hove, H.A.; van Moorsel, G.W.N.M.; Hoeksema, B.W. Nocturnal predation of Christmas treeworms by a batwing coral crab at Bonaire (Southern Caribbean). *Diversity* **2020**, *12*, 455. [CrossRef]

24. Frahm, J.L.; Brooks, W.R. The use of chemical cues by *Sargassum* shrimps *Latreutes fucorum* and *Leander tenuicornis* in establishing and maintaining a symbiosis with the host *Sargassum* algae. *Diversity* **2021**, *13*, 305. [CrossRef]

25. Canizales-Flores, H.M.; Rodríguez-Troncoso, A.P.; Rodríguez-Zaragoza, F.A.; Cupul-Magaña, A.L. A long-term symbiotic relationship: Recruitment and fidelity of the crab *Trapezia* on its coral host *Pocillopora*. *Diversity* **2021**, *13*, 450. [CrossRef]

26. Cruz, R.; Torres, M.T.; Santana, J.V.M.; Cintra, I.H.A. Lobster distribution and biodiversity on the continental shelf of Brazil: A review. *Diversity* **2021**, *13*, 507. [CrossRef]

27. Bianchi, C.N.; Gerovasileiou, V.; Morri, C.; Froglia, C. Distribution and ecology of decapod crustaceans in Mediterranean marine caves: A review. *Diversity* **2022**, *14*, 176. [CrossRef]

28. Simões, M.V.P.; Saeedi, H.; Cobos, M.E.; Brandt, A. Environmental matching reveals non-uniform range-shift patterns in benthic marine Crustacea. *Clim. Chang.* **2021**, *168*, 31. [CrossRef]

29. Briones-Fourzán, P.; Álvarez-Filip, L.; Barradas-Ortiz, C.; Morillo-Velarde, M.P.; Negrete-Soto, F.; Segura-García, I.; Sánchez-González, A.; Lozano-Álvarez, E. Coral reef degradation differentially alters feeding ecology of co-occurring congeneric spiny lobsters. *Front. Mar. Sci.* **2019**, *5*, 516. [CrossRef]

30. Shields, J.D. Climate change enhances disease processes in crustaceans: Case studies in lobsters, crabs, and shrimps. *J. Crust. Biol.* **2019**, *39*, 673–683. [CrossRef]

 *diversity*

Article

Unexpected Absence of Population Structure and High Genetic Diversity of the Western Atlantic Hermit Crab *Clibanarius antillensis* Stimpson, 1859 (Decapoda: Diogenidae) Based on Mitochondrial Markers and Morphological Data

Keity S. Nishikawa, Mariana Negri and Fernando L. Mantelatto *

Laboratory of Bioecology and Crustacean Systematics (LBSC), Department of Biology, Faculty of Philosophy, Sciences and Letters at Ribeirão Preto (FFCLRP), University of São Paulo (USP), Av. Bandeirantes 3900, Ribeirão Preto 14040-901, São Paulo, Brazil; keity_nishikawa@hotmail.com (K.S.N.); ma_negri90@hotmail.com (M.N.)
* Correspondence: flmantel@usp.br

Citation: Nishikawa, K.S.; Negri, M.; Mantelatto, F.L. Unexpected Absence of Population Structure and High Genetic Diversity of the Western Atlantic Hermit Crab *Clibanarius antillensis* Stimpson, 1859 (Decapoda: Diogenidae) Based on Mitochondrial Markers and Morphological Data. *Diversity* **2021**, *13*, 56. https://doi.org/10.3390/d13020056

Academic Editors: Michael Wink, Patricia Briones-Fourzán and Michel E. Hendrickx

Received: 11 December 2020
Accepted: 27 January 2021
Published: 1 February 2021

Abstract: Recent studies on genetic variability have revealed different patterns of genetic structure among populations of marine decapod species with wide geographical distribution. The hermit crab *Clibanarius antillensis* has a broad distribution along the western Atlantic Ocean, from south Florida (United States) to Santa Catarina (Brazil). This factor, in addition to differences in larval morphology and in adult coloration, makes this species a good model for studies on intraspecific variations. Therefore, we evaluated the molecular and morphological variability of *C. antillensis* along its distribution in order to check the levels of population structure. The results were based on the morphological analyses of 187 individuals and 38 partial sequences of the mitochondrial gene 16S rRNA and 46 of cytochrome c oxidase subunit I (COI) from specimens whose locations covered the whole species distribution. The molecular analyses did not show any apparent population structure of *C. antillensis*. This result was corroborated by the morphological analyses since the characters analyzed did not show any pattern of variation. Our results may be explained by a set of factors, such as the dispersive potential of the species and the absence of barriers that could prevent gene flow. In addition, high genetic diversity was observed, mainly for COI, which may be explained by the historical processes of the species, which seem to be in almost constant expansion in the last 700,000 years and experienced no genetic bottleneck. Apparently, this species was little affected by the climate fluctuations of Pleistocene. Additionally, our morphological analyses allowed us to present herein a redescription of the studied species since we noted differences from the characters in the diagnosis.

Keywords: cytochrome c oxidase subunit I (COI); larval dispersal; mitochondrial genes; molecular data; 16S rRNA; redescription

1. Introduction

Species and their populations are constantly changing. Their history, as well as details from their current stage of genetic structure, are a combination of different past events [1], which may be understood by investigating their genetic processes [2]. Gene flow, for example, is essential to maintain genetic homogeneity or heterogeneity among populations of a species [3,4].

For most marine invertebrate species, planktonic larvae and their life span influence their dispersion process which allows them to interconnect populations by reaching long distances [5,6]. Therefore, long larval stages are usually related to high dispersal capacity and levels of gene flow and reduced population genetic structure [7–9]. However, some studies revealed that high levels of connectivity and genetic homogeneity were not necessarily related to the duration of planktonic stages [10–12]. In addition, gene flow may also be influenced by marine currents circulation, local oceanic conditions, physical barriers, food availability, ecological interactions, as well as past geological events and recent history [8,10,13–22].

Therefore, each marine species has its own patterns of gene flow and genetic differences along its distribution [14,23]. This occurs because each individual has a unique way to respond to different factors at specific moments [14]. Gene flow patterns may be revealed by studies on genetic variability of populations [24], which might show different levels of geographic structure and genetic diversity [25].

Many studies have revealed geographic structure on marine decapod crustaceans with wide distribution. As examples, the hermit crabs *Calcinus tibicen* Herbst, 1791 [26] and *Clibanarius vittatus* Bosc, 1802 [27] exhibited different patterns of population structure along their distribution in the western Atlantic Ocean.

The hermit crab *Clibanarius antillensis* Stimpson, 1859 (Figure 1) occurs in Bermuda, Florida (US), Gulf of Mexico, Belize, Costa Rica, Panama, Antilles, north of South America and Brazil (in Atol das Rocas and from the state of Piauí to Santa Catarina) (Figure 2) [28–32]. It is found in intertidal zones, shallow waters, over rocks, coral reefs, and banks of *Halodule* [30,33]. The species has a larval development of five to six stages that require at least 43 days to complete [34,35].

Figure 1. *Clibanarius antillensis* Stimpson, 1859. Preserved female specimen, CCDB 5061. Photo by Buranelli, R.C.

Figure 2. Distribution of *Clibanarius antillensis* Stimpson, 1859 (blue line). Colored dots indicate collecting sites where we sampled specimens whose partial sequences of 16S rRNA and cytochrome c oxidase subunit I (COI) were obtained. * indicates only sequences of 16S rRNA obtained. Abbreviations: TM: Tamaulipas; VE: Veracruz; TB: Tabasco; QR: Quintana Roo; PI: Piauí; CE: Ceará; RN: Rio Grande do Norte; PE: Pernambuco; AL: Alagoas; BA: Bahia; ES: Espírito Santo; RJ: Rio de Janeiro; SP: São Paulo; SC: Santa Catarina.

Some morphological differences were found between larvae from Brazil and Panama and Mexico populations, such as the number of antennular aesthetascs, the number of denticles of crista dentate of the third maxilliped, the development of the external lobe of the maxillule and endopod of the maxilla [34–36]. Additionally, distinct coloration patterns were found among adults from different localities [37].

Additional investigations on genetic variability may contribute to a better comprehension of biogeographic processes, population differentiation and biodiversity along groups/families of hermit crabs in the western Atlantic. Additionally, it may allow checking if there is an evolutive signal among them. Based on the reported scenario, the wide distribution of *C. antillensis*, its larval stage duration and the context previously described, this hermit crab is a suitable species for investigations on genetic variability and morphological analyses. Therefore, the aim of this study was to: (1) check the levels of population structure along *C. antillensis* distribution; (2) analyze morphological and molecular variations, and (3) analyze, preliminarily, demographic factors related to its current diversification pattern.

2. Materials and Methods

2.1. Sample Collection

Most individuals were obtained from the Crustacean Collection of the Department of Biology, University of São Paulo, Brazil (CCDB). In order to cover most part of the species distribution (Figure 2), we also analyzed specimens obtained by means of loans or donations from the following collections: University of Louisiana at Lafayette Zoological Collection, LA, USA (ULLZ—recently transferred to the National Museum of Natural History, Smithsonian Institution, Washington, DC, USA. USNM; both catalog numbers are used, as specimens are now permanently cross-referenced under both numbers at the USNM); Florida Museum of Natural History, University of Florida, FL, USA (UF); United States National Museum, Smithsonian National Museum of Natural History, Washington, DC, USA (USNM); American Museum of Natural History, NY, USA (AMNH); Natural History Museum of Los Angeles County, CA, USA (NHMLA); Colección Nacional de Crustáceos, Universidad Autónoma de Mexico, Mexico (CNCR). Before the analyses, the identification of specimens was confirmed based on previous morphological characters established in the literature [28,30,31,38,39].

2.2. DNA Extraction, Amplification and Sequencing

For DNA extraction, we used muscle tissue from pereiopods or abdomen and followed saline protocols described by Schubart et al. [40], with modifications from Mantelatto et al. [41], and Chelex® resin [42]. Some adaptations were made to suit our material.

The fragments were amplified by polymerase chain reaction (PCR) [43] in a Veriti 96-Well Thermal Cycler® (Applied Biosystems, Foster City, CA, USA). The molecular markers 16S rRNA and cytochrome c oxidase subunit I (COI) were chosen, since these mitochondrial genes have been widely used and effective on studies that contribute to our comprehension of Decapod diversity [40,41,44,45]. Of all primers used in this study (Table 1), we designed one pair of each marker in Primer-Blast (National Center for Biotechnology Information, Bethesda, MD, USA) [46]: 16SLClib and 16SHClib; COILClib and COIHClib, due to amplification difficulties. For this purpose, we based the design on the alignment of two 16S rRNA GenBank sequences (KF182529 and DQ369941) and new sequences of 16S rRNA and COI.

Table 1. Sequences of primers used for amplification of 16S rRNA and cytochrome c oxidase subunit I (COI) by means of PCR.

Gene	Primers	Sequence
16S rRNA	16SL2	5′-TGCCTGTTTATCAAAAACAT-3′ [40]
	16SH2	5′-AGATAGAAACCAACCTGG-3′ [40]
	16SLClib	5′-TTTGACCTGCCCACTGAA-3′ [Present study]
	16SHClib	5′-GAAACCAACCTGGCT CACG-3′ [Present study]
COI	COL6b	5′-ACAAATCATAAAGATATYGG-3′ [47]
	COL6b2	5′-ACWAAYCAYAAAGAYATYGG-3′ [48]
	COIAL2o	5′-ACGCAACGATGATTATTTTCTAC-3′ [48]
	COIAL1m	5′-GAGCTTGAGCYGGRATAGTAGG-3′ [48]
	COH6	5′-TADACTTCDGGRTGDCCAARAAYCA-3′ [47]
	COIAH2m	5′-GACCRAAAAATCARAATAAATGTTG-3′ [48]
	COIAH1m	5′-CTCCWGCRGGGTCAAAGAAAGA-3′ [48]
	COILClib	5′-GCGTGAGCAGGAATAGTAGGT T-3′ [Present study]
	COIHClib	5′-AAAACAGGGTCTCCTCCTC-3′ [Present study]

Each PCR was performed with 25 µL total volume, containing ultrapure water, betaine (5 M), DNTPs (10 mM), PCR *Buffer* (10×), MgCl$_2$ (25 mM), bovine serum albumin (BSA) 1% solution, primers (10 µM each), *Thermus aquaticus* (Taq) DNA polimerase (5 U/µL) and previously calculated extracted DNA. The thermal cycle consisted of: 16S rRNA—initial denaturing for 4 min at 95 °C; annealing for 40 cycles of 45 s at 95 °C, 45 s at 54 °C and 1 min at 72 °C; final extension for 6 min at 72 °C; COI—initial denaturing for 5 min at

95 °C; annealing for 40 cycles of 1 min at 95 °C, 1 min at 38–48 °C and 75 s at 72 °C; final extension for 6 min at 72 °C. PCR products were electrophoresed on 1.5% agarose gel for confirmation, purified using the SureClean Plus® kit (Bioline, Tauton, MA, USA), following the manufacturer's instructions, and sequenced with the ABI BigDye Terminator Mix (Applied Biosystems, Foster City, CA, USA) in an ABI 3730 XL DNA Analyzer (Applied Biosystems automated sequencer, Foster City, CA, USA), following the manufacturers' protocol.

The forward and reverse obtained sequences were edited and used to construct a consensus sequence in BioEdit 7.2.5 (Ibis Therapeutics, Carlsbad, CA, USA) [49]. The identity of the consensus was confirmed with BLAST (Basic Local Alignment Search Tool) [50] by comparisons to accessioned sequences of GenBank database. COI consensus were checked for the occurrence of pseudogenes at the online Translate tool on SIB ExPASy [51]. Multiple sequences were aligned for each gene using MUSCLE (Multiple Sequence Comparison by Log-Expectation, European Molecular Biology Laboratory–The European Bioinformatics Institute, Hinxton, UK) [52].

Besides the sequences we obtained, which were all submitted to GenBank (National Center for Biotechnology Information, Bethesda, MD, USA), we also included two 16S rRNA sequences of *C. antillensis* retrieved from GenBank (Table 2). For genetic distance and phylogenetic analyses, we added five 16S rRNA and 13 COI sequences of other species of the genus *Clibanarius* (Table 3); the following outgroup sequences, based on Bracken-Grissom et al. [53]: *Calcinus laevimanus* Randall, 1840 (GenBank: 16S rRNA–FJ620175; COI–FJ620271), *C. osbcurus* Stimpson, 1859 (GenBank: 16S rRNA–FJ620216; COI–FJ620314), *C. tibicen* (GenBank: 16S rRNA–FJ620220; COI–FJ620318), *Isocheles pilosus* Holmes, 1900 (GenBank: 16S rRNA–AF436057), *I. sawayai* Forest and Saint Laurent, 1968 (GenBank: 16S rRNA–DQ369938), and *I. wurdemanni* Stimpson, 1859 (GenBank: 16S rRNA–KF182530).

Table 2. Specimens of *Clibanarius antillensis* Stimpson, 1859 used in molecular analyses, sampling localities, museum catalog number, and GenBank accession numbers. New sequences are in bold. AMNH: American Museum of Natural History. CCDB: Crustacean Collection of the Department of Biology—Faculty of Philosophy, Sciences and Letters at Ribeirão Preto, University of São Paulo. CNCR: Colección Nacional de Crustáceos, Universidad Autónoma de Mexico. UF: Florida Museum of Natural History, University of Florida. ULLZ: University of Louisiana at Lafayette Zoological Collection. USNM: United States National Museum, Smithsonian National Museum of Natural History. (-): missing sequences.

Locality	Catalog Number	GenBank	
		16S rRNA	**COI**
Florida, United States of America	ULLZ 4683–USNM 1540491	DQ369941	–
Florida, United States of America	ULLZ 9433–USNM 1544313	KF182529 **MG264431** **MG264432** **MG264433**	– – **MG264468** –
Florida, United States of America	CCDB 6267	**MG264434**	**MG264469**
Andros Island, Bahamas	AMNH 18726	**MG264435**	**MG264470**
Barra del Tordo, Mexico	ULLZ 15019–USNM 1548156	**MG264436**	–
Veracruz, Mexico	CNCR 24702	**MG264438** **MG264439**	**MG264471** **MT740091**
Veracruz, Mexico	CNCR 22223	**MG264437**	**MG264472**
Tabasco, Mexico	CNCR 18624	**MG264440**	–
Quintana Roo, Mexico	CNCR 3729	**MG264441**	–
Carrie Bow Cay, Belize	USNM 1277880	**MG264442**	**MG264473**

Table 2. *Cont.*

Locality	Catalog Number	GenBank	
		16S rRNA	**COI**
Tortola Island, British Virgin Islands	USNM 1277883	MG264444	–
Saint Martin, French Antilles	UF 32041	MG264443	–
Grande-Terre, Guadeloupe	USNM 1277879	MG264445	–
Playa Puerto Viejo, Costa Rica	CCDB 550	MG264446 MG264447 –	MG264474 MG264475 MG264476
Bocas del Toro, Panama	CCDB 3578	MG264448 MG264449	MG264477 MG264478
Isla Margarita, Venezuela	CCDB 1810	MG264450 –	MG264479 MG264480
Luís Correia, Piauí, Brazil	CCDB 4158	MG264451 –	MG264481 MG264482
Trairi, Ceará, Brazil	CCDB 2651	MG264452 MG264453	MG264483 MG264484
Fortaleza, Ceará, Brazil	CCDB 4274	–	MG264485
Touros, Rio Grande do Norte, Brazil	CCDB 3366	MG264454	MG264488
Touros, Rio Grande do Norte, Brazil	CCDB 3367	–	MG264486
Touros, Rio Grande do Norte, Brazil	CCDB 3373	–	MG264487
Ipojuca, Pernambuco, Brazil	CCDB 1727	MG264455 MG264456 – – –	MG264489 MG264490 MG264491 MG264492 MG264493
Maragogi, Alagoas, Brazil	CCDB 4920	MG264457	MG264494 MG264495 MG264496
Ilhéus, Bahia, Brazil	CCDB 2597	–	MG264498
Ilhéus, Bahia, Brazil	CCDB 2610	–	MG264500
Porto Seguro, Bahia, Brazil	CCDB 585	MG264458 MG264459 –	MG264497 – MG264499
Guarapari, Espírito Santo, Brazil	CCDB 2243	MG264460 MG264461 –	MG264501 MG264502 MG264503
Búzios, Rio de Janeiro, Brazil	CCDB 497	MG264462	MG264504
Búzios, Rio de Janeiro, Brazil	CCDB 761	MG264463	–
Búzios, Rio de Janeiro, Brazil	CCDB 5656	– –	MG264505 MG264506
Ubatuba, São Paulo, Brazil	CCDB 2906	MG264464	MG264508
São Sebastião, São Paulo, Brazil	CCDB 5061	MG264465	–
São Sebastião, São Paulo, Brazil	CCDB 5062	– –	MG264507 MG264509
Itajaí, Santa Catarina, Brazil	CCDB 1876	MG264466 – –	MG264510 MG264511 MG264512

Table 3. Specimens of *Clibanarius* spp. used in molecular analyses, sampling locality, museum catalog number, and GenBank accession numbers. New sequence is in bold. CBM-ZC: Natural History Museum and Institute, Zoology Crustacea. CCDB: Crustacean Collection of the Department of Biology—Faculty of Philosophy, Sciences and Letters at Ribeirão Preto, University of São Paulo. ULLZ: University of Louisiana at Lafayette Zoological Collection. USNM: United States National Museum, Smithsonian National Museum of Natural History. (−): missing data.

Species	Locality	CatalogNumber	Gen Bank	
			16S rRNA	COI
Clibanarius albidigitus Nobili, 1901	Panama City, Panama	−	AF425323	−
	Punta Morales, Costa Rica	CCDB 1711	−	JN671591
Clibanarius clibanarius Herbst, 1791	−	−	-	JX676177
Clibanarius corallinus H. Milne Edwards, 1848	Tuamotus, French Polynesian	ULLZ 10121–USNM 1544831	KF182528	−
	Okinawa, Japan	CBM-ZC 9622	−	AB507374
Clibanarius erythropus Latreille, 1818	Cádiz, Spain	CCDB 488	−	JN671592
Clibanarius lineatus H. Milne Edwards, 1848	Porosi, Nicaragua	CCDB 2444	−	JN671594
Clibanarius longitarsus De Haan, 1849	Okinawa, Japan	CBM-ZC 9583	−	AB496944
Clibanarius rhabdodactylus Forest, 1953	Okinawa, Japan	CBM-ZC 9593	−	AB496946
Clibanarius sclopetarius Herbst, 1796	São Sebastião, SP, Brazil	CCDB 2961	JN671523	JN671584
Clibanarius signatus Heller, 1861	Iran	CCDB 3694	−	JN671590
Clibanarius symmetricus Randall, 1840	Paraty, RJ, Brazil	CCDB 2237	JN671529	JN671548
Clibanarius tricolor Gibbes, 1850	Quintana Roo, Mexico	CCDB 504	**MG264467**	JN671593
Clibanarius virescens Krauss, 1843	Okinawa, Japan	CBM-ZC 9587	−	AB496948
Clibanarius vittatus Bosc, 1802	Florida, United States of America	CCDB 3783	−	JX238506
	Texas, United States of America	CCDB 1185	JN671527	−

2.3. Genetic Distance Analyses

Genetic distances were calculated to determine intra and interspecific variation rates with the software MEGA 6.06 [54], using the Kimura 2-parameters substitution model [55]. Two genetic distances histograms were constructed in Microsoft Excel 2010, with interval ranges of 0.2%.

2.4. Phylogenetic Analyses

Maximum likelihood (ML) analyses [56] were conducted in RAxML—HPC Black Box 8.2.4 (Randomized Axelerated Maximum Likelihood, Heidelberg Institute for Theoretical Studies, Heidelberg, Germany) [57], implemented at the online platform Cyber Infrastructure for Phylogenetic Research (CIPRES). We used the default parameters for RAxML and the evolution model GTR + Γ + I [General Time Reversible [58] + Gama + Invariables sites] and the consistency of the topologies was measured by bootstrap method (1000 replicates). The topologies were visualized and edited using FigTree 1.4.2 (University of Edinburgh, Edinburgh, UK) [59]; only values >50% were reported.

2.5. Genetic Variability Analyses

The genetic variability analyses were conducted for both 16S rRNA and COI. The genetic diversity indexes, such as number of haplotypes (H), haplotype diversity (Hd), nucleotide diversity (π) and average number of nucleotide differences (K), were calculated in DnaSP 5.10.1 [60]. Haplotype networks were constructed using statistical parsimony method with TCS 1.21 [61]. In case of ambiguous connections, the criteria proposed by Excoffier and Langaney [62] were considered. Analyses of Molecular Variance (AMOVA) were conducted using the software Arlequin 3.5.2.2 (University of Bern, Bern, Switzerland) [63] to calculate the variance within and between previously established groups and the fixation index values (FST).

2.6. Demographic Analyses

Demographic analyses were conducted for both 16S rRNA and COI. Demographic history was inferred by the neutrality tests Tajimas' D [64] and Fu's Fs [65] using Arlequin 3.5.2.2 (University of Bern, Bern, Switzerland) [63]. In addition, pairwise mismatch distribution were analyzed to test population expansion [66]. The graphic was created in DnaSp 5.10.1 [60] and the sum of squared deviations (SSD) [67] and Harpending raggedness index (HRI) [68] were calculated using Arlequin 3.5.2.2 [63].

The Bayesian skyline plot (BSP) [69] analyses was conduct only for COI and it was used to infer the demographic history of the species under coalescent model. First, the substitution model HKY + I + G [Hasegawa-Kishino-Yano [70] + Invariable sites + Gama] was selected using jModelTest 2.1.10 (Free Software Foundation, Inc., Boston, MA, USA) [71] with Bayesian information criterion (BIC). Afterwards, some parameters were selected in BEAUti (Bayesian Evolutionary Analysis Utility, University of Auckland, Auckland, New Zealand) to create the input file in BEAST 1.8.4 (Bayesian Evolutionary Analysis Sampling Trees, University of Auckland, Auckland, New Zealand). The divergence rate was 1.4% per million years [72], the number of Markov chain Monte Carlo interactions was 10 million, at every 1000 chains, with a 10% burn-in. Then, the output was analyzed using Tracer [69], and a graphic was created.

2.7. Morphological Assessment

Morphological data was accessed to compare specimens of *C. antillensis* from different localities. We adopted all diagnostic characters found in the taxonomic literature [28,30,31,38,39]. Therefore, we measured length of shield (sl), rostrum, lateral projections, left ocular peduncle, right chelae, dactyl, propodus, carpus, merus and ischium of the left second pereiopod; width of front and right chelae. We also analyzed shape and disposition of tufts of setae of shield; shape of rostrum, front and telson lobes; shape and number of spines of ocular acicle; number and disposition of spines of antennal acicle; number and disposition of spines and tufts of setae of right cheliped; coloration and number and disposition of spines of second and third pair of pereiopods. A redescription of the species was made, since we noted differences between some characters observed in this study in comparison to literature descriptions.

3. Results

3.1. Genetic Distance Analyses

The automated sequencing protocols to obtain two fragments of mitochondrial genes resulted in ~1170 base pairs (bp). The alignment of 16S rRNA with 530 bp included 38 sequences of *C. antillensis* and 12 sequences from other species of Diogenidae. The intraspecific divergence for *C. antillensis* varied from 0–0.99%, whereas interspecific values ranged from 1.48–24.98%, with the first value corresponding to the divergence between sequences of *C. vittatus* and *C. symmetricus* (Figure 3a). An interspecific gap was not evident for this marker. The alignment of COI with 640 bp included 46 sequences of *C. antillensis* and 16 sequences from other species of Diogenidae. In this case, the interspecific gap was evident, since the intraspecific divergence for *C. antillensis* varied from 0–2.90% and the interspecific values ranged from 5.80–22.80% (Figure 3b).

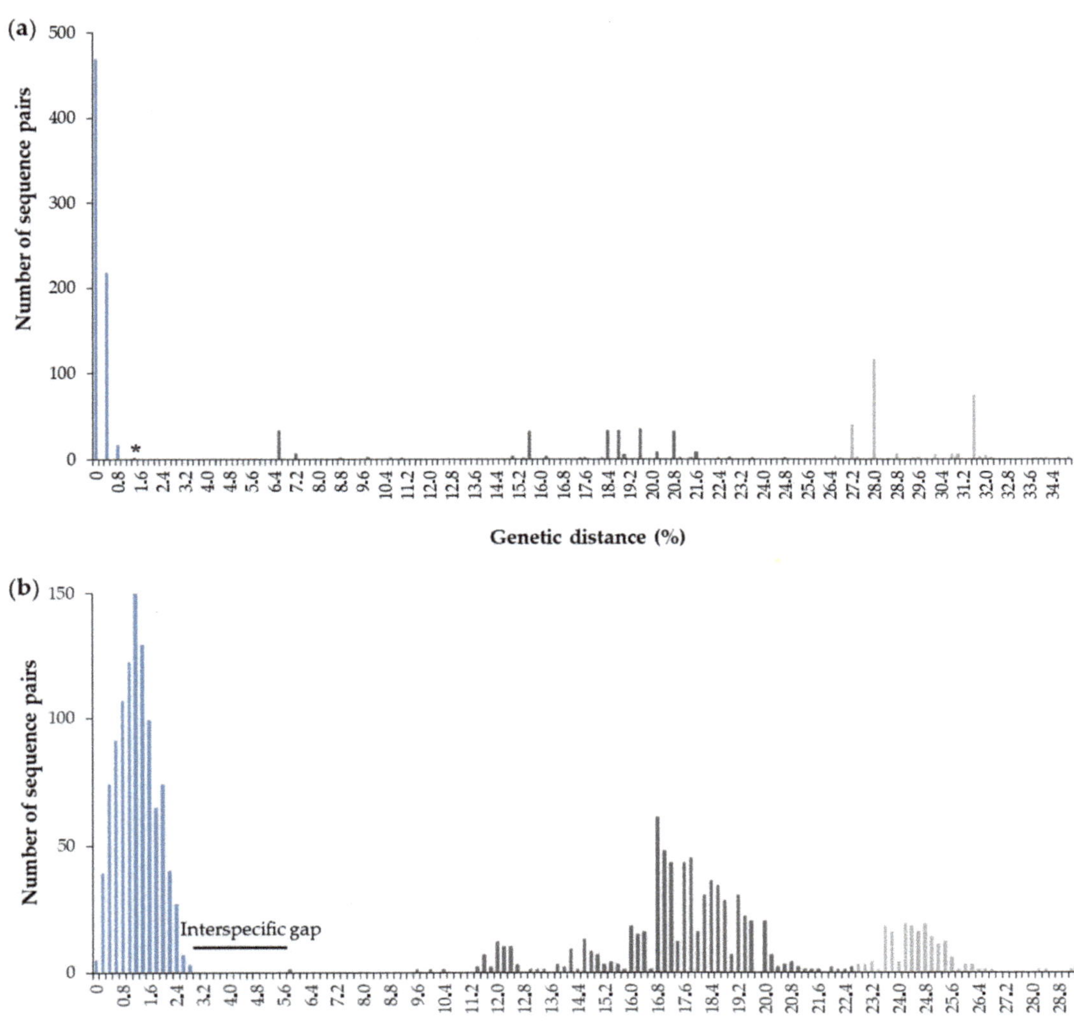

Figure 3. Histograms of Kimura 2-parameters genetic distances calculated for each pair of sequences of *Clibanarius antillensis* Stimpson, 1959 and other species of Diogenidae for 16S rRNA (**a**) and cytochrome c oxidase subunit I (COI) (**b**). * indicates the divergence between sequences of *Clibanarius vittatus* and *Clibanarius symmetricus*.

3.2. Phylogenetic Analyses

Both phylogenetic trees, generated by ML analyses, indicated the monophyly of *C. antillensis* in clades with bootstrap values of 87% for 16S rRNA (Figure 4) and 79% for COI (Figure 5). There were no pattern dividing groups that could reveal genetic structure. Additionally, in both trees, *C. tricolor* was closer to *C. antillensis* than other congeneric species.

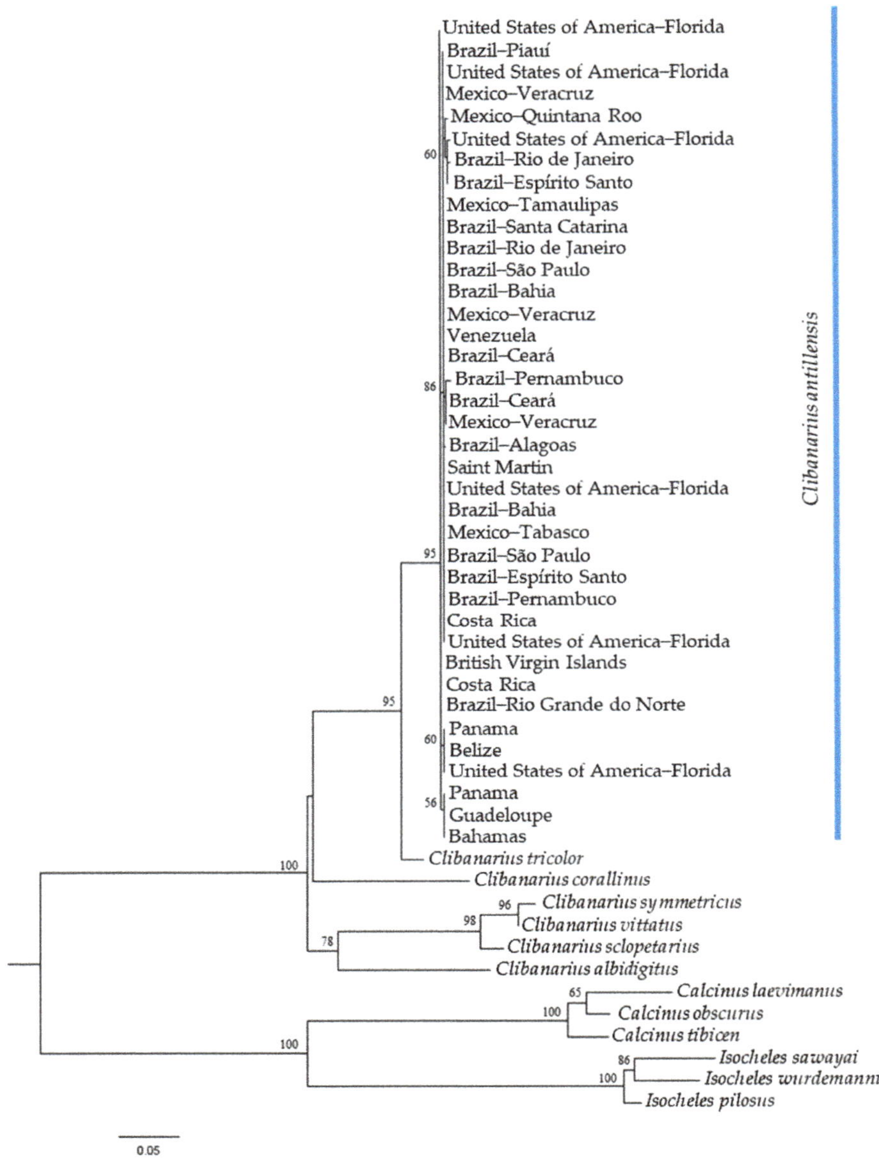

Figure 4. Maximum likelihood phylogram obtained for 16S rRNA sequences of *Clibanarius antillensis* Stimpson, 1859 specimens (blue bar) and other species of Diogenidae. Numbers represent bootstrap values (1002 replicates) and only bootstrap values >50% were shown.

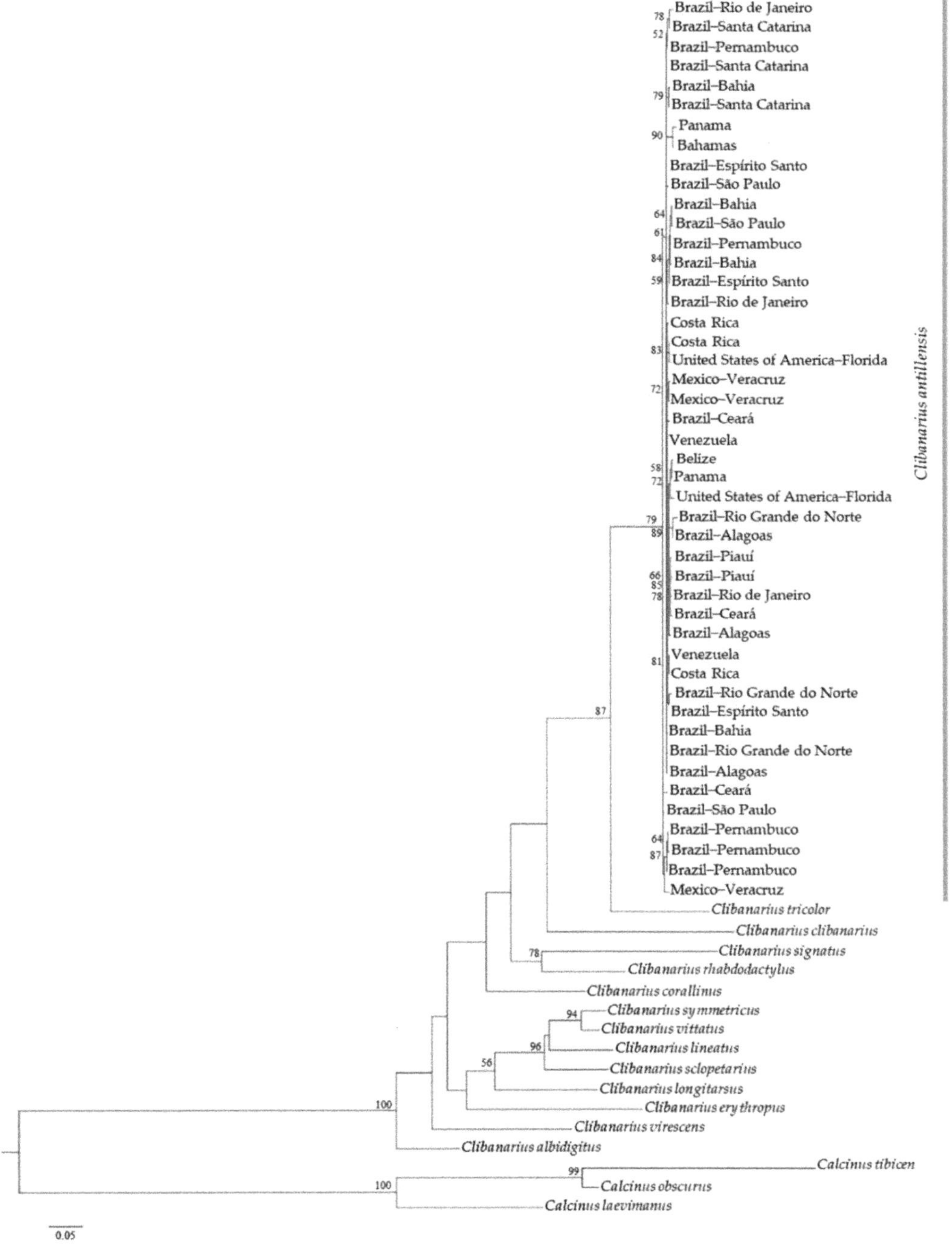

Figure 5. Maximum likelihood phylogram obtained for cytochrome c oxidase subunit I (COI) sequences of *Clibanarius antillensis* Stimpson, 1859 specimens (blue bar) and other species of Diogenidae. Numbers represent bootstrap values (1002 replicates) and only bootstrap values > 50% were shown.

3.3. Genetic Variability Analyses

The population alignment of 16S rRNA consisted of 396 bp with 38 specimens from 20 localities. Seven haplotypes were detected, haplotype diversity was 0.417, total nucleotide diversity was 0.00241, the average number of nucleotide differences was 0.506, and the number of polymorphic sites was five.

The COI alignment had 524 bp with 46 specimens from 17 localities. Forty-two haplotypes were detected, haplotype diversity was 0.995, total nucleotide diversity was 0.01253, the average number of nucleotide differences was 6.564, and the number of polymorphic sites was 63. Genetic diversity index for each locality are in Table 4.

Table 4. Values of number of specimens (N), polymorphic sites (S), number of haplotypes (H), haplotype diversity (Hd), nucleotide diversity (π), and average number of nucleotide differences (K) for each sampled locality of *Clibanarius antillensis* Stimpson, 1859 distribution for 16S rRNA and cytochrome c oxidase subunit I (COI) mitochondrial genes.

Locality	16S rRNA						COI					
	N	S	H	Hd	π	K	N	S	H	Hd	π	K
United States	6	2	3	0.6	0.00168	0.66667	2	5	2	1.0	0.00954	5.00000
Bahamas	1	–	–	–	–	–	1	–	–	–	–	–
Mexico	6	2	3	0.6	0.00189	0.66667	3	10	3	1.0	0.01272	6.66667
Belize	1	–	–	–	–	–	1	–	–	–	–	–
Antilles	3	0	1	0.0	0.00000	0.00000	–	–	–	–	–	–
Costa Rica	2	0	1	0.0	0.00000	0.00000	3	2	3	1.0	0.00254	1.33333
Panama	2	1	2	1.0	0.00252	1.00000	2	12	2	1.0	0.02290	12.0000
Venezuela	1	–	–	–	–	–	2	0	1	0.0	0.00000	0.00000
Brazil–Piauí	1	–	–	–	–	–	2	2	2	1.0	0.00382	2.00000
Brazil–Ceará	2	1	2	1.0	0.00281	1.00000	3	14	3	1.0	0.01781	9.33333
Brazil–Rio Grande do Norte	1	–	–	–	–	–	3	11	3	1.0	0.01399	7.33333
Brazil–Pernambuco	2	3	2	1.0	0.00758	3.00000	5	12	5	1.0	0.01202	6.30000
Brazil–Alagoas	1	–	–	–	–	–	3	9	3	1.0	0.01145	6.00000
Brazil–Bahia	2	0	1	0.0	0.00000	0.00000	4	10	4	1.0	0.01081	5.66667
Brazil–Espírito Santo	2	0	1	0.0	0.00000	0.00000	3	5	3	1.0	0.00636	3.33333
Brazil–Rio de Janeiro	2	0	1	0.0	0.00000	0.00000	3	13	3	1.0	0.01654	8.66667
Brazil–São Paulo	2	0	1	0.0	0.00000	0.00000	3	11	3	1.0	0.01399	7.33333
Brazil–Santa Catarina	1	–	–	–	–	–	3	5	3	1.0	0.00636	3.33333

For 16S rRNA, a central haplotype (H1) was shared by 29 individuals from 18 localities, two haplotypes were shared by two (H2) and three (H3) individuals from different localities, and four were singletons. This network did not show any genetic structure (Figure 6a). For COI, two of 41 detected haplotypes were shared by two individuals from different localities of Brazil (H2 and H3); one (H1) by two specimens from Venezuela and one from Costa Rica; the others were singletons. The network did not show any genetic structure; however, there was high genetic diversity for this gene (Figure 6b).

For 16S rRNA, AMOVA revealed that the variance component within localities (102.02%) exceeded the variance component among localities (-2.02%), with negative and no significant F_{ST}-value ($F_{ST} = -0.0202$; $p > 0.05$), which suggested low or absence of genetic differentiation between localities. For COI, even though F_{ST}-value was positive, moderate ($0.05 < F_{ST} < 0.15$) [73] and significant ($F_{ST} = 0.1231$; $p < 0.05$), which suggested genetic differentiation among localities, variance component within localities (87.69%) was higher than that found among localities (12.31%).

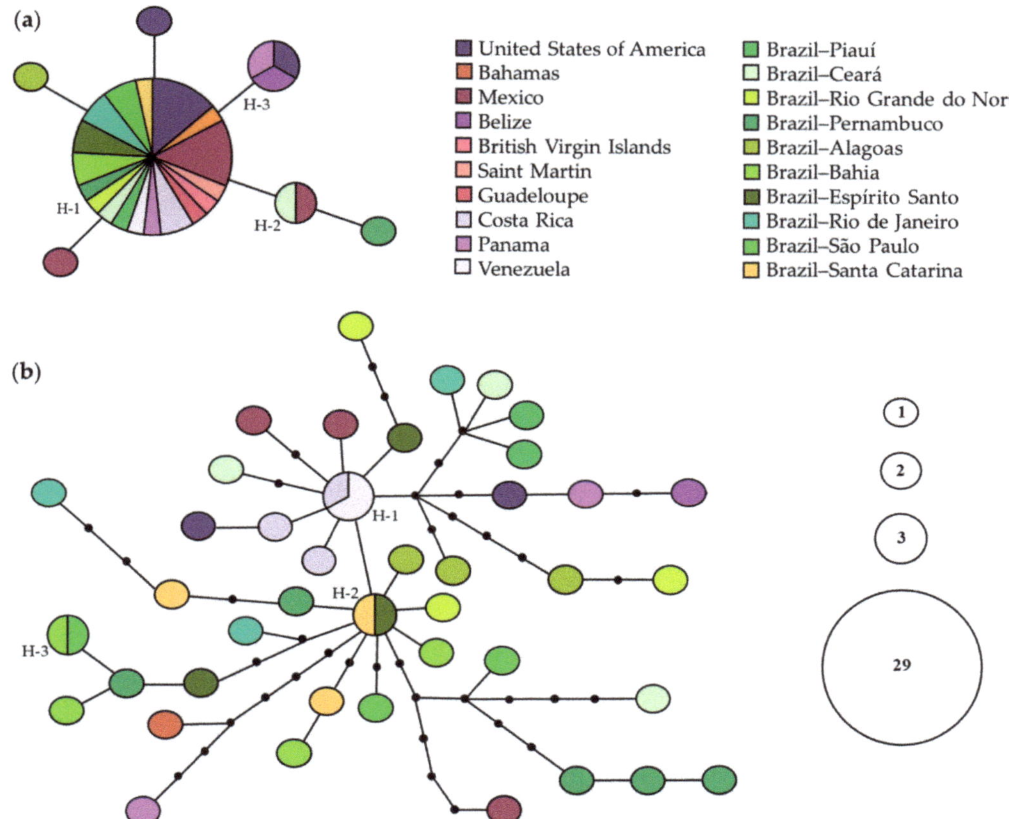

Figure 6. Statistical parsimony haplotype networks of *Clibanarius antillensis* Stimpson, 1859 for 16S rRNA (**a**) and cytochrome c oxidase subunit I (COI) (**b**). Only shared haplotypes are numbered (H). Size of each circle is proportional to haplotype frequency, according to the caption. Black dots indicate median vectors. Each line indicates one mutation step.

3.4. Demographic Analyses

Tajima's D and and Fu's Fs values were significant and negative for both 16S rRNA (D = −1.64246, $p < 0.05$; Fs = −4.64238, $p < 0.02$) and COI (D = −1.91472, $p < 0.05$; Fs = −25.14949, $p < 0.02$) genes, which indicated the rejection of the null hypothesis of population neutrality. Mismatch distribution graphics revealed a unimodal distribution pattern for both genes, which were compatible with the sudden population expansion model (p values for SSD and HRI statistics > 0.05) (Figure 7). Therefore, the null hypothesis of population expansion may not be rejected. The BSP for COI gene showed an increase in effective population size, suggesting that the species had expanded over the past 700,000 years, with a period of stabilization between 450,000 and 250,000 years ago, yet there was no evidence of genetic bottleneck (Figure 8).

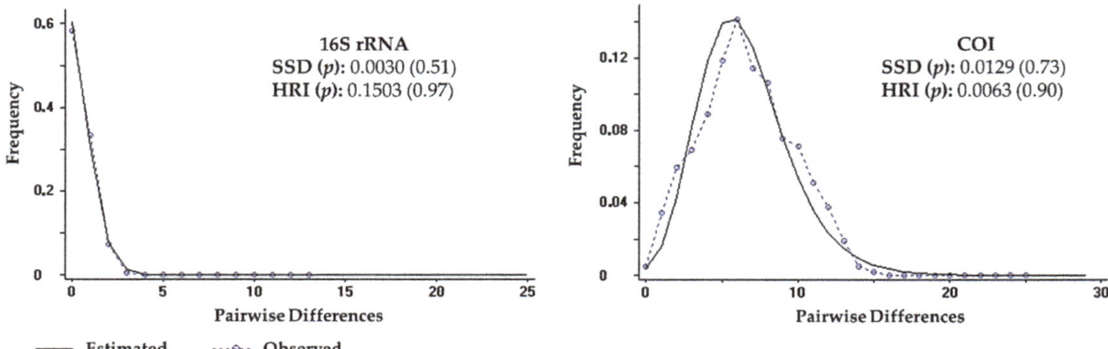

Figure 7. Mismatch distribution for 16S rRNA and cytochrome c oxidase subunit I (COI) sequences of *Clibanarius antillensis* Stimpson, 1859. Statistics for sum of squared deviations (SSD) and Harpending raggedness index (HRI) with *p* values are indicated.

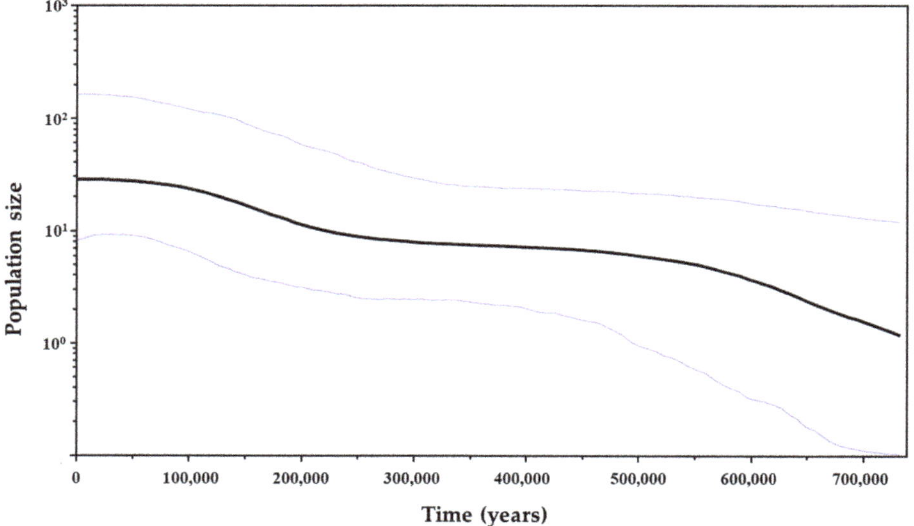

Figure 8. Bayesian skyline plot for cytochrome c oxidase subunit I gene showing demographic history of *Clibanarius antillensis* Stimpson, 1859. Black line represents the variation of average effective population size over time (years ago). Blue lines represent highest posterior density (95%). The population size (*y*-axis) is the product of the effective population size and the generation time.

3.5. Morphological Assessment

We analyzed 187 specimens of *C. antillensis* (121 males, 33 females and 33 ovigerous females) from 17 localities (covering the entire distribution) and with sl ranging from 1.40 to 5.65 mm. We found some variations on the number of spines of ocular acicles, antennal acicles and dorsaldistal surface of carpus of second pereiopod (Table 5). These variations did not show any pattern of morphological distinction between geographic groups; however, they differed from literature descriptions. Therefore, we made a redescription of the species, as follows.

Taxonomy

Family Diogenidae Ortmann, 1892
Genus *Clibanarius* Dana, 1852
Clibanarius antillensis Stimpson, 1859 (Figures 1 and 9)
Clibanarius brasiliensis Dana, 1852: 467 [75].
Clibanarius antillensis Stimpson, 1858: 235 [*nomen nudum*] [76]; 1859: 85 [38].
Clibanarius antillensis.—Smith, 1869: 18, 39 [77].—Nobili, 1897: 4 [78].—Rathbun, 1900: 144 [79].—Benedict, 1901: 142 [74].—Moreira, 1901: 29, 87 [80].—Schmitt, 1924: 79 [81]; 1935: 199 [82]; 1936: 375 [83].—Provenzano, 1959: 368 [39]; 1960: 119 [84]; 1961: 152 [85].—Forest and Saint Laurent, 1967: 99 [28].—Coelho and Ramos-Porto, 1972: 169 [86]; 1987: 51 [29].—Rieger, 1998 [87]: 421.—Melo, 1999: 48 [31].—McLaughlin et al., 2010: 19 [88].—Nucci and Melo, 2015: 331 [31].

Type locality—Barbados.

Table 5. Comparison of analyzed characters of *Clibanarius antillensis* Stimpson, 1859 that differed from literature descriptions [30,31,39,74].

Characters	Literature	Present Study
Ocular acicles: spines	up to 6	3–9
Antennal acicles: spines	up to 7	5–9
Second pereiopod: carpus spines	1 or 2	1–4

Material Examined

UNITED STATES OF AMERICA: <u>Florida</u>—Miami, Biscayne Bay, CCDB 6267, 15 June 2017, coll. H. Bracken-Grissom, 1 ovigerous female (sl 3.55 mm). BAHAMAS: <u>Andros Island</u>—Blanked Sound, Forfar Field Station, AMNH 18726, 29 August 2000, coll. P.M. Mikkelsen, G. Hendler and C.B. Boyko, 1 male (sl 3.20 mm). MEXICO: <u>Veracruz</u>—Actopan, CNCR 24702, 20 April 2006, coll. Y. de los Santos, 2 males (sl 1.56 mm; 2.15 mm)—San Andrés Tuxtla, CNCR 22223, 15 July 2002, coll. A. Argüelles and M. Maldonado, 5 males (sl 1.70–3.37 mm), 1 ovigerous female (sl 1.90 mm). <u>Tabasco</u>—Tacotalpa, CNCR 18624, 14 May 1996, coll. F. Alvarez and R. Robles, 3 males (sl 2.61–2.70 mm). <u>Quintana Roo</u>—Felipe Carrillo Puerto, CNCR 3729, 16 January 1985, coll. J.C. Nates, J.L. Villalobos and A. Cantu, 5 males (sl 2.10–3.03 mm), 1 ovigerous female (sl 2.20 mm). SAINT MARTIN: <u>Le Galion</u>, UF 32041, 15 April 2012, coll. J. Slapcinsky, M. Bernis and A. Anger, 1 male (sl 2.80 mm). COSTA RICA: <u>Talamanca</u>—Cahuita, NHMLA 555-3, 15 July 1977, coll. K. Nelson and D. Hedgecock, 2 males (sl 3.17 mm; 3.19 mm)—Puerto Vargas, CCDB 4131, 14 February 2009, coll. F.L. Mantelatto and I. Wehrtmann, 1 male (sl 1.72 mm), 2 females (sl 1.58 mm; 2.56 mm), 1 ovigerous female (sl 2.27 mm)—Puerto Vargas, CCDB 4160, 23 May 2010, coll. F.L. Mantelatto, M. Terossi, D.F. Peiró and I. Wehrtmann, 2 males (sl 2.24 mm; 3.03 mm), 1 female (sl 2.48 mm)—Puerto Viejo, Playa Puerto Viejo, CCDB 550, 05 April 2007, coll. F.L. Mantelatto et al., 4 males (sl 2.30–3.39 mm), 1 female (sl 2.95 mm). PANAMA: <u>Bocas del Toro</u>—Bocas del Toro, CCDB 3578, 03 August 2011, coll. F.L. Mantelatto, 3 males (sl 2.67–2.96 mm), 2 females (sl 2.35 mm; 3.25 mm), 1 ovigerous female (sl 2.85 mm)—Bocas del Toro, Playa Paunch, CCDB 3575, 05 August 2011, coll. F.L. Mantelatto, M.P. Negri, N. Rossi and T. Magalhães, 4 males (sl 2.07 –2.81 mm), 2 ovigerous females (sl 2.29 mm; 2.52 mm)—Bocas del Toro, Playa Bluff, CCDB 4164, 17 February 2009, coll. F.L. Mantelatto, M. Terossi, I. Miranda and A. Baeza, 1 male (sl 3.23 mm). VENEZUELA: Nueva Esparta—Porlamar, Isla Margarita, Playa Valdez, CCDB 1810, 27 August 2006, coll. F.L. Mantelatto and L.G. Pileggi, 3 males (sl 3.56–4.77 mm), 1 female (sl 2.96 mm), 3 ovigerous females (sl 2.88–3.20 mm). BRAZIL: <u>Piauí</u>—Luís Correia, Praia do Coqueiro, CCDB 4158, 01 July 2006, coll. J.M. Góes, 2 males (sl 3.87 mm; 4.40 mm), 1 ovigerous female (sl 3.55 mm). <u>Ceará</u>—Trairi, Praia Flecheiras, CCDB 2651, 20 May 2008, coll. M. Terossi and I. Miranda, 8 males (sl 1.94–5.65 mm)—Trairi, Praia Flecheiras, CCDB 4273, 20 May 2008, coll. M. Terossi, 2 males (sl 1.44 mm; 2.01 mm), 1 female (sl 2.53 mm)—Paracuru, Praia da Pedra Rachada,

CCDB 5923, 14 November 2015, coll. F.L. Mantelatto, L. Bezerra and A. Almeida, 6 males (sl 3.10–4.51 mm)— Caucaia, Praia do Pacheco, CCDB 4503, 12 February 2013, coll. F.L. Mantelatto, L. Bezerra and F. Bezerra, 4 males (sl 2.35–4.41 mm), 1 female (sl 2.73 mm)— Fortaleza, Praia Meireles, CCDB 4274, 22 May 2008, coll. M. Terossi and I. Miranda, 3 males (sl 2.47–3.51 mm), 2 females (sl 2.60 mm; 2.84 mm). <u>Rio Grande do Norte</u>—Touros, Praia de Perobas, CCDB 3366, 10 June 2011, coll. L.G. Pileggi and R. Robles, 1 male (sl 2.92 mm)— Touros, Praia de Perobas, CCDB 3367, 10 June 2011, coll. L.G. Pileggi and R. Robles, 1 male (sl 4.64 mm)—Touros, Praia de Perobas, CCDB 3373, 10 June 2011, coll. L.G. Pileggi and R. Robles, 1 female (sl 2.40 mm). <u>Pernambuco</u>—Ipojuca, Praia de Serrambi, CCDB 1727, 25 July 2012, coll. F.L. Mantelatto and F.B. Mantelatto, 5 males (sl 2.53–5.12 mm), 2 females (sl 2.23 mm; 2.98 mm), 4 ovigerous females (sl 2.11–2.04 mm)—Ipojuca, Praia de Serrambi, CCDB 5762, 21 July 2015, coll. F.L. Mantelatto, F.B. Mantelatto and R.B. Mantelatto, 4 males (sl 2.84–3.13 mm), 2 females (sl 1.91 mm; 2.53 mm). <u>Alagoas</u>—Maragogi, CCDB 4920, 5 October 2013, coll. F.L. Mantelatto and F.B. Mantelatto, 9 males (sl 2.80–3.71 mm), 1 ovigerous female (sl 3.30 mm)—Maragogi, Praia do Bitingui, CCDB 5586, 10 January 2015, coll. F.L. Mantelatto, F.B. Mantelatto, R.B. Mantelatto and H. Mantelatto, 4 males (sl 2.20– 3.60 mm), 1 female (sl 2.30 mm), 3 ovigerous females (sl 2.46 mm–2.51 mm). <u>Bahia</u>—Lauro de Freitas, Praia do Ipitanga, 22 July 2011, coll. F.L. Carvalho and E.A. Souza-Carvalho, 1 male (sl 3.30 mm), 1 female (sl 2.70 mm)—Salvador, Ilha dos Frades, CCDB 4139, 17 July 2003, coll. M. Terossi, 1 male (sl 2.91 mm)—Salvador, Praia do Forte, CCDB 4133, 18 December 2003, coll. M. Terossi, 1 male (sl 2.21 mm)—Salvador, Praia do Forte, CCDB 4137, 18 December 2003, coll. M. Terossi, 1 male (sl 2.62 mm)—Salvador, Praia do Forte, CCDB 4138, 18 December 2003, coll. M. Terossi, 1 male (sl 1.94 mm)—Salvador, Praia de Ondina, CCDB 4135, 14 December 2003, coll. M. Terossi, 2 males (sl 3.63 mm; 4.10 mm), 1 female (sl 3.02 mm), 2 ovigerous females (sl 2.22 mm; 2.93 mm)—Ilhéus, Olivença, Praia Back Door, CCDB 2610, 18 July 2003, coll. A.O. Almeida and J.T.A. Santos, 2 males (sl 4.26 mm; 4.29 mm)—Ilhéus, Praia da Maramata, CCDB 2597, 31 March 2009, coll. F.L. Mantelatto and A.O. Almeida, 2 males (sl 3.05 mm; 4.44 mm), 1 ovigerous female (sl 2.96 mm)—Porto Seguro, Arraial D'Ajuda, CCDB 4193, 08 January 2012, coll. F.L. Carvalho and E.A. Souza-Carvalho,1 female (sl 2.50 mm)—Porto Seguro, Praia da Pitinga, CCDB 585, 29 January 2001, coll. F.L. Mantelatto and R. Garcia, 5 males (sl 2.90–3.67 mm), 2 ovigerous females (sl 3.40 mm; 3.67 mm). <u>Espírito Santo</u>—Vitória, Ilha do Frade, CCDB 4118, 21 June 2012, coll. F.L. Carvalho, R. Robles and D.F. Peiró, 1 male (sl 3.91 mm), 3 females (sl 3.40–3.74 mm), 2 ovigerous females (sl 3.01 mm; 3.32 mm)—Guarapari, Canal de Guarapari, CCDB 2243, 04 November 2006, coll. F.L. Mantelatto, D.F. Peiró, and E.C. Mossolin, 2 males (sl 3.30 mm; 3.91 mm), 3 ovigerous females (sl 3.28 mm–3.63 mm)—Anchieta, Praia de Iriri, CCDB 4012, 19 June 2012, coll. F.L. Carvalho, R. Robles and D.F. Peiró, 1 male (sl 3.74 mm), 1 female (sl 3.32 mm)—Piúma, Praia de Piúma, CCDB 4072, 15 June 2012, coll. F.L. Carvalho, R. Robles and D.F. Peiró, 1 male (sl 1.82 mm), 1 female (sl 3.90 mm), 1 ovigerous female (sl 3.33 mm). <u>Rio de Janeiro</u>—Búzios, Praia da Tartaruga, CCDB 5655, 20 May 2015, coll. N. Rossi, 1 male (sl 4.97 mm)—Búzios, Porto da Barra, CCDB 5902, 24 April 2006, coll. R. Bispo, R. Johnsson, W. Santana and F. Faria, 3 males (sl 2.95–5.34 mm), 2 females (sl 2.74 mm; 3.11 mm). <u>São Paulo</u>—Ubatuba, Praia do Perequê Mirim, CCDB 2906, 19 November 2002, coll. F.L. Mantelatto, 3 males (sl 4.16–4.90 mm), 1 female (sl 3.92 mm)—Ubatuba, Saco do Codó, CCDB 2813, 01 May 2002, coll. F.L. Mantelatto, 1 male (sl 2.68 mm)—São Sebastião, Mangue do Araçá, CCDB 1462, 18 July 2004, coll. F.L. Mantelatto, 1 male (sl 3.60 mm)—São Sebastião, Mangue do Araçá, CCDB 5061, 10 September 2013, coll. F.L. Mantelatto et al., 1 male (sl 4.24 mm)—São Sebastião, Mangue do Araçá, CCDB 5062, 10 September 2013, coll., F.L. Mantelatto et al., 2 males (sl 3.29 mm; 3.76 mm), 5 females (sl 2.76–3.76 mm), 2 ovigerous females (sl 2.81 mm; 3.53 mm). <u>Santa Catarina</u>—Itajaí, Praia Cabeçudas, CCDB 1876, 19 June 2007, coll. F.L. Mantelatto, L.G. Pileggi, L.S. Torati and E.C. Mossolin, 2 males (sl 4.89 mm; 4.90 mm), 1 ovigerous female (sl 4.40 mm).

Diagnosis

Shield subrectangular. Second and third pair of pereiopods with dactyl shorter than propodus, lateral surface of merus with a dark stripe on light background and lateral surfaces of carpus, propodus and dactyl with a light stripe on dark background; dactyl with orange distal region.

Description

Shield (Figures 1 and 9a) subrectangular, longer than broad, with cervical suture and *linea transversalis* well developed; anterior margin between rostrum and lateral projections straight; lateral margins slightly sloping; dorsal surface plain, lateral region with 2–5 tufts of long setae and anterior region with few scattered setae. Rostrum triangular, twice as long as lateral projections.

Ocular peduncles (Figures 1 and 9a) as long as frontal width, cylindrical, slightly broader at the base, left slightly longer than right; dorsal surface with scattered tufts of short setae. Corneas slightly dilated. Ocular acicles (Figure 9a,b) subrectangular, long, closely set; dorsodistal margin with 39 spines, spines shorter in middle region; dorsal surface plain and slightly concave; dorsodistal margin with few setae.

Antennular peduncles (Figure 9a) long, occasionally exceeding distal margin of left cornea when extended. Last segment with short, scattered dorsal setae. Penultimate segment with long, scattered dorsal setae.

Antennal peduncles (Figures 1 and 9a,c) barely reaching distal margin of cornea. Fifth segment dorsal surface with tufts of short setae, lateral margin with tufts of long setae. Fourth segment dorsolateral region of distal margin with one spine and setae. Third segment ventrodistal margin with one spine and setae. Second segment dorsodistal and laterodistal margins with tufts of setae; laterodistal margin with one spine; lateral margins occasionally with projections. First segment unarmed. Flagella long, slender, reaching to dactyl of first pair of pereiopods, with short setae. Antennal acicle lateral and dorsal surfaces with long, scattered setae; lateral and dorsal surfaces with 59 spines.

Chelipeds subequal, right slightly larger than left. Chela (Figures 1 and 9d) twice as long as broad; dorsal surface with short spines; ventral surface with tubercles and tufts of setae; palm and fixed finger with scattered setae; fixed finger lateral surface with tufts of short setae; fixed finger and dactyl ending in spoon-shaped corneous tip. Carpus short, lateral and mesial surfaces with scattered tubercles and long setae, similar to chela; dorsal surface mesial angle with row of spines and long setae; few dorsodistal spines; ventral surface unarmed. Merus long, dorsal surface with small tubercles, long setae and some dorsodistal spines; ventromesial margin with row of short spines; ventral surface with few tufts of setae and few lines. Ischium unarmed.

Second and third pereiopods (Figures 1 and 9e) similar, long and slender. Dactyl about 0.8 length of propodus, ending in a sharp, curved corneous claw; dorsal and ventral surfaces with tufts of setae; ventral surface with row of spines; left third pereiopod flattened, with dorsolateral ridge. Propodus about 1.5 as long as carpus; surfaces with tufts of scattered setae; laterodistal and ventrodistal margins with short spines; left third pereiopod flattened, with dorsolateral ridge. Carpus about 0.7 length of merus; second pair of pereiopods with dorsal row of tufts of short setae; third pair of pereiopods with row of tufts of long setae; lateral and mesial surfaces with few tufts of setae; second pair of pereiopod with 1–4 dorsodistal spines, third pair with one dorsodistal spine. Merus ventral and dorsal surfaces, and dorsodistal and dorsoventral margins with row of tufts of setae, with distoventral short spines and one distolateral spine. Ischium with tufts of setae.

Fourth pereiopod (Figures 1 and 9f) semichelate; dorsal and ventral surfaces with long setae. Dactyl ending in corneous claw, ventrolateral row of small spines. Propodal rasp well developed. Carpus with dorsodistal spine.

Fifth pereiopod (Figures 1 and 9g) chelate, with scattered tufts of long setae. Propodal rasp well developed, covering about one third of propodus lateral surface.

Uropods asymmetrical, left larger than right. Endopodal and exopodal rasps well developed, dorsolateral margins with setae.

Telson (Figure 9h) asymmetrical, left lobe larger than right. Distal margin of posterior lobes rounded, with row of short spines and long setae; lobes separated by distinct median cleft; lateral margins with long setae and indentations distinct.

Color (Fresh Specimen)

Shield with small white spots and darker anterior region. Ocular peduncles greenish-blue with a brown area on dorsal surface. Antennular peduncles orange with a bluish color on distal region of the segment; antennular flagella orange. Antennal peduncles orange with a yellowish color on first two segments; Antennal flagella orange. Chelipeds olive to rusty brown with white spines and white tubercles; chela with a lighter color. Second and third pair of pereiopods with a dark stripe on light background on lateral surface of merus; a light stripe on dark background on lateral surfaces of carpus, propodus and dactyl; dactyl with orange distal region. Figure 1 shows a preserved specimen with the original color pattern and supplements the above description.

Distribution

Western Atlantic: Bermuda, Florida, Gulf of Mexico, Belize, Costa Rica, Panama, Antilles, north of South America, and Brazil (Atol das Rocas, Piauí, Ceará, Rio Grande do Norte, Paraíba, Pernambuco, Alagoas, Bahia, Espirito Santo, Rio de Janeiro, São Paulo, and Santa Catarina). Usually intertidal and found in shallow waters, over rocks, coral reefs and banks of *Halodule* [28–32].

Remarks

The similarity between *C. brasiliensis* and *C. antillensis* was first noted by Stimpson [38]; however, Forest and Saint Laurent [28] later stated that the description of *C. brasiliensis* and the original figure of Dana [75] corresponded to *C. antillensis*. Although the name *C. brasiliensis* had priority, Forest and Saint Laurent [28] did not reestablish it, because it was not mentioned since Moreira [80]; therefore, the valid name is *C. antillensis*. The holotype of *C. antillensis* was collected by Theo Gill and it should be at the National Museum of Natural History, Smithsonian Institution, at United States of America; however, it seems to be lost, according to Provenzano [39]; according to the database of WoRMS edited by Lemaitre and McLaughlin [89], the syntype is deposited in the Naturhistorisches Museum, Switzerland (catalogue NHM 61.44), but not checked by us.

Clibanarius tricolor and *C. antillensis* are very close morphologically and it is hard to distinguish them when they are preserved and lost their original color. *C. antillensis* is found from the USA (Florida) to the south of Brazil (Santa Catarina) and *C. tricolor* is found from the USA (Florida) to the southeast of Brazil (Espírito Santo) [29,30,32]. Both are the only species of the genus *Clibanarius* from the Western Atlantic that have dactyls of second and third pair of pereiopods shorter than propodi. They are easily distinguishable by their second and third pair of pereiopods original color pattern, once *C. tricolor* has transverse orange bands on proximal margins of segments, which, except for white or yellow background dactyl, is otherwise blue with dark punctae; *C. antillensis* has broad longitudinal stripes on dark background, as described above. When preserved in alcohol, the blue on pereiopods of *C. tricolor* fades, remaining only orange bands and punctae; on *C. antillensis*, they become orange with lighter stripes [39].

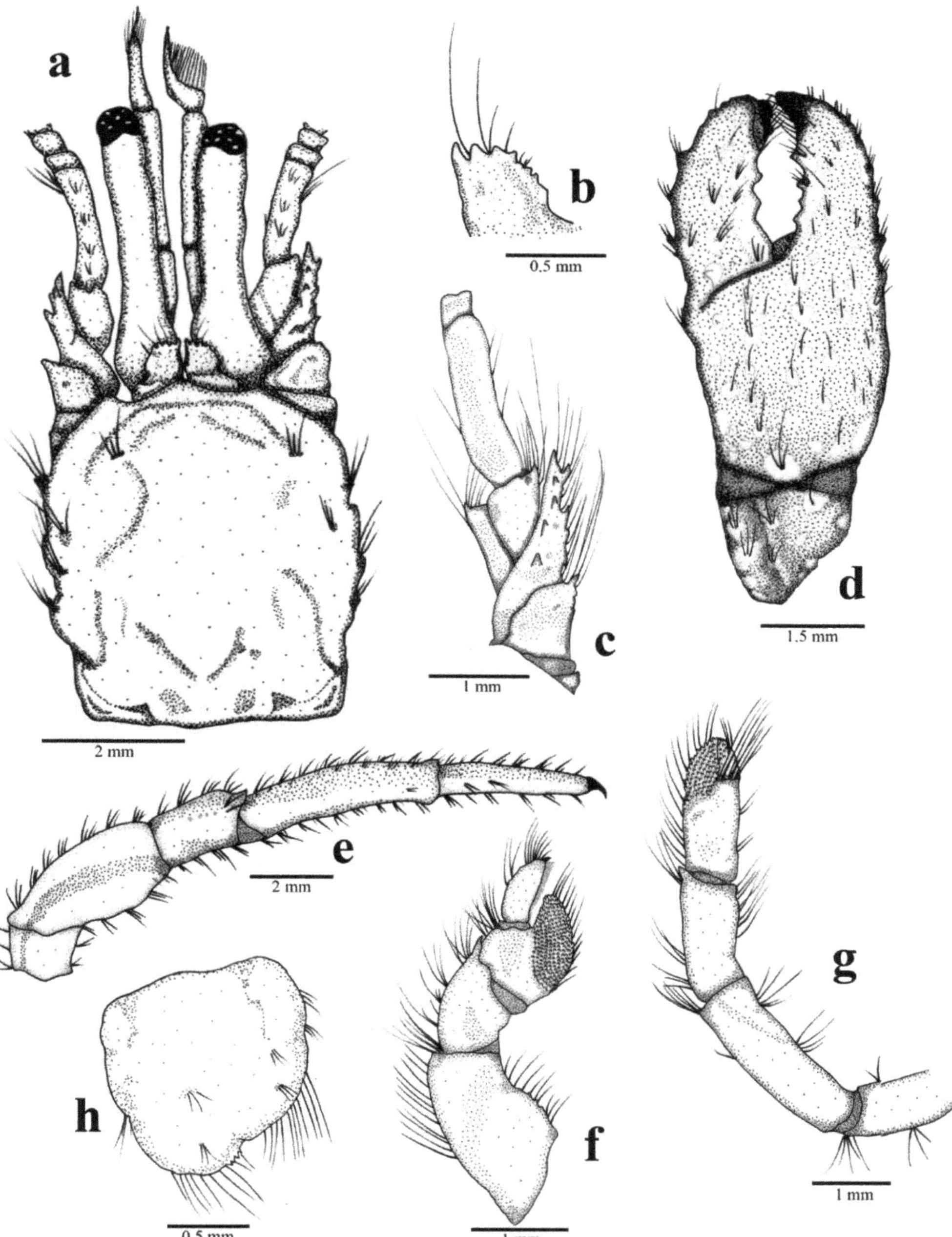

Figure 9. *Clibanarius antillensis* Stimpson, 1859: male, sl 4.97 mm. CCDB 5655, Praia das Tartarugas, Búzios, Rio de Janeiro, Brazil. (**a**): cephalothoracic shield and cephalic region, dorsal view; (**b**): right ocular acicle, dorsal view; (**c**): right antennal peduncle, dorso-mesial view; (**d**): right chela, dorsal view; (**e**): right third pereiopod, lateral view; (**f**): right fourth pereiopod, lateral view; (**g**): right fifth pereiopod, lateral view; (**h**): telson, dorsal view. Some setae were removed for a better observation of structures.

4. Discussion

4.1. Genetic Structure

Based on our analyses for 16S rRNA and COI genes, we found no genetic structure for *C. antillensis* along its distribution. The intraspecific divergence was lower than the interspecific variability for both genes, without an evident interspecific gap for 16S rRNA (Figure 3). This occurred due to the proximity between the intraspecific divergence of *C. antillensis* (0.99%) and the interspecific divergence between *Clibanarius vitattus* and *C. symmetricus* (1.48%), two species that possibly went through recent divergence processes [44]. Besides that, there was no gap within intraspecific variability of *C. antillensis*, which indicated the absence of population structure. Additionally, the phylogenetic trees, as well as the haplotype networks, did not show any grouping pattern that may indicate genetic structure (Figures 4–6). This was also evidenced by AMOVA, as within localities variance components were higher than among localities.

Many marine species have populations widely distributed with low genetic differentiation and habitats interconnected by gene flow [8]. Some examples among decapods distributed along the western Atlantic can be mentioned: the slipper lobsters *Scyllarides brasiliensis* Rathbun, 1906 [90], the mangrove crab *Ucides cordatus* Linnaeus, 1763 [24,91], the swimming crab *Callinectes danae* Smith, 1869 [22] and the congeneric species *Clibanarius sclopetarius* [27]. The absence of genetic structure within these species, as well as among specimens of *C. antillensis* from different localities, may be explained by the lack of physical barriers restricting gene flow and by their larval dispersive capacity [92].

In general, many marine species have planktonic larval stages, and their wide dispersal may happen during their first development weeks. In this period, a large number of larvae are released and passively transported by marine currents system, through which individuals might reach long distances and promote genetic and demographic connectivity among populations [18,93–95]. Along the western Atlantic, there is the South Equatorial Current, which reaches the Brazilian coast (9–15° S) and bifurcates into north (Northern Brazilian Current) and south (Brazil Current) [96]. This bifurcation has different effects on the genetic structure of many marine species, acting as a barrier to gene flow [97,98] or not [45,90,91]. These currents may not prevent gene flow of *C. antillensis*, in fact, they may facilitate the dispersion of its larvae. Current systems are associated with long-distance connectivity and long duration of larval stages [3,99]. *C. antillensis* larvae go through five to six stages of development, which altogether take at least 43 days [34,35].

Salinity is another feature limiting the dispersal of species. It is relevant especially in estuarine areas and other coastal environments, since it presents high and constant variations, which affect the physiology and ecology of organisms [19]. It may influence biochemical composition, growth, survival and development of larvae [100,101], feeding activity [19], carbon accumulation rates [102], as well as osmoregulatory activities [103]. In fact, salinity has been described as a barrier for dispersion and gene flow of some decapod's species [22,26,104,105]. In those studies, the absence of gene flow between populations resulted from the incapacity of their larvae to traverse the Amazon River plume at the Atlantic Ocean, where the volume of water discharged changes the local salinity.

On the other hand, the outflow of the Amazon River has not been a barrier for the dispersal of other decapods [22,91] since the larvae may be more tolerant to low salinity. This might be the case of *C. antillensis* larvae, which can develop at salinity levels of 29–35 ppt [34,35]. However, adults probably are not able to establish on conditions where the salinity is reduced, since there is a gap along their distribution [28,30,31], which corresponds to the north region of Brazil, where the Amazon river ends and promote salinity influence.

Genetic connectivity may also be influenced by behavioral site fidelity and local retention of larvae [11,20]. Hence, even if the larvae present high dispersal potential, if they are retained next to their natal populations for many generations, populations might undergo through enough differentiation, resulting on genetic structure [11]. In addition, there are many other features interfering on genetic structure, such as the biology and life cycle, habitat, local oceanic conditions, local adaptation, ecological and geographic limitations, past geological events, and recent history. Together, they may influence gene flow at specific directions or moments [8,10,14,106–108]. Therefore, even if different species have similar dispersal capacity, it is not easy to establish genetic structure patterns, since they are influenced by different factors at the same time [8,23,109].

4.2. Genetic Diversity

In addition to the lack of genetic structure in *C. antillensis*, a high genetic diversity was found, especially for the COI gene, which presented a total nucleotide diversity of 0.01253 and haplotype diversity of 0.995. The former value is considered high when $\pi > 0.005$ [110] and the closeness of the latter to 1 indicates high number of singletons, which corresponds to an individual sequence of certain gene [111]—as observed in the haplotype network for COI (Figure 6b). The high number of low-frequency haplotypes, as well as high values of nucleotide and haplotype diversities might be related to large and stable populations with long evolutionary history and high mutation rates or with secondary contact between different lineages [110,112]. High diversity indexes using the same gene were also found on studies of *Opecarcinus hypostegus* (Hd = 0.9994, π = 0.02558) [113], *U. cordatus* (Hd = 0.9820, π = 0.005862) [91] and *Callinectes ornatus* (Hd = 0.9570, π = 0.01360) [22], which may indicate that their populations have been stable through time or undergone through a slightly recent expansion [22,91]. For *C. antillensis*, the BSP recovered a long demographic history from 700,000 years ago, with periods of stabilization and small population expansion (Figure 8). Demographic expansion was also evidenced by significant and negative values of neutrality tests for both genes [114] and by mismatch distribution, with a unimodal distribution pattern and non-significant SSD and HRI values [66] (Figure 7).

These results may reflect historical processes, such as past geological events and the demographic history, which could influence current geographical distribution and genetic variation of marine individuals. The high genetic diversity is common to many marine species [115–117]. It might be preserved by long-distance dispersal during expansion [116], or many migrations among close areas, generating a higher number of new haplotypes than others that are lost [118,119].

The demographic history and distribution of many marine species were influenced by climatic fluctuations that occurred during Pleistocene (~2.6 million–10,000 years ago) [120, 121]. During glacial periods, many populations of marine species used to refuge on low latitude regions, possibly resulting in genetic drift. When the climate became warmer, they would recolonize other areas and reestablish populations that had disappeared during the previous glacial event [122–125]. Consequently, genetic diversity would be considerably higher in areas where colonizers came from different refuges compared to those originated from a single population source [126,127]. Some haplotypes were exclusive because they may have not participated in recolonization [124].

Nonetheless, each species had a unique response to climatic oscillations during glaciation [17]. There are many marine species that were not strongly affected by glacial periods, consequently, their populations probably continued to expand during these periods, resulting in lack of genetic structure [89,125,128,129]. This might be the case of *C. antillensis*. The almost continuous expansion with periods of stabilization indicated by the BSP analyses for the last 700,000 years, associated with high values of haplotype and nucleotide diversities, do not indicate that this species went through any genetic bottleneck followed by expansion. Therefore, if the species had refuged during Pleistocene, there would not have been sufficient isolation to cause a reduction on genetic diversity or its populations would have isolated themselves in many refuges, maintaining the diversity on periods when population growth was more stable.

4.3. Morphological Variations

Morphological analyzes have also corroborated the absence of genetic structure. Although some features presented variability, they have not shown any pattern related to geographic groups, and in some cases, specimens from the same locality presented differences in characters. Intraspecific morphological variations without any pattern have been reported for other decapods with wide distribution [130–132]. These differences, as well as the distinct coloration patterns found among adults [37], may be related to environmental conditions (habitat, wave action, food supply, and salinity), local selection pressures and intra or interspecific interactions that may affect each organism differently [133–135].

Among the characters analyzed, the number of spines of the ocular acicles, antennal acicles and second pereiopod carpus presented the largest variation. They also differed from the literature descriptions (Table 5).

In the present study, ocular acicles had three to nine spines on dorsodistal margin, in which seven to nine spines were found on 30 specimens, mainly on male, followed by ovigerous females and females from different localities. According to the literature, the number of spines was: three or four [30,74], six [39], and up to six [31].

In the present study, antennal acicles had five to nine spines on lateral and dorsal surfaces, in which eight or nine spines were found on 34 specimens, mainly on male, followed by female and ovigerous females from different localities. According to the literature, the number of spines was up to seven [31].

In the present study, the carpus of the second pereiopod had one to four spines, in which three or four spines were found only on five males and one ovigerous female from Mexico and four Brazilian states. According to the literature, the number of spines was one or two [39].

The importance of including detailed variations on a redescription of the species is to assure that some traits are not neglected and to facilitate the differentiation of closely related taxon [130]. Hermit crabs are usually hard to be distinguished by a unique character, especially if they have lost their original color [130]. *C. vittatus* and *C. symmetricus*, for example, only differ by the color pattern of their pereiopods [44], as well as *C. antillensis* and *C. tricolor*. Therefore, if preserved specimens have lost their color, the availability of a set of characters is required to facilitate their distinction.

It is important to define the genetic diversity of marine species once it allows us to understand how historical processes and contemporary environmental conditions have influenced their populations along their distribution. In addition, they may reveal aspects of gene flow, evolution, genetic differentiation, and spatial population boundaries [136]. Such studies, consequently, provide information about biodiversity and conservation strategies of species [137,138]. The present study enables the understanding of marine phylogeographic patterns along the western Atlantic Ocean. Overall, our mitochondrial data for both 16S rRNA and COI genes and morphological comparisons did not reveal structure patterns, related or not to geographical patterns, among populations of *C. antillensis*. These results may be explained by a set of factors including planktonic larval duration of the species and the absence of effective barriers to gene flow. Besides, there were high genetic diversity for COI gene and signs of population expansion in neutrality tests, mismatch distribution and Bayesian skyline plot. This last analysis revealed small population effective size expansion in the last 700,000 years, with some periods of stabilization, and no evidence of bottleneck effect. Therefore, the species might not have been strongly influenced by Pleistocene climatic oscillations.

Author Contributions: Conceptualization, F.L.M.; methodology, K.S.N., M.N. and F.L.M.; formal analysis, K.S.N. and M.N.; investigation, K.S.N. and M.N.; resources, F.L.M.; writing—original draft preparation, K.S.N., M.N. and F.L.M.; writing—review and editing, K.S.N., M.N. and F.L.M.; visualization, K.S.N., M.N. and F.L.M.; supervision, M.N. and F.L.M.; project administration, F.L.M.; funding acquisition, F.L.M. All authors have read and agreed to the published version of the manuscript.

Funding: The present study is part of a long-term project to evaluate the taxonomy and genetic variability of decapods in the Western Atlantic, and was supported by scientific grants provided to FLM. This research is part of a Bachelor's thesis by KSN, supported by scientific fellowships from Coordenação de Aperfeiçoamento de Pessoal de Nível Superior (CAPES—Ciências do Mar II) and Fundação de Amparo à Pesquisa do Estado de São Paulo (FAPESP—Proc. 2016/22448-1).

MN received support from a post-doctoral fellowship from Conselho Nacional de Desenvolvimento Científico e Tecnológico (CNPq—PROTAX Proc. 152377/2016-6). Major financial support for this project was provided by FAPESP (Temáticos Biota 2010/50188-8 and INTERCRUSTA 2018/13685-5; Coleções Científicas 2009/54931-0; PROTAX 2016/50376-5), CNPq (301359/2007-5; 473050/2007-2; 302748/2010-5; 471011/2011-8; PQ 304968/2014-5 and 302253/2019-0) and CAPES—Código de financiamento 001 (Ciências do Mar II Proc. 2005/2014—23038.004308/2014-14) to FLM.

Institutional Review Board Statement: Not applicable.

Informed Consent Statement: Not applicable.

Data Availability Statement: Sequences of this study are available from GenBank (accession numbers MG264431–MG264512, MT740091).

Acknowledgments: We are grateful to Darryl Felder and Catherine Craig (ULLZ), Fernando Álvarez and José Luis Villalobos (CNCR), Gustav Paulay and John Slapcinskyl (UF), Rafael Lemaitre and Karen Reed (USNM), Estefania Rodriguez and Lily Berniker (AMNH), Kathy Omura, Regina Wetzer, Dean Pentcheff and Adam Wall (NHMLA), and Heather Bracken-Grisson (FIU) for loans and donations of material; Natália Rossi, Rafael Robles, Raquel Buranelli and Tatiana Magalhães (LBSC) for their help with materials from collections; all members of LBSC for their contributions during the development of this research, and the anonymous reviewers for their important corrections during review process. The collections of species conducted in this study complied with current applicable state and federal laws of Brazil (permanent license to FLM for collection of Zoological Material No. 11777-1 MMA/IBAMA/SISBIO).

Conflicts of Interest: The authors declare no conflict of interest.

References

1. Futuyma, D.J. *Evolutionary Biology*, 2nd ed.; Sinauer Associates: Sunderland, MA, USA, 1986; ISBN 978-0-87893-188-0.
2. Heywood, V.H.; Watson, R.T. *Global Biodiversity Assessment: Summary for Policy-Makers*; Watson, R.T., Heywood, V.H., Baste, I., Dias, B., Gámez, R., Reid, W., Ruark, G., Eds.; Cambridge University Press: Cambridge, UK, 1995; ISBN 978-0-521-56481-6.
3. Slatkin, M. Gene flow and the geographic structure of natural populations. *Science* **1987**, *236*, 787–792. [CrossRef] [PubMed]
4. Slatkin, M.; Barton, N.H. A comparison of three methods for estimating average levels of gene flow. *Evolution* **1989**, *43*, 1349–1368. [CrossRef] [PubMed]
5. Palumbi, S.R. Macrospatial genetic structure and speciation in marine taxa with high dispersal abilities. In *Molecular Zoology: Advances, Strategies, and Protocols*; Ferraris, J.D., Palumbi, S.R., Eds.; Wiley-Liss: New York, NY, USA, 1996; pp. 101–113. ISBN 978-0-471-14449-6.
6. Collin, R. The effects of mode of development on phylogeography and population structure of north Atlantic *Crepidula* (Gastropoda: Calyptraeidae). *Mol. Ecol.* **2001**, *10*, 2249–2262. [CrossRef] [PubMed]
7. Avise, J.C. *Molecular Markers, Natural History and Evolution*; Springer: New York, NY, USA, 1994; ISBN 978-1-4615-2381-9.
8. Palumbi, S.R. Genetic divergence, reproductive isolation, and marine speciation. *Annu. Rev. Ecol. Syst.* **1994**, *25*, 547–572. [CrossRef]
9. Silva, I.C.; Mesquita, N.; Paula, J. Lack of population structure in the fiddler crab *Uca annulipes* along an East African latitudinal gradient: Genetic and morphometric evidence. *Mar. Biol.* **2010**, *157*, 1113–1126. [CrossRef]
10. Hedgecock, D. Is gene flow from pelagic larval dispersal important in the adaptation and evolution of marine invertebrates? *Bull. Mar. Sci.* **1986**, *39*, 550–564.
11. Taylor, M.S.; Hellberg, M.E. Genetic evidence for local retention of pelagic larvae in a Caribbean reef fish. *Science* **2003**, *299*, 107–109. [CrossRef]
12. Baums, I.B.; Paris, C.B.; Chérubin, L.M. A bio-oceanographic filter to larval dispersal in a reef-building coral. *Limnol. Oceanogr.* **2006**, *51*, 1969–1981. [CrossRef]
13. Burton, R.S. Protein polymorphisms and genetic differentiation of marine invertebrate populations. *Mar. Biol. Lett.* **1983**, *4*, 193–206.
14. Scheltema, R.S. On dispersal and planktonic larvae of benthic invertebrates: An eclectic overview and summary of problems. *Bull. Mar. Sci.* **1986**, *39*, 290–322.
15. Morgan, S.G. Predation by planktonic and benthic invertebrates on larvae of estuarine crabs. *J. Exp. Mar. Biol. Ecol.* **1992**, *163*, 91–110. [CrossRef]
16. Morgan, S.G. Life and death in the plankton: Larval mortality and adaptation. In *Ecology of Marine Invertebrate Larvae*; McEdward, L., Ed.; CRC Press: Boca Raton, FL, USA, 1995; pp. 279–321. ISBN 978-0-13-875895-0.
17. Hewitt, G. Some genetic consequences of ice ages, and their role in divergence and speciation. *Biol. J. Linn. Soc. Lond.* **1996**, *58*, 247–276. [CrossRef]

18. Gaylord, B.; Gaines, S.D. Temperature or transport? Range limits in marine species mediated solely by flow. *Am. Nat.* **2000**, *155*, 769–789. [CrossRef] [PubMed]

19. Anger, K. *The Biology of Decapod Crustacean Larvae*; A.A. Balkema Publishers: Lisse, The Netherlands, 2001; ISBN 978-90-265-1828-7.

20. Avise, J.C. Phylogeography: Retrospect and prospect. *J. Biogeogr.* **2009**, *36*, 3–15. [CrossRef]

21. Boehm, J.T.; Woodall, L.; Teske, P.R.; Lourie, S.A.; Baldwin, C.; Waldman, J.; Hickerson, M. Marine dispersal and barriers drive atlantic seahorse diversification. *J. Biogeogr.* **2013**, *40*, 1839–1849. [CrossRef]

22. Peres, P.A.; Mantelatto, F.L. Salinity tolerance explains the contrasting phylogeographic patterns of two swimming crabs species along the tropical western Atlantic. *Evol. Ecol.* **2020**, *34*, 589–609. [CrossRef]

23. Gopurenko, D.; Hughes, J.M. Regional patterns of genetic structure among Australian populations of the mud crab, *Scylla serrata* (Crustacea: Decapoda): Evidence from mitochondrial DNA. *Mar. Freshw. Res.* **2002**, *53*, 849–857. [CrossRef]

24. Oliveira-Neto, J.F.; Pie, M.R.; Boeger, W.A.; Ostrensky, A.; Baggio, R.A. Population genetics and evolutionary demography of *Ucides cordatus* (Decapoda: Ocypodidae). *Mar. Ecol.* **2007**, *28*, 460–469. [CrossRef]

25. Mayr, E. *Populações, Espécies e Evolução*; Cia Ed Nacional: São Paulo, Brasil, 1977.

26. Mandai, S.S.; Buranelli, R.C.; Schubart, C.D.; Mantelatto, F.L. Phylogenetic and phylogeographic inferences based on two DNA markers reveal geographic structure of the orange claw hermit crab *Calcinus tibicen* (Anomura: Diogenidae) in the western Atlantic. *Mar. Biol. Res.* **2018**, *14*, 565–580. [CrossRef]

27. Negri, M.; Pileggi, L.G.; Mantelatto, F.L. Molecular barcode and morphological analyses reveal the taxonomic and biogeographical status of the striped-legged hermit crab species *Clibanarius sclopetarius* (Herbst, 1796) and *Clibanarius vittatus* (Bosc, 1802) (Decapoda: Diogenidae). *Invert. Syst.* **2012**, *26*, 561–571. [CrossRef]

28. Forest, J.; Saint Laurent, M. Compagne de la calypso au large des côtes Atlantiques de l'Amérique Du Sud (1961–1962). 6. Crustacés Décapodes: Pagurides. *Ann. Inst. Oceanogr.* **1968**, *45*, 47–169.

29. Coelho, P.A.; Ramos-Porto, M. Sinopse dos crustáceos decápodos Brasileiros (Família Callianassidae, Callianideidae, Upogebiidae, Parapaguridae, Paguridae, Diogenidae). *Trop. Ocean.* **1987**, *19*, 27–53. [CrossRef]

30. De Melo, G.A.S. *Manual de identificação dos Crustacea Decapoda do Litoral Brasileiro: Anomura, Thalassinidea, Palinuridea, Astacidea*; Editora Plêiade: São Paulo, Brasil, 1999; ISBN 978-85-85795-08-5.

31. Nucci, P.R.; De Melo, G.A.S. Hermit crabs from Brazil: Family Diogenidae (Crustacea: Decapoda: Paguroidea), except *Paguristes*. *Zootaxa* **2015**, *3947*, 327–346. [CrossRef] [PubMed]

32. Mantelatto, F.L.; Miranda, I.; Vera-Silva, A.L.; Negri, M.; Buranelli, R.C.; Terossi, M.; Magalhães, T.; Costa, R.C.; Zara, F.J.; Castilho, A.L. Checklist of decapod crustaceans from the coast of the São Paulo State (Brazil) supported by integrative molecular and morphological data: IV. Infraorder Anomura: Superfamilies Chirostyloidea, Galatheoidea, Hippoidea and Paguroidea. *Zootaxa* **2021**, in press.

33. Hazlett, B.A. The behavioral ecology of hermit crabs. *Annu. Rev. Ecol. Syst.* **1981**, *12*, 1–22. [CrossRef]

34. Brossi-Garcia, A.L.; Hebling, N.J. Desenvolvimento pós-embrionário de *Clibanarius antillensis* Stimpson, 1859 (Crustacea, Diogenidae), em laboratório. *Bol. Zool.* **1983**, *6*, 89–111. [CrossRef]

35. Siddiqui, F.A.; McLaughlin, P.A.; Crain, J.A. Larval development of *Clibanarius antillensis* Stimpson, 1859 (Crustacea: Anomura: Diogenidae) reared under laboratory conditions: A comparison between Panamanian and Brazilian populations. *J. Nat. Hist.* **1991**, *25*, 917–932. [CrossRef]

36. Cházaro-Olvera, S.; Robles, R.; Montoya-Mendoza, J.; Herrera-López, J.A. Intraspecific variation in Megalopae of *Clibanarius antillensis* (Anomura, Diogenidae) among western Atlantic populations. *Nauplius* **2018**, *26*, e2018031. [CrossRef]

37. Mantelatto, F.L.; (University of São Paulo Ribeirão Preto, São Paulo, Brazil). Personal Communication, 2014.

38. Stimpson, W. XI.-Notes on North American Crustacea, No. 1. *Ann. Lyceum Nat. Hist. N. Y.* **1859**, *7*, 49–93. [CrossRef]

39. Provenzano, A.J. The shallow-water hermit crabs of Florida. *Bull. Mar. Sci.* **1959**, *9*, 349–420.

40. Schubart, C.; Neigel, J.; Felder, D. The use of the mitochondrial 16S rRNA gene for phylogenetic and biogeographic studies of crustacea. *Crustac. Issues* **2000**, *12*, 817–830.

41. Mantelatto, F.L.; Robles, R.; Felder, D.L. Molecular phylogeny of the western Atlantic species of the Genus *Portunus* (Crustacea, Brachyura, Portunidae). *Zool. J. Linn. Soc.* **2007**, *150*, 211–220. [CrossRef]

42. Estoup, A.; Largiader, C.; Perrot, E.; Chourrout, D. Rapid one-tube DNA extraction for reliable PCR detection of fish polymorphic markers and transgenes. *Mol. Mar. Biol. Biotechnol.* **1996**, *5*, 295–298.

43. Sambrook, J.; Fritsch, E.F.; Maniatis, T. In vitro amplification of DNA by the polymerase chain reaction. In *Molecular Cloning: A Laboratory Manual*; Sambrook, J., Fritsch, E.F., Maniatis, T., Eds.; Cold Spring Harbor Laboratory Press: New York, NY, USA, 1989; Volume 2, pp. 2–35.

44. Negri, M.; Lemaitre, R.; Mantelatto, F.L. Molecular and morphological resurrection of *Clibanarius symmetricus* (Randall, 1840), a cryptic species hiding under the name for the "Thinstripe" hermit crab *C. vittatus* (Bosc, 1802) (Decapoda: Anomura: Diogenidae). *J. Crust. Biol.* **2014**, *34*, 848–861. [CrossRef]

45. Buranelli, R.C.; Mantelatto, F.L. Broad-ranging low genetic diversity among populations of the yellow finger marsh crab *Sesarma rectum* Randall, 1840 (Sesarmidae) Revealed by DNA Barcode. *Crustaceana* **2017**, *90*, 845–864. [CrossRef]

46. Ye, J.; Coulouris, G.; Zaretskaya, I.; Cutcutache, I.; Rozen, S.; Madden, T.L. Primer-BLAST: A tool to design target-specific primers for polymerase chain reaction. *BMC Bioinform.* **2012**, *13*, 134. [CrossRef] [PubMed]

47. Schubart, C.D.; Huber, M.G.J. Genetic comparisons of German populations of the stone crayfish, *Austropotamobius torrentium* (Crustacea: Astacidae). *Bull. Fr. Pêche Piscic.* **2006**, *318*, 1019–1028. [CrossRef]

48. Mantelatto, F.L.; Carvalho, F.L.; Simões, S.M.; Negri, M.; Souza-Carvalho, E.A.; Terossi, M. New primers for amplification of cytochrome c oxidase subunit I barcode region designed for species of Decapoda (Crustacea). *Nauplius* **2016**, *24*, e2016030. [CrossRef]

49. Hall, T.A. BioEdit: A user-friendly biological sequence alignment editor and analysis program for windows 95/98/NT. *Nucl. Acids. Symp. Ser.* **1999**, *41*, 95–98.

50. Altschul, S.F.; Gish, W.; Miller, W.; Myers, E.W.; Lipman, D.J. Basic local alignment search tool. *J. Mol. Biol.* **1990**, *215*, 403–410. [CrossRef]

51. Artimo, P.; Jonnalagedda, M.; Arnold, K.; Baratin, D.; Csardi, G.; de Castro, E.; Duvaud, S.; Flegel, V.; Fortier, A.; Gasteiger, E.; et al. ExPASy: SIB Bioinformatics Resource Portal. *Nucleic Acids Res.* **2012**, *40*, W597–W603. [CrossRef] [PubMed]

52. Edgar, R.C. MUSCLE: Multiple sequence alignment with high accuracy and high throughput. *Nucleic Acids Res.* **2004**, *32*, 1792–1797. [CrossRef] [PubMed]

53. Bracken-Grissom, H.D.; Cannon, M.E.; Cabezas, P.; Feldmann, R.M.; Schweitzer, C.E.; Ahyong, S.T.; Felder, D.L.; Lemaitre, R.; Crandall, K.A. A Comprehensive and Integrative reconstruction of evolutionary history for Anomura (Crustacea: Decapoda). *BMC Evol. Biol.* **2013**, *13*, 128. [CrossRef] [PubMed]

54. Tamura, K.; Stecher, G.; Peterson, D.; Filipski, A.; Kumar, S. MEGA6: Molecular evolutionary genetics analysis version 6.0. *Mol. Biol. Evol.* **2013**, *30*, 2725–2729. [CrossRef] [PubMed]

55. Kimura, M. A simple method for estimating evolutionary rates of base substitutions through comparative studies of nucleotide sequences. *J. Mol. Evol.* **1980**, *16*, 111–120. [CrossRef] [PubMed]

56. Felsenstein, J. Evolutionary trees from DNA sequences: A maximum likelihood approach. *J. Mol. Evol.* **1981**, *17*, 368–376. [CrossRef]

57. Stamatakis, A. RAxML-VI-HPC: Maximum likelihood-based phylogenetic analyses with thousands of taxa and mixed models. *Bioinformatics* **2006**, *22*, 2688–2690. [CrossRef]

58. Tavaré, S. Some probabilistic and statistical problems in the analysis of DNA sequences. In *Lectures on Mathematics in the Life Sciences*; Miura, R.M., Ed.; American Mathematical Society: Providence, RI, USA, 1986; Volume 17, pp. 57–86.

59. Rambaut, A.S.M.; Suchard, M.A.; Xie, D.; Drummond, A.J. Tracer 1.6. Available online: http://beast.bio.ed.ac.uk/Tracer (accessed on 15 November 2019).

60. Rozas, J.; Rozas, R. DnaSP Version 3: An integrated program for molecular population genetics and molecular evolution analysis. *Bioinformatics* **1999**, *15*, 174–175. [CrossRef]

61. Clement, M.; Posada, D.; Crandall, K.A. TCS: A computer program to estimate gene genealogies. *Mol. Ecol.* **2000**, *9*, 1657–1659. [CrossRef]

62. Excoffier, L.; Langaney, A. Origin and differentiation of human mitochondrial DNA. *Am. J. Hum. Genet.* **1989**, *44*, 73–85.

63. Excoffier, L.; Lischer, H.E.L. Arlequin suite version 3.5: A new series of programs to perform population genetics analyses under Linux and Windows. *Mol. Ecol. Resour.* **2010**, *10*, 564–567. [CrossRef] [PubMed]

64. Tajima, F. Statistical method for testing the neutral mutation hypothesis by DNA polymorphism. *Genetics* **1989**, *123*, 585–595. [CrossRef] [PubMed]

65. Fu, Y.X. Statistical tests of neutrality of mutations against population growth, hitchhiking and background selection. *Genetics* **1997**, *147*, 915–925. [CrossRef] [PubMed]

66. Rogers, A.R.; Harpending, H. Population growth makes waves in the distribution of pairwise genetic differences. *Mol. Biol. Evol.* **1992**, *9*, 552–569. [CrossRef] [PubMed]

67. Schneider, S.; Excoffier, L. Estimation of past demographic parameters from the distribution of pairwise differences when the mutation rates vary among sites: Application to human mitochondrial DNA. *Genetics* **1999**, *152*, 1079–1089.

68. Harpending, H.C. Signature of ancient population growth in a low-resolution mitochondrial DNA mismatch distribution. *Hum. Biol.* **1994**, *66*, 591–600.

69. Drummond, A.J.; Rambaut, A.; Shapiro, B.; Pybus, O.G. Bayesian coalescent inference of past population dynamics from molecular sequences. *Mol. Biol. Evol.* **2005**, *22*, 1185–1192. [CrossRef]

70. Hasegawa, M.; Kishino, H.; Yano, T. Dating of the human-ape splitting by a molecular clock of mitochondrial DNA. *J. Mol. Evol.* **1985**, *22*, 160–174. [CrossRef]

71. Darriba, D.; Taboada, G.L.; Doallo, R.; Posada, D. JModelTest 2: More models, new heuristics and parallel computing. *Nat. Methods* **2012**, *9*, 772. [CrossRef]

72. Knowlton, N.; Weigt, L.A. New dates and new rates for divergence across the Isthmus of Panama. *Proc. R. Soc. Lond. B* **1998**, *265*, 2257–2263. [CrossRef]

73. Wright, S. *Evolution and the Genetics of Populations: A Treatise*; University of Chicago Press: Chicago, IL, USA, 1978; ISBN 978-0-226-91049-9.

74. Benedict, J.E. Four new symmetrical hermit crabs (Pagurids) from the West India Region. *Proc. USA Natl. Mus.* **1901**, *23*, 771–778. [CrossRef]

75. Dana, J.D. Crustacea. Part I. In *United States Exploring Expedition during the Year 1838, 1839, 1840, 1841, 1842. under the Command of Charles Wilkes, U.S.N.*; C. Sherman: Philadelphia, PA, USA, 1852; Volume 13, p. 685.

76. Stimpson, W. Prodomus descriptionis animalium evertebratorum, quae in expeditione ad oceanum Pacificum Septentrionalem, a Republica Federate Missa, Cadevaladero Ringgold et Johanne Rodgers ducibus, observavit et descripsit. VII. Crustacea Anomura. *Proc. Acad. Nat. Sci. Phila.* **1858**, *10*, 225–252.

77. Smith, S.I. Notice of the Crustacea collected by prof. C.F. Hart on the coast of Brazil in 1867. *Trans. Conn. Acad. Arts Sci.* **1869**, *2*, 1–41.

78. Nobili, G. Decapodi i stomatopodi raccolti dal dr. Enrico Festa Nel Darien, Curaçao, La Guayra, Porto Cabello, Colon, Panama. *Boll. Musei Zool. Anat. Comp. R Univ. Torino* **1897**, *12*, 1–8.

79. Rathbun, M.J. Results of the Branner-Agassiz expedition to Brazil. I. The Decapod and Stomatopod Crustacea. *Proc. Wash. Acad. Sci. USA* **1900**, *2*, 133–156.

80. Moreira, C. Contribuições para o conhecimento da fauna Brasileira. Crustáceos do Brasil. *Arch. Mus. Nac.* **1901**, *11*, 1–151.

81. Schmitt, W.L. Report on the Macrura, Anomura and Stomatopoda collected by the Barbados-Antigua expedition from the university of Iowa in 1918. *Univ. Iowa Stud. Nat. Hist.* **1924**, *10*, 65–99.

82. Schmitt, W.L. Crustacea Macrura and Anomura of Porto Rico and the Virgin Islands. Scientific survey of Porto Rico and the Virgin Islands. *N. Y. Acad. Sci.* **1935**, *15*, 125–227.

83. Schmitt, W.L. Zoologische ergebnisse einer reise nach bonaire, Curaçao und Aruba im jahre 1930. No. 16. Macruran and Anomuran Crustacea from Bonaire, Curaçao und Aruba. *Zool. Jahrb. Abt. Syst. Geog. Biol. Tiere* **1936**, *67*, 363–378.

84. Provenzano, A.J. Notes on Bermuda hermit crabs (Crustacea: Anomura). *Bull. Mar. Sci.* **1960**, *10*, 117–124.

85. Provenzano, A.J. Pagurid Crabs (Decapoda Anomura) from St. John, Virgin Islands, with descriptions of three new species. *Crustaceana* **1961**, *3*, 151–166. [CrossRef]

86. Coelho, P.A.; Ramos, M.D.A. A constituição e a distribuição da fauna de decápodos do litoral leste da América do sul entre as latitudes de 5° N e 39° S. *Trop. Ocean.* **1972**, *13*, 133–236. [CrossRef]

87. Rieger, P.J. Malacostraca—Eucarida. Paguroidea. In *Catalogue of Crustacea of Brazil*; Young, P.S., Ed.; Museu Nacional: Rio de Janeiro, Brazil, 1998; pp. 413–429. ISBN 978-85-7427-001-2.

88. McLaughlin, P.A.; Komai, T.; Lemaitre, R.; Rahayu, D.L. annotated checklist of Anomuran Decapod Crustaceans of the world (exclusive of the Kiwaoidea and families Chirostylidae and Galatheidae of the Galatheoidea). Part 1, Lithodoidea, Lomisoidea and Paguroidea. *Raffles Bull. Zool.* **2010**, *23*, 5–107.

89. Lemaitre, R.; McLaughlin, P. World Paguroidea & Lomisoidea database. *Clibanarius antillensis* Stimpson, 1859. Available online: http://marinespecies.org/aphia.php?p=taxdetails&id=367492 (accessed on 18 January 2021).

90. Rodríguez-Rey, G.T.; Solé-Cava, A.M.; Lazoski, C. Genetic homogeneity and historical expansions of the slipper lobster, *Scyllarides brasiliensis*, in the South-West Atlantic. *Mar. Freshw. Res.* **2014**, *65*, 59–69. [CrossRef]

91. Buranelli, R.C.; Felder, D.L.; Mantelatto, F.L. Genetic diversity among populations of the western Atlantic mangrove crab *Ucides cordatus* (Linnaeus, 1763) (Decapoda: Brachyura: Ocypodidae): Evidence for Panmixia and useful data for future management and conservation. *J. Crust. Biol.* **2019**, *39*, 386–395. [CrossRef]

92. Liu, Y.; Liu, R.; Ye, L.; Liang, J.; Xuan, F.; Xu, Q. Genetic differentiation between populations of swimming crab *Portunus trituberculatus* along the coastal waters of the East China Sea. *Hydrobiology* **2009**, *618*, 125–137. [CrossRef]

93. Scheltema, R.S. larval dispersal as a means of genetic exchange between geographically separated populations of shoal-water benthic marine gastropods. *Biol. Bull.* **1971**, *140*, 284–322. [CrossRef]

94. Crisp, D.J. Genetic consequences of different reproductive strategies in marine invertebrates. In *Marine Organisms: Genetics, Ecology, and Evolution*; Battaglia, B., Beardmore, J.A., Eds.; Nato Conference Series, IV: Marine Sciences; Plenum Press: New York, NY, USA, 1978; pp. 257–273. ISBN 978-0-306-40020-9.

95. Palumbi, S.R.; Wilson, A.C. Mitochondrial DNA diversity in the sea urchins *Strongylocentrotus purpuratus* and *S. droebachiensis*. *Evolution* **1990**, *44*, 403–415. [CrossRef]

96. Cirano, M.; Mata, M.M.; Campos, E.J.D.; Deiró, N.F.R. A Circulação oceânica de larga-escala na região oeste do Atlântico sul com base no modelo de circulação global OCCAM. *Rev. Bras. Geof.* **2006**, *24*, 209–230. [CrossRef]

97. Puchnick-Legat, A.; Levy, J.A. Genetic structure of Brazilian populations of white mouth croaker *Micropogonias furnieri* (Perciformes: Sciaenidae). *Braz. Arch. Biol. Technol.* **2006**, *49*, 429–439. [CrossRef]

98. Affonso, P.; Galetti, P.M., Jr. Genetic diversity of three ornamental reef fishes (Families Pomacanthidae and Chaetodontidae) from the Brazilian Coast. *Braz. J. Biol.* **2007**, *67*, 925–933. [CrossRef] [PubMed]

99. Day, R.; McEdward, L. Aspects of the physiology and ecology of pelagic larvae of marine benthic invertebrates. In *Marine Plankton Life Cycle Strategies*; Steidinger, K.A., Walker, L.M., Eds.; CRC Press: Boca Raton, FL, USA, 1984; pp. 93–120. ISBN 978-0-8493-5222-5.

100. Lárez, M.B.; Palazón-Fernández, J.L.; Bolaños, C.J. The effect of salinity and temperature on the larval development of *Mithrax caribbaeus* Rathbun, 1920 (Brachyura: Majidae) reared in the laboratory. *J. Plankton Res.* **2000**, *22*, 1855–1869. [CrossRef]

101. Torres, G.; Giménez, L.; Anger, K. Effects of reduced salinity on the biochemical composition (lipid, protein) of Zoea 1 decapod crustacean larvae. *J. Exp. Mar. Biol. Ecol.* **2002**, *277*, 43–60. [CrossRef]

102. Anger, K.; Spivak, E.; Luppi, T. Effects of reduced salinities on development and bioenergetics of early larval shore crab, *Carcinus maenas*. *J. Exp. Mar. Biol. Ecol.* **1998**, *220*, 287–304. [CrossRef]

103. Gonçalves, R.R.; Masui, D.C.; McNamara, J.C.; Mantelatto, F.L.M.; Garçon, D.P.; Furriel, R.P.M.; Leone, F.A. A kinetic study of the gill (Na+, K+)-ATPase, and its role in ammonia excretion in the intertidal hermit crab, *Clibanarius vittatus*. *Comp. Biochem. Physiol. Part A Mol. Integr. Physiol.* **2006**, *145*, 346–356. [CrossRef]

104. Terossi, M.; Mantelatto, F.L. Morphological and genetic variability in *Hippolyte obliquimanus* Dana, 1852 (Decapoda, Caridea, Hippolytidae) from Brazil and the Caribbean Sea. *Crustaceana* **2012**, *85*, 685–712. [CrossRef]

105. Tourinho, J.L.; Solé-Cava, A.M.; Lazoski, C. Cryptic species within the commercially most important lobster in the tropical Atlantic, the spiny lobster *Panulirus argus*. *Mar. Biol.* **2012**, *159*, 1897–1906. [CrossRef]

106. Duffy, J.E. Resource-associated population subdivision in a symbiotic coral-reef shrimp. *Evolution* **1996**, *50*, 360–373. [CrossRef]

107. Palumbi, S.R.; Grabowsky, G.; Duda, T.; Geyer, L.; Tachino, N. Speciation and population genetic structure in tropical pacific sea urchins. *Evolution* **1997**, *51*, 1506–1517. [CrossRef]

108. Benzie, J.A.H. Major genetic differences between crown-ofthorns starfish (*Acanthaster planci*) populations in the Indian and Pacific Oceans. *Evolution* **1999**, *53*, 1782–1795. [CrossRef]

109. Wieman, A.C.; Berendzen, P.B.; Hampton, K.R.; Jang, J.; Hopkins, M.J.; Jurgenson, J.; McNamara, J.C.; Thurman, C.L. A panmictic fiddler crab from the coast of Brazil? Impact of divergent ocean currents and larval dispersal potential on genetic and morphological variation in *Uca maracoani*. *Mar. Biol.* **2014**, *161*, 173–185. [CrossRef]

110. Grant, W.; Bowen, B.W. Shallow population histories in deep evolutionary lineages of marine fishes: Insights from sardines and anchovies and lessons for conservation. *J. Hered.* **1998**, *89*, 415–426. [CrossRef]

111. Freeland, J. *Molecular Ecology*; John Wiley & Sons: Chichester, UK; Hoboken, NJ, USA, 2005; ISBN 978-0-470-09061-9.

112. Cassone, B.J.; Boulding, E.G. Genetic structure and phylogeography of the lined shore crab, *Pachygrapsus crassipes*, along the northeastern and western Pacific coasts. *Mar. Biol.* **2006**, *149*, 213–226. [CrossRef]

113. Van Tienderen, K.M.; van der Meij, S.E.T. Extreme mitochondrial variation in the Atlantic gall crab *Opecarcinus hypostegus* (Decapoda: Cryptochiridae) reveals adaptive genetic divergence over *Agaricia* coral hosts. *Sci. Rep.* **2017**, *7*, 39461. [CrossRef] [PubMed]

114. Aris-Brosou, S.; Excoffier, L. The impact of population expansion and mutation rate heterogeneity on DNA sequence polymorphism. *Mol. Biol. Evol.* **1996**, *13*, 494–504. [CrossRef]

115. Bucklin, A.; Wiebe, P.H. Low mitochondrial diversity and small effective population sizes of the copepods *Calanus finmarchicus* and *Nannocalanus minor*: Possible impact of climatic variation during recent glaciation. *J. Hered.* **1998**, *89*, 383–392. [CrossRef] [PubMed]

116. Duran, S.; Palacín, C.; Becerro, M.A.; Turon, X.; Giribet, G. Genetic diversity and population structure of the commercially harvested sea urchin *Paracentrotus lividus* (Echinodermata, Echinoidea). *Mol. Ecol.* **2004**, *13*, 3317–3328. [CrossRef]

117. Stamatis, C.; Triantafyllidis, A.; Moutou, K.A.; Mamuris, Z. Mitochondrial DNA variation in northeast Atlantic and Mediterranean populations of Norway lobster, *Nephrops norvegicus*. *Mol. Ecol.* **2004**, *13*, 1377–1390. [CrossRef]

118. Ray, N.; Currat, M.; Excoffier, L. Intra-deme molecular diversity in spatially expanding populations. *Mol. Biol. Evol.* **2003**, *20*, 76–86. [CrossRef]

119. Excoffier, L. Patterns of DNA sequence diversity and genetic structure after a range expansion: Lessons from the infinite-island model. *Mol. Ecol.* **2004**, *13*, 853–864. [CrossRef]

120. Imbrie, J.; Boyle, E.A.; Clemens, S.C.; Duffy, A.; Howard, W.R.; Kukla, G.; Kutzbach, J.; Martinson, D.G.; McIntyre, A.; Mix, A.C.; et al. On the structure and origin of major glaciation cycles 1. Linear responses to Milankovitch forcing. *Paleoceanography* **1992**, *7*, 701–738. [CrossRef]

121. Hellberg, M.E. Genetic approaches to understanding marine metapopulation dynamics. In *Marine Metapopulations*; Kritzer, J.P., Sale, P.F., Eds.; Elsevier: Amsterdam, The Netherlands, 2006; pp. 431–455. ISBN 978-0-12-088781-1.

122. Roy, K.; Valentine, J.W.; Jablonski, D.; Kidwell, S.M. Scales of climatic variability and time averaging in pleistocene biotas: Implications for ecology and evolution. *Trends Ecol. Evol.* **1996**, *11*, 458–463. [CrossRef]

123. Hewitt, G.M. Post-glacial re-colonization of European Biota. *Biol. J. Linn. Soc. Lond.* **1999**, *68*, 87–112. [CrossRef]

124. Provan, J.; Bennett, K. Phylogeographic insights into cryptic glacial refugia. *Trends Ecol. Evol.* **2008**, *23*, 564–571. [CrossRef]

125. Marko, P.B.; Hoffman, J.M.; Emme, S.A.; Mcgovern, T.M.; Keever, C.C.; Nicole Cox, L. The 'Expansion-Contraction' model of Pleistocene biogeography: Rocky shores suffer a sea change? *Mol. Ecol.* **2010**, *19*, 146–169. [CrossRef]

126. Austerlitz, F.; Jung-Muller, B.; Godelle, B.; Gouyon, P.-H. Evolution of coalescence times, genetic diversity and structure during colonization. *Popul. Biol.* **1997**, *51*, 148–164. [CrossRef]

127. Petit, R.J.; Aguinagalde, I.; de Beaulieu, J.-L.; Bittkau, C.; Brewer, S.; Cheddadi, R.; Ennos, R.; Fineschi, S.; Grivet, D.; Lascoux, M.; et al. Glacial Refugia: Hotspots but not melting pots of genetic diversity. *Science* **2003**, *300*, 1563–1565. [CrossRef]

128. Cárdenas, L.; Castilla, J.C.; Viard, F. A phylogeographical analysis across three biogeographical provinces of the south-eastern Pacific: The case of the marine gastropod *Concholepas concholepas*. *J. Biogeogr.* **2009**, *36*, 969–981. [CrossRef]

129. Ibáñez, C.M.; Argüelles, J.; Yamashiro, C.; Adasme, L.; Céspedes, R.; Poulin, E. Spatial genetic structure and demographic inference of the Patagonian squid *Doryteuthis gahi* in the south-eastern Pacific Ocean. *J. Mar. Biol. Ass.* **2012**, *92*, 197–203. [CrossRef]

130. Mclaughlin, P.A. *Paguristes puniceus* Henderson, 1896 (Decapoda: Anomura: Paguroidea: Diogenidae): A study in intraspecific variability. *Zootaxa* **2004**, *742*, 1. [CrossRef]

131. Hermoso-Salazar, M.; Solís-Weiss, V. Distribution and morphological variation of *Synalpheus superus* Abele and Kim, 1989 and notes on the distribution of *S. fritzmuelleri* Coutière, 1909 (Decapoda: Caridea: Alpheidae). *Zootaxa* **2010**, *2505*, 65. [CrossRef]
132. Negri, M.; Schubart, C.D.; Mantelatto, F.L. Tracing the introduction history of the invasive swimming crab *Charybdis hellerii* (A. Milne-Edwards, 1867) in the western Atlantic: Evidences of high genetic diversity and multiple introductions. *Biol. Invasions* **2018**, *20*, 1771–1798. [CrossRef]
133. Schluter, D.; Price, T. Conflicting selection pressures and life history trade-offs. *Proc. R. Soc. Lond. B* **1991**, *246*, 11–17. [CrossRef]
134. Debuse, V.J.; Addison, J.T.; Reynolds, J.D. Morphometric variability in UK populations of the European lobster. *J. Mar. Biol. Assoc.* **2001**, *81*, 469–474. [CrossRef]
135. Brian, J.V.; Fernandes, T.; Ladle, R.J.; Todd, P.A. Patterns of morphological and genetic variability in UK Populations of the shore crab, *Carcinus maenas* Linnaeus, 1758 (Crustacea: Decapoda: Brachyura). *J. Exp. Mar. Biol. Ecol.* **2006**, *329*, 47–54. [CrossRef]
136. Rousset, F. Genetic approaches to the estimation of dispersal rates. In *Dispersal*; Clobert, J., Danchin, E., Dhondt, A.A., Nichols, J.D., Eds.; Oxford University Press: Oxford, UK, 2001; ISBN 978-0-19-850660-7.
137. Avise, J.C. Molecular population structure and the biogeographic history of a regional fauna: A case history with lessons for conservation biology. *Oikos* **1992**, *63*, 62. [CrossRef]
138. Bickford, D.; Lohman, D.J.; Sodhi, N.S.; Ng, P.K.L.; Meier, R.; Winker, K.; Ingram, K.K.; Das, I. Cryptic species as a window on diversity and conservation. *Trends Ecol. Evol.* **2007**, *22*, 148–155. [CrossRef]

Article

Diversity and Molecular Phylogeny of Pagurid Hermit Crabs (Anomura: Paguridae: *Pagurus*)

Zakea Sultana [1,*], **Isaac Adeyemi Babarinde** [2] **and Akira Asakura** [3]

[1] Japan International Research Center for Agricultural Sciences, 1-1 Ohwashi, Ibaraki 305-8686, Japan
[2] Shenzhen Key Laboratory of Gene Regulation and Systems Biology, Department of Biological Sciences, Southern University of Science and Technology, Shenzhen 518055, China; babarindeia@sustech.edu.cn
[3] Faculty of Agriculture (Main Building), Kyoto University (Yoshida North Campus), Kitashirakawa Oiwake-cho, Sakyo-ku, Kyoto 606-8502, Japan; asakura.akira.6w@kyoto-u.ac.jp
* Correspondence: zadabor@gmail.com; Tel.: +81-80-3829-0625

Abstract: Species of the genus *Pagurus* have diversified into a wide variety of marine habitats across the world. Despite their worldwide abundance, the genus diversity and biogeographical relationship are relatively less understood at species-level. We evaluated the phylogenetic relationship and genetic diversity among the *Pagurus* species based on publicly available mitochondrial and nuclear markers. While independent analyses of different markers allowed for larger coverage of taxa and produced largely consistent results, the concatenation of 16S and COI partial sequences led to higher confidence in the phylogenetic relationships. Our analyses established several monophyletic species clusters, substantially corresponding to the previously established morphology-based species groups. The comprehensive species inclusion in the molecular phylogeny resolved the taxonomic position of a number of recently described species that had not been assigned to any morpho-group. In mitochondrial markers-based phylogenies, the *"Provenzanoi"* group was identified as the basal lineage of *Pagurus*. The divergence time estimation of the major groups of *Pagurus* revealed that the Pacific species originated and diversified from the Atlantic lineages around 25–51 MYA. The molecular results suggested a higher inter-regional species diversity and complex phylogenetic relationships within the diverse and heterogeneous members of the genus *Pagurus*. The study presents a comprehensive snapshot of the diversity of pagurid hermit crabs across multiple geographic regions.

Keywords: hermit crab; Paguridae; diversity; molecular phylogeny

Citation: Sultana, Z.; Babarinde, I.A.; Asakura, A. Diversity and Molecular Phylogeny of Pagurid Hermit Crabs (Anomura: Paguridae: *Pagurus*). *Diversity* **2022**, *14*, 141. https://doi.org/10.3390/d14020141

Academic Editors: Patricia Briones-Fourzán, Michel E. Hendrickx and Michael Wink

Received: 26 November 2021
Accepted: 11 February 2022
Published: 16 February 2022

Publisher's Note: MDPI stays neutral with regard to jurisdictional claims in published maps and institutional affiliations.

1. Introduction

Hermit crabs (Anomura) are a morphologically and ecologically highly diverse group of decapod crustaceans, inhabiting various dimensions of elevations from inland to deep-ocean. To cope with versatile ecological biotopes, species of hermit crabs have evolved a variety of specialized body forms and behaviors. In fact, species of this group are well-recognized for their asymmetric, decalcified, and coiled hind body (pleon) which can be looped into gastropod shells for protective shelter [1,2]. These attributes have long attracted the interests of evolutionary biologists for addressing various questions in the evolution of asymmetrical body form and phenotypic diversity. Moreover, being the ancestor of complete carcinized king crabs (Lithodidae) [3], these species are potentially excellent models for investigating adaptive ecology, comparative population biology, speciation, and biogeographic process [4]. While several aspects of phylogenetic relationship of anomuran have been extensively studied at higher taxonomic levels [5–16], a large-scale species level phylogeny and diversity using molecular data has not been exhaustively investigated.

Among hermit crabs, the genus *Pagurus* Fabricius, 1775 [17] is one of the most diverse groups with high morphological and ecological variability among the species [18]. The genus encompasses about 176 species worldwide [19]. The most obvious feature for differentiating *Pagurus* from other hermit crabs is the larger right cheliped, hence they are also known as right-handed hermit crabs. They are very common and important members among the macro-zoobenthic groups in the intertidal and subtidal communities. The pelagic larval stages (typically four stages) of *Pagurus* facilitate their wide distribution. However, due to the reliance on the abandoned shells of gastropods for protection; establishment of *Pagurus* populations in a particular environment is directly linked to the availability of marine gastropod shells and their quality and quantity [20]. Furthermore, the interspecific differences in the patterns of shell utilization also may limit the range of biogeographical distribution of these species [21]. Therefore, species composition and distribution of a sympatric community of the intertidal hermit crabs are largely associated with the features of resource (gastropod shell) partitioning [22]. Consequently, these species can potentially be used as an index for the study of biological interactions (i.e., competition for shell or other resources) and adaptive mechanisms of successful colonization of a variety of habitats ranging from costal to deep sea.

Over several decades, the classification of *Pagurus* hermit crabs has largely been carried out by reviewing both adult and larval morphology extensively, leading to a list of hypothetical species relationships [6–16,23–29]. For example, considering the characters of males (presence or absences of a pleopod on the second abdominal somite), Bouvier (1940) divided the Northeast Atlantic species into Group I (presence of a pleopod) and Group II (absences of a pleopod) [23]. After describing the larval and adult characters of the European species (northeastern Atlantic-Mediterranean regions), Ingle (1985) categorized the *Pagurus* species into eight groups under two subdivisions: Subdivision I (three groups) and Subdivision II (five groups) [28]. Later, McLaughlin et al. (1988) designated four larval groups and 12 adult groups in the eastern Pacific species (with two "unnamed group") of *Pagurus* [29]. However, many species of the "unnamed group" (Group XII) were previously assigned to Ingle's Subdivision I (Table 1). In the Ingle (1985) and McLaughlin et al. (1988) species arrangements, there are three groups (*"Bernhardus"*, *"Anachoretus"*, and *"Trigonocheirus"*) in common. As a result, there have been 15 groups recognized in the genus *Pagurus* based on adult morphology up until now (Table 1). However, these lists were geographically specific to the Atlantic-Mediterranean species [30], many species from the Indo-Pacific region were not included. Further, numerous new species have been described in more than three decades since the publications of Ingle (1985) and McLaughlin et al. (1988). Therefore, a number of species are yet to be assigned to the morpho-groups.

Table 1. Species groups of the genus *Pagurus* based on adults' morphological characters.

Species groups of the genus *Pagurus*	Ingle (1985) [28] — Subdivision I — Group 1 "Alatus"	Subdivision I — Group 2 "Prideaux"	Subdivision I — Group 3 "Cuanensis"	Subdivision II — Group 1 (Group V) "Bernhardus"	Subdivision II — Group 2 (Group VIII) "Trigonocheirus"	Subdivision II — Group 3 (Group XI) "Anachoretus"	Subdivision II — Group 4 "Chevreuxi"	Subdivision II — Group 5 "Carneus"	McLaughlin et al. (1988) [29] — Group I "Provenzanoi"	Group II "Exilis"	Group III "Comptus"	Group IV "Gaudichaudii"	Group VI "Capillatus"	Group VII "Confragosus"	Group X "Smithi"	Group IX "Unnamed"	Group XII "Unnamed"
Species assigned in the groups	*P. alatus*	*P. prideaux*	*P. cuanensis*	*P. aleuticus*	*P. brandti*	*P. anachoretoides*	*P. chevreuxi*	*P. carneus*	*P. annexus**	*P. albus**	*P. beringianus*	*P. gaudichaudii*	*P. arcuatus*	*P. confragosus*	*P. impressus*	*P. delsolari*	*P. alatus*
	P. excavatus	*P. pycnacanthus*	*P. forbesii*	*P. armatus*	*P. dalli*	*P. anachoretus*			*P. arenisaxatilis**	*P. gladius*	*P. caurinus*		*P. capillatus*	*P. cornutus*	*P. pollicaris*	*P. imarpe*	*P. cuanensis*
	P. pubescentulus			*P. bernhardus*	*P. pubescens*	*P. gordonae*			*P. benedicti*	*P. limatulus*	*P. comptus*		*P. kennerlyi*	*P. tanneri*	*P. smithi*		*P. ruber*
	P. pulchellus			*P. ochotensis*	*P. stevensae*	*P. laurentae*			*P. brevidactylus*	*P. longicarpus*	*P. edwardsii*		*P. setosus*		*P. spilocarpus*		*P. sculptimanus*
					P. trigonocheirus	*P. souriei*			*P. carolinensis*	*P. mertensii*	*P. forceps*						*P. pubescentulus*
					P. undosus	*P. triangularis*			*P. criniticornis*	*P. perlatus**	*P. granosimanus*						*P. prideaux*
									P. galapagensis		*P. hemphilli*						*P. variailis*
									P. gymnodactylus		*P. hirsutiusculus*						
									P. lepidus		*P. middendorffii*						
									P. leptonyx		*P. samuelis*						
									P. maclaughlinae		*P. quaylei*						
									*P. nanodes**								
									*P. nesiotes**								
									P. protuberocarpus								
									P. provenzanoi								
									P. redondoensis								
									*P. retrorsimanus**								
									*P. rhabdotus**								
									*P. spighti**								
									P. stimpsoni								
									P. trichocerus								
									*P. vetaultae**								
									P. villosus								
									*P. virgulatus**								

Table 1. *Cont.*

References								Ingle (1985) [28]								McLaughlin et al. (1988) [29]	
Compiled species groups of the genus *Pagurus*	Group 1 "Alatus"	Group 2 "Prideaux"	Group 3 "Cuanensis"	Group 4 "Bernhardus"	Group 5 "Trigonocheirus"	Group 6 "Anachoretus"	Group 7 "Chevreuxi"	Group 8 "Carneus"	Group 9 "Provenzanoi"	Group 10 "Exilis"	Group 11 "Comptus"	Group 12 "Gaudichaudii"	Group 13 "Capillatus"	Group 14 "Confragosus"	Group 15 "Smithi"	Group 16 "Unnamed" (combined of Group IX and XII)	Ungrouped Species
Species assigned in the groups	*P. alatus* *P. excavatus* *P. pubescentulus* *P. pulchellus*	*P. prideaux* *P. pycnacanthus*	*P. cuanensis* *P. forbesii*	*P. aleuticus* *P. armatus* *P. bernhardus* *P. ochotensis*	*P. brandti* *P. dalli* *P. pubescens* *P. stevensae* *P. trigonocheirus* *P. undosus*	*P. anachoretoides* *P. anachoretus* *P. gordonae* *P. laurentae* *P. souriei* *P. triangularis*	*P. chevreuxi*	*P. carneus*	*P. annexus** *P. arenisaxatilis** *P. benedicti* *P. brevidactylus* *P. carolinensis* *P. criniticornis* *P. galapagensis* *P. gymnodactylus* *P. lepidus* *P. leptonyx* *P. maclaughlinae* *P. nanodes** *P. nesiotes** *P. protuberocarpus* *P. provenzanoi* *P. redondoensis* *P. retrorsimanus** *P. rhabdotus** *P. spighti** *P. stimpsoni* *P. trichocerus* *P. vetaultae** *P. villosus* *P. virgulatus**	*P. albus** *P. gladius* *P. limatulus* *P. longicarpus* *P. mertensii* *P. perlatus**	*P. beringianus* *P. caurinus* *P. comptus* *P. edwardsii* *P. forceps* *P. granosimanus* *P. hemphilli* *P. hirsutiusculus* *P. middendorffii* *P. samuelis* *P. quaylei*	*P. gaudichaudii*	*P. arcuatus* *P. capillatus* *P. kennerlyi* *P. setosus*	*P. confragosus* *P. cornutus* *P. tanneri*	*P. impressus* *P. pollicaris* *P. smithi* *P. spilocarpus*	*P. delsolari* *P. hartae** *P. imarpe* *P. meloi** *P. rathbuni** *P. retrorsimanus** *P. ruber* *P. sculptimanus* *P. townsendi** *P. variailis* *P. venturensis**	*Pagurus acadianus* *Pagurus boriaustraliensis* *Pagurus brachiomastus* *Pagurus bullisi* *Pagurus constans* *Pagurus emmersoni* *Pagurus filholi* *Pagurus fraserorum* *Pagurus gracilipes* *Pagurus heblingi* *Pagurus japonicus* *Pagurus kulkarni* *Pagurus lanuginosus* *Pagurus liochele* *Pagurus maculosus* *Pagurus mbizi* *Pagurus minutus* *Pagurus nigrivittatus* *Pagurus nigrofascia* *Pagurus pectinatus* *Pagurus pitagsaleei* *Pagurus proximus* *Pagurus pseudosculptimanus* *Pagurus quinquelineatus* *Pagurus rectidactylus* *Pagurus rubrior* *Pagurus similis* *Pagurus simulans* *Pagurus spina*

"*" According to Lemaitre and Castaño (2004) [30].

Identification of taxon based on morphological characteristics is a common method in the classical taxonomy, and the accuracy of such method is broadly accepted [31]. However, many challenging issues remain in the identification and delimitation of species that are phenotypically similar, but reproductively isolated. Furthermore, sibling and cryptic species are sometimes not properly identified by morphological characteristics alone [32,33]. On the other hand, convergent evolution of morphological features in distantly related species may also lead to wrong identification [34]. The availability and easy access to ever increasing DNA sequences and the development of the necessary computational tools are being employed in phylogenetic studies. The analyses of these sequences have proved useful for identification and delimitation of species that have less morphological distinctness in particular [35,36]. As a result, the molecular taxonomy or the integration of the morphology and molecular (morpho-molecular) data in taxonomy has become a more frequent and widely accepted approach to evaluate the evolutionary relationships among species at various taxonomic levels.

To resolve the complexity in taxonomic relationships among the morphologically convergent species in *Pagurus*, a considerable number of molecular phylogenetic studies have been conducted in recent years [37–44]. As a consequence, the paraphyly or polyphyly of the genus *Pagurus* was also suggested in some studies [15,39,40,44]. However, most of these reports are limited to a restricted geographic scope and provide only a fragmented picture of the taxonomic relationships. Despite the world-wide distribution of the *Pagurus* hermit crabs, information on the molecular phylogeny among the biogeographical lineages, intra/interspecific variations, and the regional species diversities remain deficient and unresolved. The increasing discoveries of new species from various geographical localities in recent years provides an opportunity to investigate the molecular phylogeny and geographical distribution patterns of the genus *Pagurus* from the major geographical localities.

In this study, we investigated the phylogenetic relationships, and the patterns of species diversity among the species of the genus *Pagurus* based on multiple genetic markers. The genetic diversity was determined by estimating both intra and interspecific genetic variations among the sequences. Next, the association between the morphology-based species groups and molecular phylogenetic relationships was assessed to evaluate the genetic evidence for supporting morpho-groups. Finally, the geographical distributions of the *Pagurus* species were investigated for the identification of the regional patterns of species diversity. The inclusion of a large number of species from a wide range of habitats and the use of multiple molecular data sets allowed us to present the most complete evolutionary picture for the genus *Pagurus* to date.

2. Materials and Methods

2.1. Taxon Sampling

To elucidate the phylogenetic relationships among the species of the genus *Pagurus*, the publicly available molecular data were retrieved from the GenBank with the aim of sampling as many species as possible. The specimens were chosen to obtain molecular data from a wide range of species across every major habitat, providing a complete geographical coverage throughout all possible distributions of *Pagurus* species (Figure 1). Species were assigned a location code defined by their sampling regions as Arctic Ocean (AO), North Sea (NS), North Atlantic (NA), northeast Atlantic (NEA), northwestern Atlantic (NWA), Caribbean Sea (CS), Mediterranean Sea (MS), Alboran Sea (AO), Gulf of Mexico (GoM), South Atlantic (SA), southwest Atlantic (SWA), North Pacific (NP), East Pacific (EP), northeast Pacific (NEP), northwest Pacific (NWP), southeast Pacific (SEP), Yellow Sea China (YSC), Yellow Sea South Korea (YSSK), East China Sea (ECS), South China Sea (SCS), Sea of Japan (SOJ), and Indian Ocean (IO).

Figure 1. Map shows geographical locations of collection sites, adopted from the source of sequences in NCBI. Abbreviations are herein referred to as Arctic Ocean (AO), North Sea (NS), North Atlantic (NA), northeast Atlantic (NEA), northwestern Atlantic (NWA), Caribbean Sea (CS), Mediterranean Sea (MS), Gulf of Mexico (GoM), South Atlantic (SA), southwest Atlantic (SWA), North Pacific (NP), East Pacific (EP), northeast Pacific (NEP), northwest Pacific (NWP), southeast Pacific (SEP), Yellow Sea China (YSC), Yellow Sea South Korea (YSSK), East China Sea (ECS), South China Sea (SCS), Sea of Japan (SOJ), and Indian Ocean (IO).

2.2. Retrieval of Molecular Data

Appropriate genetic markers for species-level phylogenetic analyses are those that mutate quickly enough for difference to be observed between species. Though mitochondrial (mt) DNA markers have been proven useful in elucidating phylogenetic problem at species level, some factors derived from intergenomic co-adaptation can quicken or slow the rate of base substitution at various mtDNA loci and cause evolutionary rate variation between sites. Therefore, inclusion of more than two coding genes and nuclear markers are suggested to depict the evolutionary relationships between different species [45]. We therefore incorporated both mitochondrial and nuclear markers in our analyses.

2.3. Characteristics of Analysed Sequences Data

The resulting data set of complete mt genome consisted of 15 *Pagurus* species (https://www.ncbi.nlm.nih.gov/nuccore, accessed on 15 March 2021). However, among them, two sequences are duplicates (Table S1). Excluding the duplicates, 13 sequences of 10 species were left in the data set for final molecular analyses. Additonally, sequences of the partial mitochondrial gene regions, including 30 of the small subunit ribosomal gene (12S), 288 of the large subunit ribosomal gene (16S), and 1010 of the protein coding cytochrome c oxidase gene (COI), were obtained from both NCBI and UNIPROT databases (https://www.uniprot.org/uniprot/, accessed on 18 March 2021). In addition, sequences of three available nuclear genes, including 21 of the small subunit ribosomal gene (18S), 35 of the large subunit ribosomal gene (28S), and 19 of the Histone 3 (H3) were retrieved from the NCBI database (https://www.ncbi.nlm.nih.gov/nuccore, accessed on 23 March 2021).

Each individual marker was exploited for subsequent analyses. However, because of the numbers of sequences available, the inference of the molecular phylogenetic relationship and species diversity of the genus *Pagurus* was mainly based on 16S and COI gene trees. Moreover, because of the genetic relatedness of species from the same regions, only one species per location was included in the subsequent analyses. Therefore, the resulting pruned data sets for subsequent analyses included 73 sequences of 50 species (Table S3-1) and 127 sequences of 58 species (Table S4-1) for the 16S and COI genes, respectively. Furthermore, to obtain a more robust phylogenetic tree with increased branch support values, 16S and COI sequences were concatenated for the individuals from the same localities. A total of 50 sequences of 35 species were left in the concatenated (16S and COI) data set (Table S8). Despite the relatively small proportion of remaining data analyzed, the samples collectively encompass the breadth of morphological and ecological diversity of the genus *Pagurus*. For the analyses of the complete mitochondrial genomes, crayfish, *Procambarus clarkii* (NC_016926), was used as an outgroup taxon.

2.4. Phylogenetic Analyses

Different alignment methods were employed. The alignment of the ribosomal gene sequences was performed using ClustalW [46] in MEGA 7.0.16 [47] with the default settings: pairwise parameters, gap opening 15.0, gap extension 6.66, multi-alignment parameters, gap opening 15.0, and gap extension 6.66. On the other hand, aliment for the coding genes was performed based on the translated amino acid sequences using ClustalW (codons) with the default settings: pairwise parameters, gap opening 10.0, gap extension 0.1, multi-alignment parameters, gap opening 10.0, and gap extension 0.2. Ambiguous regions of the alignment were removed. Additionally, all alignments were visually examined for mismatches, and gaps/missing data were treated as complete deletions.

Phylogenetic trees were inferred by both the maximum likelihood (ML) [48] and neighbor joining (NJ) [49] methods. In ML analyses, the best-fit model under the "Find Best DNA/Protein Models" was searched for each gene. On the other hand, Kimura 2-parameter (K2P) model [50] was used as a model of substitution in NJ analyses. To determine confidence values for the resulting trees, for both NJ and ML methods, non-parametric bootstrap analyses were performed on 1000 replicates [51]. In case of identical topologies and bootstrap supporting values of two methods, the ML tree is shown. On the molecular trees, confidence values $\geq$50% were reported. Patterns of interspecific and intraspecific pairwise distances were calculated using the K2P substitution model implemented in MEGA.

2.5. Estimating Diversification Rates from Phylogenetic Information

We dated the ML phylogenetic tree derived from the combined 16S and COI sequences of 35 *Pagurus* species, which included one individual per taxon from a location, with *Procambarus clarkii* as an outgroup. While *P. clarkii* has been estimated to have diverged from Anomura about 357–385 millions of years ago (MYA) [15,52,53], *P. brevidactylus* and *P. bernhardus* diverged relatively recently at about 70.4 MYA [14]. We therefore eliminated *Procambarus clarkii* from the alignment and phylogenetic tree to focus the estimations of divergence times to the last 80 million years. The phylogenetic tree and aligned sequences were then used for the estimation of divergence times using MCMC tree in PAML package [54], following previously reported steps [55,56]. Briefly, the overall substitution rate was computed with *baseml* in PAML package, using <80 MYA divergence between *P. brevidactylus* and *P. bernhardus*. MCMC tree was then used to compute the gradient and Hessian using the overall substitution rate. The final steps involved the computation of the actual divergence times using the computed gradient and Hessian. The calibrations used were 60–80 MYA for the divergence of *P. brevidactylus* and *P. bernhardus* and 14–24 MYA for *P. brevidactylus* and *P. maclaughlinae* split [15]. The tree was then visualized in Fig Tree v 1.4.4 (http://tree.bio.ed.ac.uk/software/figtree/, accessed on 19 November 2021).

2.6. Morphological Characters Analyses

The morphological characteristics of *Pagurus* species were retrieved from the taxonomic literatures [35,36]. The diagnostic characters include the shape of chelipeds, number of spines/tubercles on its surface, shape of ambulatory pereopods, maxillipeds, telson, and so on. The morphological features were investigated with reference to the phylogeny of the morpho-based species groups.

3. Results

3.1. Species Composition

The localities of obtained sequences data may reflect the species richness and pattern of geographical distribution. Most of the species in our data set tend to be geographically restricted lineages; very few species show wider distributions (Table S9). The NWP species (Chinese, Korean and Japanese adjacent water) are restricted to the Kuroshio and Tsushima warm currents. Similarly, species of the NA region are confined to the North Atlantic gyre. While most of species maintain an exclusive limited distribution, a few species (*P. longicarpus*, *P. prideaux*, and *P. pubescens*) extend their ranges of geographical distibution widely along the North Atlantic and eastern Pacific regions. Some South African species (*P. emmersoni* and *P. liochele*) are also found in the Indian Ocean regions (Table S9).

3.2. Phylogenetic Aspects

The complete mt genome, and partial 12S, 16S, COI, 18S, 28S, and H3 data sets consisted of sequences of 10, 27, 50, 58, 19, 22, and 16 species, respectively (Table S1–S7). Several sequences were excluded from the analysis due to suspicious topologies after constructing phylogenetic trees (data derived from specimens may have been erroneously linked to species name on NCBI database due to misidentification, i.e., *Pagurus japonicus*: MG214647; *Pagurus fraserorum*: MF695071; *Pagurus pollicaris*: AF483159, etc.). The best fit models for each gene found in MEGA were as follows: (GTR + G) mt genome nucleotide, (mtREV + G + F) mt genome amino acid; (GTR + G + I) COI, combined 16S + COI; (T92 + G) 12S, 16S, 18S, 28S, and H3.

3.2.1. Complete mt Genome Sequences

Most of the taxa with complete mitochondrial genome sequences were exclusive to the NWP regions, except one species from the NA region. In the ML phylogenetic tree obtained from complete mt genome sequences, the NWP species formed a distinct clade that was further separated into three distinct sub-clades (Figure 2). The basal sub-clade consisted of two individuals of *P. nigrofascia* from the Yellow Sea, China (YSC), and Japanese waters. The genetic distance between them was very small (0.0003) (Table S10). *Pagurus lanuginosus* and *P. maculosus* were distinctly clustered together as sister species with 0.04 genetic diversities. *Pagurus japonicus* and *P. filholi* showed genetic closeness with 0.005 genetic distances, and distinctly separated from *P. gracilipes*. *Pagurus minutus*, and *P. similis* were clustered together as sister species. A *Pagurus* species (*Pagurus* sp.) collected at a 20 m depth from the northwester coast of Japan, showed genetic closeness with the species from the NA region (*P. longicarpus*). The genetic distances between the *Pagurus* species from the NWP regions and *P. longicarpus* ranged from 0.19–0.20, (Table S10).

3.2.2. Partial mtDNA Markers

In the 12S ML tree (Figure S1), species belonging to the "*Provenzanoi*" group (*P. maclaughlinae*, *P. brevidactylus*, and *P. stimpsoni*) were nested together. Similar to mt genome-based tree, species from the NWP regions clustered together and formed a distinct clade with one representative species of the "*Smithi*" and "*Exilis*" groups. However, unlike the mt genome-based tree, *P. similis* from the NWP region showed genetic closeness to an unassigned group species *P. bullisi* from the IO region and formed a distinct clade. On the other hand, most of the species from the Atlantic regions nested together in a distinct clade, and showed long genetic distance to the species from other regions. Representative species of the "*Bernhardus*" and "*Comptus*" groups were bifurcated, and positioned separately with low support values.

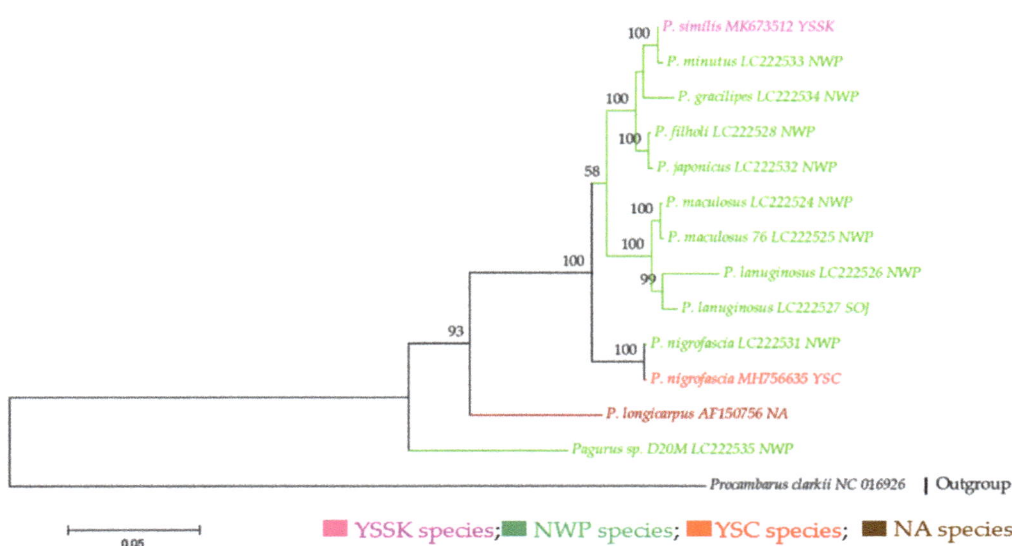

Figure 2. Maximum-likelihood (ML) tree based on mitochondrial genome (2709 amino acid) of *Pagurus* species with 1000 bootstrap replicates. Each sample was represented by species name, sequence accession number and the sampling location. The sampling locations included Yellow Sea South Korea (YSSK), northwest Pacific (NWP), Sea of Japan (SOJ), Yellow Sea China (YSC), and North Atlantic (NA). Different colors represent the localities of species analyzed.

In the 16S ML tree, a single representative of the *"Chevereuxi"* group (belonging to the Ingle 1985's Subdivision II), *P. chevreuxi*, positioned at the base (Figure 3). Similar to the result of 12S tree, species of the *"Provenzanoi"* group clustered together and formed a distinct clade. However, unlike the 12S tree, *P. stimpsoni* diverged at the base of this clade, and *P. brevidactylus* showed close genetic relationship with *P. provenzanoi* instead of *P. maclaughlinae*. *P. criniticornis*, and *P. nr. criniticornis* were likely to be the same species as they clustered together and formed a clade. On the other hand, *P. nr. maclaughlinae* and *P. maclaughlinae* were likely to be different species, since they positioned separately in two distinct clades. *Pagurus nr. maclaughlinae* was genetically close to *P. nr. carolinensis*, whereas, *P. maclaughlinae* was genetically close to *P. nr. criniticornis / P. criniticornis* species. *Pagurus leptonyx* and *P. villosus* were found genetically closed and nested with the *"P. maclaughlinae-P. nr. criniticornis / P. criniticornis"* clade. The results demonstrated the existence of genetically and geographically distinct populations in *P. brevidactylus* and *P. criniticornis* species. *Pagurus meloi* and *P. heblingi*, two unassigned species to any morpho-group, nested together in a clade with high bootstraps support, and positioned as sister-taxon to the *"Provenzanoi"* clade.

Species of the *"Alatus"*, *"Prideaux"*, and *"Cuanensis"* groups (belonging to the Ingle 1985's Subdivision I) were positioned together in a distinct clade. The *"Alatus"* group was bifurcated into two distinct clades, since *P. excavates* was isolated distinctly from other two members of this group. Similar pattern of isolation was also observed between the species of the *"Cuanensis"* group that two members of this groups, *P. forbesii* and *P. cuanensis* were positioned distantly. An unassigned morpho-group species, *P. pseudosculptimanus* clustered with the Subdivision I clade. On the other hand, among the five species groups of the Ingle 1985's Subdivision II, species of only three groups were included in the analyses. A single representative of the *"Trigonocheirus"* group, *P. pubescens* formed a distinct clade and showed genetic closeness with species of the *"Capillatus"* group, *P. kennerlyi*. An unassigned morpho-group species, *P. pectinatus* was nested with *P. kennerlyi*. Species belonging to the

"*Bernhardus*" group were nested together. However, *P. aleuticus* was positioned distantly and showed genetic closeness with *P. pubescens*, *P. kennerlyi*, and *P. pectinatus*.

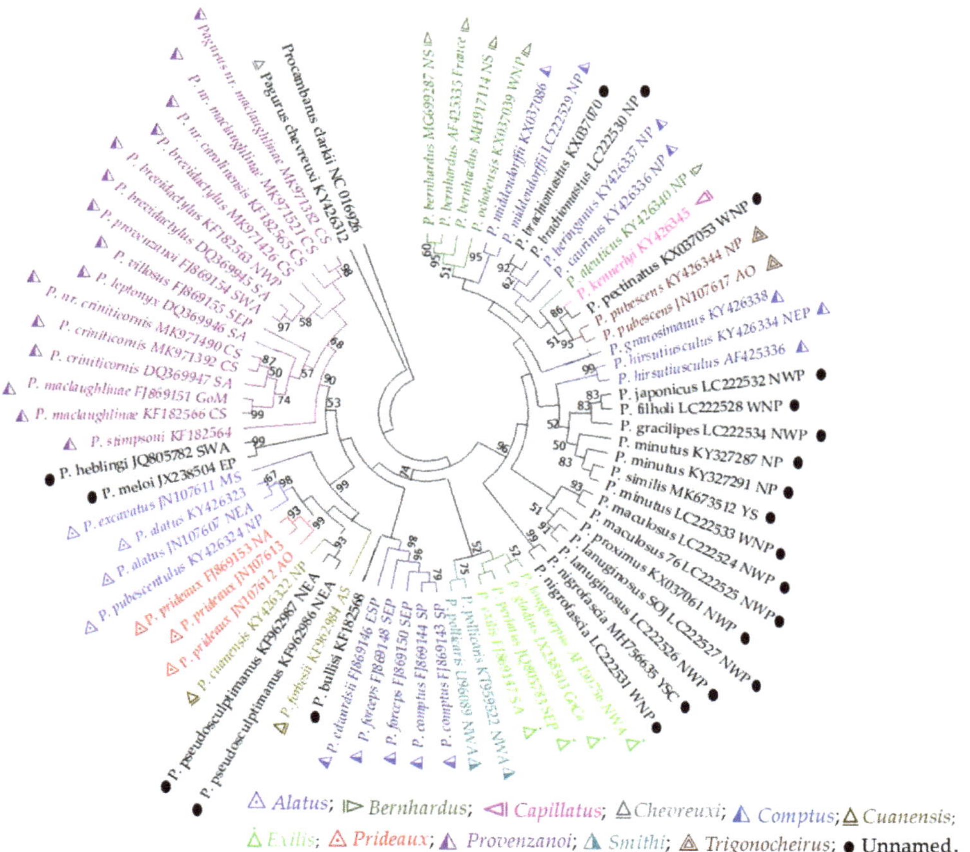

Figure 3. Phylogenetic tree of *Pagurus* species based on the 16S sequence inferred with the maximum-likelihood (ML) method using T92 + G model. Branch lengths are proportional to inferred nucleotide substitutions. Numbers at nodes represent bootstrap value in percentages; value below 50% are not shown. The species name, accession number, and sampling location are shown for each sequence. Samples were collected from Arctic Ocean (AO), North Sea (NS), North Atlantic (NA), northeast Atlantic (NEA), northwestern Atlantic (NWA), Caribbean Sea (CS), Mediterranean Sea (MS), Gulf of Mexico (GoM), South Atlantic (SA), southwest Atlantic (SWA), North Pacific (NP), East Pacific (EP), northeast Pacific (NEP), northwest Pacific (NWP), southeast Pacific (SEP), Yellow Sea China (YSC), Yellow Sea South Korea (YSSK), East China Sea (ECS), South China Sea (SCS), Sea of Japan (SOJ), and Indian Ocean (IO). Different colors represent the morpho-based species groups.

Species of the "*Comptus*" group were split into several clades suggesting a polyphyletic group. *Pagurus forceps*, *P. comptus*, and *P. edwardsii* were clustered together and isolated from other members such as *P. beringanus*, *P. caurinus*, *P. granosimanus*, and *P. middendorffii*. Unlike the 12S tree, *P. bullisi* from the IO regions clustered with the "*Comptus*" species from the SP regions, whereas *P. similis* clustered with *P. minutus* as was in the complete mt genome tree (Figure 2). The species of "*Exilis*" group (*P. exilis*, *P. gladius*, *P. longicapus*, and *P. perlatus*) showed a close genetic relationship with a species of the *"Smithi"* group (*P. pollicaris*) (Figure 3).

The NWP regions species comprised several clusters with high bootstrap support values. Two individuals of *P. nigrofascia* clustered together. *Pagurus maculosus* and *P. lanuginosus* were nested with *P. proximus* from Russian Far East. *Pagurus hirsutiusculus* was clustered with the NWP species. Consistent with the result of the complete mt genome tree, *P. minutus* and *P. similis* were closely related species. Unlike 12S phylogeny, but similar to the complete mt genome tree, *P. filholi*, *P. japonicus*, and *P. gracillipes* showed genetic closeness. An affinity between *P. brachiomastus* and *P. beringanus* (species of the *"Comptus"* group) was found.

In the most species-rich COI based ML tree (Figure 4), the species clustering was mostly consistent with the results of other mtDNA markers. However, a few inconsistencies were found in species topologies. Three representative species of the *"Provenzanoi"* group were divided into two clades. Similar to the 16S tree, *P. brevidactylus* and *P. nr. maclaughlinae* clustered together in a distinct clade, and *P. criniticornis*/*P. nr. criniticornis* formed another clade, though the bootstrap support values were low (<50%).

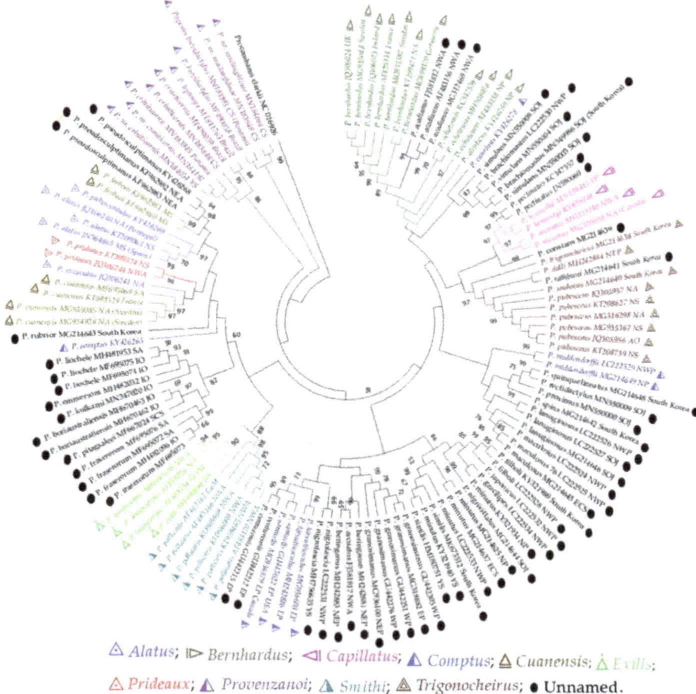

Figure 4. Phylogenetic tree of *Pagurus* species based on the COI sequence inferred with the maximum-likelihood (ML) method using GTR + G + I model. Branch lengths are proportional to inferred nucleotide substitutions. The species name, accession number and sampling location are shown for each sequence. Samples were collected from Arctic Ocean (AO), North Sea (NS), North Atlantic (NA), northeast Atlantic (NEA), northwestern Atlantic (NWA), Caribbean Sea (CS), Mediterranean Sea (MS), Gulf of Mexico (GoM), South Atlantic (SA), southwest Atlantic (SWA), North Pacific (NP), East Pacific (EP), northeast Pacific (NEP), northwest Pacific (NWP), southeast Pacific (SEP), Yellow Sea China (YSC), Yellow Sea South Korea (YSSK), East China Sea (ECS), South China Sea (SCS), Sea of Japan (SOJ), and Indian Ocean (IO). Different colors represent the morpho-based species groups.

Similar to the 16S tree, species of the Ingle 1985's Subdivision I ("*Alatus*", "*Prideaux*", and "*Cuanensis*") groups formed a distinct clade. Phylogenetic relationship of some unassigned morpho-group species was found to correlate with geographical distributions. We observed that species from the IO regions clustered together, and formed a distinct clade near to the Subdivision I clade. The genetic relationship among the species of the "*Bernhardus*" group was clearly reflected by the COI tree; species of this group were clustered together in a distinct clade. However, an unassigned morpho-group species *P. acadianus* was nested in this clade. Species of the "*Trigonocheirus*" group formed a distinct monophyletic clade with an unassigned morpho-group species, *P. rathbuni*. The close genetic relationship between the species of the "*Capillatus*" group and *P. pectinatus* was also proven by the COI tree. An unassigned morpho-group species, *P. constans* showed close genetic relationship with the "*Capillatus*" group. A close genetic relationship between the species of the "*Trigonocheirus-Capillatus*" and "*Exilis-Smithi*" groups were also proven by the COI tree.

Consistent with other mtDNA marker phylogenies, species from the NWP regions were divided into several sub-clades with higher bootstrap support values. *Pagurus filholi*, *P. japonicus*, *P. minutus*, and *P. similis* were nested in a distinct clade with *P. nigrivittatus*. The taxonomic position of *P. middendorffii*, *P. nigrofascia*, and *P. brachiomastus* was inconsistent among mtDNA markers (Figures 3,4 and Figure S1). Unlike 16S tree, species of the "*Comptus*" group clustered together, however, *P. caurinus* was isolated into different clade. An unassigned morpho-group species, *P. venturensis* showed genetic closeness with the "*Comptus*" group.

To obtain more informative sites, the aligned sequences of 16S and COI were concatenated for species with both markers. The phylogenetic tree obtained from the concatenated sequences (Figure 5) had a better resolution of ancestral relationships among the morpho-based *Pagurus* groups, depicted by three monophyletic linages (L 1–3) with higher bootstrap support. These three major linages can be further divided into five clades (clade I–V), though some clades have low bootstrap supports (Figure 5). Consistent with both 16S and COI trees, species of the "*Provenzanoi*" group clustered together as the basal clade (clade I). Species of the "*Alatus-Prideaux-Cuanensis*" groups formed a distinct clade II. However, *P. alatus* and *P. pubescentulus* showed genetic closeness with *P. forbesii* of the "*Cuanensis*" group. Species of the "*Exilis*" and "*Smithi*" groups were genetically the closest and formed the clade III with *P. comptus*. The hairy hermit crab, *P. hirsutiusculus*, was diverged at the base of clades IV and V. The NWP species around the Japanese adjacent waters clustered together and formed the clade IV. Species of the "*Trigonocheirus*" group nested with that of the "*Capillatus*" group, and formed the clade V with three unassigned morpho-group species *P. brachiomastus* and *P. nigrofascia* and *P. pectinatus*. On the other hand, species of the "*Bernhardus*" and "*Comptus*" groups were also nested together with this clade V.

3.2.3. Partial Nuclear DNA Markers

The sequences of all three nuclear markers were not available for the same analyzed taxa, therefore, the comparison or concatenation of the sequences were difficult. Nevertheless, the phylogenetic relationships established using 18S, 28S, and H3 ML trees (Figure 6A–C, respectively), were congruent with those inferred from mtDNA markers. However, a few inconsistencies were found in general tree topologies. Species of the "*Provenzanoi*" group were always clustered together in a distinct clade in all three markers. Similar to the 16S tree, in the H3 tree *P. stimpsoni* diverged at the base, and *P. maclaughlinae*, *P. criniticornis*, *P. leptonyx*, and *P. villosus* were found to be genetically closely related species. However, *P. heblingi* and *P. meloi* that were close to the "*Provenzanoi*" group in 16S tree, were positioned distantly in H3 tree.

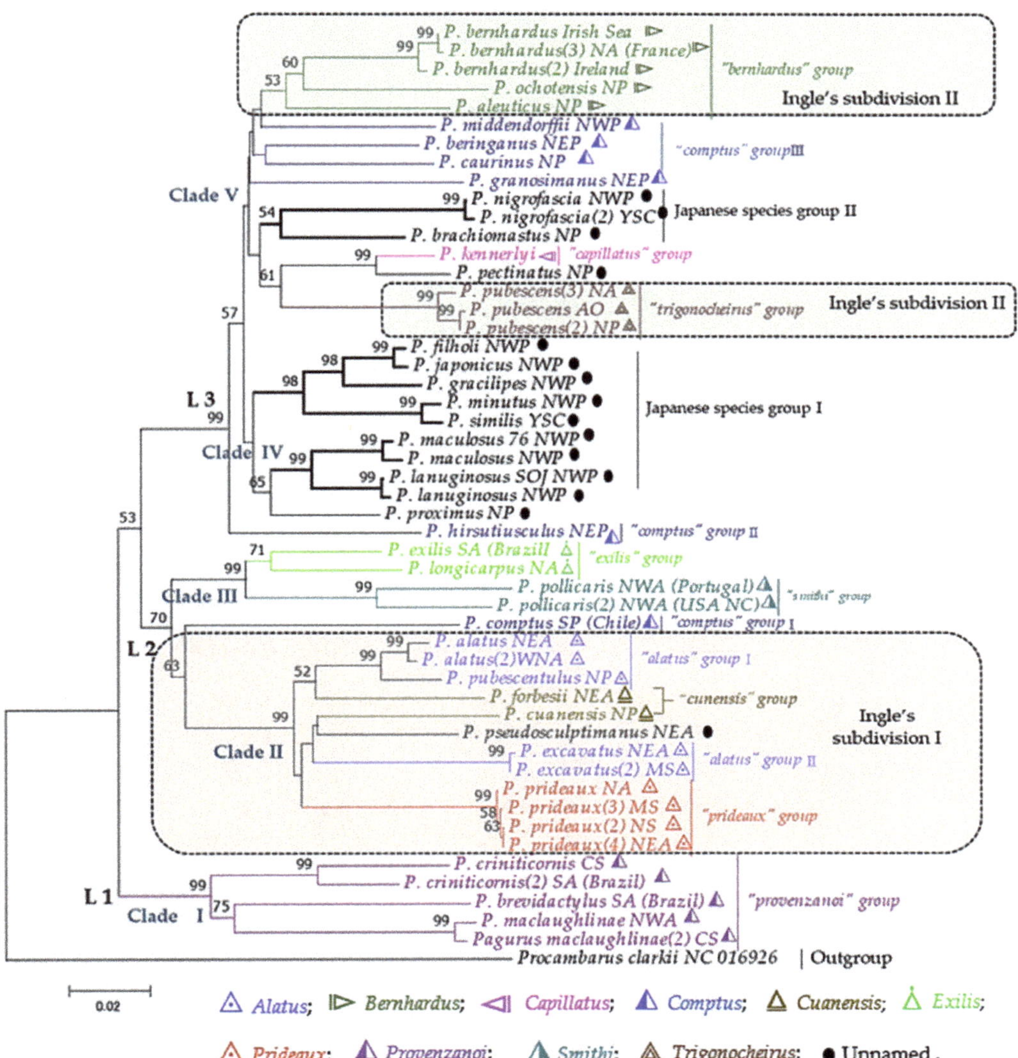

Figure 5. Phylogenetic tree of *Pagurus* species based on the combined gene (16S & COI) sequence inferred with the maximum-likelihood (ML) method using GTR + G + I model. Branch lengths are proportional to inferred nucleotide substitutions. Numbers at nodes represent bootstrap value in percentages; value below 50% are not shown. One concatenated sequence set from a species was included per location. Sampling locations included Arctic Ocean (AO), North Sea (NS), North Atlantic (NA), northeast Atlantic (NEA), northwestern Atlantic (NWA), Caribbean Sea (CS), Mediterranean Sea (MS), Gulf of Mexico (GoM), South Atlantic (SA), southwest Atlantic (SWA), North Pacific (NP), East Pacific (EP), northeast Pacific (NEP), northwest Pacific (NWP), southeast Pacific (SEP), Yellow Sea China (YSC), Yellow Sea South Korea (YSSK), East China Sea (ECS), South China Sea (SCS), Sea of Japan (SOJ), and Indian Ocean (IO). Different colors represent the morpho-based species groups.

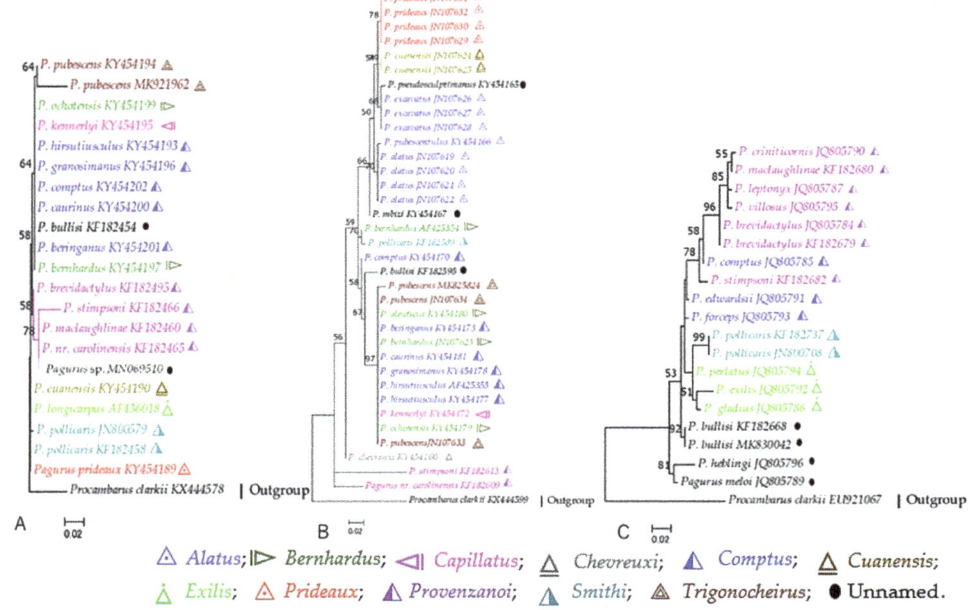

Figure 6. Individual gene trees based on maximum-likelihood (ML) method with 1000 bootstraps constructed from nuclear markers of Pagurus species: (**A**) (18S), (**B**) (28S), and (**C**) (H3). Numbers at nodes represent bootstrap value in percentages; value below 50% are not shown. The species name and sequence accession number are shown for each sample. Different colors represent the morpho-based species groups.

Consistent with the results of mtDNA markers, in both 18S and H3 ML trees (Figure 6A,C), species of the *"Exilis"* and *"Smithi"* groups were nested together showing genetic closeness to each other. Similar to the results of mt-based trees, species of the *'Comptus'* and *"Bernhardus"* groups were scattered in several clades, and showed inconsistencies in topologies. *Pagurus beringanus* and *P. bernhardus* clustered together in both 18S and 28S trees. However, two individuals of *P. bernhardus* were isolated distantly in the 28S tree. *P. comptus* clustered with the *"Comptus"* group in the 18S tree, was positioned separately in both 28S and H3 trees. Similarly, in 16S tree, *P. comptus* clustered with *P. foceps* and *P. edwardsii*, however, it isolated from these two species and clustered with the *"Provenzanoi"* group in the H3 tree. Unlike in the mtDNA marker trees, *P. hirsutiusculus* and *P. granosimanus* were clustered together in both 18S and 28S trees. In the 28S tree, a single representative of the *"chevereuxi"* group, *P. chevereuxi*, positioned separately. Similar to the results of mtDNA markers, species of the *"Alatus"* group formed two distinct clades, however, *P. excavatus* was clustered with *P. pseudosculptimanus*, instead of *P. cuanensis* as in the COI tree. Consistent with the results from mtDNA markers, the clustering and formation of morpho-based species *"Prideaux-Alatus-Cuanensis"* clade was also supported by the nuclear marker 28S (Figure 6B).

3.3. Genetic Variability among Geographical Regions

Pagurus species are most likely to be restricted to narrow geographical localities, with the largest number of endemic species confined to the Northwest Pacific, and the western and eastern North Atlantic regions. A few species inhabit the South Atlantic, the Indian Ocean, and the China Sea regions. Analysis of interspecific variations reflected the genetic diversifications among the *Pagurus* species from various geographical localities (Tables 2 and S10–S13). Based on 16S, the interspecific genetic distances between *P. provenzanoi* and *P. brevidactylus* ranged from 0.003–0.06. These values were smaller than the intraspecific genetic diversity of *P. brevidactylus* individuals that ranged from 0.06–0.07 (Table S11), and the highest intraspecific genetic variability (0.07) was found between NWP and CS populations. *Pagurus nr. maclaughlinae* and *P. maclaughlinae* are seemed to be different species, as they had 0.09–0.10 K2P genetic distances based on 16S data, larger than that the interspecific genetic variation between *P. criniticornis* and *P. maclaughlinae* (0.07), or between *P. nr. maclaughlinae* and *P. nr. carolinensis* (0.08–0.09) (Table S11). Based on the third codon of COI gene, the genetic distances between the species of "*Alatus*" group (*P. alatus* and *P. pubescentulus*) and the species of "*Cuanensis*" group (*P. forbesii*) ranged from 0.38–0.43, smaller than those 0.75–0.82 with another species of the "*Alatus*" group, *P. excavate* (Table S12).

Table 2. Pairwise genetic distance Kimura's two-parameter model (K2P) values (per site) among *Pagurus* species between geographic groups based on combined (16S + COI) genes.

	NEA	NWA	NP	NEP	NA	SA	SP	CS	MS	NWP	NWA	YSC	NS	AO
NEA														
NWA	0.06													
NP	0.14	0.12												
NEP	0.16	0.15	0.11											
NA	0.14	0.13	0.11	0.12										
SA	0.15	0.14	0.15	0.15	0.15									
SP	0.14	0.13	0.15	0.15	0.15	0.15								
CS	0.17	0.16	0.16	0.17	0.16	0.12	0.16							
MS	0.08	0.09	0.15	0.17	0.14	0.16	0.14	0.17						
NWP	0.16	0.14	0.12	0.11	0.12	0.15	0.16	0.17	0.16					
NWA	0.16	0.16	0.16	0.16	0.16	0.14	0.15	0.14	0.16	0.17				
YSC	0.17	0.15	0.11	0.10	0.12	0.15	0.15	0.16	0.17	0.10	0.16			
NS	0.08	0.09	0.15	0.17	0.13	0.16	0.15	0.18	0.05	0.16	0.16	0.17		
AO	0.17	0.15	0.10	0.10	0.10	0.15	0.17	0.17	0.18	0.11	0.17	0.10	0.17	

Samples were collected from Arctic Ocean (AO), North Sea (NS), North Atlantic (NA), northeast Atlantic (NEA), northwestern Atlantic (NWA), Caribbean Sea (CS), Mediterranean Sea (MS), Gulf of Mexico (GoM), South Atlantic (SA), southwest Atlantic (SWA), North Pacific (NP), East Pacific (EP), northeast Pacific (NEP), northwest Pacific (NWP), southeast Pacific (SEP), Yellow Sea China (YSC), Yellow Sea South Korea (YSSK), East China Sea (ECS), South China Sea (SCS), Sea of Japan (SOJ), and Indian Ocean (IO). The cells are shown in heatmaps with green representing lower values and red representing higher values.

Species from the Indian Ocean, which formed a monophyletic clade in the COI tree, showed low level of interspecific genetic distances. The least interspecific diversity (0.12) was found between *P. emmersoni* and *P. liochele*, whereas the highest interspecific diversity (0.59) was found between *P. emmersoni* and *P. fraserorum*. Surprisingly, that was 0.01 between *P. pitagsaleei* from SCS and *P. boriaustraliensis* from IO (Table S12), which was even smaller than the intraspecific genetic variation among two different populations of *P. liochele* from IO and SA (0.02). The North Atlantic species, *P. pollicaris* and *P. longicarpus*, each formed a monophyletic clade (Figure 4), with a range of intraspecific genetic distances of 0.01–0.84 and 0.02–0.12, respectively (Table S12). The higher genetic divergence of the gulf (GoM) populations of *P. longicarpus* (Figure 4) may indicate the less genetic connectivity and less expansion of population compare to the open waters. A similar case was found between two populations of *P. beringanus* from the eastern North Atlantic region that showed 0.03 intraspecific genetic diversity. Likewise, two population of *P. arcuatus* showed 0.02 intraspecific genetic diversity. Similarly, higher inter-regional intraspecific diversity (0.17) was also found between populations of *P. cuanensis* from the North and South Atlantic regions. Among the populations of *P. bernhardus*, the highest (0.04) intraspecific genetic diversity was found between the North Sea and North Atlantic individuals.

Based on the combined 16S + COI data (Table 2), the highest interregional genetic diversity (0.18) was found between the species from MS and AO, CS, and NS, whereas the lowest genetic diversity 0.05 was found between the species from MS and NS. On the other hand, the highest intraregional genetic diversity was found among the species from NWA (0.14), followed by species from the SA (0.13), CS (0.12), and NP (0.10), whereas the lowest was found among the species from NEA (0.08) and NEP (0.09) (results were not shown).

3.4. Geographical Diversification

To investigate the timescale for biogeographical connection, we dated the divergence among *Pagurus* species (Figure 7). Divergence times estimated in the present study revealed that *Pagurus* species from the Atlantic and Pacific diverged between 25–51 MYA. The main diversification of extant lineages took places between 4–25 MYA. The major NWP lineages seemed to have diverged from NP ancestor between 16–18 MYA.

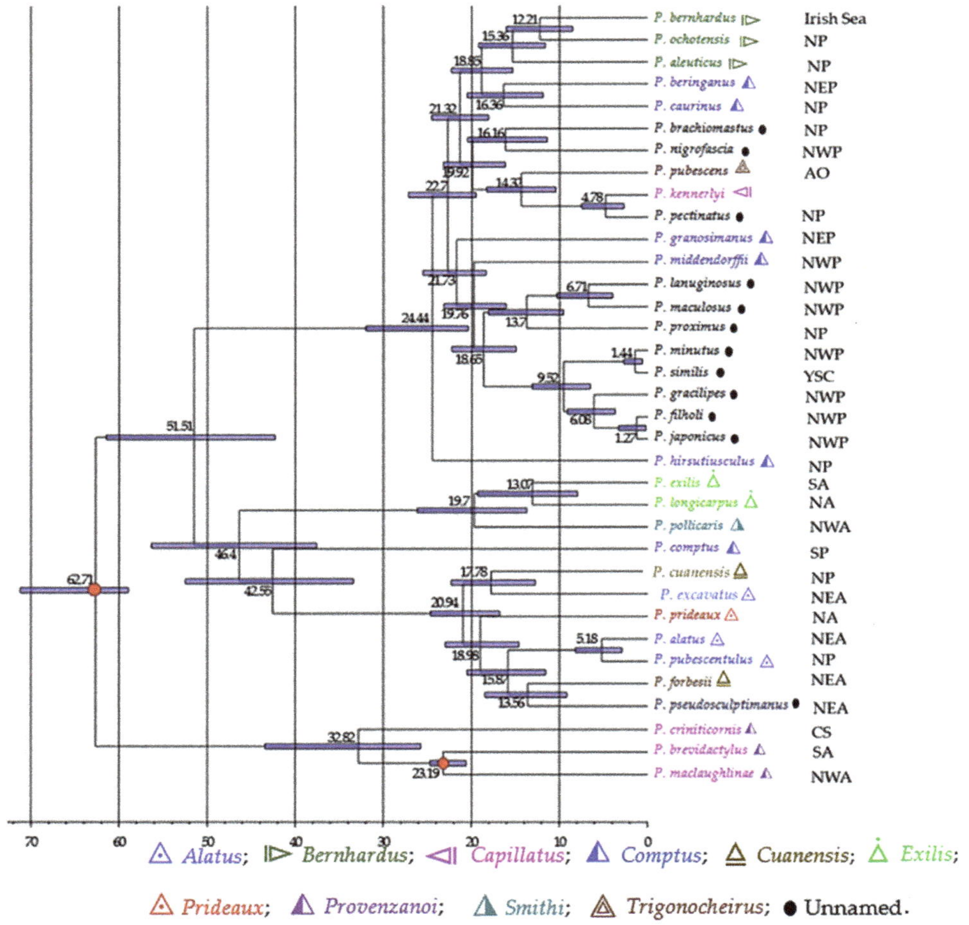

Figure 7. Dated phylogenetic tree of *Pagurus* species obtained from analysis of combined 16 and COI sequences. The evolutionary time-scale in millions of years ago (MYA) is shown below. The calibration points are shown in red circles. Different colors represent the morpho-based species groups.

3.5. Morphological Characteristics within Pagurus Species

In molecular phylogeny, the patterns of distinct separation among species may interpret the variations in morpho-anatomical traits. The diagnostic characteristics among the species group were shown in Table 3. The genetic closeness between *P. provenzannoi* and *P. brevidactylus* is consistent with their common diagnostic character of bearing of a row of more than tree terminal spines [26]. The genetic closeness between *P. meloi* and *P. heblingi* in both 16S and H3 trees may correspond to several similar morphological characteristics, e.g., shape of chelipeds, long second and third pereopods, slender dactyls, and shape of gill [35,40].

Table 3. Variabilities in morphological features among species groups of the genus *Pagurus*.

Common features	Morpho-traits	Ingle (1985) Subdivision I	Ingle (1985) Subdivision II
	Maxillule (endopod outer distal margin)	rounded or obtusely produced	produced as a conspicuous lobe or papilla, sometimes recurved, rerely segmented
	Pereiopod 3 (anterior lobe of sternite)	triangular to subtriangular, and setose with acute to subacute spines	broadly subquadrate to suboval, and setos, smooth
	Telson (posterior margin)	armed with teeth that extend on to lateral margin	armed with teeth but that not extend on to lateral margin
	Male's pleopod (number)	0-4 unpaired	3 unpaired

Diagnostic features	Species groups	Group 1 "Alatus"	Group 2 "Pride aux"	Group 3 "Cuanen-sis"	Group 4 "Bern-hardus"	Group 5 "Trigono cheirus"	Group 6 "Anachore-tus"	Group 7 "Chevr euxi"	Group 8 "Carn-eus"
	ocular peduncles	stout, noticeably subcylindrical,	stout, noticeably subcylindrical,	long and cylindrical	short, moderately-stout	subcylindrical	/	/	/
	ocular acicles	spatulate to subtriangular, dorsal surface concave	spatulate, dorsal surface relatively flat	spatulate to subtriangular, dorsal surface slightly concave	spatulate, dorsal surface concave	subtriangular, dorsal surface flat to slightly concave	subtriangular, dorsal surface flat	subtriangular, dorsal surface flat	narrowly subtriangular, dorsal surface slightly concave
	Maxilliped 1 (peduncle of exopod)	moderately broadened proximally	slightly broadened proximally	broadened proximally	broadened proximally	moderately broadened proximally	slightly broadened proximally	/	broadened proximally
	Maxilliped 3	outer distal margin of corpus and merus with a spine, crista dentata with many acute teeth	outer distal margin of corpus and merus with a spine; crista dentata with moderate number of acute teeth	outer distal margin of corpus and merus without spines; crista dentata with many teeth	endopod, inner margin and outer distal margin of merus each with a spine; crista dentata with moderate number of teeth	endopod, inner margin and outer distal margin of merus each with a spine; crista dentata with many teeth	endopod, outer distal margin of merus without or with a spine; crista dentata with moderate number of teeth	endopod, inner and outer distal margin of merus with a spine; crista dentata with numerous small teeth	outer margin of merus with a spine; crista dentata with numerous small teeth

Species groups	Group 9 "Proven-zanoi"	Group 10 "Exilis"	Group 11 "Com-ptus"	Group 12 "Gaudi-chaudii"	Group 13 "Capilla-tus"	Group 14 "Con-frago-sus"	Group 15 "Smithi"
ocular peduncles	subcylindrical	strong	narrow, subcylindrical	/	cylindrical	moderately short, stout	/
ocular acicles	broad, denticulate terminally	triangular, dorsal surface concave	triangular, with subterminal spine inserted ventrally	/	acutely or ovately triangular	ovately triangular	/
Maxilliped 1 (peduncle of exopod)	very narrow	strongly inflated proximally	/	/	slender	moderately inflated	/
Maxilliped 3	outer margin of merus unarmed; crista dentata with one tooth	outer margin of merus with small spines; crista dentata with few teeth	/	/	outer margin of merus with spines; crista dentata with 1 or 2 teeth	merus and carpus each with spin, crista dentata with 1 tooth	/

Table 3. *Cont.*

Morpho-traits	Ingle (1985) Subdivision I	Ingle (1985) Subdivision II
Common features		
Maxillule (endopod outer distal margin)	rounded or obtusely produced	produced as a conspicuous lobe or papilla, sometimes recurved, rerely segmented
Pereiopod 3 (anterior lobe of sternite)	triangular to sub-triangular, and setose with acute to subacute spines	broadly subquadrate to suboval, and setos, smooth
Telson (posterior margin)	armed with teeth that extend on to lateral margin	armed with teeth but that not extend on to lateral margin
Male's pleopod (number)	0-4 unpaired	3 unpaired

Species groups	Group 1 "Alatus"	Group 2 "Prideaux"	Group 3 "Cuanensis"	Group 4 "Bernhardus"	Group 5 "Trigonocheirus"	Group 6 "Anachoretus"	Group 7 "Chevreuxi"	Group 8 "Carneus"	Group 9 "Provenzanoi"	Group 10 "Exilis"	Group 11 "Comptus"	Group 12 "Gaudichaudii"	Group 13 "Capillatus"	Group 14 "Confragosus"	Group 15 "Smithi"
Diagnostic features															
Left cheliped (propodal outer surface)	medially elevated and with prominent longitudinal row of acute tubercles	slightly elevated and, with scattered tubercles	medially elevated and, with acute to subacute tubercles	gently convex and tuberculate	medially elevated with spiniform tubercles	convex and relatively smooth	relat-ively flat	a longitudinal median elevation bearing tubercles	with spiny tubercles and numerus bristles	compressed laterally.	/	/	oblique, not produced into prominent ridge	strongly oblique, concave, elevated into prominent ridge	/
pereiopods 2, 3 (dactyls)	slightly obliquely compressed distally	obliquely compressed distally	slightly compressed distally not obliquely	obliquely compressed distally	obliquely compressed distally	not obliquely compressed distally	not obliquely compressed distally	slightly obliquely compressed distally	terminating in sharp, corneous claw	long, arched, ventral margin with spiniform bristles	short, or-namented by strong, spiniform bristles on ventral margin	/	straight or twisted, with setos	moderately long, slender, usually twisted	/
Telson (posterior margin)	bicon-cave, many teeth extending on to lateral margin	concave, many teeth extending on to lateral margin	bicon-cave, and small to med-ium size teeth extending on to lateral margin	concave, with a few small teeth	bicon-cave, with a few large, recurved teeth	bicon-cave, with many or a few small teeth	bicon-cave, with numerous small teeth	concave, with large subacute teeth	lest uropod bordered with long, ribbon-like bristles	difurcated lobes, spinous, without bristles	separated by fine median incision, denticulated	/	oblique, with 2-8 strong spines	subtriangular, oblique, denticulate or with small spinules	/
Male's pleopod (number)	3	4	4	3	3	/	/	/	3	3	/	/	3	3	/
Reference Literature	Ingle (1985) [28]											Mclaughlin (1974) [26]			

" / " means information unavailable.

In the "Ingle's Subdivision I" clade, the relationship between the species of "*Alatus*" and "*Cuanensis*" groups remains ambiguous. Generally, *P. excavatus* was separated from other members of the "*Alatus*" group. Instead, it positioned at the base of the "Ingle's Subdivision I" clade in 16S, nested with *P. cuanensis* in COI, *P. pseudosculptimanus* in 28S, and with both in combined 16S + COI trees. These relationships are consistent with the adult morphology that is quite unlike the "*Alatus*" group but similar to *P. cuanensis* and *P. pseudosculptimanus*, with the male *P. excavatus* having four unpaired pleopods [38,57]. The main differences in morphological traits between *P. alatus* and *P. excavatus* were described as follow; segment 1 of antennular peduncle with 1-3 spines, cheliped palm not striking concave, male with three unpaired pleopods in *P. alatus*; whereas in *P. excavatus*, segment 1 of antennular peduncle without or with small obtuse spines, cheliped palm strikingly concave, male with four unpaired pleopods [38]. The distinct genetic separation of the two species of the "*Cuanensis*" group, *P. forbesii* and *P. cuanensis* can be explained by their morphological differences that in outer surface of right cheliped with long, dense setae in *P. cuanensis*, which is either absent or present in a few numbers in *P. forbesii*. The lower margin of dactyl of pereiopod 3 is relatively smooth in *P. cuanensis* but with subacute teeth and spines in *P. forbesii* [28].

Though the suggested closely related species of *P. proximus* was *P. brachiomastus* [58], *P. proximus* clustered with *P. maculosus* and *P. lanuginosus* in the molecular phylogeny. This relationship may reflect an accord with a common morphological feature among these species having very dense hairs on cheliped/pereopods. *Pagurus maculosus* can be distinguished from *P. lanuginosus* by the number of calcareous teeth on the dactyl of and presence or absence of a large tubercle of the merus of the right cheliped, and distinct chromatophore color (white vs. black- spotted, respectively) across the carapace and walking legs [59]. On the other hand, *P. proximus* differs from the other two species by the color of antenna and antennules (former have bright red in color) and horizontal striped pattern instead of spotted chromatophore [58].

On the contrary, *P. brachiomastus* showed close genetic relationship with *P. beinganus* (in 16S tree), *P. simulans* (in COI tree), and *P. nigrofascia* (in combined 16S + COI tree). *Pagurus brachiomastus* is recognizable by the distally red colored chelae and dactyli of the ambulatory pereopods, whereas *P. nigrofascia* is characterized by the olive colored chelipeds with scattered dark brown spots in general, and orangish tips; the dactyls are banded proximally with dark brown or gray, without stripes. *Pagurus simulans* differs from these two species by brown chelipeds and ambulatory pereopods [60]. *Pagurus beinganus* is diagnosed by chelipeds with flame scarlet spines, distal end of merus with a distinct scarlet band; ambulatory legs with large irregular spots of claret brown, antennae claret brown to brussels brown [26].

4. Discussion

The scarcity of genetic data for a significant number of species covering broader geographic locations is a major challenge in biogeographical studies of non-model organisms. In this study, however, an attempt was made to evaluate extensive molecular phylogeny of the genus *Pagurus* from a wide range of taxa currently available from a wide range of geographical locations. Many recent molecular studies have dramatically increased our knowledge on ancestral relationship of *Pagurus* species, however, the exploration of many new species in recent years highlights the necessity for integrated and continued molecular phylogenetic study of this group.

Of the molecular data sets analyzed, the COI gene consisted of the largest number of taxa representing wider locations. The phylogenetic relationship among the *Pagurus* species was mainly delineated based on the 16S, COI, and combined 16S and COI data sets. In the double-marker genetic approaches, the tree topologies showed substantially similar phylogenetic relationships among the terminal taxa, and largely consistent with the previously assigned morpho-trait based species groups. The close relationship among the morpho-groups in the phylogenetic analyses may suggest that the phenotypic traits are substantially supported by the molecular evidence. The inclusion of species from the Pacific regions especially from the Chinese, Korean and Japanese waters, and Indian Ocean in the present study resulted an increased clade diversity as well as regional diversity.

Despite the aim that the current study endeavored to evaluate an extensive molecular phylogeny of the genus *Pagurus* including a wide range of taxa; a few representative taxa of each morpho-based species group were obtained for the analyses. However, in terms of taxon numbers, the present study is so far the most comprehensive of *Pagurus*. The distinct cladistics forms obtained here may likely reflect interspecific variation related to unique morphological traits. Taxon composition of the "*Provenzanoi*" group were consistent among markers in 12S, 16S, COI, and 28S trees (Figures 3, 4, 6B and Figure S1). Species of this group always formed a monophyletic linage with some subdivisions, and represented the earliest diverging group of *Pagurus*.

In agreement with the morphology-based taxonomy, phylogenetic trees inferred from both mtDNA and nuclear markers established the monophyly of the Ingle's (1985) Subdivision I, consisting of "*Alatus*" group (*P. alatus*, *P. excavates*, and *P. pubescentulus*), "*Prideaux*" group (*P. prideaux*), and the "*Cuanensis*" group. However, the assemblage of species in the "*Alatus*" and "*Cuanensis*" groups were not well-resolved due to the inconsistencies in phylogenetic relationships among markers. In the 16S tree, *P. excavatus* positioned at the base of the "Ingle's Subdivision I" clade. In contrast, it nested with *P. cuanensis* in COI, *P. pseudosculptimanus* in 28S, and with both in combined 16S + COI trees. With the conclusion of morpho-taxonomy and based on the results obtained from the molecular phylogenetic analyses, the genetic closeness between *P. pseudosculptimanus* and *P. excavatus* may suggest proposing a new species group to unite these two species.

On the other hand, species of only three groups ("*Bernhardus*", "*Trigonocheirus*", and "*Chevreuxi*") among five groups of Ingle's (1985) Subdivision II were attained in the analyses. The phylogenetic relationship among three representative species of the "*Bernhardus*" group were clearly depicted by COI tree (Figure 4) than the 16S tree (*P. aleuticus* was isolated) (Figure 3). *Pagurus acadianus*, unassigned to any morpho-group, was nested with the "*Bernhardus*" group as sister species in COI tree. Species of the "*Trigonocheirus*" group nested together in a distinct monophyletic clade with *P. rathbuni* (Figure 4), and showed genetic closeness with the "*Capillatus*" group. A single representative species of "*Chevreuxi*" group, *P. chevreuxi*, showed distinct genetic isolation in both 16S and 28S trees.

The previously reported polyphyly of the "*Comptus*" group was largely congruent with the present study [15,37,39,44,61]. In most cases, the actual phylogenetic relationship could not be established because the species of the "*Comptus*" group were split into several distinct clades in both individual and combined trees. According to the description of McLaughlin et al. (1988), species of the "*Comptus*" group form the North American region has ambulatory legs with short dactyls ornamented by strong, spiniform bristles on ventral margin; and a thin chitinous layer on lateral margins of the posterior lobes of the telson. Therefore, the review of the species listed in this group has been suggested [29]. The genetic closeness between the "*Exilis*" and "*Smithi*" groups was supported by both mtDNA and nuclear markers. This genetic closeness was consistent with the previous spermiotaxonmic study that the morphology of spermatophore (an ovoid and slightly flattened capsule shaped main ampulla and similar accessory ampulla) is identical among *P. exilis*, *P. pollcaris*, and *P. longicarpus* [7,13]. The Species of the "*Capillatus*" group showed genetic relationship with *P. pectinatus* in all mtDNA markers and always nested together (Figures 3–5). Based on the results, the possible inclusion of *P. pectinatus* in "*Capillatus*" group could be proposed.

In all phylogenetic analyses, species from the NWP regions did not appear as monophyletic group, but rather appeared as more distantly related than their localities suggest. Species were split and formed several sub-clades, and positioned distantly, though the species topologies were not consistent. Each united sub-clade may correspond to a particular morphological trait. One sub-clade consists of *P. lanuginosus*, *P. maculosus*, and *P. proximus* from the Russian Far East, while the other sub-clade consists of *P. filholi*, *P. japonicus*, *P. gracilipes*, and *P. minutus* (Figure 5). *Pagurus brachiomastus*, *P. middendorffii*, and *P. nigrofascia* isolated and positioned with the "Ingle's Subdivision II" clade. The phylogenetic position and interspecific relationship among *P. brachiomastus*, *P. middendorffii*, and *P. nigrofascia* were not well-resolved due to a number of contradicting tree topologies produced by mtDNA markers (Figures 3–5 and Figure S1). *Pagurus brachiomastus* were related with *P. nigrofascia* based on the COI and combined trees. However, based on the 12S and 16S trees, they were found to be distantly related. Similar topological inconsistencies were also found for *P. middendorffii* in 16S and COI trees (Figures 3 and 4). Although in phylogenetic trees, the clustering of taxa was mainly formed based on morpho-groups; highly supported clades were also found that consisted of species collected from the same geographical localities. In most cases, species distributed in the IO, SCS, and SA regions nested together and formed a monophyletic clade (Figure 4). A representative of *P. pitagsaleei* population from the Taiwan has close relationship with *P. boriaustraliensis* from the Australian region. Indeed, previous report suggested that these two species might be two populations of the same species [62]. The relationship presented here suggests a close ancestral relationship among the species. Similarly, *P. meloi* and *P. heblingi* were likely to be genetically close species (Figures 3 and 6C). This result is consistent with previously reported findings from both molecular [39], and morphological studies that they are morphologically similar but can be differentiated from each other by the morphology of telson plate [43,48]. Interestingly, the analyses of H3 sequences support the previous report that the two species were separate from "*Provenzanoi*" group [39].

Generally, species of the genus *Pagurus* appear as geographically disjunct populations, since species distributed in the Pacific regions were rarely found in Atlantic regions, and vice versa. It may suggest that the migrations across geographical barriers are rare. The *Pagurus* hermit crabs, with the active migratory pelagic larval stages, can disperse widely via the transportation of the ocean currents. However, their ranges of expansion remain somewhat static throughout their shell-dwelling juvenile-to-adult lives. The water temperature could be one of the most important barriers for their dispersion particularly for dispersal at the larval stages. Further the biological interactions (i.e., competition for shell or other resources) may affect the observed distribution patterns. Recent discoveries of several new anomuran species in areas of the Indo-West Pacific regions support the prediction that there may be many undescribed species particularly in areas of Indo-west Pacific, the Central Indian Ocean, and the East Asian Sea [63].

The knowledge of the distribution patterns of organisms is important for the understanding of global diversity. All species occurring in either the Atlantic or the Pacific regions were likely restricted to the respective areas. The divergence time estimation indicated that the Pacific species likely diverged from the Atlantic ancestors, perhaps through the Arctic Ocean connectivity pathway (Figure 7). The "*Provenzanoi*" species represented the basal clade, and the time tree suggested that it arose at the age of 33 MYA. Species of the "*Exilis*" group diverged from the "*Smithi*" group around 20 MYA. The "*Alatus*" group appeared to have spilt at 15.87 MYA. The Japanese intertidal hermit crab species that possesses a similar geographic distribution, have likely derived from the North Pacific origin, estimated dates for the subsequent diversification ranged from 16 (*P. nigrofascia*) to 18 MYA (major Japanese species clade). The estimated clade ages suggest that *P. filholi* and *P. japonicus*, *P. minutus*, and *P. similis* clades diversified nearly at the same time, since the divergence times of these species were 1.27 MYA and 1.44 MYA.

To advance our understanding of the biogeographical history of a group, a robust and well-resolved phylogenetic tree is required [64]. Taxon sampling is thus fundamental, as limited taxon sampling may bias phylogenetic relationships as well as influence the reliability of any estimate of ancestral area of distribution. Genetic divergence and geographic distances were not compared analytically in this study due to the inability to include multiple populations for each species in the analyses. Despite that, an empirical estimation for the level of relationship was obtained by using species representatives from distant localities where possible. A high inter-regional genetic population structure was found in our analyses. The highest rate of genetic diversification (0.18) was found between the species of MS and AO regions (Table 2), suggesting more ancestral relationship and/or genetic isolation. Similar result of genetic diversification was also found between the species of CS and NS regions. It may indicate that the marine biogeographic barriers (e.g., the Mid-Atlantic Barrier that directs the straight-line flow of the major east–west current at the Atlantic zones, and the Amazon Barrier at the Amazon and Orinoco rivers) effectively influence the long-distance dispersal of the pelagic larvae of the *Pagurus* species along these regions [65]. On the other hand, the lowest genetic diversity (0.05) was found between the species from MS and NS regions. The low level of genetic differences between species from different localities may indicate close ancestral or habitats interconnected by gene flow [66]. It may also suggest weak selective pressures leading to divergence or limited localized adaptation than that in the gulf population. Our analyses revealed that the highest intraregional genetic diversity was found among the species from NWA (0.14), followed by species from the SA (0.13), CS (0.12), and NP (0.10), whereas the lowest was found among the species from NEA (0.08) and NEP (0.09) (results are not shown).

Some phylogenetic relationships could not be exhaustively resolved because of the limited genetic markers and taxonomic coverages. Nonetheless, it is unlikely that the general patterns of phylogeny observed here have been biased by such limitations. Independent analysis of different markers allowed for the inclusion of large number of species from different habitats, while the concatenation of 16S and COI regions gave higher confidence because of increased number of informative sites. The incomplete taxon inclusion may influence the estimation of spatial diversification. Therefore, future comprehensive molecular studies on the regional endemic *Pagurus* fauna, and the integration of those results would benefit from inclusion of larger number of taxa from broader geographical localities. In addition, to establish a precise taxonomic relationship among the major genetic lineages of this genus *Pagurus,* larger sample size per population and sequencing of more genetic markers across geographical ranges are necessary. These more comprehensive studies would require international collaborative studies/joint research agreements.

5. Conclusions

In this study, we have assessed the phylogenetic relationship and estimated the intra/interspecific genetic diversity among the major lineages of *Pagurus* with the currently available molecular data. The data sets analyzed here were one of the most comprehensive molecular data sets of a larger number of representative individuals of the *Pagurus* species across global bioregions. While independent analysis of each molecular marker led to the analyses of numerous taxa, the concatenation of the sequences provided results with higher confidence. The results provide a new insight into evolutionary relationships among the *Pagurus* species. We confirmed previous morphology-based taxonomic classifications from the molecular data. The taxonomic classification of a number of previously unassigned group species were resolved. We established that *"Provenzanoi"* was the basal group of *Pagurus* genus. Additionally, our analyses supported previous studies reporting the polyphyly of *"Comptus"* group. Taken together, this study provides a comprehensive phylogenetic analysis revealing the diversity of *Pagurus* species distributed across multiple geographical locations.

Supplementary Materials: The following are available online at https://www.mdpi.com/article/10.3390/d14020141/s1. Figure S1: Individual gene tree constructed from mitochondrial marker (12S) based on maximum-likelihood (ML) method with 1000 bootstraps. Different colors represent the morpho-based species groups: *"Alatus"* in light blue; *"Bernhardus"* in green; *"Capillatus"* in magenta; *"Chevereuxi"* in gray; *"Comptus"* in blue; *"Cuanensis"* in olive; *"Exilis"* in lawn green; *"Prideaux"* in red; *"Provenzanoi"* in purple; *"Smithi"* in medium turquoise; *"Trigonocheirus"* in brown; and unassigned morpho-group species are in black. Table S1: Species information and accession numbers for the complete mitochondrial genome sequences used for assessment of *Pagurus* species diversity. Table S2: Species information and accession numbers for the 12S sequences used for assessment of *Pagurus* species diversity. Table S3: Species information and accession numbers for the 16S sequences downloaded from NCBI for assessment of *Pagurus* species diversity. Table S3-1: Species information and accession numbers for the selected 16S sequences used for assessment of *Pagurus* species diversity. Table S4: Species information and accession numbers for the COI sequences downloaded from NCBI for assessment of *Pagurus* species diversity. Table S4-1: Species information and accession numbers for the selected COI sequences used for assessment of *Pagurus* species diversity. Table S5: Species information and accession numbers for the selected 18S sequences used for assessment of *Pagurus* species diversity. Table S6: Species information and accession numbers for the selected 28S sequences used for assessment of *Pagurus* species diversity. Table S7: Species information and accession numbers for the selected H3 sequences used for assessment of *Pagurus* species diversity. Table S8: Species information and accession numbers for the concatenated mitochondrial genome sequences used for assessment of *Pagurus* species diversity. Table S9: Species of the genus *Pagurus* along the major geographical localities. Table S10: Pairwise genetic distance Kimura's two-parameter model (K2P) values (per site) among *Pagurus* species based on complete mt genome. Table S11: Pairwise genetic distance Kimura's two-parameter model (K2P) values (per site) among *Pagurus* species based on the 16S gene. Table S12: Pairwise genetic distance Kimura's two-parameter model (K2P) values (per site) among *Pagurus* species based on the third codon of the COI gene. Table S13: Pairwise genetic distance Kimura's two-parameter model (K2P) values (per site) among *Pagurus* species based on the 16S + COI genes.

Author Contributions: A.A. and Z.S. conceptualized this work. Z.S. collected sequences and performed the phylogenetic analyses. I.A.B. led the additional phylogenetic analyses and provided valuable advice on other analyses. A.A. managed funds for the publication. Z.S. primarily wrote the manuscript and finalized with input from all authors. All authors have read and agreed to the published version of the manuscript.

Funding: This research received no external funding.

Institutional Review Board Statement: Not applicable.

Data Availability Statement: All sequence data used in this study have been downloaded from the NCBI database with accession numbers provided by each gene list as a supplementary file. Intermediate data used for these analyses are available upon reasonable request to the corresponding author.

Acknowledgments: The authors are deeply indebted with all the colleagues which have collaborated on the molecular phylogenetic study of *Pagurus* species over the years and made those molecular data available to the public database. Special thanks to two anonymous reviewers for their critic comments on the manuscript. Authors are grateful to Patricia Briones-Fourzán and Michel E. Hendrickx, Guest Editors of the open-access journal *Diversity*, for providing an opportunity to submit this manuscript for a special issue on the topic "Ecology and Diversity of Marine Decapods".

Conflicts of Interest: The authors declare no conflict of interest.

References

1. Walker, S. Criteria for recognizing marine hermit crabs in the fossil record using gastropod shells. *J. Paleontol.* **1992**, *66*, 535–558. [CrossRef]
2. Tsang, L.M.; Chan, T.Y.; Ahyong, S.T.; Chu, K.H. Hermit to king, or hermit to all: Multiple transitions to crab-like forms from hermit crab ancestors. *Syst. Biol.* **2011**, *60*, 616–629. [CrossRef] [PubMed]
3. Cunningham, C.W.; Blackstone, N.W.; Buss, L.W. Evolution of king crab from hermit crab ancestors. *Nature* **1992**, *355*, 539–542. [CrossRef] [PubMed]
4. Wolfe, J.M.; Luque, J.; Bracken-Grissom, H.D. How to become a crab: Phenotypic constraints on a recurring body plan. *BioEssays* **2021**, *43*, 2100020. [CrossRef]
5. McLaughlin, P.A. A review of the phylogenetic position of the Lomidae (Crustacea, Decapoda, Anomala). *J. Crust. Biol.* **1983**, *3*, 431–437. [CrossRef]
6. McLaughlin, P.A.; Lemaitre, R. Carcinization in the Anomura—Fact or fiction? I. Evidence from adult morphology. *Contrib. Zool.* **1997**, *67*, 79–123. [CrossRef]
7. Tudge, C.C. Spermatophore morphology in the hermit crab families Paguridae and Parapaguridae (Paguroidea, Anomura, Decapoda). *Invertebr. Reprod. Dev.* **1999**, *35*, 203–214. [CrossRef]
8. McLaughlin, P.A.; Lemaitre, R.; Tudge, C.C. Carcinization in the Anomura—Fact or fiction? Part II. Evidence from larval, megalopal and early juvenile morphology. *Contrib. Zool.* **2004**, *73*, 165–205. [CrossRef]
9. McLaughlin, P.A.; Lemaitre, R.; Sorhannus, U. Hermit crab phylogeny: A reappraisal and its "fall-out". *J. Crust. Biol.* **2007**, *27*, 97–115. [CrossRef]

10. Tsang, L.M.; Ma, K.Y.; Ahyong, S.T.; Chan, T.Y.; Chu, K.H. Phylogeny of Decapoda using two nuclear protein-coding genes: Origin and evolution of the Reptantia. *Mol. Phylogenet. Evol.* **2008**, *48*, 359–368. [CrossRef]

11. Lemaitre, R.; McLaughlin, P.A. Recent advances and conflicts in concepts of anomuran phylogeny (Crustacea: Malacostraca). *Arthropod Syst. Phylogeny* **2009**, *67*, 119–135.

12. Lemaitre, R.; McLaughlin, P.A.; Sorhannus, U. Phylogenetic relationships within the Pylochelidae (Decapoda: Anomura: Paguroidea): A cladistics analysis based on morphological characters. *Zootaxa* **2009**, *2022*, 1–14. [CrossRef]

13. Scelzo, M.A.; Fantucci, M.Z.; Mantelatto, F.L. Spermatophore and gonopore morphology of the Southwestern-Atlantic hermit crab *Pagurus exilis* (Benedict, 1892) (Anomura, Paguridae). *Zool. Stud.* **2010**, *49*, 421–433.

14. Tirelli, T.; Gamba, M.; Pessani, D.; Tudge, C. Spermatophore and spermatozoal ultrastructure of the Mediterranean hermit crab *Pagurus excavatus* (Paguridae: Anomura: Decapoda). *JMBA* **2013**, *93*, 1363–1371. [CrossRef]

15. Bracken-Grissom, H.D.; Cannon, M.E.; Cabezas, P.; Feldmann, R.M.; Schweitzer, C.E.; Ahyong, S.T.; Felder, D.L.; Lemaitre, R.; Crandall, K.A. A comprehensive and integrative reconstruction of evolutionary history for Anomura (Crustacea: Decapoda). *BMC Evol. Biol.* **2013**, *13*, 128. [CrossRef] [PubMed]

16. Tan, M.H.; Gan, H.M.; Lee, Y.P.; Linton, S.; Grandjean, F.; Bartholomei-Santos, M.L.; Miller, A.D.; Austin, C.M. ORDER within the chaos: Insights into phylogenetic relationships within the Anomura (Crustacea: Decapoda) from mitochondrial sequences and gene order rearrangements. *Mol. Phylogenet. Evol.* **2018**, *127*, 320–331. [CrossRef]

17. Fabricius, J.C. Systema entomologiae, sistens insectorum classes, ordines, genera, species, adiectis synonymis, locis, descriptionibus, observationibus. *Flensbg. Lipsiae Kortii* **1775**, *832*, 1745–1808.

18. Susana, M.P.; Fran, R. Hermit crabs (Decapoda: Crustacea) from deep Mauritanian waters (NW Africa) with the description of a new species. *Zootaxa* **2015**, *3926*, 151–190.

19. De Grave, S.; Pentcheff, N.D.; Ahyong, S.T.; Chan, T.-Y.; Crandall, K.A.; Dworschak, P.C.; Felder, D.L.; Feldmann, R.M.; Fransen, C.H.J.M.; Goulding, L.Y.D.; et al. A classification of living and fossil genera of decapods crustaceans. *Raffles Bull. Zool.* **2009**, *21*, 1–109.

20. Oba, T.; Goshima, S. Temporal and spatial settlement patterns of sympatric hermit crabs and the influence of shell resource availability. *Mar. Biol.* **2004**, *144*, 871–879. [CrossRef]

21. Gherardi, F.; Nardone, F. The question of coexistence in hermit crabs: Population ecology of a tropical intertidal assemblage. *Crustaceana* **1997**, *70*, 608–629.

22. Biagi, R.; Meireles, A.L.; Mantelatto, F.L. Bio-ecological aspects of the hermit crab *Paguristes calliopsis* (Crustacea, Diogenidae) from Anchieta Island, Brazil. *An. Acad. Bras. Ciênc.* **2006**, *78*, 451–462. [CrossRef] [PubMed]

23. Bouvier, E.-L. *Décapodes Marcheurs Faune de France, 37*; Paul Lechevalier: Paris, France, 1940; 404p.

24. MacDonald, J.D.; Pike, R.B.; Williamson, D.I. Larvae of the British species of *Diogenes, Pagurus, Anapagurus* and *Lithodes* (Crustacea, Decapoda). *Proc. Zool. Soc.* **1957**, *128*, 209–257. [CrossRef]

25. Pike, R.B.; Williamson, D.I. *Crustacea, Decapoda: Larvae XI. Paguridea, Coenobitidea, Dromiidea and Homolidea*; Zooplankton Identification Sheet; Conseil International pour l'Exploration de la Mer ICES: Copenhagen, Danmai, 1959; Volume 81, pp. 1–9.

26. McLaughlin, P.A. The hermit crabs (Crustacea, Decapoda, Paguridae) of northwestern North America. *Zool. Verh.* **1974**, *130*, 1–396.

27. McLaughlin, P.A. On the identity of *Pagurus brevidactylus* (Stimpson) (Decapoda: Paguridae), with the description of a new species of *Pagurus* from the western Atlantic. *Bull. Mar. Sci.* **1975**, *25*, 359–376.

28. Ingle, R.W. Northeastern Atlantic and Mediterranean hermit crabs (Crustacea: Anomura: Paguroidea: Paguridae). I. The genus *Pagurus* Fabricius, 1775. *J. Nat. Hist.* **1985**, *19*, 745–769. [CrossRef]

29. McLaughlin, P.A.; Gore, R.H.; Crain, J.A. Studies on the Provenzanoi and other pagurid groups: II. A reexamination of the larval stages of *Pagurus hirsutiusculus* (Dana) (Decapoda: Anomura: Paguridae) reared in the laboratory. *J. Crust. Biol.* **1988**, *8*, 430–450. [CrossRef]

30. Lemaitre, R.; Castaño, N.C. A new species of *Pagurus* Fabricius, 1775 from the Pacific coast of Colombia, with a checklist of the eastern Pacific species of the genus. *Nauplius* **2004**, *12*, 71–82.

31. Scotland, R.W.; Olmstead, R.G.; Bennett, J.R. Phylogeny reconstruction: The role of morphology. *Syst. Biol.* **2003**, *52*, 539–548. [CrossRef]

32. Machordom, A.; Macpherson, E. Rapid radiation and cryptic speciation in squat lobsters of the genus *Munida* (Crustacea, Decapoda) and related genera in the south west Pacific: Molecular and morphological evidence. *Mol. Phylogenet. Evol.* **2004**, *33*, 259–279. [CrossRef] [PubMed]

33. Negri, M.; Pileggi, L.G.; Mantelatto, F.L. Molecular barcode and morphological analyses reveal the taxonomic and biogeographical status of the striped-legged hermit crab species *Clibanarius sclopetarius* (Herbst, 1796) and *Clibanarius vittatus* (Bosc, 1802) (Decapoda: Diogenidae). *Invert. Syst.* **2012**, *26*, 561–571. [CrossRef]

34. Hultgren, K.M.; Stachowicz, J.J. Alternative camouflage strategies mediate predation risk among closely related co-occurring kelp crabs. *Oecologia* **2008**, *155*, 519–528. [CrossRef] [PubMed]

35. Bickford, D.; Lohman, D.J.; Sodhi, N.S.; Ng, P.K.L.; Meier, R.; Winker, K.; Ingram, K.K.; Das, I. Cryptic species as a window on diversity and conservation. *Trends Ecol. Evol.* **2007**, *22*, 148–155. [CrossRef] [PubMed]

36. Piffaretti, J.; Vanlerberghe-Masutti, F.; Tayeh, A.; Clamens, A.L.; D'Acier, A.C.; Jousselin, E. Molecular phylogeny reveals the existence of two sibling species in the aphid pest *Brachycaudus helichrysi* (Hemiptera: Aphididae). *Zool. Scr.* **2012**, *41*, 266–280. [CrossRef]

37. Mantelatto, F.L.; Pardo, L.M.; Pileggi, L.G.; Felder, D.L. Taxonomic re-examination of the hermit crab species *Pagurus forceps* and *Pagurus comptus* (Decapoda: Paguridae) by molecular analysis. *Zootaxa* **2009**, *2133*, 20–32. [CrossRef]

38. da Silva, M.J.; dos Santos, A.; Cunha, M.R.; Costa, F.O.; Creer, S.; Carvalho, G.R. Multigene molecular systematics confirm species status of morphologically convergent *Pagurus* hermit crabs. *PLoS ONE* **2011**, *6*, e28233.

39. Olguín, N.; Mantelatto, F.L. Molecular analysis validates of some informal morphological groups of *Pagurus* (Fabricius, 1775) (Anomura: Paguridae) from South America. *Zootaxa* **2013**, *3666*, 436–448. [CrossRef]

40. Lemaitre, R.; Tavares, M. New taxonomic and distributional information on hermit crabs (Crustacea: Anomura: Paguroidea) from the Gulf of Mexico, Caribbean Sea, and Atlantic coast of South America. *Zootaxa* **2015**, *3994*, 451–506. [CrossRef]

41. Sultana, Z.; Asakura, A.; Kinjo, S.; Nozawa, M.; Nakano, T.; Ikeo, K. Molecular phylogeny of ten intertidal hermit crabs of the genus *Pagurus* inferred from multiple mitochondrial genes, with special emphasis on the evolutionary relationship of *Pagurus lanuginosus* and *Pagurus maculosus*. *Genetica* **2018**, *146*, 369–381. [CrossRef]

42. Hwang, J.-Y.; Haque, N.; Lee, D.-H.; Kim, B.-M.; Rhee, J.-S. Complete mitochondrial genome of the intertidal hermit crab, *Pagurus similis* (Crustacea, Anomura). *Mitochondrial DNA B* **2019**, *4*, 1861–1862. [CrossRef]

43. Gong, L.; Jiang, H.; Zhu, K.; Lu, X.; Liu, L.; Liu, B.; Jiang, L.; Ye, Y.; Lü, Z. Large-scale mitochondrial genere arrangements in the hermit crab *Pagurus nigrofascia* and phylogenetic analysis of the Anomura. *Gene* **2019**, *695*, 75–83. [CrossRef] [PubMed]

44. Landschoff, J.; Gouws, G. DNA barcoding as a tool to facilitate the taxonomy of hermit crabs (Decapoda: Anomura: Paguroidea). *J. Crust. Biol.* **2018**, *38*, 780–793. [CrossRef]

45. Takezaki, N.; Gojobori, T. Correct and incorrect vertebrate phylogenies obtained by the entire mitochondrial DNA sequences. *Mol. Biol. Evol.* **1999**, *15*, 727–737. [CrossRef] [PubMed]

46. Thompson, J.D.; Higgins, D.G.; Gibson, T.J. CLUSTAL W: Improving the sensitivity of progressive multiple sequence alignment through sequence weighting, position-specific gap penalties and weight matrix choice. *Nucleic Acids Res.* **1994**, *22*, 4673–4680. [CrossRef]

47. Kumar, S.; Stecher, G.; Tamura, K. MEGA7: Molecular evolutionary genetics analysis version 7.0 for bigger datasets. *Mol. Biol. Evol.* **2016**, *33*, 1870–1874. [CrossRef]

48. Felsenstein, J. Evolutionary trees from DNA sequences: A maximum likelihood approach. *J. Mol. Evol.* **1981**, *17*, 368–376. [CrossRef]

49. Saitou, N.; Nei, M. The neighbor-joining method: A new method for reconstructing phylogenetic trees. *Mol. Biol. Evol.* **1987**, *4*, 406–442.

50. Kimura, M.A. simple method for estimating evolutionary rate of base substitutions through comparative studies of nucleotide sequences. *J. Mol. Evol.* **1980**, *16*, 111–120. [CrossRef]

51. Felsenstein, J. Confidence limits on phylogenies: An approach using the bootstrap. *Evolution* **1985**, *39*, 783–791. [CrossRef]

52. Hedges, S.B.; Kumar, S. *The Timetree of Life*; Oxford University Press: Oxford, UK, 2009.

53. Yang, J.-S.; Lu, B.; Chen, D.-F.; Yu, Y.-Q.; Yang, F.; Nagasawa, H.; Tsuchida, S.; Fujiwara, Y.; Yang, W.-J. When did decapods invade hydrothermal vents? Clues from the western Pacific and Indian oceans. *Mol. Biol. Evol.* **2013**, *30*, 305–309. [CrossRef] [PubMed]

54. Yang, Z. PAML 4: A program package for phylogenetic analysis by maximum likelihood. *Mol. Biol. Evol.* **2007**, *24*, 1586–1591. [CrossRef] [PubMed]

55. Inoue, J.G.; Miya, M.; Lam, K.; Tay, B.H.; Danks, J.A.; Bell, J.; Walker, T.I.; Venkatesh, B. Evolutionary origin and phylogeny of the modern holocephalans (Chondrichthyes: Chimaeriformes): A mitogenomic perspective. *Mol. Biol. Evol.* **2010**, *27*, 2576–2586. [CrossRef] [PubMed]

56. Babarinde, I.A.; Saitou, N. The dynamics, causes, and impacts of mammalian evolutionary rates revealed by the analyses of Capybara draft genome sequences. *Genome Biol. Evol.* **2020**, *12*, 1444–1458. [CrossRef] [PubMed]

57. Muñoz, J.E.; Cuesta, J.A.; Raso, J.E. Taxonomic study of the *Pagurus forbesii* "complex" (Crustacea: Decapoda: Paguridae). Description of *Pagurus pseudosculptimanus* sp. nov. from Alborán Sea (Southern Spain, Western Mediterranean Sea). *Zootaxa* **2014**, *3753*, 25–46. [CrossRef] [PubMed]

58. Komai, T. The identity of *Pagurus brachiomastus* and descriptions of two new species of *Pagurus* (Crustacea: Decapoda: Anomura: Paguridae) from the Northwestern Pacific. *Species Divers.* **2000**, *5*, 229–265. [CrossRef]

59. Komai, T.; Imafuku, M. Redescription of *Pagurus lanuginosus* with the establishment of a neotype, and description of a new closely related species (Decapoda: Anomura: Paguridae). *J. Crust. Biol.* **1996**, *16*, 782–796. [CrossRef]

60. Komai, T. Identities of *Pagurus japonicus* (Stimpson, 1858) and *P. similis* (Ortmann, 1892), with description of a new species of *Pagurus*. *Zoosystema* **2003**, *25*, 377–411.

61. Noever, C.; Glenner, H. The origin of king crabs: Hermit crab ancestry under the magnifying glass. *Zool. J. Linn. Soc.* **2017**, *182*, 300–318. [CrossRef]

62. Landschoff, J.; Komai, T.; du Plessis, A.; Gouws, G.; Griffiths, C.L. MicroCT imaging applied to description of a new species of *Pagurus Fabricius*, 1775 (Crustacea: Decapoda: Anomura: Paguridae), with selection of three-dimensional type data. *PLoS ONE* **2018**, *13*, e0203107.

63. Trivedi, J.N.; Vachhrajani, K.D. An annotated checklist of hermit crabs (Crustacea, Decapoda, Anomura) of Indian waters with three new records. *J. Asia Pac. Biodivers.* **2017**, *10*, 175–182. [CrossRef]

64. Ree, R.H.; Smith, S.A. Maximum likelihood inference of geographic range evolution by dispersal, local extinction, and cladogenesis. *Syst. Biol.* **2008**, *57*, 4–14. [CrossRef] [PubMed]

65. Luiz, O.J.; Madin, J.S.; Robertson, D.R.; Rocha, L.; Wirtz, P.; Floeter, S.R. Ecological traits influencing range expansion across large oceanic dispersal barriers: Insights from tropical Atlantic reef fishes. *Proc. Biol. Sci.* **2012**, *279*, 1033–1040. [CrossRef] [PubMed]
66. Hellberg, M.E. Gene flow and isolation among populations of marine animals. *Annu. Rev. Ecol. Evol. Syst.* **2009**, *40*, 291–310. [CrossRef]

Article

Integrative Taxonomy of New Zealand Stenopodidea (Crustacea: Decapoda) with New Species and Records for the Region

Kareen E. Schnabel [1,], Qi Kou [2,3] and Peng Xu [4]

1 Coasts and Oceans Centre, National Institute of Water & Atmospheric Research, Private Bag 14901 Kilbirnie, Wellington 6241, New Zealand
2 Institute of Oceanology, Chinese Academy of Sciences, Qingdao 266071, China; kouqi@qdio.ac.cn
3 College of Marine Science, University of Chinese Academy of Sciences, Beijing 100049, China
4 Key Laboratory of Marine Ecosystem Dynamics, Second Institute of Oceanography, Ministry of Natural Resources, Hangzhou 310012, China; xupeng@sio.org.cn
* Correspondence: kareen.schnabel@niwa.co.nz; Tel.: +64-4-386-0862

Abstract: The New Zealand fauna of the crustacean infraorder Stenopodidea, the coral and sponge shrimps, is reviewed using both classical taxonomic and molecular tools. In addition to the three species so far recorded in the region, we report *Spongicola goyi* for the first time, and formally describe three new species of Spongicolidae. Following the morphological review and DNA sequencing of type specimens, we propose the synonymy of *Spongiocaris yaldwyni* with *S. neocaledonensis* and review a proposed broad Indo-West Pacific distribution range of *Spongicoloides novaezelandiae*. New records for the latter at nearly 54° South on the Macquarie Ridge provide the southernmost record for stenopodidean shrimp known to date.

Keywords: sponge shrimp; coral cleaner shrimp; taxonomy; cytochrome oxidase 1; 16S ribosomal RNA; association; southwest Pacific Ocean

Citation: Schnabel, K.E.; Kou, Q.; Xu, P. Integrative Taxonomy of New Zealand Stenopodidea (Crustacea: Decapoda) with New Species and Records for the Region. *Diversity* **2021**, *13*, 343. https://doi.org/10.3390/d13080343

Academic Editors: Michael Wink, Patricia Briones-Fourzán and Michel E. Hendrickx

Received: 24 June 2021
Accepted: 11 July 2021
Published: 27 July 2021

Publisher's Note: MDPI stays neutral with regard to jurisdictional claims in published maps and institutional affiliations.

1. Introduction

The unique group of coral shrimp and Venus or sponge shrimp, united in the infraorder Stenopodidea Spence Bate, 1888 [1], is a small group of marine decapod crustaceans with 92 species, 13 genera and three families currently recognized [2]. While the infraorder is well-defined considering shared morphological synapomorphies [3], the placement of the Stenopodidea within the Decapoda remains unresolved [4–6]. Similarly, the internal classification remains in flux with a recently erected family [7], four new genera [7–10] and over one-third of the current species diversity described since 2006, e.g., [11–14].

The Stenopodidea currently contains three families: (1) The shallow-water, free-living Macromaxillocarididae Alvarez, Iliffe & Villalobos, 2006 [7] that includes the anchialine species *Macromaxillocaris bahamaensis* Alvarez, Iliffe & Villalobos, 2006 [7,10]; (2) the Stenopodidae, as defined presently, include the colorful and popular ornamental shrimps in the aquarium trade [3,15] and are almost exclusively free-living shallow-water taxa inhabiting coral reefs down to about 50 m; (3) the more diverse Spongicolidae represent primarily deep-water taxa, typically associated with hexactinellid sponges or octocorals, which can extend beyond 2300 m depth [3,8,12,16]. Comparative morphology and molecular phylogenetics more recently called for internal taxonomic revisions, e.g., a molecular phylogeny provided by Chen et al. [17] refuted both the family and genus level classification and the authors suggested to unite all species in a single family Stenopodidae. More recently, Bochini et al. [10] provided additional molecular evidence and hypotheses of current taxonomic delimitations. Work is underway to increase taxon sampling and to settle the classification for this group by e.g., Kou and Goy with colleagues (unpubl.).

Stenopodidean shrimps inhabit all oceans, with an overall pan-tropical distribution and the highest diversity centered in the Indo-West Pacific [3]. The most northern record was provided by Hansen [18] off Iceland, in the Atlantic, the most northern record in the Pacific is Japan, while South Africa, Tasmania and New Zealand represent the southern latitudinal boundaries of this group [12,19–21]. However, the New Zealand stenopodideans have historically received limited attention. The known fauna currently only comprises three species: the first record was provided by Yaldwyn [22] for the banded coral shrimp *Stenopus hispidus* Olivier, 1911 [23] from northern New Zealand, followed by *Spongiocaris yaldwyni* Bruce & Baba, 1973 [21] from the Bay of Plenty off the central North Island, which remained the only specimen record of this species to date (gene sequences were for one sample were recently presented by Chen et al. [17] from Tonga), and *Spongicoloides novaezelandiae* Baba, 1979 [20] from the Chatham Rise off the South Island. The latter species was recently reported by Goy [13] with a broad Indo-Pacific distribution, but records are either referred to other species or called into doubt in this study.

Here, we review historical and new specimens of stenopodidean shrimps collected in the New Zealand region by combining DNA sequencing and morphological classification. We provide evidence of a higher diversity for the region than previously reported, with one new record and three new species presented, including the most southern record known to date.

2. Materials and Methods

2.1. Specimen Collections

The primary study area encompasses the New Zealand charting area [24] which includes portions of the Australian Exclusive Economic Zone that surround Norfolk, Lord Howe and Macquarie Islands (Figure 1). Samples were collected between the years of 1962–2017 and from depths ranging from 0–1998 m. Most of the recent samples were provided by the following RV *Tangaroa* surveys: the 2003 NORFANZ voyage (TAN0308); 2008 "MacRidge 2" Macquarie Ridge voyage (TAN0803); two "Impact of resource use on vulnerable deep-sea communities" project voyages (TAN1104 and TAN1206); the 2016 "Biodiversity of the Kermadec Islands and offshore waters of the Kermadec Ridge—a coastal, marine mammal and deep-sea survey" (TAN1612); and the 2017 *PoribacNewZ* voyage using the ROV *KIEL 6000* on the German RV *Sonne* (voyage SO254). Please see the Acknowledgments section for details. Specimens examined are deposited at the National Institute of Water & Atmospheric Research, Wellington (NIWA), Museum of New Zealand Te Papa Tongarewa, Wellington (NMNZ) and Tāmaki Paenga Hira Auckland War Memorial Museum, Auckland (AWMM).

2.2. Morphological Examination

Morphological terminology and measurements follow Goy [3]. Measurements of specimens are given in millimeters (mm). Postrostral carapace length (PCL) is measured along the dorsal midline, from the posterior end of the orbit to the posterior margin of the carapace; total carapace length (CL) is measured from the anterior end of the rostrum to the posterior end of the carapace; total body length (TL) is measured from the tip of the rostrum to the posterior end of the telson. Specimens were measured and illustrated using a MZ9.5 (KS, Leica, Heerbrugg, Switzerland) and SteREO Discovery V8 (QK Zeiss, Oberkochen, Germany) stereomicroscope. Line drawings and color plates were made using a Intuous Pro Graphics Tablet (WACOM, Saitama, Japan), Adobe Illustrator CS6 and Adobe Photoshop 2020 (KS) and CS4 (QK) (Adobe, San Jose, CA, USA); sample records were mapped using ArcGIS Pro version 2.6.1 (ESRI, Redlands, CA, USA) and a NIWA Basemap (Mercator 41 Projection, NIWA 2018) using maps based on NIWA Regional Bathymetry data (https://niwa.co.nz/our-science/oceans/bathymetry/download-the-data (accessed on 12 July 2021)).

2.3. DNA Extraction and Analysis

DNA was extracted from muscle or branchial tissue of recently collected specimens. Extraction using the DNeasy Blood & Tissue Kit (QIAGEN, Germantown, MD, USA) followed the manufacturer's protocols. Genomic DNA was eluted in 50 µL of sterile distilled H$_2$O (RNase free) and stored at −20 °C until processed further. A partial sequence of the mitochondrial cytochrome c oxidase I (COI) gene was amplified using the universal primer pairs LCO1490/HCO2198 [25] or the newly-designed stenopodidean-specific primer pairs: COI-stenF (5′-TTTATTTTYGGWRCWTGARSAGG-3′) and COI-stenR (5′-TAACTGAYCGWAATMTTAAYACTTC-3′). The 16S ribosomal RNA gene was amplified using the universal primer pair 16S-arL/brH [26] or the newly-designed stenopodidean-specific primer pairs: 16S-stenF (5′-TTGAYGARARATADTCTGTC-3′) and 16S-stenR (5′-CGGTBTGAACTCAAATCAT-3′). Polymerase chain reaction (PCR) amplifications were performed in a reaction mix containing 25 µL of Premix Taq™ (Takara, Otsu, Shiga, Japan), 2−5 µL of template DNA, 1 µL of each primer (10 mM), and sterile distilled H$_2$O to a total volume of 50 µL. The PCR protocol was run on a 2720 Thermal Cycler (Applied Biosystems, Waltham, MA, USA) as follows: the reactions were processed with an initial denaturation step (95 °C, 3 min), followed by 35 cycles of denaturation (95 °C, 30 s), annealing (48 °C, 30 s) and extension (72 °C, 45 s), with a final extension of 5 min at 72 °C. PCR products were assessed by agarose gel electrophoresis, cleaned using ExoSAP-IT reagent (USB, Cleveland, OH, USA) and commercially sequenced (Macrogen Inc., Seoul, Korea) using the same primers used for the PCR.

The 16S rRNA and COI gene sequence chromatograms were checked using CHRO-MAS 2.23 (Technelysium Pty Ltd., Brisbane, Australia) by eye. The forward and reverse sequence fragments were assembled by CONTIG EXPRESS (a component of Vector NTI Suite 6.0, Life Technologies, Carlsbad, CA, USA). Sequences were checked for potential contamination using the Basic Local Alignment Search Tool (BLAST) through GenBank. The homologous sequence alignment was conducted with MAFFT version 7 webserver [27] using the default parameters, and manually trimmed to the same length. The Kimura's 2-parameter genetic distances of COI and 16S rRNA genes were calculated using MEGA 6.06 [28].

The 16S phylogenetic tree was reconstructed using maximum likelihood (ML) method. The best-fit nucleotide substitution model was selected by ModelFinder [29] implemented in IQ-TREE 2.1.2 [30]. Then the ML analysis was conducted using IQ-TREE 2.1.2 and the node supports were evaluated by performing Bayesian-like transformation of aLRT (aBayes) test [31], as well as SH-like approximate likelihood ratio test (SH-aLRT) [32] and ultrafast bootstrap (UFBoot) with 1000 replicates [33]. The phylogenetic trees were visualized using FigTree 1.4.3 [34] and annotated with Adobe Photoshop CS4®. Finally, all the new sequences were submitted to the GenBank database (Table 1).

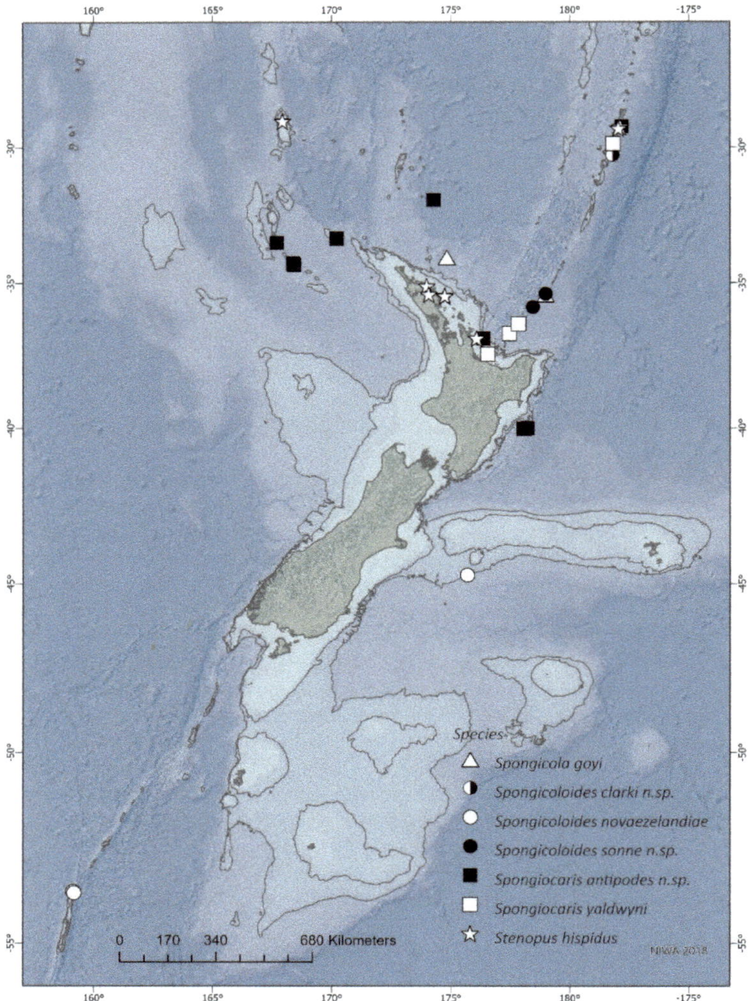

Figure 1. Map of the New Zealand region showing sampling locations for New Zealand Stenopodidea. 1000 and 2000 m bathymetric contours are highlighted.

Table 1. List of all Stenopodidea sequences used in phylogenetic analyses, collection locality and their respective GenBank accession numbers. NA = not available. Collection catalogue numbers in bold indicate misidentified specimens. GenBank accession numbers in bold indicate new sequences presented in this study.

Family	Genus	Species	Collection Cat. No.	Collection Locality	CO1	16S rRNA	Sequence References
Macromaxillocarididae	Chicosciencea	Chicosciencea pernambucensis	DZ/UFRGS-6779	Pernambuco, Brazil	NA	MN482688	[10]
	Macromaxillocaris	Macromaxillocaris bahamaensis	ULLZ-11769	Eleuthera, Bahamas	NA	KX086378	[17]
	Microprosthema	Microprosthema semilaeve	MNHN-IU-2013-4307	Guadeloupe	NA	KX086382	[17]
		Microprosthema takedai	MNHN-IU-2014-6693	Vanuatu	NA	KX086404	[17]
Spongicolidae	Engystenopus	Engystenopus palmipes	NTOU-M01900	The Philippines	NA	KX086379	[17]
	Globospongicola	Globospongicola spinulatus	NTOU-M01877	Taiwan	KU188326 #	KU188326 #	[5]
	Paraspongicola	Paraspongicola inflatus	MNHN-IU-2014-12066	New Caledonia	NA	KX086392	[17]
	Spongicola	Spongicola andamanicus	NTOU-M01903	South China Sea	NA	KX086415	[17]
		Spongicola goyi	NIWA-127123	New Zealand	**MZ539859**	**MZ531909**	this study
			NTOU-M01905	South China Sea	NA	KX086426	[17]
		Spongicola levigatus	NTOU-M01876	South China Sea	KU188325 #	KU188325 #	[5]
		Spongicola robustus	MNHN-IU-2010-85	Madagascar	NA	KX086395	[17]
	Spongicoloides	Spongicoloides clarki sp. nov.	NIWA-118650	New Zealand	**MZ539860**	**MZ531910**	this study
		Spongicoloides corbitellus	MBM-286006	Seamount near the Mariana Trench	MG386936	MG429776	[35]
		Spongicoloides iheyaensis	NTOU-M01908	Taiwan	NA	KX086424	[17]
		Spongicoloides novaezelandiae	NMNZ-CR.1889	New Zealand	**MZ539861**	**MZ531911**	This study
			NIWA-40567	New Zealand	**MZ539862**	**MZ531912**	This study
			NIWA-40638	New Zealand	**MZ539863**	**MZ531913**	This study
			MNHN-IU-2013-19488	New Caledonia	**MZ539867**	**MZ531918**	This study
			MNHN-IU-2013-19622	Vanuatu	NA	**MZ531917**	This study
			MNHN-IU-2013-19625	New Caledonia	NA	**MZ531919**	This study
			MNHN-IU-2013-19627	New Caledonia	NA	**MZ531920**	This study
			MNHN-IU-2013-19630	New Caledonia	NA	**MZ53192**	This study

Table 1. *Cont.*

Family	Genus	Species	Collection Cat. No.	Collection Locality	CO1	16S rRNA	Sequence References
		Spongicoloides sonne sp. nov.	NIWA-127110	New Zealand	MZ539864	MZ531914	This study
			NIWA-127111	New Zealand	MZ539865	MZ531915	This study
			NMNZ-CR.019650	New Zealand	MZ539866	MZ531916	This study
		Spongicoloides weijiaensis	SRSIO-16050001	Weijia Guyot, NW Pacific	KY404238	KY404237	[36]
			SRSIO-17050301	South China Sea	MZ539868	MZ531922	This study
	Spongiocaris	*Spongiocaris antipodes* sp. nov.	NIWA-88622	New Zealand	MZ539869	MZ531923	This study
			NIWA-127133	New Zealand	MZ539870	MZ531924	This study
			NMNZ-CR.005952	New Zealand	MZ539871	MZ531925	This study
			NMNZ-CR.019259	New Zealand	MZ539872	MZ531926	This study
			NMNZ-CR.019491	New Zealand	MZ539873	MZ531927	This study
			NMNZ-CR.019492	New Zealand	NA	MZ531928	This study
			NMNZ-CR.025704	New Zealand	MZ539874	MZ531929	This study
			NMNZ-CR.019494	New Zealand	MZ539875	MZ531930	This study
		Spongiocaris japonicus	NTOU-M01906	Japan	NA	KX086416	[17]
		Spongiocaris koehleri	MNHN-IU-2014-12841	Seamount near Bermuda	NA	KX086425	[17]
		Spongiocaris neocaledonensis	MNHN-IU-2014-23852	New Caledonia	MZ539876	MZ531931	This study
			MNHN-IU-2014-23853	New Caledonia	MZ539877	MZ531932	This study
			MNHN-IU-2014-23855	New Caledonia	MZ539878	MZ531933	This study
			MNHN-IU-2014-23856	New Caledonia	NA	MZ531934	This study
			MNHN-IU-2014-23857	Loyalty Islands	NA	MZ531935	This study
			MNHN-IU-2014-23858	New Caledonia	MZ539879	MZ531936	This study
			MNHN-IU-2014-23859	New Caledonia	NA	MZ531937	This study
			MNHN-IU-2014-23860	New Caledonia	NA	MZ531938	This study
			MNHN-IU-2014-23861	New Caledonia	MZ539880	MZ531939	This study

Table 1. *Cont.*

Family	Genus	Species	Collection Cat. No.	Collection Locality	CO1	16S rRNA	Sequence References
		Spongiocaris panglao	NTOU-M01909	The Philippines	NA	KX086429	[17]
			Sun et al. (2018) *	Seamount near the Yap Trench	MG812382 [#]	MG812382 [#]	[37]
		Spongiocaris yaldwyni	NMNZ-CR.1888	New Zealand	NA	**MZ531940**	This study
			NIWA-82821	New Zealand	**MZ539881**	**MZ531941**	This study
			NIWA-83102	New Zealand	**MZ539882**	**MZ531942**	This study
			MNHN-IU-2014-12842	Tonga	NA	KX086439	[17]
Stenopodidae	Juxtastenopus	*Juxtastenopus spinulatus*	MNHN-IU-2014–6692	Vanuatu	NA	KX086403	[17]
	Odontozona	*Odontozona meloi*	MNHN-IU-2013-2887	French Guinea	NA	KX086387	[17]
	Richardina	*Richardina spinicincta*	MNHN-IU-2013–19177	Guadeloupe	NA	KX086442	[17]
	Stenopus	*Stenopus goyi*	NTOU-M01912	The Philippines	NA	KX086422	[17]
		Stenopus hispidus	NIWA-118168	New Zealand	NA	**MZ531943**	This study
			AM-MA-168313	New Zealand	**MZ539883**	**MZ531944**	This study
			AM-MA-182117	New Zealand	**MZ539884**	**MZ531945**	This study
			MNHN-IU-2013-233	Papua New Guinea	NA	KX086420	[17]
		Stenopus pyrsonotus	MNHN-IU-2011-8952	French Polynesia	NA	KX086431	[17]
		Stenopus scutellatus	MNHN-IU-2013-4378	Guadeloupe	NA	KX086432	[17]
		Stenopus spinosus	CCDB-1525	Sao Paulo, Brazil	MF490042	MF490145	[38]
		Stenopus tenuirostris	MNHN-IU-2013-10246	Papua New Guinea	NA	KX086434	[17]
		Stenopus zanzibaricus	MNHN-IU-2013-10243	Papua New Guinea	NA	KX086441	[17]
Outgroups		*Alvinocaris chelys*	NTOU-M01671	Taiwan	NA	KX086443	[17]
		Gnathophyllum americanum	MSLKHC-CA02	Aquarium shop, Hong Kong	NA	KX086444	[17]
		Palaemon serratus	MSLKHC-CA98	Hawaii	NA	KX086445	[17]
		Procaris hawaiiana	OUMNH.ZC.2010-13-007	Hawaii	NA	KX086446	[17]

* Collection catalogue number was not provided. [#] Sequences are extracted from complete mitochondrial genome sequence.

3. Results

3.1. Molecular Taxonomy

A total of 32 CO1 and 70 16S rRNA sequences were included in our analyses. The final CO1 and 16S rRNA dataset for phylogeny inference were trimmed to the same length after the alignments and consisted of 590 bp and 390 bp, respectively. The best-fit models for 16S rRNA dataset selected by ModelFinder is TIM3 + F + I + G4. The Kimura's 2-parameter pair-wise genetic distances between individuals for the CO1 and 16S rRNA datasets are shown in Tables S1 and S2.

For the CO1 dataset, the inter-generic genetic divergences of Stenopodidea ranged from 8.5% (between *Spongiocaris yaldwyni* and *Spongicoloides sonne* sp. nov.) to 26.6% (between *Spongicola goyi* and *Spongicoloides sonne* sp. nov.). The intra-generic genetic divergences ranged from 3.8% (between *Spongicoloides novaezelandiae* and *S. clarki* sp. nov.) to 16.8% (between *Stenopus hispidus* and *S. spinosus*). The intra-specific genetic divergences ranged from 0.0% to 4.2% (between MNHN-IU-2014-23855 and NIWA-83102 in *Spongiocaris yaldwyni*). These high levels of intraspecific divergences represent substantial internal genetic structure and may warrant further investigation.

For the 16S rRNA dataset, the inter-generic genetic divergences of Stenopodidea ranged from 4.0% (between *Juxtastenopus spinulatus* and *Stenopus goyi*) to 27.5% (between *Microprosthema takedai* and *Spongicola levigatus*). The intra-generic genetic divergences ranged from 0.6% (between *Spongiocaris japonica* and *S. panglao*) to 17.1% (between *Spongicoloides iheyaensis* and *S. sonne* sp. nov.). The intra-specific genetic divergences ranged from 0.0% to 2.6% (between MNHN-IU-2014-12842 and MNHN-IU-2014-23855, in *Spongiocaris yaldwyni*).

The phylogenetic trees inferred from ML analysis was generally well supported (Figure 2). Our findings show the ingroup (Stenopodidea) was recognized as a monophyletic group with high support values. At the familial level, Macromaxillocarididae was at the basal position and monophyletic, but with weak support. The monophyly of the other two families, Spongicolidae and Stenopodidae was not supported, as the spongicolid genus *Globospongicola* Komai & Saito, 2006 [8] was embedded in the Stenopodidae clade. At the generic level, *Stenopus*, *Spongicola*, *Spongiocaris* and *Spongicoloides* were suggested to be non-monophyletic, which was generally in accord with the results of previous phylogenetic studies [10,17]. Specifically, *Stenopus*, *Spongicola* and *Spongiocaris* are paraphyletic, while the species of *Spongicoloides* formed a polyphyletic assemblage. At the specific level, all the species were well separated. Significantly, our tree suggested that *Spongiocaris neocaledonensis* should be the synonym of *S. yaldwyni*, and some specimens previously identified as *Spongicoloides novaezelandiae* in Goy [13] are actually *S. weijiaensis* Xu, Zhou & Wang, 2017 [36] or *S. sonne* sp. nov. These findings are discussed further under the species below.

3.2. Systematics

Order DECAPODA Latreille, 1802 [39]
Infraorder STENOPODIDEA Spence Bate, 1888 [1]
Family SPONGICOLIDAE Schram, 1986 [40]
Genus ***Spongicola*** de Haan, 1844 [41]

3.2.1. *Spongicola goyi* Saito & Komai, 2008

In (Figures 1–4). *Spongicola goyi* Saito & Komai, 2008 [12]: 107.—Goy 2010 [3]: 224.—De Grave & Fransen 2011 [42]: 251.—Goy 2015 [13]: 307.

Material examined. Northland. 1 M (PCL: 5.7 mm); Seamount #441, 34.043° S, 174.817° E, 880–792 m; 19 Apr 2002; R/V *Kaharoa* Stn. KAH0204/47; 'seamount' sled; NIWA 3621. **Southern Kermadec Ridge.** 1 F ov. (PCL: 5.9 mm); Southern Kermadec Ridge, 35.380° S, 178.980° E, 1184.1 m; 7 February 2017; RV *Sonne* Stn. SO254/33ROV08, Remote Operated Vehicle; NIWA 127111; found inside *Corbitella* sp.; NIWA 127123.

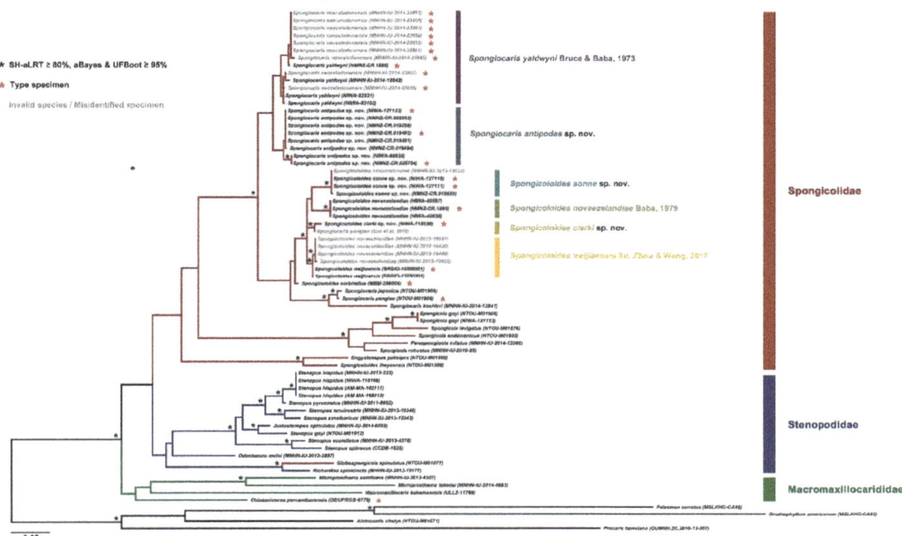

Figure 2. Maximum likelihood tree inferred from the 16S rRNA gene sequences. The colors of the branches indicate the familial affiliation proposed by Bochini et al. [10]. * indicate high nodal support and a star highlights a type specimen.

Figure 3. *Spongicola goyi* Saito & Komai, 2008 [12] (NIWA127123) sponge host, most likely *Pheronema conicum* Lévi & Lévi, 1982 [43] (det. Michelle Kelly, NIWA), (Hexactinellida, Amphidiscosida, Pheronematidae) immediately prior to (top) and during collection (bottom) on the southern Kermadec Ridge (1168 m). Image courtesy of ROV *Kiel 6000* GEOMAR, SO254 voyage PoriBac-NewZ, ICBM.

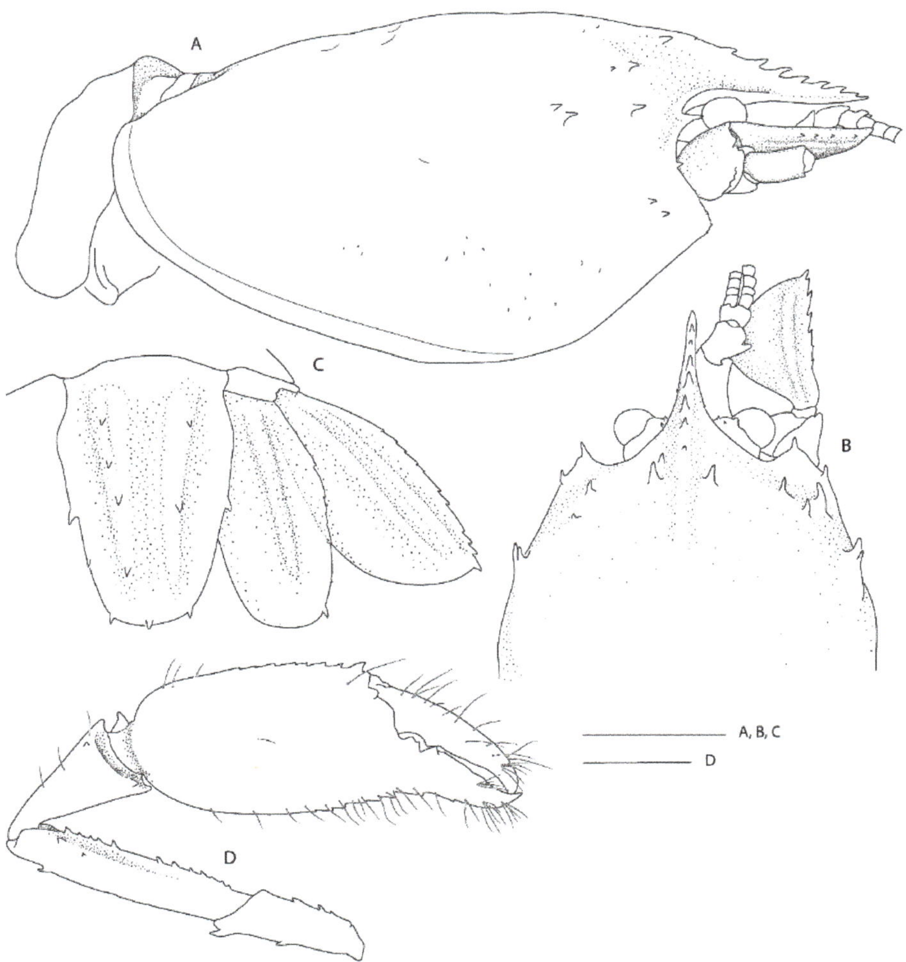

Figure 4. *Spongicola goyi* Saito & Komai, 2008 [12], F ov. (NIWA 127123). (**A**) Carapace and anterior pleonal segments, lateral. (**B**) Anterior portion of carapace and cephalic appendages, dorsal. (**C**) Telson and uropods, dorsal. (**D**) Third pereiopod, lateral. Scale bars = 2 mm.

Diagnosis. Carapace with postrostral submedian and anterolateral spines; branchial region with fine scattered spinules or granules. Merus of third pereiopod unarmed or armed with row of small teeth or denticles on lateral margin; palm distinctly carinate on dorsal margin. Lateral margin of uropodal endopod serrate.

Measurements. CL: 7.6 mm (female), 7.9 mm (male), PCL: 5.9 mm (female), 5.7 mm (male), TL: 26.0 mm (female), 19.4 mm (male).

Distribution. Japan, Indonesia, Vanuatu, New Caledonia, Australia, Madagascar, and now northern New Zealand; 315–1184 m.

Hosts. Previously reported on *Demospongia* sp., Hexactinellidae sp., *Hyalonema* sp., *Euplectella* sp. and *Pheronema semiglobosum* Lévi & Lévi, 1982 [43] by Saito & Komai [12] and *Euhyalonema* sp. by Goy [13]. The female (NIWA 127123) reported here was collected with a sponge host most likely *Pheronema conicum* Lévi & Lévi, 1982 [43] (NIWA 126108, det. Michelle Kelly, Figure 3). The male (NIWA 3621) included a collection note "with hexactinellid", of which eight were collected at the same station, including three *Pheronema conicum* but also *Chonelasma* sp. and *Euryplegma auriculare* Schulze, 1886 [44].

Colour in life. Unknown. Collection note with NIWA 127123 refers to 'bright blue eggs'.

Remarks. This is the first record for this genus and species in the New Zealand region, although the type series included specimens from New Caledonia and the Norfolk Ridge just north of New Zealand. The two specimens presented here extend the known range slightly southward to seamounts off the Northland Plateau and the southern Kermadec Ridge (Figure 1). They were both collected from within hexactinellid sponges, possibly both from the amphidiscosid *Pheronema conicum*. Genetically, the female (NIWA 127123) aligns with conspecific reference sequences (see below). Morphologically, both specimens match the type description in nearly all aspects. Slight variation for the ovigerous female compared to the original description is an additional large postorbital carapace spine, the first pleonite being more distinctly ridged (Figure 4A), the posterior margin of the telson being distinctly convex (rather than slightly convex or nearly truncate) (Figure 4C), the dactylus of the third pereiopod is smooth on the dorsal margin (not with row of teeth or denticles) (Figure 4D) and the subdivision of the propodi of pereiopods 4 and 5 is indistinct. The specimen bears approximately 100 spherical eggs of a diameter 0.8 mm each. The male (NIWA 3621) matches the paratypes illustrated by Saito & Komai [12], the third pereiopod is distinctly spinose along the ischium and merus and serrate along the margins of the palm and fingers. A slight difference is apparent in the shape of the eighth thoracic sternite that is anteriorly projected to a distinct, round tooth (instead of subrectangular).

The species resembles *S. andamanicus* Alcock, 1901 [45] (widely distributed throughout the Indo-West Pacific) but can be chiefly distinguished by the lack of a large tooth on the lateral margin of the third pereiopod merus. The merus is instead furnished with a row of small teeth or denticles only.

DNA sequence data. DNA sequences for the large female (NIWA 127123) aligned well with previously published sequences, with intraspecific levels of divergences for 16S (≤0.8%) compared to four specimens from South China Sea, Papua New Guinea and New Caledonia [17] (Figure 2).

Genus *Spongicoloides* Hansen, 1908 [18].

3.2.2. *Spongicoloides clarki* sp. nov.

In (Figures 1, 2 and 5–7). *Spongiocaris panglao*.—Sun, Sha & Wang 2018 [37]: 124 (complete mitochondrial genome) (not *Spongiocaris panglao* Komai, De Grave & Saito, 2016 [15]).

Material examined. Holotype: F ov. (PCL: 11.8 mm); Macauley Island, Kermadec Islands, 30.28–30.29° S, 178.20–178.20° W, 1431–1426 m; 29 Oct 2016; RV *Tangaroa* Stn. TAN1612/71, beam trawl; found inside *Regadrella okinoseana*; NIWA 118650.

Diagnosis. Carapace with faint hepatic groove; surface with only scattered small postrostral, hepatic and anterolateral spines, cervical groove not lined with distinct spines; anterior margins without antennal spine, few small anterolateral spines. Rostrum reaching distal margin of basal article of antennular peduncle; with nine small dorsal teeth, laterally unarmed. Epistome anteriorly straight, nearly smooth. Second and third pleonite with acute articular knob; fifth and sixth pleonites smooth on dorsal midline, unarmed posteroventral margin. Telson broadly rectangular, about twice as long as wide, small teeth along posterior margin. Antennular peduncle unarmed except for minute distomesial spine on second article, stout stylocerite. Antenna with small ventromedian spines on basicerite; scale with 7–8 small lateral teeth. First maxilliped with single arthrobranch. Second maxilliped with single arthrobranch, well-developed podobranch, lacking epipod. Third maxilliped with two arthrobranchs and rudimentary epipod, setiferous organ well-developed. First pereiopod with setiferous organ on propodus only. Third pereiopod nearly entirely glabrous and unarmed; fixed finger unarmed on distoventral margin; ischium with distodorsal spine and row of small ventral spines; coxa mesially unarmed. Fourth and fifth pereiopods dactyli with ventral unguis bearing a small ventral and dorsal tooth; coxa mesially unarmed; fourth pereiopod with paired arthrobranchs.

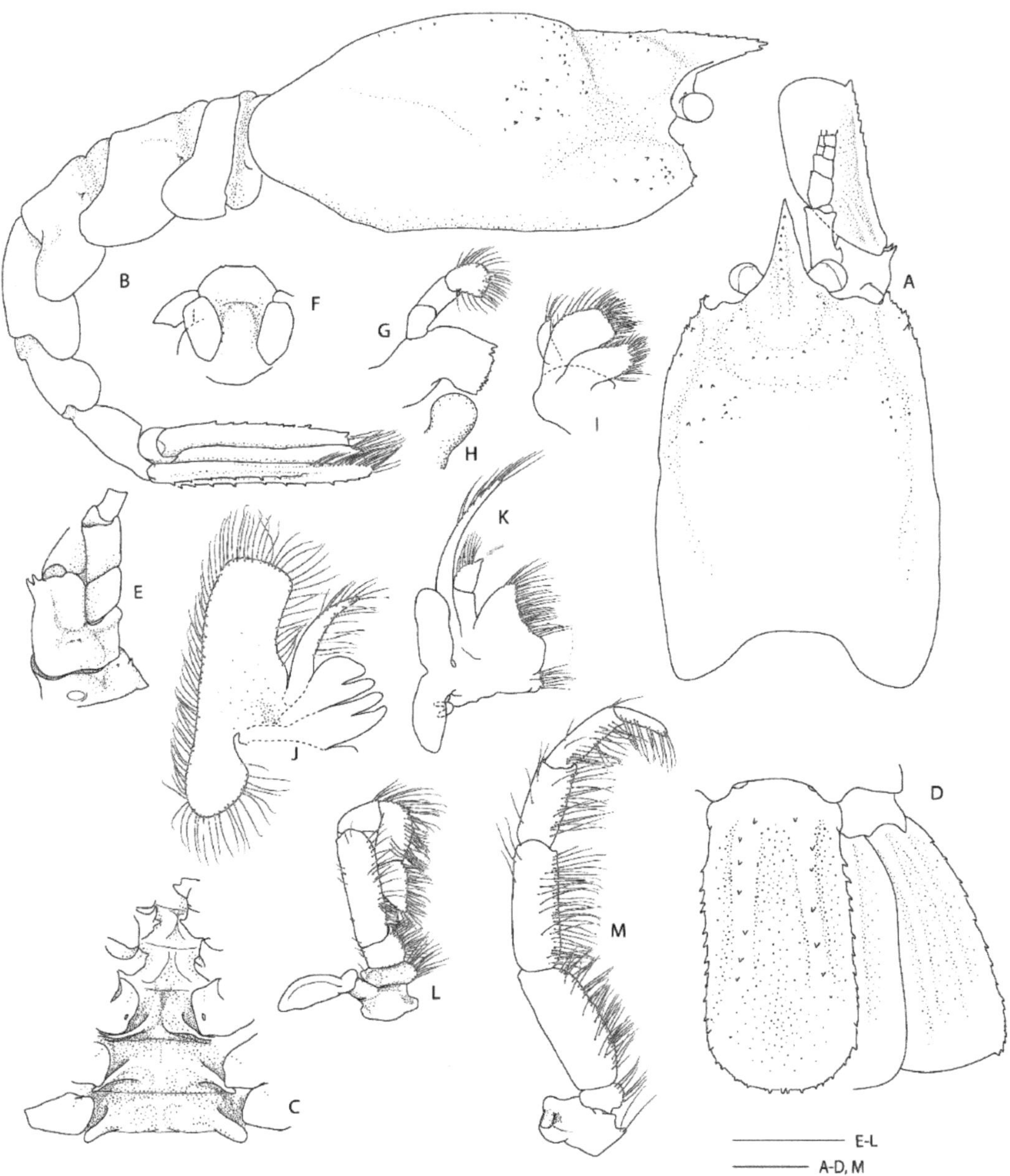

Figure 5. *Spongicoloides clarki* sp. nov., holotype F ov. (NIWA 118650). (**A**) Carapace and cephalic appendages, dorsal. (**B**) Carapace and pleon, lateral. (**C**) Thoracic sternites, ventral. (**D**) Telson and uropods, dorsal. (**E**) Right antennal peduncle, ventral. (**F**) Epistome and labrum showing the distal mandibular palp, ventral. (**G**) Right mandible, ventral. (**H**) Right paragnath, ventral. (**I**) Right maxillule. (**J**) Right maxilla. (**K**) Right Mxp1. (**L**) Right Mxp2. (**M**) Right Mxp3. Scale bars = 2 mm.

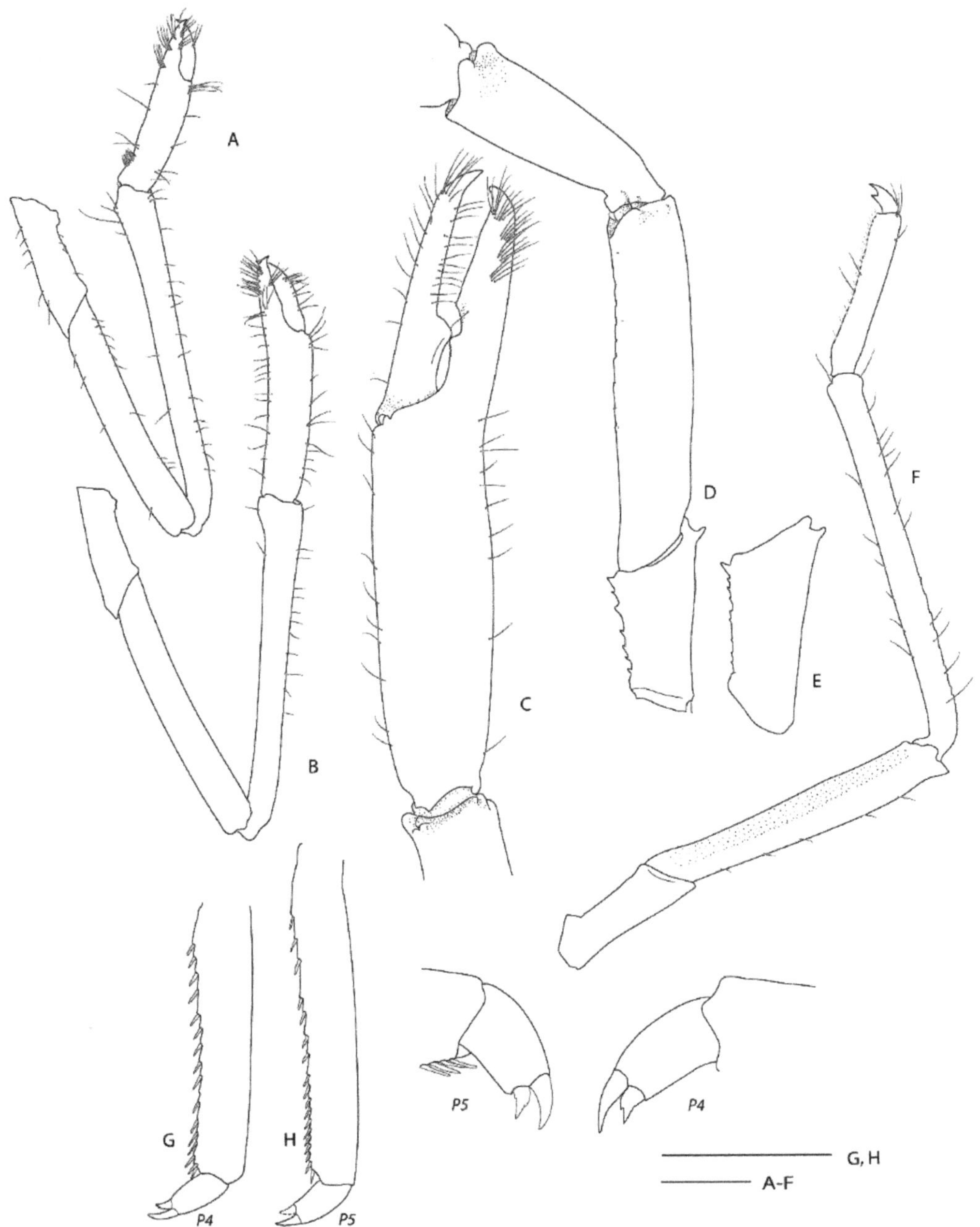

Figure 6. *Spongicoloides clarki* sp. nov., holotype F ov. (NIWA 118650). (**A**) P1, left, lateral. (**B**) P2, left, lateral. (**C**) P3, fingers, palm and distal carpus, right, outer surface. (**D**) P3, carpus, merus and ischium, right, mesial. (**E**) P3, ischium, left, lateral. (**F**) Left P4 lateral. (**G**) Right P4 propodus and dactylus, lateral. (**H**) Right P5 propodus and dactylus, lateral. Detail of P4 and P5 dactyli without scale. Scale bars = 2 mm.

Figure 7. *Spongicoloides clarki* sp. nov., holotype F ov. (NIWA 118650) with euplectellid sponge host (NIWA 118649) collected off Macauley Island. Scale bar = 5 cm. Image credits: Rob Stewart, NIWA.

Etymology. Named after Malcolm Clark, Principle Scientist at NIWA, for his contribution to deep-sea, specifically seamounts, research, and who led the voyage that collected the holotype (Kermadec-Rangitahua, TAN1612), with KS's gratitude for his mentoring and guidance.

Description of holotype female (Figures 5 and 6).

Body large, robust, surface generally glabrous. Rostrum narrowly triangular in dorsal view, one-third (0.3) as long as PCL, horizontal, slightly over-reaching distal margin of first article of antennular peduncle; dorsal margin with nine small teeth and posterior blunt eminence at level of posterior orbital margin; ventral margin with two small teeth close to rostral tip; ventrolateral ridges unarmed.

Carapace glabrous, slightly inflated. Cervical and branchiostegal grooves distinct. Small spines on anterolateral, anterior branchial surface, smaller spines and granules scattered loosely on postrostral, anterior cardiac and hepatic surfaces. Orbital margin concave, antennal margin unarmed, small branchiostegal spine present, pterygostomian angle round, with series of three small spines on margin.

Thoracic sternites 6–8 anteriorly rounded, minutely serrate but unarmed; sternite 6 posterior margins evenly concave.

Pleonal somites glabrous, surfaces smooth. First pleonite shortest; divided into two sections by distinct transverse carina; ventral margin continuous, straight. Second and third pleonite subequal in length, with feeble transverse carina and acute articular knob; ventral margin rounded. Fourth and fifth pleonites with round posteroventral margin, minute granule on fifth tergite; dorsally smooth and unarmed. Sixth pleonite dorsally smooth; posterolateral corner angular, with very small granule; posterior margin smooth. Pleonal sternites unarmed.

Telson about 2.2 times as long as broad, subquadrangular, posterior margin shallowly convex, with long plumose setae; dorsolateral ridges distinct, with 7–8 spines each, proximalmost largest and placed mesially; lateral margins nearly straight, slightly constricted proximally, each side armed with a subproximal spine and 10–11 small, lateral spines; posterior margin with 11 small spines (4–5 lateral and a pair of median spines).

Eye well developed; cornea semiglobular, about half length of ocular peduncle in dorsal view, distinctly inflated, unpigmented; minute mesial distal spine on both eye stalks, and 1–2 small spines on dorsal surface.

Antennular peduncle reaching to about middle of antennal scale, not reaching spinose lateral margin of antennal scale. Basal article stout, length-width ratio of 2.8 at midlength, without statocyst, about twice as long as second article; lateral margin concave, distally produced to small, rounded lobe, stylocerite short but distinct, acute, barely reaching proximal quarter of basal article; mesial margin almost straight, unarmed other than row of serration where plumose seta are inserted. Second article about 1.5 times longer than distal article, distomesially with small spine. Distal article as long as wide, unarmed. Flagella slender, about twice as long as peduncle.

Antenna with first article (coxa, bearing antennal gland) mesially glabrous, not carinate, two small mesial spines. Basicerite stout; mesially unarmed, paired spines distolaterally; dorsal and lateral surfaces smooth, one (left) or two (right) central small spines on ventral surface. Antennal scale broad, 2.6 × as long as wide; lateral margin nearly straight, armed with 7–8 small teeth along the distal 0.4 portion, including the most prominent distal tooth; dorsal surface with two distinct longitudinal ridges. Carpocerite reaching to distal end of first article of antennular peduncle, minute dorsal and ventral spine distally. Flagellum at about 1.3 times the CL.

Epistome broadly convex, minutely granulate at anterior margin. Labrum smooth. Paragnaths bilobed, with deep median fissure, distodorsally spatulate.

Mandible robust, fused molar and incisor processes. Molar surface with three teeth; incisor bearing two distal teeth with regular row of small proximal teeth. Palp well-developed, 3-segmented; proximal article shortest, without setae; middle article with few setae on flexor and more, longer setae on extensor margins; distal article suboval, slightly longer than intermediate article (measured along extensor margin), densely setose.

Maxillule with simple palp, with four terminal setae; distal endite broad, round, with numerous simple setae and eight slender spines; proximal endite oval, with simple setae distally.

Maxilla with palp slender, tapering, with plumose setae, nearly reaching end of scaphognathite; distal and proximal endites both deeply bilobed, with numerous plumose setae; scaphognathite well developed, anterior portion longer than posterior portion, about three times longer than broad, with dense fringe of plumose setae along entire margin.

Branchial formula summarized in Table 2.

First maxilliped with 2-segmented palp, bearing long plumose setae; with distal blunt spine, bearing distal plumose seta; proximal article stout, slightly longer than distal article in length. Distal endite large, subtriangular, densely setose; proximal endite deeply bilobed, with distal setae. Exopod slender, with long plumose setae. Epipod well developed, subequally bilobed. Arthrobranch small.

Second maxilliped with 5-segmented endopod, unarmed; dactylus sub-oval, tapering distally, about twice as long as broad, with dense setae on flexor margin; propodus slightly longer than dactylus, densely setose along flexor margin; carpus triangular, about two-third length of propodus measured at mid-line, with long distodorsal setae; merus nearly straight, about 1.6 × length of propodus, nearly four times longer than broad, with a row of setae along the mesial margin and sparse short setae on surfaces; ischium and basis not fused, each about 0.2 × meral length and long setae along mesial margin. Coxa with blunt mesial process; epipod absent, podobranch and arthrobranch. Exopod absent.

Table 2. Branchial formulae of New Zealand species of *Spongicoloides* and *Spongiocaris* (r = rudimentary).

Spongicoloides clarki **sp. nov.**							
	Maxillipeds			Pereiopods			
	I	II	III	I	II	III	IV	V
Pleurobranchs	–	–	1	1	1	1	1	1
Arthrobranchs	1	1	2	2	2	2	2	–
Podobranchs	–	1	–	–	–	–	–	–
Epipods	1	–	r	–	–	–	–	–
Exopods	1	–	–	–	–	–	–	–

Spongicoloides novaezelandiae **Baba, 1979** [20]							
	Maxillipeds			Pereiopods			
	I	II	III	I	II	III	IV	V
Pleurobranchs	–	–	1	1	1	1	1	1
Arthrobranchs	1	2, 1 or r	2	2	2	2	2	–
Podobranchs	–	1	–	–	–	–	–	–
Epipods	1	1	1	–	–	–	–	–
Exopods	1	–	–	–	–	–	–	–

Spongicoloides sonne **sp. nov.**							
	Maxillipeds			Pereiopods			
	I	II	III	I	II	III	IV	V
Pleurobranchs	–	–	1	1	1	1	1	1
Arthrobranchs	1	1	2	2	2	2	2	–
Podobranchs	–	1	–	–	–	–	–	–
Epipods	1	1	–	–	–	–	–	–
Exopods	1	–	–	–	–	–	–	–

Spongiocaris antipodes **sp. nov.**							
	Maxillipeds			Pereiopods			
	I	II	III	I	II	III	IV	V
Pleurobranchs	–	–	1	1	1	1	1	1
Arthrobranchs	1	1	2	2	2	2	2	–
Podobranchs	–	1	–	–	–	–	–	–
Epipods	1	1	–	–	–	–	–	–
Exopods	1	1	–	–	–	–	–	–

Spongiocaris yaldwyni **Bruce & Baba, 1973** [2]							
	Maxillipeds			Pereiopods			
	I	II	III	I	II	III	IV	V
Pleurobranchs	–	–	1	1	1	1	1	1
Arthrobranchs	1	1	2	2	2	2	2	–
Podobranchs	–	1	–	–	–	–	–	–
Epipods	1	1	1/–	–	–	–	–	–
Exopods	1	1	–	–	–	–	–	–

Third maxilliped endopod 5-segmented, slender, unarmed; dactylus narrow, gently tapering, about 4 times longer than broad (at mid-length), setose; propodus slightly less than twice as long as dactylus, with setiferous organ along entire length of flexor margin; carpus slightly longer than propodus; merus longest, about 1.5 × carpal length; ischium broadest, subequal in length to merus; all segments with long setal fringe along flexor margin; basis short. Coxa with rudimentary epipod, arthrobranchs present. Exopod absent.

First pereiopod slender, glabrous and sparsely setose; dactylus about 0.5 × palm length, distally setose; palm subcylindrical, tufts of long setae distally and along ventral margin of fixed finger; carpus 3 × palm length, ventral carpo-propodal setiferous organ pronounced on propodus, absent on carpus; merus about 0.7 × carpal length; ischium half meral length. Basis and coxa short, coxa with sharp mesial spine. Epipod absent.

Second pereiopod similar to first, 1.5 × longer and stronger, sparsely setose; dactylus 0.4 × palm length, distal tip formed into a strong corneous spine, tips of fingers cross when chela closed, cutting edges entire; propodus with a few tufts of setae distally and along fixed finger; carpus about twice palm length; merus 0.8 × carpal length; ischium 0.4 × length of merus. Basis and coxa short; coxa mesially produced to angular process. Epipod absent.

Third pereiopod largest, subequal and similar, about 2 × CL, overreaching the tip of antennal scale by length of chela, very sparsely setose except for a few distal tufts of setae along fingers; dactylus slightly less than 0.5 × palm length, ending in strong, hooked corneous tip, cutting edge with narrow trench along distal half, with broadly rounded tooth at mid-length, otherwise smooth; propodal cutting edge with sharp corneous spine distally, followed by narrow trench along distal half, with shallow trianguloid process at mid-length and distinct notch to accommodate dactylar tooth, proximal quarter straight, with numerous tiny teeth; outer margin of fixed finger unarmed, bearing tufts of long setae; palm sub-cylindrical, 3.1–3.3 times as long as broad, nearly entirely smooth, with few small spines on inner ventrodistal surfaces; carpus about 0.5 × palm length, narrowing proximally, without spines, with two distinct distodorsal lobes; merus 0.9 × length of palm, nearly 5 times longer than broad in lateral view, proximally compressed laterally, distoventral corners blunt, smooth other than ventral ridge with row of minute granules; ischium about three-fourth length of carpus, laterally compressed, with small, sharp distodorsal spine and small ventrodistal spine, preceded by ventral row of granules or small spines, distalmost strongest. Basis and coxa short; coxa mesially densely setose and finely granulate. Epipod absent.

Fourth and fifth pereiopods long and slender, similar, sparsely setose; dactylus about one-fifth the length of propodus, biungulate, both unguis clearly demarcated, both P5 with small accessory tooth on dorsal margin, right P5 with obsolescent accessory tooth on ventral margin proximally (P4 dactyli missing or damaged); propodus not subdivided, about 0.5 × carpal length, P5 propodus as wide as P4 propodus, with single row of 19–21 (P4) and 21 (P5) movable spine along entire flexor margin, both margins with few long, simple setae; carpus longest, not subdivided, with movable spine at distoventral angle; merus about 0.7 (P4)–0.7 (P5) × carpal length, unarmed; ischium about half length of merus, unarmed. Basis and coxa short, coxa mesially with dense setae, unarmed. Epipod absent.

First pleopod uniramous, second to fifth biramous, all lacking appendices, unarmed.

Uropod well developed, about as long as telson. Protopod stout, with sharp posterolateral spine. Exopod broader than endopod; lateral margin slightly convex, with 13 (left) and 16 (right) teeth along distal 0.8 portion of margin, closely spaced in distal quarter, distal margin nearly straight, dorsal surface with two distinct longitudinal ridges. Endopod simple, unarmed, surface with one longitudinal ridge. Exopod and endopod fringed with dense, plumose seta.

Eggs: The ovigerous female incubated around 100 eggs of diameters 1.4 × 2.1 mm.

Measurements. Holotype: CL: 15.2 mm, PCL: 11.8 mm, TL: 42.0 mm.

Distribution. Only known from holotype with certainty, Macauley Island, 1426–1431 m. Specimens that are displaying intraspecific levels of genetic divergence were collected on Yap Seamount, central-western Pacific at 1452.5 m ('*Spongiocaris panglao*' [37]) and Caroline Seamount, Micronesia, at a depth of 1205 m (Q. Kou pers. comm.)

Coloration. Unpigmented, most of the integument appears transparent (Figure 7)

Hosts. Found inside euplectellid glass sponge *Regadrella okinoseana* Ijima, 1896 [46] (Hexactinellida, Lyssacinosida, Euplectellidae) (NIWA 118649, det. Michelle Kelly, Figure 7). This host species is one of the most common euplectellids in the New Zealand region [47].

Remarks. *Spongiocoloides clarki* sp. nov. most closely resembles the two other New Zealand congeners *S. novaezelandiae* and *S. sonne* sp. nov.; the third maxilliped and the first four pereiopods have two arthrobranchs each, the pleonites are dorsally smooth, the telson is sub-rectangular and the fixed finger of the third pereiopod is unarmed on the distoventral margin. The new species is distinct in its branchial formula with the second maxilliped lacking an epipod (usually present in both *S. novaezelandiae* and *S. sonne*) and a rudimentary epipod on the third maxilliped (present in *S. novaezelandiae* and absent in *S. sonne*) (Table 2). Furthermore, the second and third pleonite has a distinctly acute articular knob (round in both other species, although this can be allometric), the labrum is nearly smooth anteriorly (furnished with granules or small spines in both other species), and the coxa of pereiopods 3–5 are mesially smooth (*S. sonne* has distinct and *S. novaezelandiae* small mesial spines). *Spongiocoloides clarki* shares with *S. novaezelandiae* the smoother carapace surface with only scattered small spines but it lacks the distinct spines along the cervical groove which are small in *S. novaezelandiae* and pronounced in *S. sonne*, the shorter rostrum, barely overreaching the first antennular segment (reaching the end of the second antennular segment in *S. sonne*). Conversely, *S. clarki* shares with *S. sonne* the spinose ischium of the third pereiopod (ventrally smooth in *S. novaezelandiae*).

The characteristics of the pereiopods 4–5 dactylar spination warrants further investigation, both available intact pereiopods of *S. clarki* have a minute dorsal accessory tooth on the ventral unguis and lacked a distinct ventral accessory tooth or teeth. The latter are pronounced in both *S. novaezelandiae* and *S. sonne*. The collection of further intact specimens might prove this character to be fixed and diagnostic.

DNA sequence data. Closest allies with intraspecific levels of divergences in all cases are a GenBank sequence presented as '*Spongiocaris panglao*' by Sun et al. [37] (MG812382) and a specimen of *Spongicoloides* sp. (Qi Kou pers. comm.).

CO1: intra-specific divergences between 1.9% ('*Spongiocaris panglao*' [37]) and 2.2%, *Spongicoloides* sp. (Q.Kou, unpubl.); intra-generic divergences range from 4.4% (*Spongicoloides novaezelandiae*) to 7.1% (*S. corbitellus* Kou, Gong & Li, 2018 [35]).

16S rRNA: intra-specific divergences of 1.0% ('*Spongiocaris panglao*' [37] and *Spongicoloides* sp. (Q. Kou, unpubl.); intra-generic divergences range from 2.3% (*Spongicoloides weijiaensis*) to 15.1% (*S. iheyaensis* Saito, Tsuchida & Yamamoto, 2006 [48]).

3.2.3. *Spongicoloides novaezelandiae* Baba, 1979

In (Figures 1 and 8–11). *Spongicoloides novaezelandiae* Baba, 1979 [20]: 311.—Goy 1980 [49]: 770.—Baba 1983 [50]: 477.—Saito et al. 2006 [48]: 224.—Burukovsky 2009 [51]: 498.—Goy 2010 [3]: 227.—De Grave & Fransen 2011 [42]: 252.—Goy 2015 [13]: 310, Figures 9–11 (in part).—Kou et al. 2018 [35]: 105 (table).

Material examined. Holotype: M (PCL: 14.4 mm); Chatham Rise, 44.73° S, 175.70° E, 1782–1998 m; 16 Jul 1968; RV *Kaiyo Maru* Stn. KM36, trawl; NMNZ CR.001889. **Other material. Macquarie Ridge:** 1 F (PCL: ~9.8 mm, specimen damaged); Seamount 7, 53.731–53.733° S, 159.166–159.169° E, 1150–1270 m; 12 Apr 2008; RV *Tangaroa* Stn. TAN0803/81, 'seamount' sled; NIWA 40567. 1 M (PCL: ~6.6 mm, specimen damaged); Seamount 7, 53.705–53.705° S, 159.115–159.106° E; 998–1100 m; 13 Apr 2008, 'seamount' sled; RV *Tangaroa* Stn. TAN0803/84; NIWA 40638.

Figure 8. *Spongicoloides novaezelandiae* Baba, 1979 [20], holotype M (NMNZ CR.001889). (**A**) Anterior carapace and cephalic appendages, lateral. (**B**) Rostrum, right ocular peduncle and antennule, dorsal. (**C**) Pleonites 5 and 6, urosome and telson, dorsal. (**D**) Pleonites 4–6 and proximal urosome and telson, lateral. (**E**) Epistome and labrum, ventral. (**F**) Thoracic sternites 6–8 with coxae of pereiopods, ventral. (**G**) Right antenna, ventral. (**H**) Distal portion of Mxp3 showing the setiferous organ on propodus. (**I**) Right P1, lateral. (**J**) P3 fingers and distal palm, mesial. (**K**) Right (*R*) and left (*L*) ischia of P3, lateral. (**L**) Left P4 or P5 dactylus and distal propodus. (**M**) Right P5 dactylus and distal propodus. Scale bars = 2 mm.

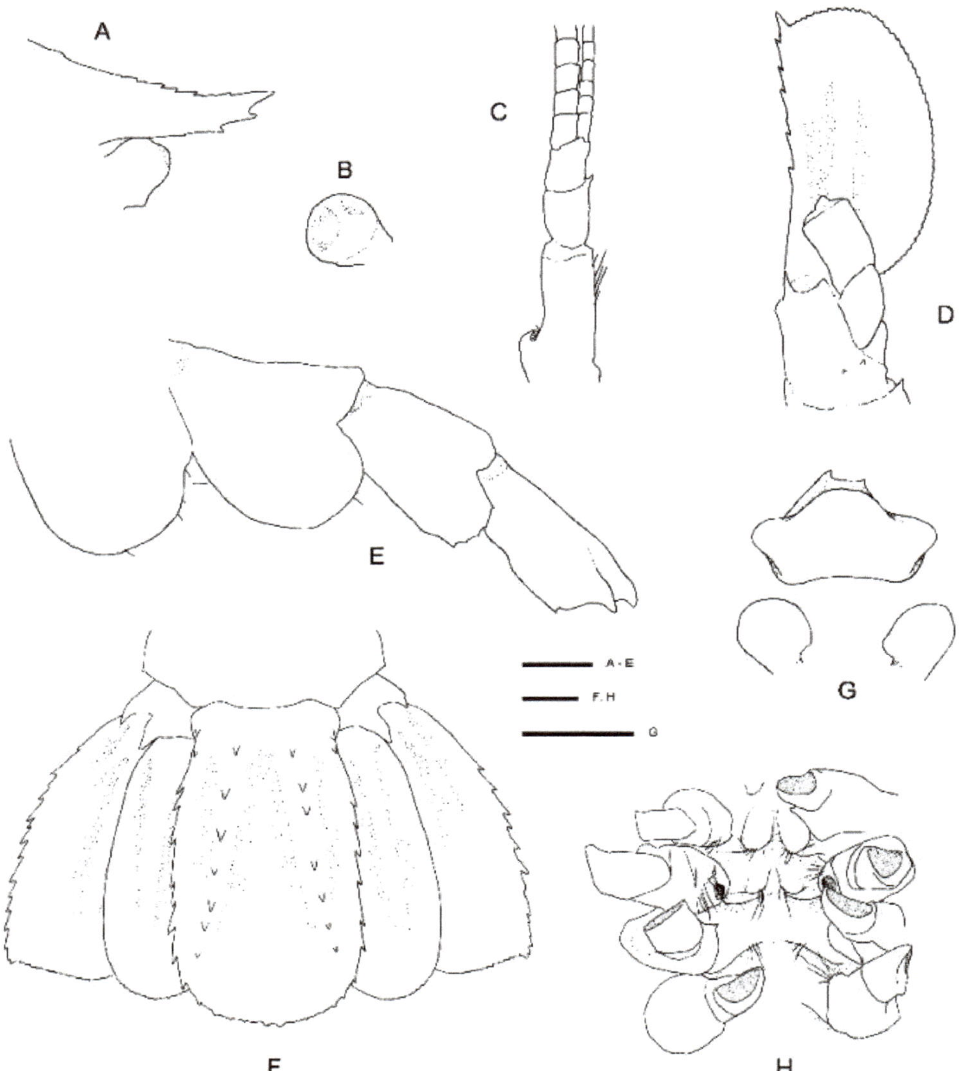

Figure 9. *Spongicoloides novaezelandiae* Baba, 1979 [20], F (NIWA 40567). (**A**) Rostrum, lateral. (**B**) Left eye, outer view. (**C**) Left antennule, dorsal. (**D**) Left antennal scale, dorsal. (**E**) Third to sixth pleura, lateral. (**F**) Telson and uropods, dorsal. (**G**) Epistome, labrum and paragnaths, ventral. (**H**) Thoracic sternites, ventral. Scale bars = 1 mm.

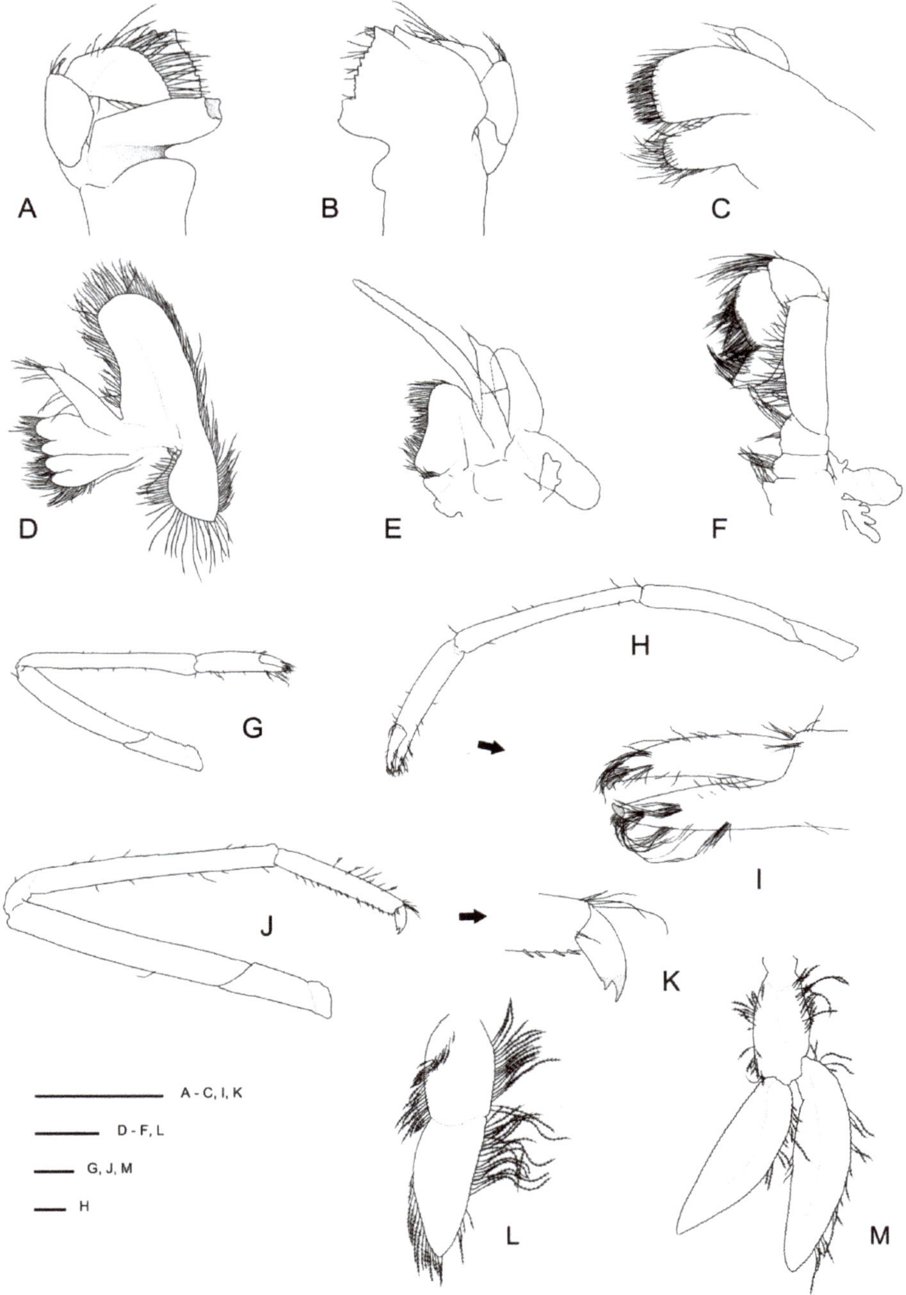

Figure 10. *Spongicoloides novaezelandiae* Baba, 1979 [20], F (NIWA 40567). (**A**) Left mandible, inner view. (**B**) Same, outer view. (**C**) Left maxillule, outer view. (**D**) Left maxilla, outer view. (**E**) Left Mxp1, outer view. (**F**) Left Mxp2, outer view. (**G**) Right P1, lateral. (**H**) Right P2, mesial. (**I**) Same, fingers, mesial. (**J**) Right P4 or P5, lateral. (**K**) Same, dactylus, lateral. (**L**) Right first pleopod, mesial. (**M**) Right second pleopod, mesial. Scale bars = 1 mm.

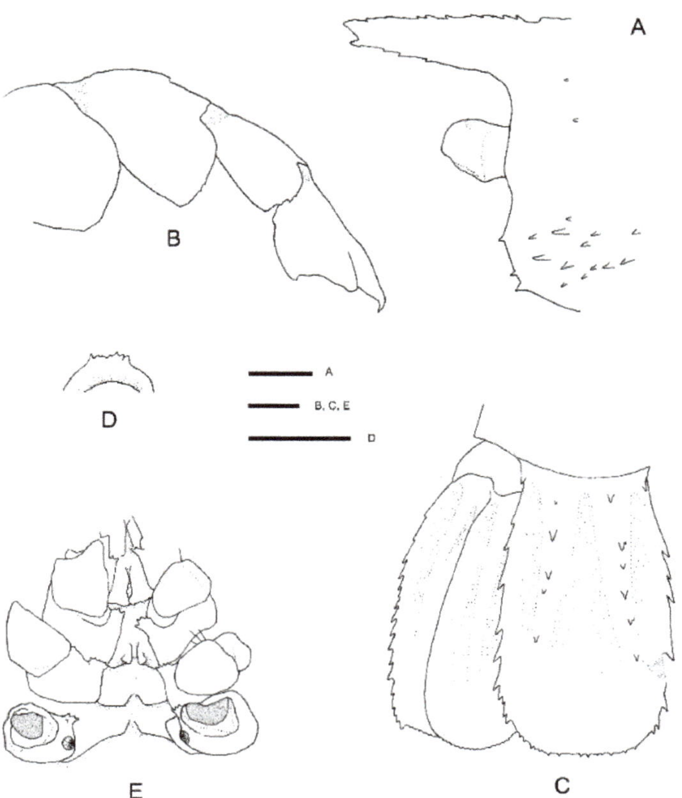

Figure 11. *Spongicoloides novaezelandiae* Baba, 1979 [20], M (NIWA 40638). (**A**) Rostrum, lateral. (**B**) third to sixth pleura, lateral. (**C**) Telson and uropods, dorsal. (**D**) Epistome, ventral. (**E**) Thoracic sternites 6–8, ventral. Scale bars = 1 mm.

Diagnosis. Carapace with shallow hepatic groove; postorbital surface smooth; minute hepatic spines; cervical groove with few distinct spines; anterior margins without antennal spine, few small anterolateral spines. Rostrum reaching distal margin of basal segment of antennular peduncle; with 8–11 small dorsal teeth, laterally unarmed. Epistome anteriorly convex, with small teeth. Pleonites smooth on dorsal midline; pleonites 5–6 with 1–2 small spines along posteroventral margin. Telson broadly rectangular, about twice as long as wide; posterior margin with median pair of spines, regular row of spines absent. Antennule unarmed, stout stylocerite. Antennal basicerite with pair of distolateral teeth, ventromedially unarmed; scale with 7–8 lateral teeth. First maxilliped with single arthrobranch developed. Second maxilliped with two arthrobranchs and well-developed podobranch. First pereiopod with setiferous organ on propodus and carpus. Third pereiopod nearly entirely glabrous and unarmed; fixed finger unarmed on distoventral margin; ischium with distodorsal spine, ventrally unarmed; coxa with or without mesial granules. Fourth and fifth pereiopod dactyli with ventral unguis bearing a small ventral and with or without slight indication of dorsal tooth; coxa with small distomesial granules; P4 with paired arthrobranchs.

Description of the holotype male (Figure 8).

Body large, robust, surface generally glabrous.

Rostrum narrowly triangular in dorsal view, one-third (0.3) as long as PCL, nearly horizontal, slightly overreaching basal segment of antennular peduncle; dorsal margin armed with 11 small, low teeth, posteriormost placed on carapace posterior to level of orbital margin; distoventral margin with two small teeth; no distinct ventrolateral ridges, lateral margins smooth, unarmed.

Carapace glabrous, not inflated. Cervical and branchiostegal grooves distinct; cervical groove lined with five small spines. Group of small spinules placed on lateral portion of gastric region; postrostral, postorbital surfaces nearly smooth, minute dorsal granules. Orbital margin concave, antennal and branchiostegal margin unarmed, anterior pterygostomian angle round, with 1–3 small spines on margin; field of several spinules just behind them.

Thoracic sternites 6–8 anteriorly rounded, unarmed; sternite six posterior margins straight.

Pleonal somites glabrous. First pleonite shortest; divided into two sections by distinct transverse carina; ventral margin continuous, not produced. Second and third pleonite subequal in length, with feeble transverse groove and blunt articular knob; ventral margin of second to fourth segments broadly rounded. Fifth and sixth pleura with acute posteroventral margin; dorsally smooth and unarmed; posterior margins smooth. Pleonal sternites unarmed.

Telson, twice as long as broad, subquadrangular, posterior margin shallowly convex, with long plumose setae; dorsolateral ridges distinct, with 8 spines each, distalmost spine very small, near posterolateral margin, distinctly separated from ordinary row. Lateral margins nearly straight, slightly constricted proximally, each side armed with a subproximal spine and 6–7 small, lateral spines. Posterior margin with two denticles equidistant between lateral groups of 3–4 similar denticles.

Eye well developed; cornea semiglobular, about half length of ocular peduncle in dorsal view, slightly inflated, unpigmented; minute distomesial spine on both eye stalks.

Antennular peduncle overreaching middle of antennal scale, reaching spinose lateral margin of antennal scale. Basal article relatively stout, length-width ratio of about 3.0 at mid-length, without statocyst, about twice as long as second article, distally unarmed but with short distodorsal carina; lateral margin concave, stylocerite small and blunt but distinct; mesial margin almost straight, unarmed other than row of serration where plumose seta are inserted. Second article about 1.5 times longer than distal article, unarmed. Distal article as long as wide, unarmed. Flagella slender, about twice as long as peduncle.

Antenna with first article (coxa, bearing antennal gland) mesially glabrous, not inflated or carinate, one small mesial spine and a row of small setiferous denticles. Basicerite stout; mesially unarmed, one (left) or two (right) outer terminal spines; surfaces smooth, unarmed. Antennal scale broad, around twice as long as broad; lateral margin almost straight, not setiferous, with 6–7 spines in distal half, distally paired on both sides; dorsal surface with two distinct longitudinal ridges; inner margin convex, inner and distal margins with long setae. Carpocerite reaching to distal end of first article of antennular peduncle, unarmed. Flagellum at least as long as PCL.

Epistome narrowly convex, with short transverse row of granules at anterior margin. Labrum smooth. Paragnaths bilobed, with deep median fissure, distodorsally spatulate.

Mandible robust, fused molar and incisor processes. Molar surface with two teeth; incisor bearing distal tooth followed by regular row of small teeth along midlength. Palp well-developed, 3-segmented; proximal article shortest, without setae; middle article with few setae on flexor and more, longer setae on extensor margins; distal article suboval, longer than intermediate article (measured along extensor margin), densely setose.

Maxillule with simple palp, with four terminal setae; distal endite broad, round, with numerous simple setae and eight slender spines; proximal endite oval, with four simple setae distally.

Maxilla with palp slender, tapering, with plumose setae, falling well short of end of scaphognathite; distal and proximal endites both deeply bilobed, with numerous plumose setae; scaphognathite well developed, anterior portion longer than posterior portion, about 3.5 times longer than broad, with dense fringe of plumose setae along entire margin.

Branchial formula summarized in Table 2.

First maxilliped with 2-segmented palp, bearing long plumose setae; distal article as long as wide, rounded, with plumose seta; proximal article stout, slightly longer than distal article in length. Distal endite large, subtriangular, densely setose; proximal endite deeply bilobed, with distal setae. Exopod slender, with long plumose setae. Epipod well developed, subequally bilobed. Arthrobranch small.

Second maxilliped with 5-segmented endopod, unarmed; dactylus sub-oval, tapering distally, about 1.5 times as long as broad, with dense setae on flexor margin; propodus slightly longer than dactylus, densely setose along flexor margin; carpus triangular, about two-third length of propodus measured at mid-line, with long distodorsal setae; merus nearly straight, about twice as long as propodus, three times longer than broad, with row of setae along mesial margin and sparse short setae on surfaces; ischium and basis not fused, each about 0.2 × meral length and long setae along mesial margin. Coxa with blunt mesial process; small epipod, podobranch and pair of arthrobranchs present. Exopod absent.

Third maxilliped endopod 5-segmented, slender, unarmed; dactylus narrow, gently tapering, about 4 times longer than broad (at mid-length), setose; propodus slightly less than twice as long as dactylus, with setiferous organ along entire length of flexor margin; carpus subequal in length to propodus; merus longest, about twice carpal length; ischium broadest, subequal in length to merus; all articles with long setal fringe along flexor margin; basis short. Coxa with small epipod, mesial margin bluntly triangular; arthrobranchs present. Exopod absent.

First pereiopod slender, glabrous and sparsely setose; fingers unarmed, half as long as palm, distally setose; palm subcylindrical, tufts of long setae distally and along ventral margin of fixed finger; carpus 2.3 times as long as palm, ventral carpo-propodal setiferous organ pronounced on propodus and carpus; merus about 0.7 × carpal length; ischium half meral length. Basis and coxa short, coxa with small mesial spine. Epipod absent.

Second pereiopod similar to first, 1.5 × longer and stronger, sparsely setose; dactylus 0.4 × palm length, distal tip formed into a strong corneous spine, tips of fingers cross when chela closed, cutting edges entire; propodus with a few tufts of setae distally and along fixed finger; carpus about 1.5 × palm length; merus 0.8 × carpal length; ischium 0.4 × length of merus. Basis and coxa short. Epipod absent.

Third pereiopod largest, subequal and similar, about 1.7 × CL, very sparsely setose except for a few distal tufts of setae along fingers; dactylus slightly less than 0.5 × palm length, ending in strong, hooked corneous tip, cutting edge with narrow trench along distal half, with broadly rounded tooth at proximal third, otherwise smooth; propodal cutting edge with sharp corneous spine distally, followed by narrow trench along distal half, mid-length barely produce, distinct proximal notch to accommodate dactylar tooth; outer margin of fixed finger unarmed, bearing tufts of long setae; palm sub-cylindrical, moderately compressed, 3.4 times as long as broad, smooth, with few minute spines on inner ventrodistal surfaces; carpus about 0.4 × palm length, narrowing proximally, without spines, with knoblike process at anterior end of inner surface; merus 0.8–0.9 × length of palm, around four times longer than broad in lateral view, proximally compressed laterally, distoventral corners blunt, smooth; ischium about three-fourth length of carpus, laterally compressed, with small, sharp distodorsal spine only, ventral margin smooth. Basis and coxa short; coxa mesially densely setose, unarmed. Epipod absent.

Fourth and fifth pereiopods long and slender, similar, sparsely setose; dactylus about one-fifth the length of propodus, biungulate; both unguis clearly demarcated, with small accessory tooth on ventral margin, distinct shoulder dorsally; propodus not subdivided, slightly less than 0.5 × carpal length, fifth pereiopod propodus as wide as fourth pereiopod propodus, with single row of 10–12 movable spine along entire flexor margin, both margins with few long, simple setae; carpus longest, not subdivided, with movable spine at distoventral angle; merus about 0.9 × carpal length, unarmed; ischium about half length of merus, unarmed. Basis and coxa short, coxa mesially with dense setae, with small granules. Epipod absent.

First pleopod uniramous, second to fifth biramous, all lacking appendices, unarmed.

Uropod well developed, about as long as telson. Protopod stout, with sharp posterolateral spine. Exopod broader than endopod; lateral margin slightly convex, with 8–10 teeth along distal 0.8 portion of margin, distal margin convex, dorsal surface with 2 distinct longitudinal ridges. Endopod simple, unarmed, surface with 1 longitudinal ridge. Exopod and endopod fringed with dense, plumose seta.

Measurements. Holotype: CL: 19.5 mm, PCL: 14.4 mm, TL: 55 mm. NIWA 40567*: CL: 12.7 mm, PCL: 9.8 mm, TL: 34.2 mm. NIWA 40638*: CL: 9.3 mm, PCL: 6.6 mm, TL: 29.3 mm. *Measurements for both NIWA specimens are approximate as the specimens are damaged.

Distribution. New Zealand (type locality), southern Chatham Rise, 990–1110 m. New records presented here extend the range southwards to Macquarie Ridge, 998–1270 m. Based on the confirmed records at hand, this appears to be the only species in this genus with a southern temperate to sub-Antarctic distribution (Figure 1). Records recently presented by Goy [13] of *S. novaezelandiae* from New Caledonia are referable to *S. sonne* sp. nov. and *S. weijiaensis* based on DNA sequence similarities (Figure 2 and see comments below). Further specimens from Tasmania, Madagascar, Indian Ocean and Fiji mentioned by Goy [13] require more detailed examination.

Coloration. Body and appendages transparent, cornea yellow [20].

Hosts. Unknown.

Remarks. Following the examination of the male holotype of *S. novaezelandiae*, the following characters were not previously presented and can be added as follows: the epistome is anteriorly rounded, bearing a row of small granules along the anterior margin; the labrum is anteriorly rounded and smooth (Figure 8D); both ocular peduncles bear a small dorsomesial granule (Figure 8A); the basal antennal articles are smooth and not bearing any ventromesial processes (Figure 8F); the coxa of pereiopods 4–5 bear small distomesial granules only (Figure 8E). The proposed apomorphy for *S. novaezelandiae* is the first maxilliped distal segment broad and rounded (not tapering) [20], this is consistent on both sides of the holotype, however, the additional specimens presented below do not share this character (Figure 10E). Baba [20] illustrates the entire dorsal and the left anterior carapace but fails to include the small but distinct series of spines that line the cervical groove. They are less distinct on the left than the right, illustrated here (Figure 8A).

Two single specimens collected at two stations on Macquarie Ridge at >53° S provide the most southerly records for the genus to date. Unfortunately, both specimens are badly damaged, both lack third maxilliped and pereiopod 3, and only one pereiopod 4 or 5 of NIWA 40567 is preserved (Figures 9–11). Genetically, these specimens clearly align with the holotype of *S. novaezelandiae* (see below). Morphologically, they share the following characters that may prove to be diagnostic: e.g., the nearly smooth postrostral carapace surface; the telson is sub-rectangular, length-width ratio is ≤ 2.0, posterior margin with few or irregular row of spines; anterior margin of epistome bearing at least two granules; maxillipeds 2 and 3 bear a small epipod each. Notably, the distal segment of the first maxilliped palp bears a distinct spine (Figure 10E).

Spongicoloides novaezelandiae was described from a single male collected on the New Zealand Chatham Rise (Figure 1) and was recently reported again from a wide geographic range and with considerable morphological variation [13]. Following the examination of New Zealand material presented here in combination with DNA sequencing, we propose that only the additional New Zealand material reported, NIWA 40567 and 40638 from Macquarie Ridge, belong to *S. novaezelandiae* and that all remaining specimens are referable to the following species:

(1). *Spongicoloides weijiaensis* Xu, Zhou & Wang, 2017 [36]

Three of four specimens reported from New Caledonia (BIOCAL specimens listed in Goy [13] as MNHN-Na-11996 [IU-2013-19630], Na-11997 [IU-2013-19488] and Na-11998 [IU-2013-19627]) were successfully sequenced and are referred to *S. weijiaensis* (Figure 2), see comments under DNA sequence section below. According to figures provided (Goy [13]: Figure 10), these all share dorsal carapace spines on parts of the postrostral, hepatic and cardiac regions, but specimens will need to be examined in more detail. In the meantime, we propose the following characteristics to separate *S. novaezelandiae* and *S. weijiaensis* to include:

(a). The postorbital carapace region is nearly entirely smooth in *S. novaezelandiae* (with at least some scattered spines in *S. weijiaensis*).

(b). The epistome bears anterior teeth or row of small granules (smooth in *S. weijiaensis*).

(c). The antennal basicerite with nearly entirely smooth ventral surface in *S. novaezelandiae* (small ventral ridge armed with 1–3 spines present in *S. weijiaensis*).

(d). P3 ischium is smooth on ventral margin in *S. novaezelandiae* (irregular, with at least some proximal granules in *S. weijiaensis*).

The presence and number of arthrobranchs on the maxilliped 1 and 2 are usually considered diagnostic and the holotype of *S. novaezelandiae* has two arthrobranchs on maxilliped 2 which would differ from the single arthrobranch in *S. weijaensis*. However, this character is unusually variable in *S. novaezelandiae* with only one arthrobranch on maxilliped 2 of NIWA 40567 and one rudimentary arthrobranch on NIWA 40638.

The distribution of *S. weijiaensis* now extends from the northwestern Pacific to New Caledonia, north of the New Zealand region.

(2). *Spongicoloides sonne* sp. nov.

The figured CALSUB specimen (MNHN-NA 11999 [IU-2013-19487]) (Goy [13]: Figures 9A and 10A) shows a distinct antennal spine on the anterolateral margin of the carapace (rounded in *S. novaezelandiae*) and a distinctly spinose ischium of the third pereiopod (smooth in *S. novaezelandiae*). Additionally, the rostrum reaches at least to the distal end of the second antennular article (reaching the end of the first article in *S. novaezelandiae*). These characteristics match those of *S. sonne* sp. nov. described and discussed below.

The material from Tasmania, Madagascar, Indian Ocean and Fiji referred to by Goy [13] will need to be examined in more detail in light of the review presented here.

Spongicoloides novaezelandiae most closely resembles *S. clarki* sp. nov. and *S. sonne* sp. nov., both reported from the northern New Zealand region (Figure 1). Morphological differences are discussed under those species below.

DNA sequence data. The holotype of *S. novaezelandiae* (NMNZ CR.001889) could be sequenced for both genes and aligned with the two specimens from Macquarie Ridge (NIWA 40567, 40638). The specimen reported as *S. novaezelandiae* by Chen et al. ([17] MNHN-IU-2014-6347) from the Solomon Islands belongs to a different species and aligns more closely with an undescribed species of *Spongicoloides* from the northwest Pacific (Zhao et al., in press.). Three of the four specimens presented as *S. novaezelandiae* by Goy [13] were successfully sequenced and align with the holotype sequence of *S. weijiaensis* (SRSIO-16050001, Figure 2).

CO1: intra-specific divergences between 0.6–1.7%; intra-generic divergences range from 4.4% (*Spongicoloides clarki* sp. nov.) to 8.5% (*S. corbitellus*).

16S rRNA: intra-specific divergences were 0.0%; intra-generic divergences range from 2.6% (*Spongicoloides clarki* sp. nov.) to 16.0% (*S. iheyaensis*).

3.2.4. *Spongicoloides sonne* sp. nov.

In (Figures 1, 2 and 12–17). *Spongicoloides novaezelandiae*.— Goy 2015 [13]: 310, Figures 9–11 (in part, New Caledonia, specimen MNHN-NA 11999 [IU-2013-19487]).

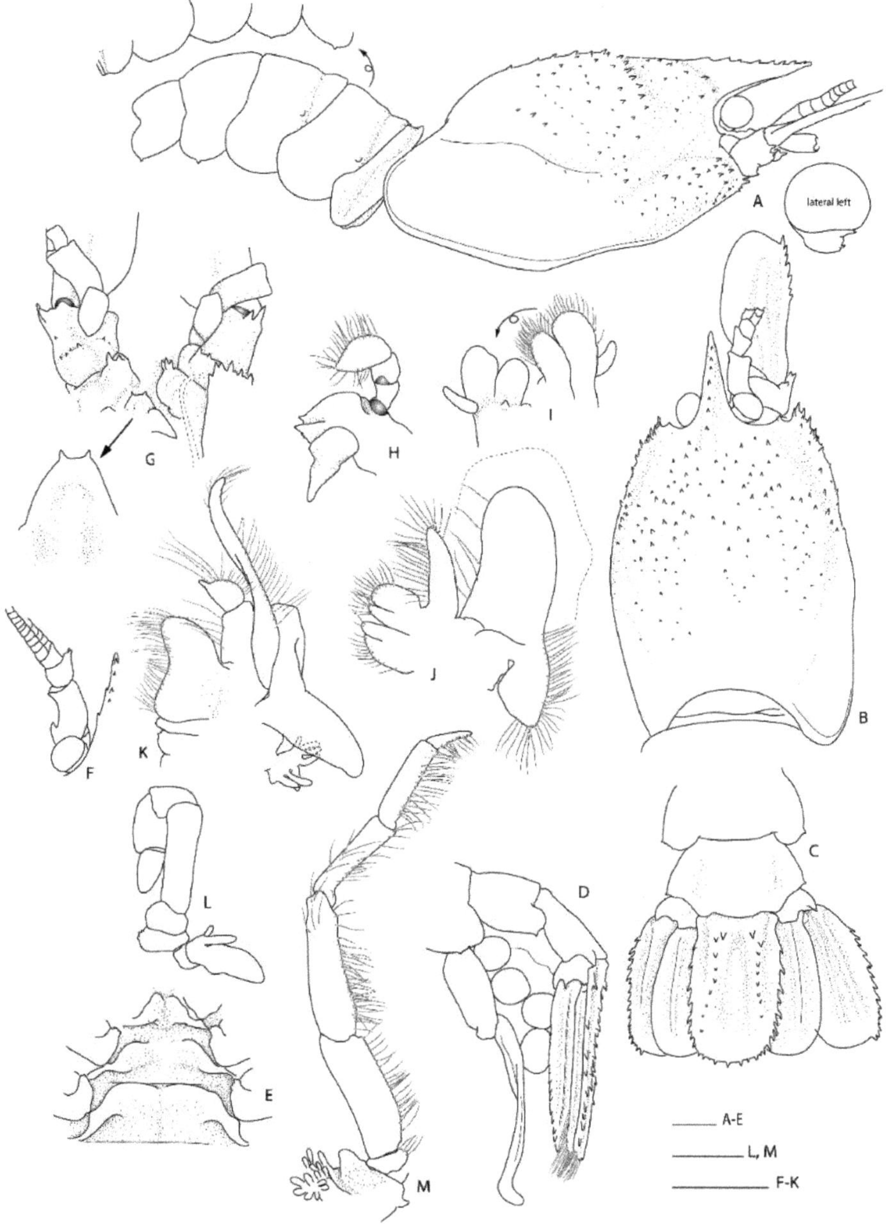

Figure 12. *Spongicoloides sonne* sp. nov., holotype F ov. (NIWA 127111). (**A**) Carapace and pleon, lateral, with inset of left pleural margins and enlarged lateral left ocular peduncle. (**B**) Carapace and cephalic appendages, dorsal. (**C**) Pleonites 5 and 6, urosome and telson, dorsal.(**D**) Pleura 4–6 with pleopod, urosome and telson, lateral. (**E**) Thoracic sternites 6–8 with coxae of pereiopods, ventral. (**F**) Rostrum, left antennule and ocular peduncle, dorsal. (**G**) Left and right antennae, left pterygostomian margin, epistome and labrum, ventral. (**H**) Left mandible with paragnath, inner and outer view. (**I**) Left maxillule, inner and outer view. (**J**) Left maxilla. (**K**) Left Mxp1. (**L**) Left Mxp2. (**M**) Right Mxp3. Scale bars = 2 mm.

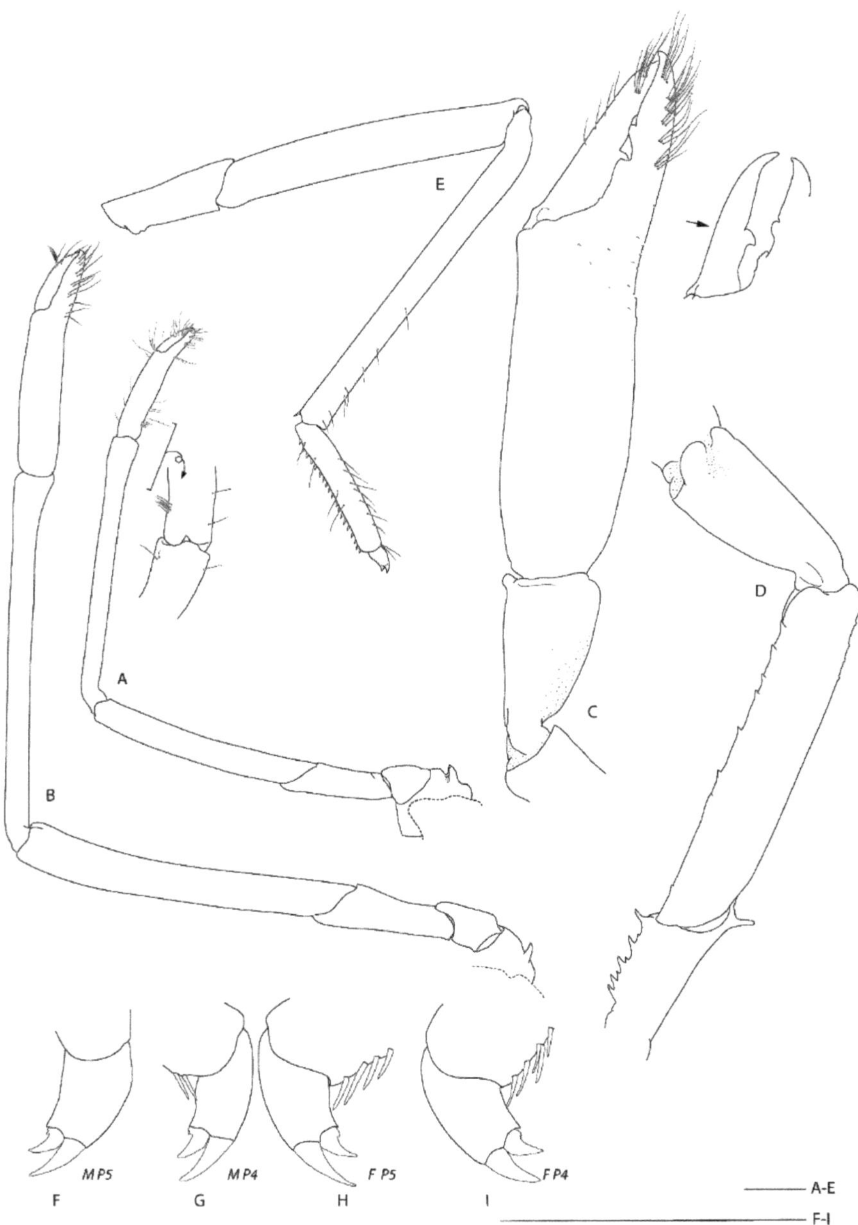

Figure 13. *Spongicoloides sonne* sp. nov., (**A–D,H,I**): holotype F ov. (NIWA 127110). F, G: allotype M (NIWA127111). (**A**) Right P1, lateral, with rotated inset of merocarpal setiferous organ. (**B**) Right P2, lateral. (**C**) Distal P3, right, lateral, with inset detail of incisor margin of fingers. (**D**) Proximal P3, right, mesial. (**E**) Right P4, lateral. (**F–I**) Dactyli and distal propodi of P4–5 for male (*M*) allotype and female (*F*) holotype. Scale = 2 mm.

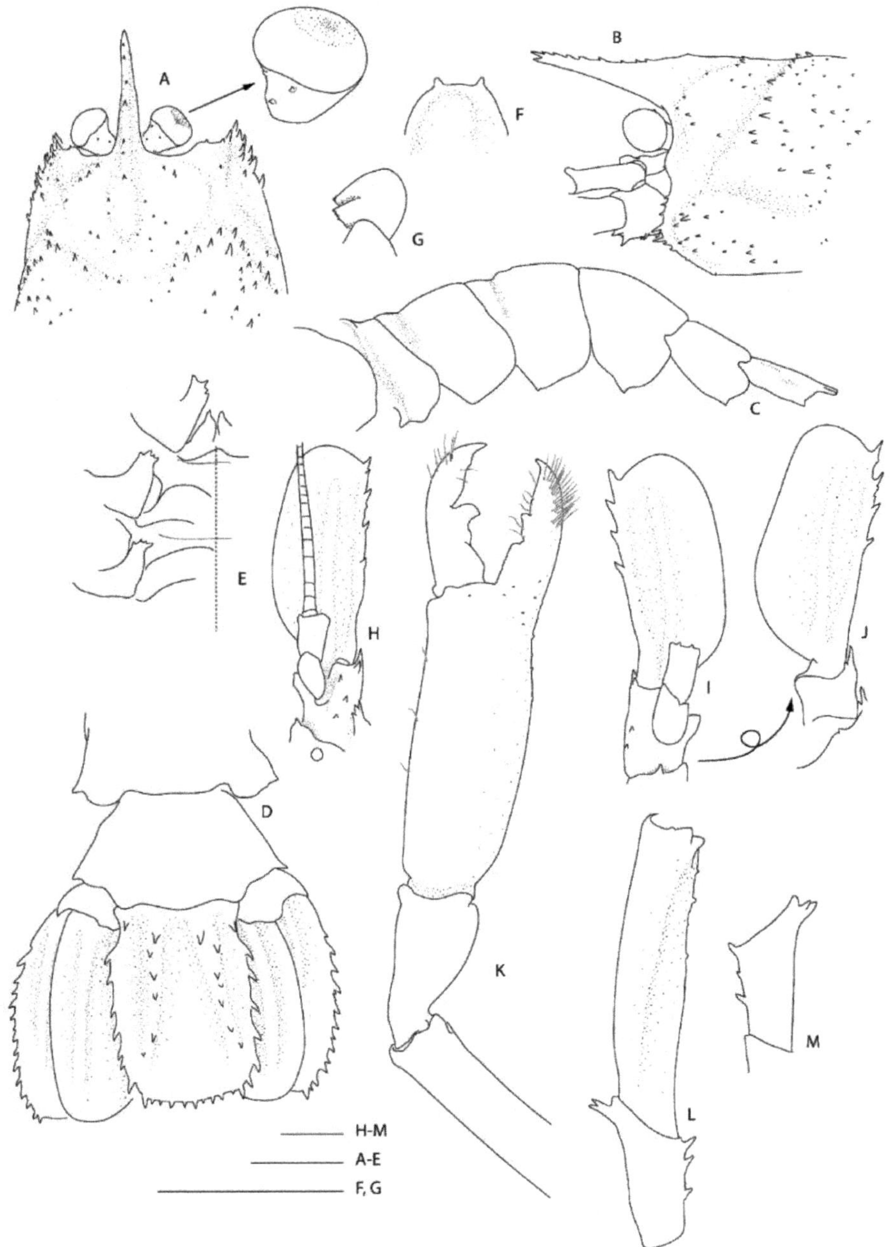

Figure 14. *Spongicoloides sonne* sp. nov., allotype M (NIWA 127110). (**A**) Anterior carapace, dorsal, with inset of enlarged right ocular peduncle. (**B**) anterior carapace and cephalic appendages, lateral. (**C**) Pleon, lateral. (**D**) Pleonites 5–6, urosome and telson, dorsal. (**E**) Thoracic sternites 6–8 with coxae of pereiopods, ventral. (**F**) Anterior margin of epistome, ventral. (**G**) Left mandible with paragnath, outer view. (**H**) Left antennae, ventral. (**I**) Right antenna, ventral. (**J**) Right antenna, dorsal. (**K**) Distal P3, lateral, right. (**L**) Proximal P3, lateral, right. (**M**) P3 ischium, left, lateral. Scale bars = 2 mm.

Figure 15. *Spongicoloides sonne* sp. nov. (pair NIWA 127110 and NIWA 127111) sponge host *Corbitella* sp. (det. Henry Reiswig, University of Victoria, Victoria, BC, Canada), Lyssacinosida, immediately prior to collection on the southern Kermadec Ridge (1168 m). Image courtesy of ROV *Kiel 6000* GEOMAR, SO254 voyage PoriBacNewZ, ICBM.

Figure 16. *Spongicoloides sonne* sp. nov. (NMNZ CR.019650). (**A**) Carapace and cephalic appendages, lateral. (**B**) Pleon, lateral. (**C**) Habitus, dorsal. (**D**) Pleonites 5–6, urosome and telson, dorsal. (**E**) Thoracic sternites with coxae of pereiopods, ventral. (**F**) Right antennule, ventral. (**G**) Right antenna, epistome and labrum, ventral. (**H**) Right mandible with paragnath, outer view, inset with rotated mandibular incisor margin. (**I**) Right maxillule. (**J**) Right maxilla. (**K**) Right Mxp1. (**L**) Right Mxp2. (**M**) Right (*R*) Mxp3 with inset left (*L*) ischium. Scale bars = 2 mm.

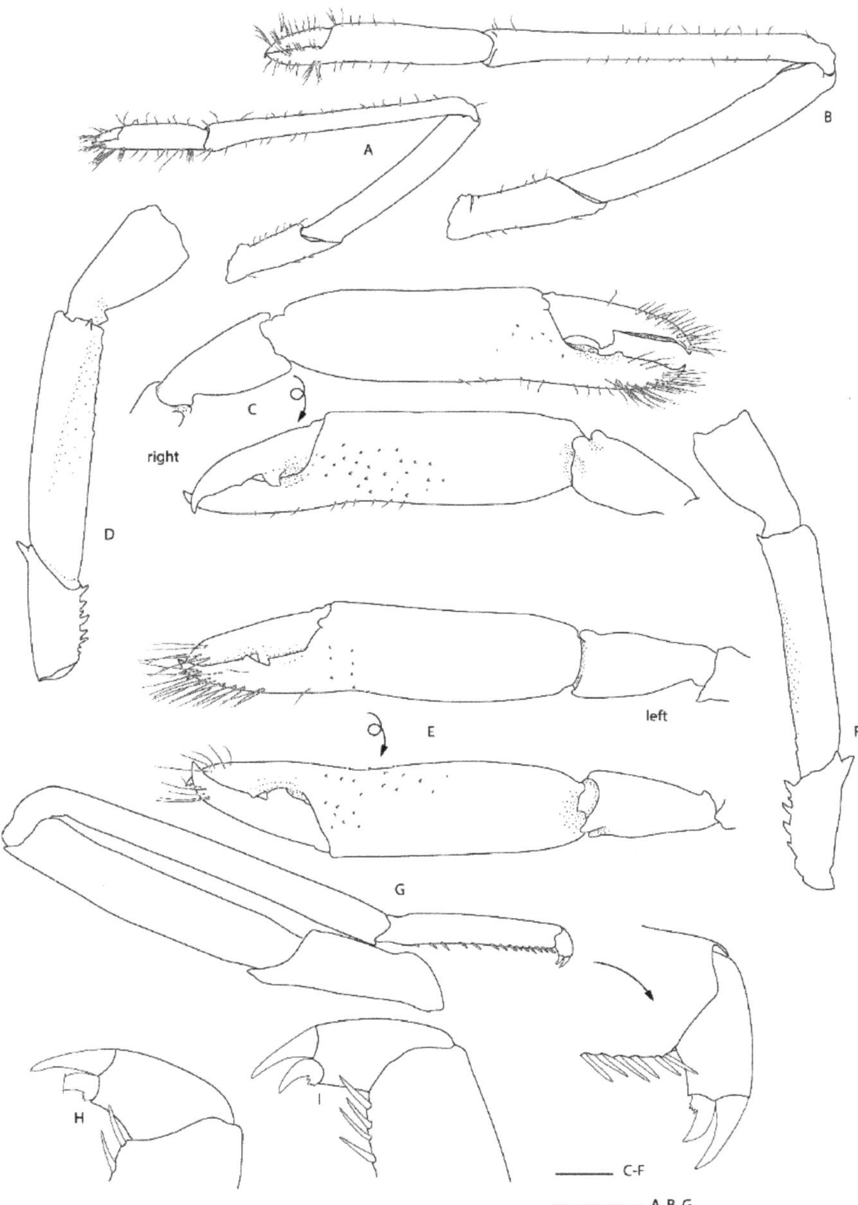

Figure 17. *Spongicoloides sonne* sp. nov. (NMNZ CR.019650). (**A**) Left P1, lateral. (**B**) Left P2, lateral. **C**. Right P3, distal portion, lateral (top) and mesial (bottom). (**D**) Right P3, proximal portion, lateral. (**E**) Left P3, distal portion, lateral (top) and mesial (bottom). (**F**) Left P3, proximal portion, lateral. (**G**) Right P4, lateral, inset detail of dactylar spination. (**H,I**) Dactyli and distal propodi of loose left P4 or P5. Scale bars = 2 mm.

Material examined. Holotype: F ov. (PCL: 15.0 mm); Southern Kermadec Ridge, 35.380° S, 178.980° E, 1184.1 m; 7 February 2017; RV *Sonne* Stn. SO254/33ROV08, Remote

Operated Vehicle; NIWA 127111; found inside *Corbitella* sp. **Allotype:** M (PCL: 11.4 mm); Southern Kermadec Ridge, 35.380° S, 178.980° E, 1184.1 m; 7 February 2017; RV *Sonne* Stn. SO254/33ROV08, Remote Operated Vehicle; NIWA 127110; found inside *Corbitella* sp. **Other material.** 1 F (PCL: 11.0 mm); West Norfolk Ridge, 34.28° S, 168.41° E, 1246–1249 m; 2 June 2003; TAN0308/142, Orange Roughy trawl; NMNZ CR.019650. 1 M (PCL: 12.0 mm); Lillie Seamount, Kermadec Ridge, 35.857–35.857° S, 178.448–178.443° E, 1237–1460 m; 19 March 2011; RV *Tangaroa* Stn. TAN1104/124, 'seamount' sled; NIWA 72929.

Diagnosis. Carapace with distinct hepatic groove; scattered small spines on postrostral, cardiac, hepatic and branchial surfaces; cervical groove lined with distinct spines; antennal spine and anterolateral spines small but distinct. Rostrum at least reaching distal margin of basal article of antennular peduncle; with 8–10 dorsal teeth. Epistome anteriorly produced, with small anterior teeth. Second and third pleonite with blunt articular knob. Fourth to sixth pleonites smooth on dorsal midline; one or more small spine along each posteroventral margin. Telson broadly rectangular, about twice as long as wide; regular row of teeth along posterior margin. Ocular peduncle with 2–3 granules on dorsal surface. Antennule basal article unarmed on mesial margin; stout stylocerite. Antenna basal article with distinct ventral spines; basicerite with scattered spines on ventral surface; scale with 4–9 lateral teeth along distal half of margin. First maxilliped distal article with sharp distal spine; single arthrobranch developed. Second maxilliped with single arthrobranch and well-developed podobranch; epipod present. Third maxilliped with well-developed setiferous organ; lacking epipod; with paired arthrobranchs. First pereiopod with setiferous organ on propodus only. Second pereiopod similar to first, 1.5 × longer and stronger. Third pereiopod nearly entirely glabrous and smooth; fixed finger unarmed on distoventral margin; palm with few to many minute granules scattered along distoventral portion; ischium with distinct distodorsal spine, with row of ventral spines; coxa mesially granulate. Fourth and fifth pereiopods dactyli with ventral unguis bearing a number of small ventral teeth. First four pereiopods with paired arthrobranchs.

Etymology. Named after the German Research Vessel *Sonne* that collected the first specimens assigned to this new species during the 2015 *PoribacNewZ* voyage SO254. Used as a substantive in apposition.

Description of holotype female (Figures 12 and 13).

Body large, robust, surface generally glabrous.

Rostrum narrowly triangular in dorsal view, one-third (0.33) as long as PCL, horizontal, barely reaching to distal margin of second article of antennular peduncle; dorsal margin with seven small teeth and posterior blunt eminence at level of posterior orbital margin; ventral margin with one small tooth close to rostral tip; ventrolateral ridges with 1–2 (right-left) small teeth on distal half.

Carapace glabrous, slightly inflated. Cervical and branchiostegal grooves distinct. Gastric, hepatic, and anterior portions of cardiac and branchial regions armed with numerous scattered small spines. Orbital margin concave, inferior orbital angle slightly produced and furnished with 1–2 small antennal spines. Small branchiostegal spine present. Pterygostomian angle round, with series of five small spines on margin.

Thoracic sternites 6–8 anteriorly rounded, minutely serrate but unarmed; sternite six posterior margins evenly concave.

Pleonal somites glabrous, surfaces smooth. First pleonite shortest; divided into two sections by distinct transverse carina; ventral margin continuous, straight. Second and third pleonite subequal in length, with feeble transverse carina and small articular knob; ventral margin rounded, left side with slight indication of a granule. Fourth and fifth pleonites each with small sharp or blunt posteroventral tooth on broad, round margin; dorsally smooth and unarmed. Sixth pleonite dorsally smooth; posterolateral process straight, angular, with small, sharp spine at corner; posterior margin smooth. Pleonal sternites unarmed.

Telson about 1.9 times as long as broad, subquadrangular, posterior margin shallowly convex, with long plumose setae; dorsolateral ridges distinct, with nine spines each, proximalmost largest and placed mesially; lateral margins relatively straight, slightly constricted proximally, each side armed with a subproximal spine and nine lateral spines; posterior margin with 10 small pines (4–5 lateral and one median spine).

Eye well developed; cornea semiglobular, about half length of ocular peduncle in dorsal view, distinctly inflated, unpigmented; minute mesial proximal spine on both eye stalks, and small distal spine on left eye stalk only.

Antennular peduncle reaching to about middle of antennal scale, does not reach spinose lateral margin of antennal scale. Basal article stout, length-width ratio of 2.7 at midlength, without statocyst, about three times as long as second article; lateral margin concave, distally produced to small, rounded lobe, stylocerite distinct, acute, reaching 1/3 length of basal article; mesial margin almost straight, unarmed other than row of serration where plumose seta are inserted. Second article about 1.5 times longer than distal article, distomesially with two small spines. Distal article as long as wide, unarmed. Flagella slender, about twice as long as peduncle.

Antenna with first article (coxa, bearing antennal gland) mesially carinate, bearing three distinct spines. Basicerite stout; mesially bearing minute (left) or distinct (right) spine, paired spines distolaterally; dorsal and lateral surfaces smooth, minute spine at midlength mesially, scattered spines on ventral surface. Antennal scale broad, 2.6 × as long as wide; lateral margin nearly straight, armed with seven (left) or eight (right) teeth along the distal 0.4 portion, including the distal tooth; dorsal surface with two distinct longitudinal ridges. Carpocerite reaching to distal end of second article of antennular peduncle, unarmed. Flagellum at least 1.7 times the CL.

Epistome narrow, subquadrate, with two small spines at anterolateral angles. Labrum smooth. Paragnaths bilobed, with deep median fissure, distodorsally spatulate.

Mandible robust, fused molar and incisor processes. Molar surface with three teeth; incisor bearing two distal teeth with five small proximal teeth. Palp well-developed, 3-segmented; proximal article shortest, without setae; middle article with few setae on flexor and more, longer setae on extensor margins; distal article suboval, slightly longer than intermediate article (measured along extensor margin), densely setose.

Maxillule with simple palp, with few terminal setae; distal endite broad, round, with numerous simple setae and seven slender spines; proximal endite oval, with simple setae distally.

Maxilla with palp stout, tapering, with plumose setae, falling short of end of scaphognathite; distal and proximal endites both deeply bilobed, with numerous plumose setae; scaphognathite well developed, anterior portion longer than posterior portion, about three times longer than broad, with dense fringe of plumose setae along entire margin.

Branchial formula summarized in Table 2.

First maxilliped with 2-segmented palp, bearing long plumose setae; distal segment broad, with distal blunt spine; proximal article stout, about 1.6 × distal article in length. Distal endite large, subtriangular, densely setose; proximal endite deeply bilobed, with distal setae. Exopod slender, with long plumose setae. Epipod well developed, subequally bilobed. Arthrobranch present.

Second maxilliped with 5-segmented endopod; dactylus sub-oval, tapering distally, about twice as long as broad, with dense setae on flexor margin; propodus subequal in length to dactylus, densely setose along flexor margin; carpus triangular, about three-forth length of propodus measured at mid-line, with long distodorsal setae; merus nearly straight, about 1.8 × length of propodus, nearly 4 times longer than broad, with row of setae along mesial margin and sparse short setae on surfaces; ischium and basis not fused, each about 0.2 × meral length and long setae along mesial margin. Coxa with mesial spine; oval epipod, podobranch and arthrobranch. Exopod absent.

Third maxilliped endopod 5-segmented, slender; dactylus narrow, about 4.5 times longer than broad (at mid-length), setose; propodus about twice as long as dactylus, with setiferous organ along distal two-thirds of flexor margin; carpus subequal in length to propodus; merus longest, about 1.5 × carpal length, minute distolateral spine on left, absent on right; ischium broadest, 0.8 × meral length, small distolateral spine on both sides;

all segments with long setal fringe along flexor margin; basis short. Coxa without epipod, arthrobranchs present. Exopod absent.

First pereiopod slender, just overreaching antennal scale when extended, glabrous and sparsely setose; dactylus about 0.5 × palm length; palm subcylindrical, tufts of long setae dorsodistally and along ventral margin of fixed finger; carpus 3 × palm length, ventral carpo-propodal setiferous organ more pronounced on propodus, very weak on carpus; merus about 0.8 × carpal length; ischium one-third meral length. Basis and coxa short, coxa with sharp mesial spine. Epipod absent.

Second pereiopod similar to first, 1.5 × longer and stronger, sparsely setose; dactylus 0.4 × palm length, distal tip formed into a strong corneous spine, tips of fingers cross when chela closed, cutting edges entire; propodus with a few tufts of setae distally and along fixed finger; carpus about 2.2 × palm length; merus 0.8 × carpal length; ischium one-third length of merus. Basis and coxa short; coxa with mesial spine. Epipod absent.

Third pereiopod largest, subequal and similar, about 2 × CL, overreaching the tip of antennal scale by length of chela, very sparsely setose except for a few distal tufts of setae along fingers; dactylus about 0.5 × palm length, ending in strong, hooked tip, cutting edge with narrow trench along distal half, with strong, hook-shaped tooth at mid-length, otherwise smooth; propodal cutting edge with sharp corneous spine distally, followed by narrow trench along distal half, rounded tooth at mid-length and molariform process with numerous tiny teeth at proximal third; outer margin of fixed finger unarmed, bearing tufts of long setae; palm sub-cylindrical, 3.3 times as long as broad, dorsal margin smooth, with few scattered small granules on surface, along ventral margin and more dense across inside palm surface; carpus about 0.5 × palm length, narrowing proximally, without spines; merus subequal in length to palm, five times longer than broad in lateral view, proximally compressed laterally, with small distoventral tooth, dorsally with few distal granules and ventral ridge with row of small teeth and granules, otherwise smooth; ischium about three-fourth length of carpus, laterally compressed, with sharp distodorsal spine and ventral row of spines, distalmost strongest. Basis and coxa short; coxa mesially granulate. Epipod absent.

Fourth and fifth pereiopods long and slender, similar, sparsely setose; dactylus about one-fourth the length of propodus, biungulate, both unguis clearly demarcated, with small, irregularly shaped accessory tooth on ventral margin proximally; propodus not subdivided, about 0.4 × carpal length, pereiopod 5 propodus 1.1 × wider than pereiopod 4 propodus, with single row of 21–22 (P4) and 24–25 (P5) movable spines along entire flexor margin, both margins with few long, simple and very few plumose setae; carpus longest, not subdivided, with movable spine at distoventral angle; merus about 0.8 × carpal length, unarmed; ischium less than half length of merus, unarmed. Basis and coxa short, coxa with blunt, serrated distomesial tooth. Epipod absent.

First pleopod uniramous, second to fifth biramous, all lacking appendices, unarmed.

Uropod well developed, about as long as telson. Protopod stout, with sharp postero-lateral spine. Exopod broader than endopod; lateral margin slightly convex, with 17 (left) and 19 (right) teeth along distal three-fourth of margin, distal margin shallowly convex, dorsal surface with two distinct longitudinal ridges. Endopod simple, unarmed, surface with 1 longitudinal ridge. Exopod and endopod fringed with dense, plumose seta.

Eggs: The ovigerous female incubates around 200 eggs of diameters 1.8 × 2.2 mm.

Comparison of female holotype to male allotype (Figures 13 and 14).

Body smaller than female holotype.

Rostrum proportionally longer, 0.4 × PCL, dorsal margin with eight small teeth in addition to proximal shallow rounded eminence at level of posterior orbital margin; ventral margin with two teeth; laterally unarmed.

The carapace less inflated, distribution of spines similar, but spines generally longer with exception of antennal spine smaller, otherwise similar to holotype.

Sixth thoracic sternites with narrowly triangular process, anteriorly furnished with sharp spine each. Seventh and eight sternites broadly rounded, unarmed.

Pleonal somites similar to holotype; first pleonite with pronounced anteroventral spine; ventral margins of tergites 3–6 more angular and with sharper distal spine each; dorsally smooth. Fifth pleomere with small median spine distally, directed posteriorly.

Telson similar, dorsal ridges with 6–7 spines, laterally with 7–8 spines; distal margin with 9 small spines (median spine with four lateral pairs).

Eye similar, eye stalk with 2–3 dorsal spines.

Antennule similar, second article with single, small distomesial granule only.

Antenna with first article (coxa) bearing strong mesial tooth (not carinate). Basicerite similar. Antennal scale with 4–5 lateral spines.

Epistome with two pairs of small spines and smooth median margin.

Mouthparts as for female holotype. Left third maxilliped merus with sharp distolateral spine, merus unarmed (right third maxilliped missing).

Pereiopod 1, 2, 4 and 5 as for female holotype. Pereiopod 3 slightly longer, 2.5 × CL; the palm is slightly more massive at 2.7–2.8 times longer than wide (left–right); ischium with double distodorsal spine, otherwise as holotype.

Uropodal exopod with 14–15 teeth along lateral margins.

Measurements. CL: 16.3–18.5 mm (M), 16.6–[20.0] mm (F), PCL: 12.0–11.4 mm (M), 11–[15] mm (F), TL: 38.5–46.0 mm (M), 40.2–[53.2] mm (F). Holotype measurements are given in square brackets.

Distribution. Confirmed in New Zealand, 1184–1460 m. Sequence only, pending morphological examination: Vanuatu, 1210–1250 m (MNHN-IU 2013-19622).

Coloration. Collection notes are retained as "white" (NIWA 127110) and "pink" (NIWA 127111).

Hosts. The pair of type specimens collected by ROV from the Southern Kermadec Ridge was taken with a sponge host of *Corbitella* sp. nov. (det. Henry Reiswig, Figure 15). This genus of glass sponges was recently reported as a host for the first time by Kou et al. [35] and this is the second record of a *Spongicoloides* species with *Corbitella*. An unidentified species of *Farrea* glass sponge is the only sponge collected at the same station with NIWA 72929 and is the most likely host at present.

Variation and remarks. Two further specimens are reported in addition to the pair of type specimens, displaying a limited amount of variation: NMNZ CR.019650 appears more spinose both considering the surface of the carapace and rostrum, the antennal peduncle and scale and the uropodal endopod (Figures 16 and 17). The epistome of the holotype is distinctly bispinose anteriorly, but all other specimens have more, indistinct spines distally. The palp of the first maxilliped bears a long and slender distal spine in this female specimen, which is shorter and blunter in all others, including the female holotype. Saito and colleagues [48] indicated this might be sexually dimorphic when describing *S. iheyaensis*. The armature of the third maxilliped ischium and merus varies slightly across specimens but all have a small or distinct distal spine on at least one side or article (NIWA 72929 had a small spine on both merus and ischium of the one remaining third maxilliped). Some (sexually dimorphic and/or allometric) variation is apparent in the proportions of the third pereiopod: it is slightly shorter for the two females (1.7–2.0 × CL) and longer (2.5 × CL) for the male allotype (pereiopod 3 are missing in the male NIWA 72929). Typical sexual dimorphism is evident with the palm more slender in the large holotype female (length-width ratio of 3.3) and proportionally broader (2.7–2.9) for the male allotype and female NMNZ CR.019650 that are both approximately the same size.

Spongicoloides sonne sp. nov. closely resembles the group of species with two arthrobranchs on third maxilliped and pereiopods 1–4, dorsally unarmed uropodal endopod, dorsally smooth pleonites and smooth distoventral margin of pereiopod 3 fixed finger. This includes *S. clarki* sp. nov., *S. novaezelandiae*, *S. hawaiiensis* Baba, 1983 [49], *S. weijiaensis* and *S. corbitellus*. *Spongicoloides sonne* sp. nov. can be distinguished from these species by the following fixed characters:

- The postrostral, cardiac and at least the anterior branchial carapace surface is covered with small spines; *S. hawaiiensis* has a nearly smooth carapace and the remaining species only have a few scattered small spinules.
- Third maxilliped is lacking an epipod; which all others have present (rudimentary in *S. clarki*).
- Third maxilliped endopod ischium and/or merus with small distolateral spine (typically varies from left to right but present on at least one of the articles); the endopod is unarmed in all other species.
- Coxa of third maxilliped and pereiopods 1–2 with distinct mesial spine or process; rounded in all other species.
- Rostrum overreaches the basal antennular article in *S. sonne*, *S. novaezelandiae*, *S. weijiaensis* and *S. clarki* but it falls short of the distal end of the article in *S. corbitellus* (the rostrum in *S. hawaiiensis* remains unknown).
- Pereiopod 3 ischium with distinct distodorsal spine and row of sharp spines along ventral margin, the distodorsal spine is small in *S. novaezelandiae*, *S. weijiaensis*, *S. clarki* and *S. corbitellus* and absent in *S. hawaiiensis*; the ventral margin smooth or irregular but not furnished with sharp spines in all other species.

A notable characteristic that has in the past been rarely considered is the ventromesial ornamentation of the first antennal segment (coxa, bearing the antennal gland) which is distinctly ridged and carinate (as in the holotype), always furnished with at least one distinct spine. In *S. novaezelandiae* the first antennal article is simple, with only a small spine. This character might hold additional phylogenetic value and could be considered further across the remaining species.

At least one of the specimens presented by Goy [13] as *S. novaezelandiae* from New Caledonia most likely belong to this new species based on the clearly spinose carapace surface and the row of distinct spines illustrated on the third pereiopod ischium of MNHN Na-11999 (Goy [13]: Figure 9; MNHN-IU-2013-19487). Most of the remaining material presented by Goy [13] are referable to *S. weijiaensis* (see comments under *S. novaezelandiae* above).

DNA sequence data. A clade formed by 3–4 sequences were monophyletic for all genes considered (Figure 2). This included a sequence from a specimen from Vanuatu (1210–1250 m) held at the MNHN (IU-2013-19622) which is considered conspecific and pending more detailed examination.

CO1: intra-specific divergences between 0–1.2%; intra-generic divergences range from 4.2% (*Spongicoloides clarki* sp. nov.) to 7.3% (*S. weijiaensis*).

16S rRNA: intra-specific divergences between 0–0.3%; intra-generic divergences range from 3.3% (*Spongicoloides weijiaensis*) to 16.7% (*S. iheyaensis*).

Genus *Spongiocaris* Bruce & Baba, 1973 [21].

3.2.5. *Spongiocaris antipodes* sp. nov.

In (Figures 1, 2 and 18–22). **Material examined. Holotype:** F ov. (PCL: 7.7 mm); West Norfolk Ridge, 34.342° S, 168.387° E, 382–390 m; 2 June 2003; NORFANZ Stn. TAN0308/139; NMNZ CR.025704. **Allotype:** M (PCL: 5.2 mm); West Norfolk Ridge, 34.285° S, 168.358° E, 785–800 m; 2 June 2003; NORFANZ Stn. TAN0308/141; NMNZ CR.019492. **Paratype:** 1 M (poor condition, PCL: ~6.2 mm); station details as for holotype; NMNZ CR.019493. 1 F (PCL: 8.0 mm); Seamount No. 986, off Hawkes Bay shelf, 39.991° S, 178.215° E, 792 m; 9 February 2017; method: manipulator arm; R/V Sonne, ROV *KIEL 6000*, cruise: SO254, dive: 36 ROV 10; NIWA 127133.

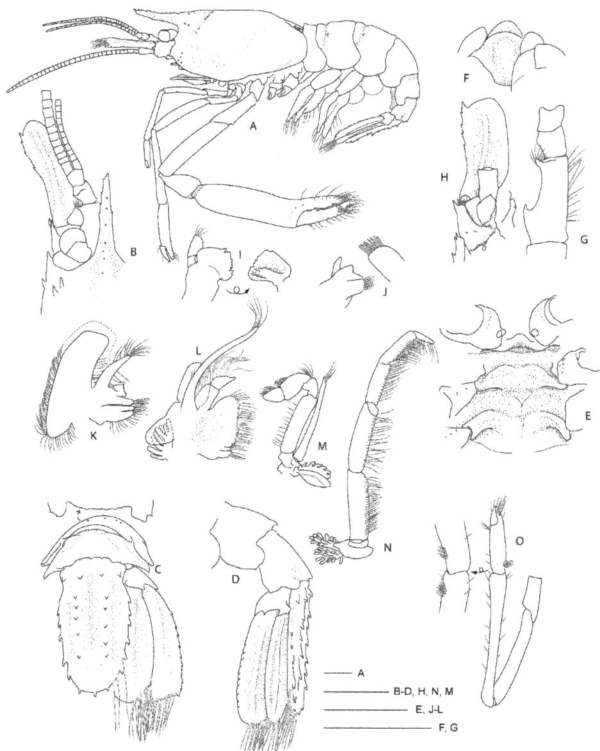

Figure 18. *Spongicaris antipodes* sp. nov., holotype F ov. (NMNZ CR.025704). (**A**) Habitus, lateral. (**B**). Anterior portion of carapace and cephalic appendages, dorsal. (**C**) Pleonites 6, urosome and telson, dorsal, with inset of dorsal view of pleonite 5. (**D**) Pleura 5–6, urosome and telson, lateral. (**E**) Thoracic sternites 6–8 with coxae of pereiopods, ventral. (**F**) Epistome, labrum partially covered by following mouthparts, ventral. (**G**) Right antennular peduncle, ventral. (**H**) Right antennae with inset view of mesial margin of basicerite. (**I**) Right mandible, inner and outer view. (**J**) Right maxillule with broken upper lacinia and left maxillule upper lacinia. (**K**) Right maxilla. (**L**) Right Mxp1. (**M**) Left Mxp2. (**N**) Right Mxp3. (**O**) Right P1, lateral, with inset detail of carpo-propodal setiferous organ. Scale bars = 2 mm.

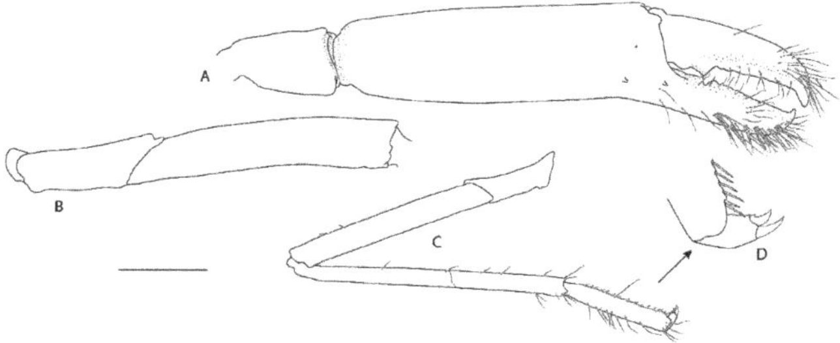

Figure 19. *Spongicaris antipodes* sp. nov., holotype F ov. (NMNZ CR.025704). (**A**) Right P3, distal portion, lateral. (**B**) Right P3, proximal portion, lateral. (**C**) Left P5, lateral, with enlarged detail of dactylar spination enlarged. Scale bar = 2 mm.

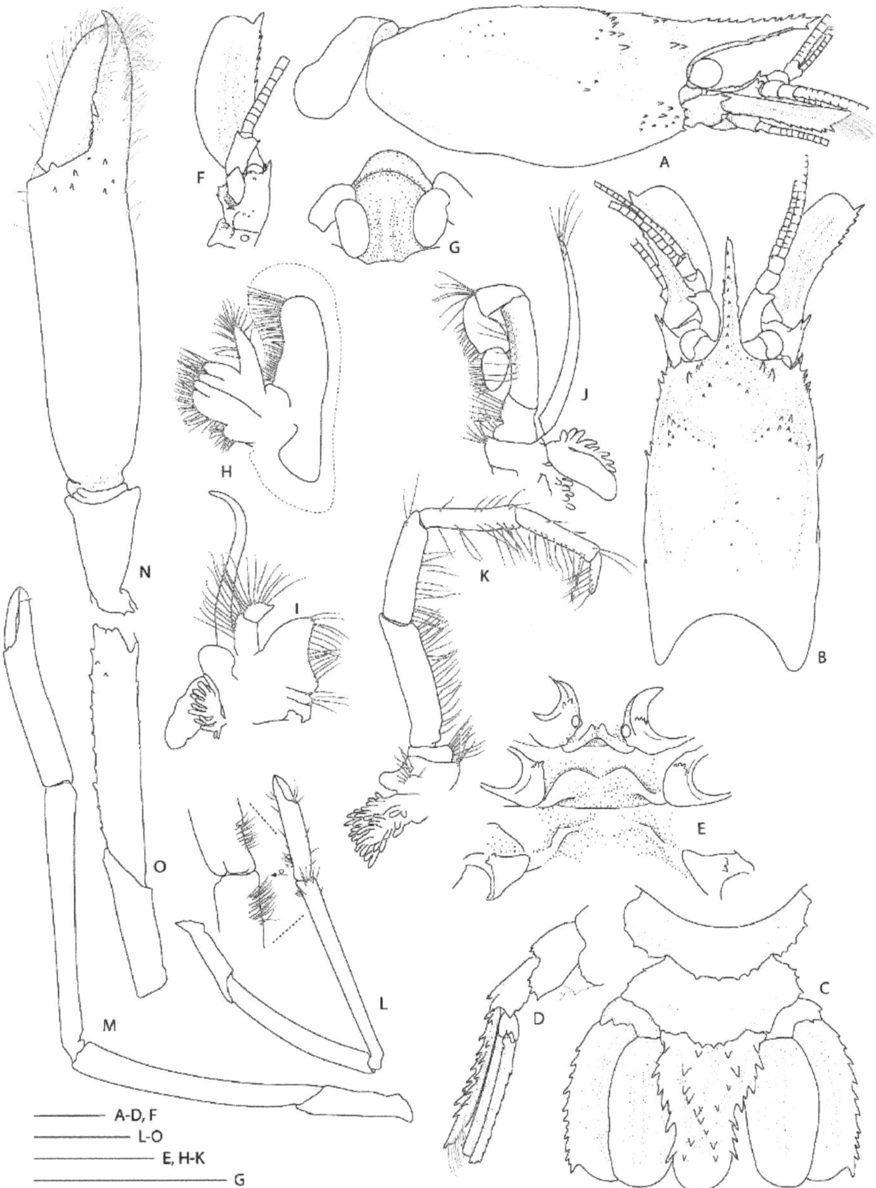

Figure 20. *Spongiocaris antipodes* sp. nov., paratype F (NIWA 127133). (**A**) Carapace and first pleonite, lateral. (**B**) Carapace and cephalic appendages, dorsal. (**C**) Pleonites 5–6, urosome and telson, dorsal. (**D**) Pleonites 5–6, urosome and telson, lateral. (**E**) Thoracic sternites 6–8 with coxae of pereiopods, ventral. (**F**) Left antenna, ventral. (**G**) Epistome, labrum partially covered by mandibular palps, ventral. (**H**) Left maxilla. (**I**) Left Mxp1. (**J**) Left Mxp2. (**K**) Right Mxp3. (**L**) Left P1 with detail of merocarpal setiferous organ. (**M**) Right P2. (**N**) Right P3 distal portion, lateral. (**O**) Right P3 proximal portion, lateral. Scale bars = 2 mm.

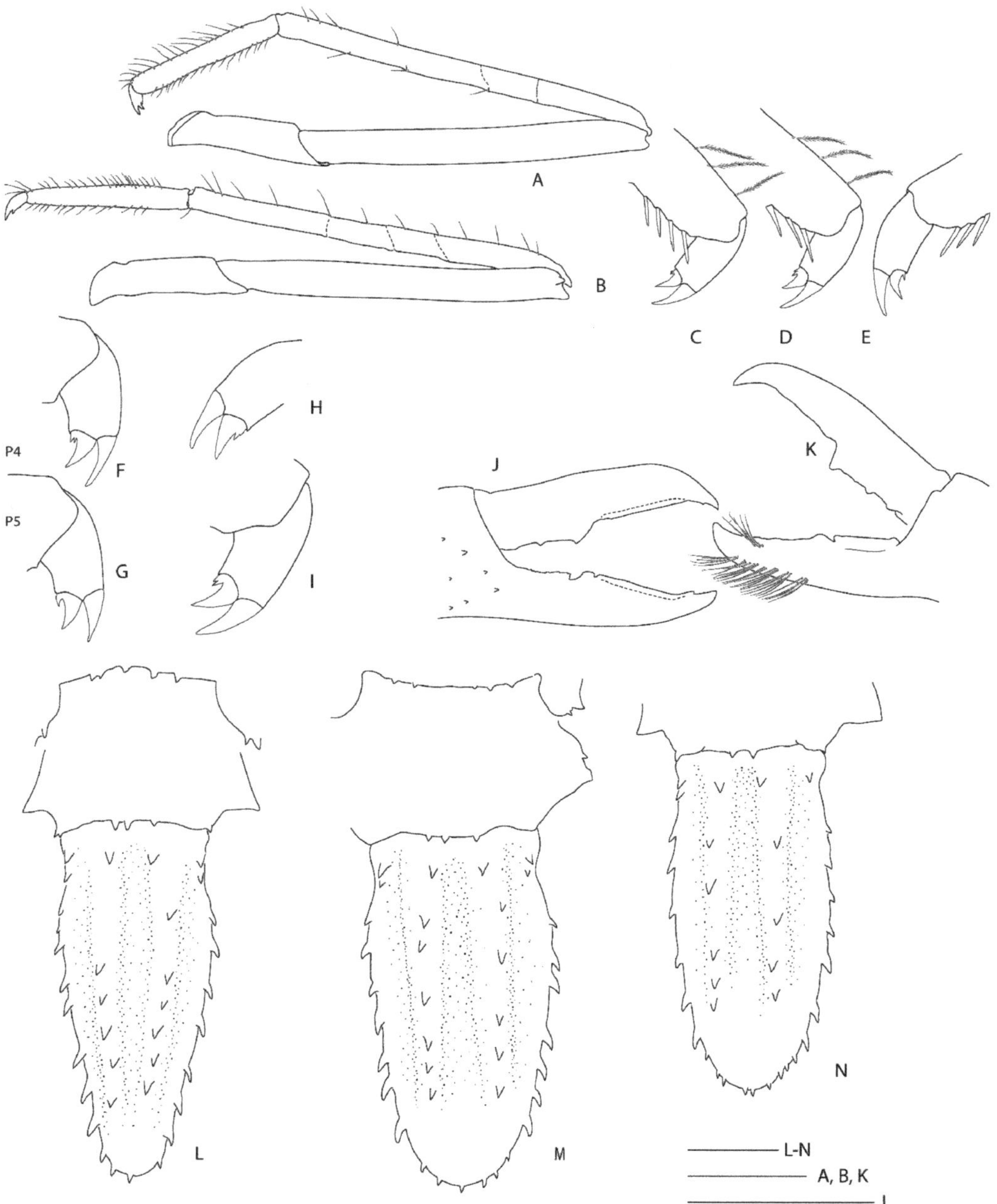

Figure 21. *Spongiocaris antipodes* sp. nov., (**A–E**): paratype F, PCL 8.0 mm (NIWA 127133); (**F,G**): F ov., PCL 7.9 mm (NMNZ CR.019259); **H,I**: lose appendages, unknown origin (NMNZ CR.019491); (**J**): M, PCL 4.6 mm (NIWA 135621); (**K,N**): M, PCL 7.3 mm (NMNZ CR.019494); (**L**): M, PCL 7.5 mm (NMNZ CR.019491); (**M**): F ov., PCL 8.0 mm (NMNZ CR.019491). (**A,B**) lose right P4 and P5. (**C,D**) Dactylus and distal propodus, lose right P4 and P5. (**E**) dactylus and propodus, presumed left P4. (**F–I**) Dactylus and distal propodus, P4 and P5. (**J,K**) fingers and distal palm, P3, outside view. (**L–N**) telson and pleonite 6, dorsal. Scale bars = 2 mm.

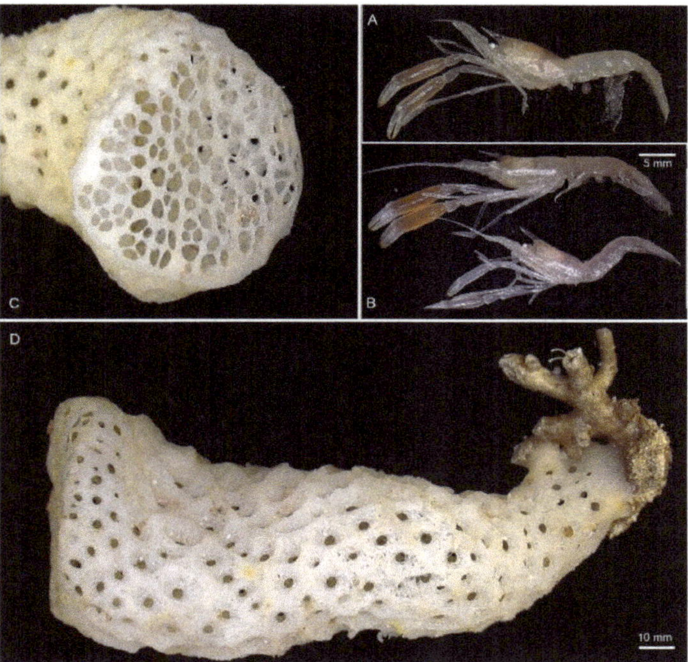

Figure 22. Live coloration of *Spongiocaris antipodes* sp. nov and euplectellid sponge host *Regadrella okinoseana* Ijima, 1896 [46] (Hexactinellida, Lyssacinosida, Euplectellidae) specimen collected on southern Norfolk Ridge during 2003 NORFANZ expedition: (**A**) NMNZ CR.019259, (**B–D**) NMNZ CR.019494, scale applies to both A and B.

Other material: Norfolk Ridge. 1 F ov. (PCL: 12.9 mm), 1 M (PCL: 8.7 mm), Wanganella bank (International Waters); 33.5–33.4° S, 167.7–167.6° E, 677–546 m; 11 November 2013; SOP Stn. TRIP3933/23; bottom longline; NIWA 88622. 2 M (PCL: 6.4, 7.3 mm); southern Norfolk Ridge, NW of Three Kings Islands, 33.390–33.391° S, 170.210–170.196° E, 469–490 m; 1 June 2003; beam trawl; NORFANZ Stn. TAN0308/136; NMNZ CR.019494 (with *Regadrella* sp.). 1 F ov. (PCL: 8.0), 2 M (PCL: 7.5, 5.2 mm); southern Norfolk Ridge, NW of Three Kings Islands, 33.390–33.391° S, 170.210–170.196° E, 469–490 m; 1 June 2003; beam trawl; NORFANZ Stn. TAN0308/136; NMNZ CR.019491. 3 F (PCL: 7.8, 7.9, 8.7 mm), 1 M (PCL: 6.3 mm); southern Norfolk Ridge, Reinga Ridge, 33.390–33.396° S, 170.190–170.203° E; 469–526 m; 31 May 2003; NORFANZ Stn. TAN0308/126; NMNZ CR.019259. **Three Kings Ridge.** 1 F (PCL: 6.0 mm), 2 M (PCL: 7.2, 4.6 mm); Seamount 148, 31.980° S, 174.265° E, 700 m; date unknown; Stn. Z9026; NIWA 135621. **Kermadec Ridge.** 1 M (PCL: 7.5 mm); Raoul Island, 6.4 km NNE of Herald Islet, 29.20° S, 177.82° W, 1189–1225 m; 5 April 1973; NMNZ Stn. 73312: NMNZ CR.016806; "inside glass sponge". **Hikurangi Margin.** 1 F ov. (damaged, TL ~ 33 mm), East Coast, E of Cape Kidnappers, 40.03° S, 178.06° E, 935 m; 28 Aug 1986; bottom trawl; RV James Cook; Stn. J10/40/86; NMNZ CR.005952 (collection note: "from *Euplectella*"). 1 F ov. (poorly preserved), 1 M (PCL: 7.0 mm); SE of Cape Kidnappers, 40.02° S, 178.08° E, 840 m; 28 August 1986; bottom trawl; RV James Cook; Stn. J10/39/86; NMNZ CR.004792.

Diagnosis. Small commensal spongicolid shrimp so far known to be associated with euplectellid glass sponges; body slightly depressed. Carapace with distinct cervical groove, each side furnished with 6–10 small spines; small branchiostegal spine and antennal spine present or absent; 2–3 pairs of postrostral spines; typically two pairs of supraorbital spines;

pterygostomian angle furnished with a few small spines; a few anterolateral spines always present; hepatic spines typically present. Rostrum length reaching to distinctly overreaching last antennular peduncle article, entire dorsal margin with spines with 5–11 dorsal, 1–3 ventral, 0–3 lateral spines; cornea unpigmented. Antennal scale subrectangular, with 4–9 spines along the lateral margin. Epistome nearly always unarmed, may have single anterior spine; endopod of maxillule unarmed. Third maxilliped setiferous organ covering nearly entire length of propodal ventral margin; epipod always absent. First pereiopod with well-developed setiferous organ. Third pereiopod robust; carpi unarmed except for low blunt distal spines; meri smooth or with dorsal row of low spines; palm with a few scattered spines along distal portion of outer surface, at least three times longer than wide; fixed finger with dense fringe of setal brushes along two-thirds of ventral margin, unarmed. Fourth and fifth pereiopods with carpi distinctly segmented by 1–3 sutures; coxa may be armed with short distomesial spinose ridge (males) or granular (females); all pereiopods lack epipods. Posterior margin of fifth pleonal tergite ending in a few spines, posterior margins of pleura irregular; sixth tergite with 3–4 posterior dorsal spines, surface smooth, posterior margin of pleura with 1–3 spines. Telson with 6–8 spines along each lateral margin. Uropodal endopodite with two terminal dorsal hairs.

Etymology. Named *antipodes*, an archaic vernacular for Australia and New Zealand, as this species is so far restricted to the southwestern Pacific region. The term is used as a noun in apposition.

Description of holotype female (Figures 18 and 19).

Body robust, surface generally glabrous.

Rostrum narrowly triangular in dorsal view, one-third (0.3) as long as PCL, horizontal, nearly reaching to distal end of antennular peduncle; dorsal margin with 7 small teeth or granules and small posterior blunt eminence at level of posterior orbital margin; ventral margin with three small teeth in distal one-third; ventrolateral ridges with 1–2 (right-left) small teeth on distal half.

Carapace glabrous, not inflated. Cervical and branchiostegal grooves distinct, seven (right)–eight (left) distinct spines line cervical groove. Postrostral region with two small submedian spines on either side; two pairs of postorbital spines; hepatic region smooth; four pairs of anterolateral spines; two small spines on anterior branchial region; otherwise carapace surface is smooth. Orbital margin concave, inferior orbital angle slightly produced and furnished with one small antennal spine (left) or granule (right); branchiostegal spine small to minute; pterygostomian angle round, with 1–2 spines on margin.

Thoracic sternites 6–8 anteriorly rounded, minutely serrate but unarmed; sternite 6 posterior margins evenly concave.

Pleonal somites glabrous, surfaces smooth. First pleonite shortest; divided into two sections by distinct transverse carina; ventral margin continuous, straight. Second to fourth pleonites subequal in length, with feeble transverse carina and blunt articular knob; ventral margin rounded. Fifth pleonite with two small posteroventral granules on broad, round margin; dorsal surface smooth, unarmed; posterior margin with four small spines. Sixth pleonite dorsally smooth; posterolateral process straight, angular, with small spine at corner; posterior margin with three small spines. Pleonal sternites unarmed.

Telson about twice as long as broad, subquadrangular, posterior margin distinctly convex, with long plumose setae; dorsolateral ridges distinct, with six spines each, proximalmost largest and placed mesially; lateral margins relatively straight, slightly constricted proximally, each side armed with a subproximal spine and six lateral spines; posterior margin with three small spines (one median spine and one lateral pair).

Eye well developed; cornea semiglobular, about half length of ocular peduncle in dorsal view, moderately inflated, unpigmented; unarmed.

Antennular peduncle nearly reaching to middle of antennal scale, does not reach spinose lateral margin of antennal scale. Basal article stout, length-width ratio of 2.0 at midlength, without statocyst, about three times as long as second article; lateral margin concave, distally produced to small, rounded lobe, stylocerite distinct, acute, reaching one-third length of basal article; mesial margin almost straight, unarmed other than row of serration where plumose seta are inserted. Second article slightly longer than distal article, unarmed. Distal article as long as wide, unarmed. Flagella slender, about twice as long as peduncle.

Antenna with first article (coxa, bearing antennal gland) with 1–2 mesial spines. Basicerite stout; 1 mesial spine proximal to base of carpocerite; 2–3 small spines on ventral surface; dorsal and lateral surfaces smooth. Margins with paired spines distolaterally; small spine below ventral angle; small spine below mesial angle (Figure 18H inset). Antennal scale broad, 2.5 × as long as wide; lateral margin slightly concave, armed with 4 (left) or 6 (right) small teeth along the distal 0.4–0.5 portion, including the distal tooth; dorsal surface with two distinct longitudinal ridges. Carpocerite not reaching to distal end of second article of antennular peduncle; first article with small mesial spine; second article with minute distoventral spine. Flagellum at least as long as CL.

Epistome narrow, inflated, subtriangular, unarmed. Labrum smooth, with median ridge. Paragnaths bilobed, with deep median fissure, distodorsally spatulate.

Mandible robust, fused molar and incisor processes. Molar surface with three blunt teeth; incisor with irregular teeth. Palp well-developed, 3-segmented; proximal article shortest, without setae; middle article with few distal setae; distal article suboval, slightly longer than intermediate article (measured along extensor margin), densely setose.

Maxillule with simple palp, with few terminal setae; distal endite broad, round, with numerous simple setae and eight slender spines; proximal endite oval, with simple setae distally.

Maxilla with palp stout, tapering, with plumose setae, falling short of end of scaphognathite; distal and proximal endites both deeply bilobed, with numerous plumose setae; scaphognathite well developed, anterior portion longer than posterior portion, about 3 times longer than broad, with dense fringe of plumose setae along entire margin.

Branchial formula summarized in Table 2.

First maxilliped with 2-segmented palp, bearing long plumose setae; distal article longer than broad, with distal spine; proximal article stout, about 1.5 × distal article in length. Distal endite large, subtriangular, densely setose; proximal endite deeply bilobed, with distal setae. Exopod slender, with long plumose setae. Epipod well developed, subequally bilobed. Arthrobranch well developed.

Second maxilliped with 5-segmented endopod; dactylus sub-oval, tapering distally, with dense setae on flexor margin; propodus twice as long as dactylus, densely setose along flexor margin; carpus triangular, about two-third length of propodus measured at mid-line, with long distodorsal setae; merus nearly straight, about 1.8 × length of propodus, 4 times longer than broad, with row of setae along mesial margin and sparse short setae on surfaces; ischium and basis not fused, ischium slightly longer, about one-fourth meral length, long setae along mesial margin. Coxa mesially produced to blunt process; exopod well developed; oval epipod present, podobranch and arthrobranch well developed.

Third maxilliped endopod 5-segmented, slender, unarmed; dactylus narrow, distally tapering, setose; propodus about twice as long as dactylus, with setiferous organ along entire flexor margin; carpus subequal in length to propodus; merus 1.5 times longer than carpus, about as long as ischium; all segments with long setal fringe along flexor margin; basis short. Coxa mesially angular, not produced to spine, without epipod, arthrobranchs well developed. Exopod absent.

First pereiopod slender, glabrous and sparsely setose, slightly overreaching antennal scale when extended. Dactylus about 0.4 × palm length; palm subcylindrical, tufts of long setae dorsodistally and along ventral margin of fixed finger; carpus about 3 times palm length, ventral carpo-propodal setiferous organ well developed on both articles; merus about 0.7 × carpal length, about 3 times longer than ischium. Basis and coxa short, unarmed. Epipod absent.

Second pereiopod similar to first, 1.5 × longer and stronger, sparsely setose; dactylus 0.4 × palm length, distal tip formed into a strong corneous spine, tips of fingers cross when chela closed, cutting edges entire; propodus with a few tufts of setae distally and along fixed finger; carpus about 1.8 × palm length; merus 0.8 × carpal length; ischium 0.4 × merus. Basis and coxa short; coxa unarmed. Epipod absent.

Third pereiopod largest, subequal and similar, about 2 × CL, sparsely setose except for a few distal tufts of setae along fingers; dactylus about 0.5 × palm length, ending in strong, hooked tip, cutting edge with narrow trench along distal half, with blunt, trianguloid tooth at distal quarter, preceding fossa not pronounced; propodal cutting edge with sharp corneous spine distally, followed by narrow trench along distal half, distinct fossa to accommodate dactylar tooth at distal third, followed proximally by flat molariform process with numerous tiny teeth; outer margin of fixed finger unarmed, distally bearing tufts of long setae; surface with small proximal spine; palm sub-cylindrical, 3.4 times as long as broad, dorsal margin smooth, with 2–3 small granules on distal surface, very few small granules on inside palm surface; carpus about 0.3 × palm length, narrowing proximally, without spines; merus three-fourth length of palm, 5 times longer than broad in lateral view, proximally compressed laterally, margins entirely smooth; ischium about as long as carpus, laterally compressed, unarmed. Basis and coxa short; coxa mesially granulate. Epipod absent.

Fourth and fifth pereiopods long and slender, similar, sparsely setose; dactylus (including spines) about one-fourth the length of propodus, biungulate, both unguis clearly demarcated, with small, simple or distally faintly bifurcate accessory tooth on proximal ventral margin; propodus not subdivided, about 0.4 × carpal length, pereiopods 4 and 5 propodi similar in width, 5.7 times longer than wide, with single row of 16–20 movable spine along entire flexor margin, both margins with few long, simple and very few plumose setae; carpus longest, 2.2–2.5 × propodus, with movable spine at distoventral angle, one median suture present; merus about 0.8 × carpal length, unarmed; ischium less than half length of merus, unarmed. Basis and coxa short, coxa mesially granular but unarmed. Epipod absent.

First pleopod uniramous, second to fifth biramous, all lacking appendices, unarmed.

Uropod well developed, about as long as telson. Protopod stout, with sharp posterolateral spine. Exopod broader than endopod; lateral margin slightly convex, with 12–13 small teeth along distal three-fourth of margin, distal margin shallowly convex, dorsal surface with two distinct longitudinal ridges. Endopod simple, unarmed, surface with one longitudinal ridge. Exopod and endopod fringed with dense, plumose seta.

Eggs: The ovigerous female retained 10 sub-oval eggs of diameters 1.5 × 2.0 mm.

Comparison of female holotype to male allotype (Figures 20 and 21).

Body smaller than female holotype.

Rostrum proportionally longer, 0.5 × PCL, distinctly overreaching the antennular peduncle, dorsal margin with 8 small teeth in addition to proximal shallow rounded eminence at level of posterior orbital margin; ventral margin with two teeth; with 3 small ventrolateral teeth.

The carapace is slightly less inflated, the distribution of spines is similar, the numbers vary slightly: 6–7 spines along the branchiostegal groove, three small postrostral spines, 3–5 anterolateral marginal spines. The two prominent postorbital spines match the holotype and are considered diagnostic.

Thoracic sternites 6–8 anteriorly produced, with spines, the coxa of the pereiopods and furnished with denticulate processes.

Pleonal somites similar to holotype; posteroventral margins of tergite 4 with 1–2 spines, tergite 5 with 3–4 small spines, posterior margin with rows of small spines; dorsal surfaces smooth. Fifth pleomere with small median spine ventrodistally, directed posteriorly.

Telson similarly shaped, dorsal ridges with 5–6 spines, laterally also with 6 spines; distal margin with 6 small spines (median spine with 2 and 3 lateral spines).

Antennule and antenna of similar shape and proportions; antennal basicerite with additional distoventral and distolateral spine, mesial angle unarmed. Antennal scale slightly more slender at 2.8 (length-width ratio), with 6 lateral spines.

Epistome similarly inflated, smooth.

Mouthparts as for female holotype.

Pereiopod 1 and 2 as for female holotype. Pereiopod 3 slightly more robust, similar length with 2.0 × CL; dactylus 0.6 × palm; palm slightly more massive at 2.8–2.9 times longer than wide; merus with distoventral spine and regular dorsal row of granules and small spines; ischium smooth as for holotype. Single loose pereiopod 4 or 5 retained for allotype has a simple, sharp accessory tooth on dactylus; propodus with 14 spines along flexor margin; carpus with two sutures.

Uropods as for holotype, except for protopod distally with pair of distolateral spines exopod with 12 and 14 teeth along lateral margins.

Measurements. CL: 9.3–12.3 [10.3] mm (females, this does not include the largest female that has a broken rostrum), 7.4–11.2 mm (males), PCL: 6.0–12.9 [7.7] mm (females) 5.2–8.7 mm (males), TL: 25–37 [25.5] mm (females), 19.5–30.7 (males). Measurements for the holotype are given in square brackets.

Distribution. Southern Norfolk Ridge, West Norfolk Ridge, eastern North Island, New Zealand; 382–1225 m (Figure 1).

Coloration. Collection note with NMNZ CR.016806 "white shrimp from inside glass sponge". Live coloration was captured for NORFANZ specimens NMNZ CR.019259 and CR.019494: the body is pale apricot, carapace and most appendages transparent, palm of third pereiopod orange (Figure 22).

Hosts. Two specimens were collected from inside the euplectellid glass sponge *Regadrella okinoseana* (Hexactinellida, Lyssacinosida, Euplectellidae) (NMNZ Cr. 09494, det. M Kelly, Figure 22). Other collection records retained are: "from *Euplectella*" (NMNZ CR.05952) and "from inside glass sponge" (NMNZ CR.016806).

Parasites. One male (NMNZ CR.016806) bears a sacculinid rhizocephalan under the pleon.

Remarks. Multi-gene sequencing revealed a clade that resolved most of the New Zealand specimens of *Spongiocaris* as an undescribed species, while it united the type specimens of both *S. yaldwyni* and *S. neocaledonensis* in a well-supported clade, requiring that the latter be synonymized with the former (see below, Figure 2). Morphologically, the variability of characters across the material examined rendered it difficult to establish fixed diagnostic characters.

Spongiocaris species are currently separated using differences in the length of the rostrum, distribution of spines on the carapace surface and anterior margin, pleomere and telson, as well as the pereiopods 3–5 [11]. Examination of a total of 21 specimens assigned to *S. antipodes* sp. nov. indicate the following variability: the rostrum is slightly shorter to longer than the antennular peduncle (not reaching to overreaching article 3), 0.3–0.5 × PCL, with 5–11 dorsal, 1–3 ventral and 0–3 lateral teeth; the carapace has 2–3 pairs of small postrostral spines always present, at least one but typically two distinct postorbital spines, 4–10 pairs of postcervical spines are always distinct, occasionally followed by a few scattered dorsal cardiac spinules (Figure 20A,B), a field of 4–10 distinct anterolateral spines, a few small hepatic and/or branchial spines are usually present (Figure 20B); an antennal spine on the anterior carapace margin may be absent, minute or well-developed and can vary from side to side (e.g., Figure 20B), a small branchiostegal spine may be present or absent and the pterygostomian angle with 1–3 distinct spines; the dorsal surfaces of the pleonites are always unarmed and smooth, the ventrodistal margins of pleura 4–6 usually bear some spines (pleuron 4 may be smooth, pleuron 5 always bears multiple spines and pleuron 6 has at least an acute angle); the telson shape is variable (Figures 18C, 20C and 21L–N), 6–8 lateral spines in addition to subproximal spine, 5–8 dorsal spines and 3–8 small spines along the distal margin, usually arranged with a median spine separated from pairs of laterally-placed spines (with the exception of the larger male, NMNZ CR.019491, that has a continuous row of 6 spines); the third cheliped is about twice as long as CL with limited sexual dimorphism apparent, the fingers are 0.5–0.6 × palm length, margins unarmed, with dense setal brushes distally, along about the distal two-thirds of the margin of the fixed finger, the palm length-width ratio ranges from 2.8 in small to 3.4 in larger specimens, the proximal portion always has 2–7 small granules, the merus is smooth (Figure 19B) or serrated (Figure 20O), the ischium is always unarmed with smooth margins; pereiopods 4–5 carpi with 1–3 sutures, propodi with 14–23 movable spines along the entire flexor margin and dactyli biungulate with ventral unguis bearing a variable accessory tooth, from simple hook-shaped (Figure 21G) to bifurcate (Figure 21F,I).

Further notable variation is the presence of an anterior spine on the epistome of a male of NMNZ CR.019491, the shape is otherwise the same as in all others, anteriorly inflated and round. The P4–5 propodi usually have between 17–22 spines along the flexor margin, but these can be as few as 12 (e.g., NMNZ CR.019491 and CR.019492)

Based on the combination of these characteristics, *S. antipodes* closely aligns with *S. yaldwyni* (to include *S. neocaledonensis*, see below). These two species appear to be difficult to separate morphologically. Diagnostic characters proposed are:

- the presence of postrostal and, typically, at least two pairs of postorbital carapace spines (the postrostral area is usually smooth in *S. yaldwyni* but see Goy ([13]: Figure 12) and the carapace only has a single postorbital spine). This character can vary from side to side and some small specimens (e.g., NMNZ CR.019494, CR.019491, NIWA 135621) have only a single spine.
- the dorsal surface of pleonal tergite 6 is always smooth (*S. yaldwyni* often has 2–3 median spines, see Figures 24C and 25A, but is smooth in holotype, Figure 23B).
- the dense setal brushes along the distal fixed finger of the third pereiopod is distinct along about two-thirds in *S. antipodes* but reach to 0.7–0.8 portion in *S. yaldwyni*.
- the fourth and fifth pereiopod dactyli are never triungulate in *S. antipodes* but can be both biungulate and triungulate in *S. yaldwyni*.

There may be a degree of overlap in these characters and DNA sequencing for confirmation of identification is advised.

DNA sequence data. Between six and nine specimens were sequenced across the genes examined (Figure 2). In all cases they formed a monophyletic clade with some indication of further intraspecific structure that might warrant further investigation in the future.

CO1: intra-specific divergences between 0.2–2.3%; intra-generic divergences between 4.9–6.7% (*Spongiocaris yaldwyni*).

16S rRNA: intra-specific divergences between 0.0–1.6%; intra-generic divergences range from 1.9% (*Spongiocaris yaldwyni*) to 11.4% (*S. koehleri*).

3.2.6. *Spongiocaris yaldwyni* Bruce & Baba, 1973

In (Figures 1, 2 and 23–25). *Spongiocaris yaldwyni* Bruce & Baba, 1973 [21]: 163, Figures 7–10.—Goy 2010 [3]: 257 (list).—Komai, de Grave & Saito 2016 [11]: 445 (key).

Spongiocaris neocaledonensis Goy, 2015 [13]: 313, Figures 12–14.

Material examined. Holotype: F (PCL 7.6 mm); Motunau Island (Plate Island), Bay of Plenty, 37.50° S, 176.55° E, 585–622 m; 29 Sep 1962; N.Z. Marine Department Haul 12; NMNZ CR.001888 (collection note: "venus flower basket sponge").

Other material: Kermadec Ridge. 1 F ov. (PCL: 7.6 mm), 1 M (PCL: 6.6 mm); 41 km NNE of Macauley Island, 29.8617 S, 178.1817 W, 965 m, 27 Jul 1974; NZOI Stn K831; Agassiz trawl; NIWA 10483. 1 F (PCL: 5.0 mm); NE of summit, Clark Seamount, Southern Kermadec Ridge, 36.445–36.442° S, 177.839–177.840° E, 850–927 m; 24 Apr 2012; NIWA Stn. TAN1206/99; Sled, epibenthic; NIWA 83102. **Bay of Plenty.** 1 M (PCL: 12.7 mm), Whakatane Seamount, 36.790–36.790° S, 177.454–177.456° E, 1160–1155 m; 23 Apr 2012; epibenthic sled; NIWA Stn. TAN1206/90; NIWA 82821. 1 F (crushed carapace, TL~25 mm); ca. 22 km E of Aldermen Islands, 36.96° S, 176.36° E, 803–846 m; 24-Jan-79; RV Tangaroa; NMNZ Stn. 79760; NMNZ CR.009840.

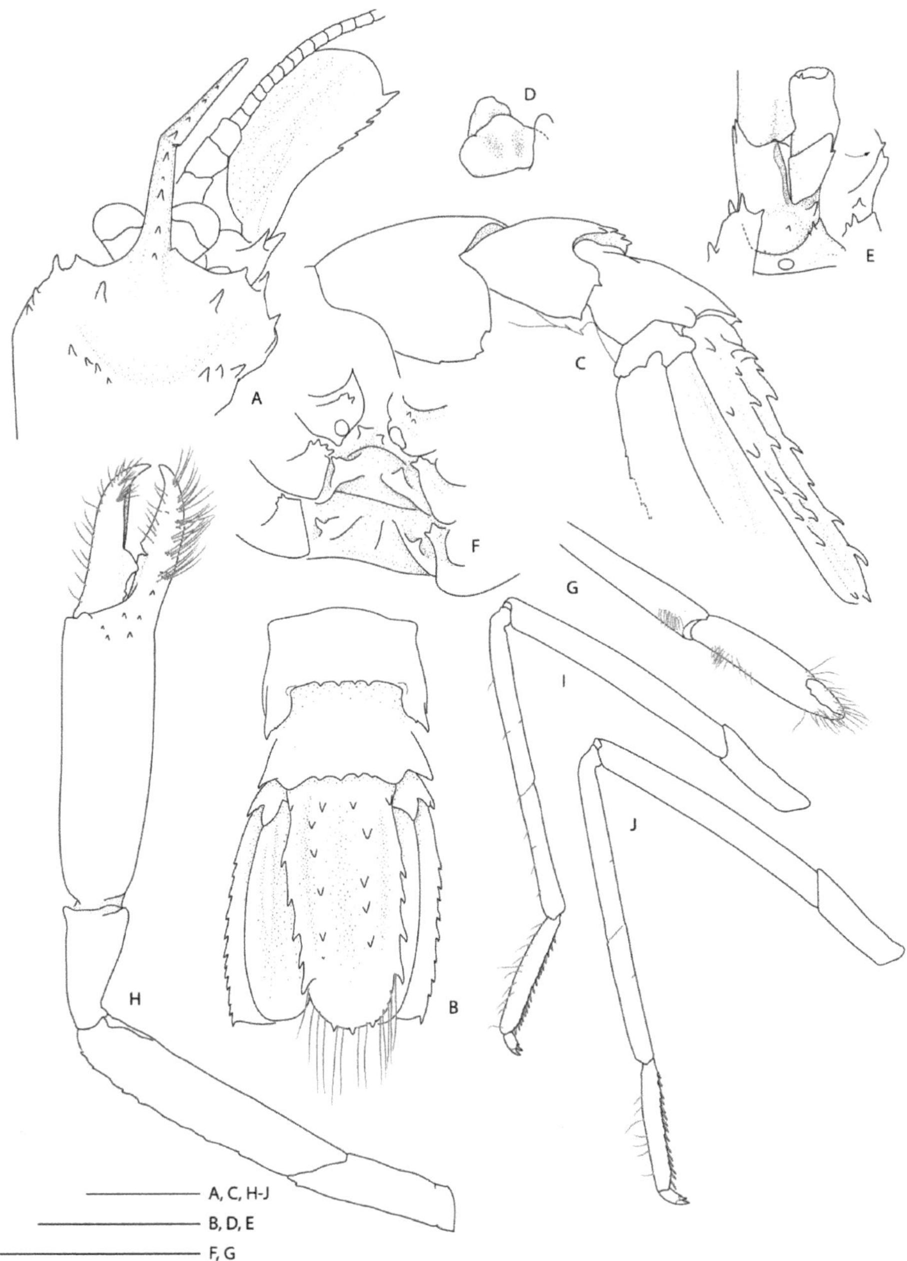

Figure 23. *Spongicaris yaldwyni* Bruce & Baba, 1973 [21], holotype F (NMNZ CR.001888). (**A**) Anterior carapace and cephalic appendages, dorsal. (**B**) Pleonites 5 and 6, urosome and telson, dorsal. (**C**) Pleonites 4–6, urosome and telson, lateral. (**D**) Epistome and labrum, oblique ventral view. (**E**) Right antenna, ventral with detail of rotated mesial margin of basicerite. (**F**) Thoracic sternites 6–8 with coxae of pereiopods, ventral. (**G**) Left distal portion of P1, mesial. (**H**) Right P1, lateral. (**I**,**J**) Lose P4 or P5, lateral. Scale bars = 2 mm.

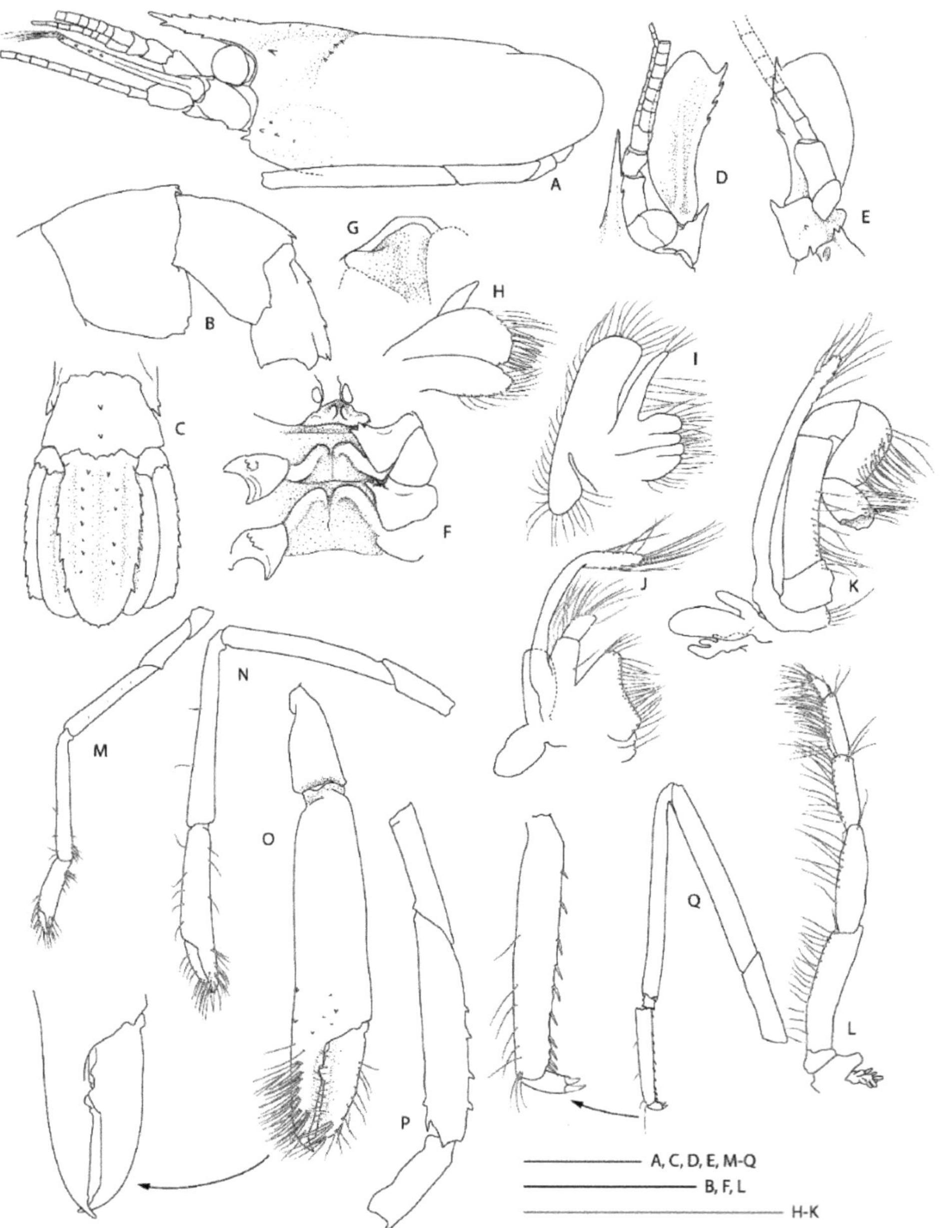

Figure 24. *Spongicaris yaldwyni* Bruce & Baba, 1973 [21], F (NIWA 83102). (**A**) Carapace and cephalic appendages, lateral. (**B**) Plenites 4–6, lateral. (**C**) Pleonites 5–6, urosome and telson, dorsal. (**D**) Rostrum with right antennule, antenna and ocular peduncle, dorsal. (**E**) Right antenna, ventral. (**F**) Thoracic sternites 6–8 with coxae of pereiopods, ventral. (**G**) Epistome, labrum partially covered by mandibular palps, ventral. (**H**) Right maxillule. (**I**) Right maxilla. (**J**) Right Mxp1. (**K**) Right Mxp2. (**L**) Mxp3. (**M**) Left P1, lateral. (**N**) Left P2, lateral. (**O**) Left P3 distal portion, lateral, with inset detail of cutting edge. (**P**) Left P3 proximal portion, lateral. (**Q**) Left P5 with enlarged propodus and dactylus. Scale bars = 2 mm.

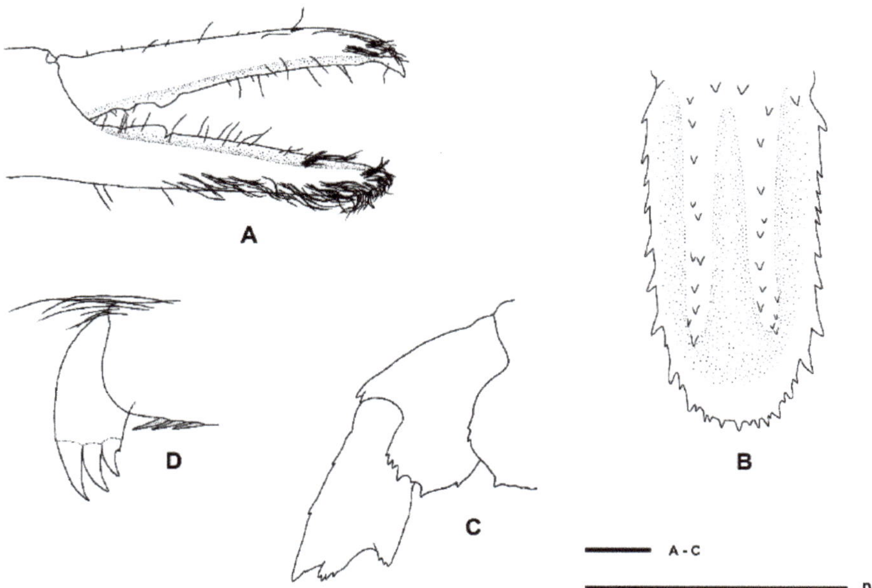

Figure 25. *Spongiocaris yaldwyni* Bruce & Baba, 1973 [21], F (NIWA 82821). (**A**) Fingers of left P3, mesial view. (**B**) Telson, dorsal view. (**C**) Fourth to sixth pleura, lateral view. (**D**) Dactylus of left P5, lateral view. Scale bars = 1 mm.

Diagnosis. Small commensal spongicolid shrimp so far known to be associated with euplectellid glass sponges and gorgonian corals; body slightly depressed. Carapace with distinct cervical groove furnished with 4–10 small spines; small branchiostegal spine and antennal spine present or absent; one pair of supraorbital spines; pterygostomian angle furnished with a few small spines; a few anterolateral spines always present; hepatic spines typically present. Rostrum length reaching to distinctly overreaching last antennular peduncle segment, entire dorsal margin with spines with 4–8 dorsal, 0–4 ventral, 0–3 lateral spines; only proximal part of cornea pigmented. Antennal scale subrectangular, with 4–7 spines along the lateral margin. Epistome unarmed; endopod of maxillule unarmed. Third maxilliped setiferous organ covering nearly entire length of propodal ventral margin; epipod present or absent. First pereiopod with well-developed setiferous organ. Third pereiopod robust; carpi unarmed; meri with row of small spines along dorsal margin; palm with a few scattered spines along distal portion of outer surface, at least 3 times longer than wide; fixed finger with dense fringe of setal brushes along more than two-thirds of ventral margin, unarmed. Fourth and fifth pereiopods with carpi distinctly or partly segmented, full suture typically present at approximately mid-point; coxa armed with short distomesial spinose ridge; all pereiopods lack epipods. Posterior margin of fifth pleonal tergite ending in four or five spines, posterior margins of pleura irregular to spinose; sixth tergite with 3–4 posterior dorsal spines, surface smooth or with two or more small spines along mid-line, posterior margin of pleura with 1–2 spines. Telson with 4–10 spines along each lateral margin. Uropodal endopodite with two terminal dorsal hairs.

Measurements. CL: 6.7–15.6 [11.8] mm, PCL: 4.6–12.7 [7.6] mm, TL: 18.0–43.0 mm. Holotype measurements are given in square brackets.

Distribution. *Spongiocaris yaldwyni* was purportedly endemic to New Zealand with a single record from the Bay of Plenty; 585–622 m. Chen et al. [17] presented gene sequences for a specimen from Tonga (627–656 m, MNHN-IU-2014-12842). New records extend the regional distribution northwards along the Kermadec Ridge, 850–1160 m (Figure 1).

Considering the synonymy of *S. neocaledonensis*, the distribution is extended to include New Caledonia, Norfolk Ridge, Loyalty Islands; 460–970 m.

Coloration. Unknown.

Hosts. The holotype was collected from within the hexactinellid sponge *Regadrella okinoseana* [21]. Goy [13] reported *S. neocaledonensis* from the hexactinellid sponge *R. okinoseana* and on a gorgonian octocoral.

Remarks. The original description of *Spongiocaris yaldwyni* is comprehensive and is not repeated here, the material presented is the first since the species was described from the single specimen [21]. The holotype (NMNZ CR.001888) is in poor condition and the body is deformed but allowed the verification of some characteristics: the epistome is unarmed, anteriorly truncate, and the labrum rounded (Figure 23D); all pleonal tergites have smooth surfaces, the posterior margins of pleonites 5–6 bear 3–4 median spines (Figure 23B), the posterior margins of the pleura bear spines (pleurae 4 and 5) or are angular (pleuron 6) (Figure 23C); sternites 6–8 are medially acute and the coxa of pereiopods 3–5 bear mesial spines along their anterior corners (Figure 23F); third maxilliped has a setiferous organ on nearly entire flexor margin of the propodus; and the first pereiopod has a well-developed setiferous organ on the proximal palm and distal carpus (Figure 23G); the third pereiopod merus is distinctly serrate along the dorsal (extensor) margin and the setal brushes extend along more than 0.6 portion of the fixed finger (Figure 23H). The fourth and fifth pereiopods carpi clearly have a suture at approximately mid-length, which was not illustrated by Bruce & Baba [21].

The type description did not mention the sex of the specimen, it bears gonopores on the coxae of the third pereiopods but also possesses the single forward curving ventral medial spine on the fifth pleomere that is typical of males (Figure 23C,F). The female (NIWA 83102), however, lacks a spine.

Three further samples are presented here for the region that all share the long rostrum (overreaching the second antennular article). The rostral ornamentation can be expanded to 6–[8] dorsal, 2–[6] ventral and [0]–3 lateral spines (holotype details are presented in square brackets); all specimens have a single pair of postorbital spines, NIWA 10483 and 83102 have 2–3 minute postrostral spines (smooth in the holotype), the carapace has a hepatic spine in the holotype and the larger female of NIWA 10483 but is smooth in NIWA 83102 (Figure 24A); the surface of pleonite 6 bears dorsal spines in all other specimens (Figures 24B and 25A), but is smooth in the holotype. The antennal scale is relatively slender in the holotype (3 × longer than wide), similar to NIWA 83102, but is relatively broader in NIWA 10483 with ratios of 2.6 and 2.8, the lateral margin has 3–7 spines; the basicerite has only a single lateral spine in NIWA 83102 (Figure 24D,E). The sternites of the females are typically rounded and the coxa are produced to distinctly projected and granulate mesial processes in all cases (e.g., Figure 24F). The third maxilliped is lacking an epipod in all specimens, the propodus is furnished with a dense fringe of seta (setiferous organ). The third pereiopod is similarly glabrous in both males and females except for the presence of small spines on the outer palm, at the base of the fixed finger, the setal brushes are distinct and extend up to 0.8 × the length of the fixed finger in NIWA 10483 and 83102 (Figure 24O). The lose walking leg of NIWA 83102 is lacking a suture on its carpus (Figure 24Q) but 1–3 sutures are present in all pereiopods 4–5 of NIWA 10483, all specimens have dactyli with small accessory teeth on the ventral unguis, either sharp, simple, or divided (Figure 24Q inset), NIWA 82821 has all pereiopod 4–5 dactyli triungulate (Figure 25D). The holotype and two most recently collected specimens (NIWA 82821 and 83102) are closely related genetically (p-distances of ≤1.6% for 16S rRNA fragment, see below) and are genetically distinct from *S. antipodes*.

DNA sequence analyses indicate that the holotype sequence for *S. yaldwyni* was in all cases near identical to the type series of *S. neocaledonensis* Goy, 2015 [13] (see Figure 2 and comments below). Goy [13] noted the close overall appearance between these two species but separated them based on the presence (*S. neocaledonensis*) and absence (*S. yaldwyni*) of an epipod on the third maxilliped and proposed *S. neocaledonensis* as distinct from *S. yaldwyni* (and all other congeners) based on the segmentation of the carpus of the last two pairs of pereiopods 4–5 (present in *S. neocaledonensis* and absent in *S. yaldwyni*), the presence of the third maxilliped's setiferous organ (absent in *S. yaldwyni*), and the dense fringe of setal brushes on the third pereiopod propodus (fixed finger) (absent in *S. yaldwyni*). As noted above, upon re-examination of the holotype and additional material of *S. yaldwyni*, characters overlap considering the presence of sutures on the pereiopod 4–5 carpi of nearly all specimens (absent on single remaining leg of NIWA 83102, Figure 24Q), presence of dense setal brushes on third maxilliped and pereiopod 3. However, none of the material examined had even rudimentary epipods on the Mxp3 and this character remains enigmatic. Nevertheless, *Spongiocaris yaldwyni* and *S. neocaledonensis* are here considered synonymous.

Most of the New Zealand specimens (the exception being the holotype) bear a median row of two small spines on pleonal tergite 6, none are as spinose as figured for *S. neocaledonensis* male holotype (Goy [13]: Figure 12). In nearly all New Zealand specimens, the pereiopod 4–5 dactyli are biungulate, with a small accessory tooth on the ventral unguis (the exception being NIWA 82821 which has a triungulate dactylus as illustrated for *S. neocaledonensis*, Figure 25D). This character has been found to be variable in other species, e.g., the stenopodid *Stenopus goyi* Saito, Okuno & Chan 2009 [52], which has the ambulatory dactyli from simple to triungulate, differing even among the pereiopods of the same specimen [52]. Variation in spination across this species is now considered as follows: the rostrum has 4–8 dorsal, 0–4 ventral and 0–3 ventrolateral spines; the carapace spination is somewhat variable with some hepatic, postrostral, minute cardiac or branchial spine, 4–10 postcervical spines are always distinct, as is one pair of postorbital spines; the telson has 4–10 spines along the lateral margin, 4–9 spines along dorsal carinae and 3–12 posterior teeth; the antennal scale has 4–7 spines along the lateral margin and the pereiopods 4–5 propodi have 8–21 movable spines along the flexor margin.

The combined distribution range now encompasses New Caledonia, the Norfolk and Kermadec Ridges, which aligns with those of other southwestern Pacific subtropical to temperate decapods (e.g., [53,54]), with an overall depth range of 460–1160 m. Both species were first described from the same hexactinellid sponge host species (*Regadrella okinoseana*).

DNA sequence data. The holotype of *S. yaldwyni* and nine type specimens of *S. neocaledonensis* were successfully sequenced, including the holotype, allotype and seven paratype samples. The sequences presented by Chen et al. [17] from Tonga are confirmed as belonging *S. yaldwyni* (Figure 2, Table 1). Both genes resolved these specimens as belonging to a single clade, distinct from *Spongiocaris antipodes* sp. nov., however, substantial internal differences remain within this group that warrant future investigation.

CO1: intra-specific divergences between 0.2–4.2%; intra-generic divergences between 4.9–6.7% (*Spongiocaris antipodes* sp. nov.).

16S rRNA: intra-specific divergences between 0.0–2.6%; intra-generic divergences range from 1.9% (*Spongiocaris antipodes* sp. nov) to 11.4% (*S. koehleri*).

Family STENOPODIDAE Claus, 1872 [55]
Genus *Stenopus* Latreille, 1819 [56]

3.2.7. *Stenopus hispidus* (Olivier, 1811)

In (Figure 26). *Palaemon hispidus* Olivier, 1811 [23]: 666.

Stenopus hispidus.—Latreille 1819 [56]: 71.—Holthuis 1946 [57]: 12 (with references and complete synonomy).—Yaldwyn 1968 [22]: 278.—Doak 1971 [58]: pl. 39D.—Yaldwyn 1974 [59]: 1044.—de Saint-Laurent & Cleva 1981 [60]: 157.—Goy 1992: 100 [61].—Poore 2004 [62]: 149.—Saito et al. 2009 [52]: 1009.—Yaldwyn & Webber 2011 [63]: 183 (list).—Webber et al. 2010 [64]: 225 (list).—Goy 2010 [3]: 215.—De Grave & Fransen 2011 [42]: 253.—Goy 2015 [13]: 336.

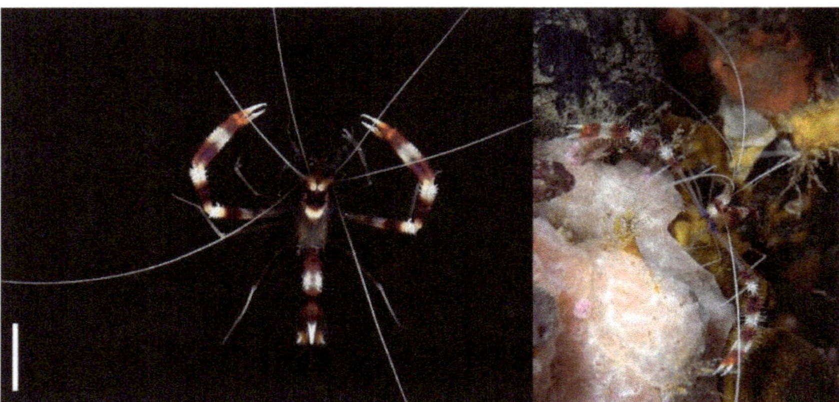

Figure 26. *Stenopus hispidus* (Olivier, 1811) [23]. (**Left**): M (AWMM MA182117). Scale bar = 2 cm (Image credit: Sadie Mills, NIWA). (**Right**): live specimen in Houhora Harbour, Northland (Image credit: Crispin Middleton, Seacology NZ).

Material examined. Kingdom of Tonga. 1 M (PCL: 11.5 mm); Tongatapu Island, Stn. I173, 21.130° S, 175.183° W, 0 m; 8 June 1976; underwater hand sampler; NIWA 10308. **Lord Howe Island (Australia EEZ).** 1 M (PCL: 14.5 mm); Stn. P31, 29.007° S, 167.915° E, 4–14 m; 28 January 1977; scuba; NIWA 10604. **Kermadec Islands.** 1 M (PCL: 9.8 mm), 1 F ov. (PCL: 12.6 mm); Raoul Island, Stn. TAN1612/37, 29.226° S, 177.880° W, 0–30 m; 25 October 2016; scuba; AWMM MA182117. 1 M (PCL: 15.8 mm); North of Napier Island, Stn. TAN1612/20, 29.230–29.231° S, 177.876–177.874° W, 0–30 m; 24 October 2016; scuba; NIWA 118168. 1 M (PCL: 10.0 mm); Raoul Island, Denham Bay, southern end, 29.288° S, 177.958° W, 3–18 m; 31 August 2017; scuba, col: Severine Hannam, Anna Murray, David Aguirre, Crispin Middleton, Carl Struthers, Matt Hill; AWMM MA168313. **Northland.** 1 M (PCL: 12.2 mm), 1 F (PCL: 14.0 mm); Bay of Islands, Tapuaetahi, 35.118° S, 173.978° E, 11 m; 4 September 1986; scuba, col: Brett Stephenson; AWMM MA6648. 1 F (poorly preserved); off northern tip of Poor Knights Islands, 35.417° S, 174.058° E, 40 m; scuba; col: Kelly Tarlton; AWMM MA8585. 1 M (PCL: 12.0); Poor Knights, Southern Island Harbour, 15–18 m; 25 May 1969; col: A.N. Baker; NMNZ CR.006006. 1 F (PCL: 14.7 mm); Poor Knights Islands, 24 m; May 1968; NMNZ CR.016812. 1 M (PCL 11.0 mm); North Island, Poor Knights Islands; 10 January 1970; col: Wade Doak; NMNZ CR.016808. 1 M (PCL: 15.3 mm); South Harbour, Poor Knights Islands, 18 m; 5 May 1969; col: B. Palmer; NMNZ CR.016811. 1 F (PCL: 13.6 mm); off Poor Knights, 18 m; 23 April 1967; col: B.B. Anderson; NMNZ CR.001779. **Bay of Plenty.** 1 M (PCL: 11.9 mm); Alderman Islands, underwater cave, several present on roof; Jan 2001; scuba; col: C.D. Roberts, J. Hope; NMNZ CR.016809.

Diagnosis. Large shrimp (total length 30–67 mm) with a slender, compressed body, densely covered with spines. Carapace and first three pleonites with curved forward-directed spines. Last three pleonites with straight posteriorly directed spines. Rostrum strong, ultimate point reaching to about middle of second article of antennular peduncle with dorsal and lateral spines but no ventral spines. Stylocerite short and pressed against basal antennular segment. Scaphocerite with outer margin entire for considerable distance before final tooth. Third maxilliped with ischium, merus and carpus provided with external row of spinules. Body white with red transverse bands on carapace, third and sixth pleonite, and third pereiopods; base of the third pereiopod deep blue, the remaining appendages are pale.

Measurements of examined specimens. CL: 13.5–21.3 mm (males), 17.5–20.0 mm (females), PCL: 9.8–15.3 mm (males), 12.6–14.7 mm (females), TL: 34.7–57.0 mm (males), 44.5–51.7 mm (females). Previously reported TL: 30–67 mm.

Distribution. *Stenopus hispidus* is the only pantropical species in the infraorder Stenopodidea. Previously recorded from the Western Atlantic from Cape Lookout, North

Carolina, throughout the Caribbean and Gulf of Mexico to the southern border of Brazil [65]; central Atlantic from Ascension Island [66]; Eastern Pacific from Taboga Island, Panama [61,67]; and throughout the tropical and warm temperate Pacific; from the intertidal to 210 m [57]. Specimens presented here are from the northern mainland New Zealand and the islands directly to the North (Lord Howe Island, Kermadec Islands); 3–40 m. Further specimens examined from the NMNZ collection were collected around the Solomon Islands, Niue and Tahiti, but are not presented in detail here.

Colour in life. Good descriptions of the color pattern for this species are previously given [22,57]. Considering the key to the species of *Stenopus* provided by Saito et al. [52], the diagnostic combination of coloration for this species are the broad red and white banding of the body and third pereiopod, not uniformly orange (as in *S. spinosus* Risso, 1827 [68]) and carapace not purple (as in *S. tenuirostris* de Man, 1888 [69]). The base of the third pereiopod deep blue, the remaining appendages are pale (the third maxilliped, fourth and fifth pereiopods are not blue as in *S. cyanoscelis* Goy, 1984 [70]) (see Figure 26).

Remarks. All specimens examined here match the keys provided by Saito et al. [52] well. *Stenopus hispidus* remains the sole pantropical species in this genus, however, recent population genetic studies indicated regional genetic divergences among locations in the Indonesian Archipelago [71] and between oceans [72]. Further studies might reveal that this supposedly widespread species indeed represents a species complex but is for now considered as *S. hispidus* sensu lato.

DNA sequence data. Three New Zealand specimens (NIWA 118168, AWMM MA182117 and MA168313, recently collected from the Kermadec Islands) unambiguously aligned with conspecific sequences deposited on GenBank with sequence divergences are 0.0% for both CO1 and 16S rRNA genes (Figure 2, Table 1). Reference sequences include specimens from across the Indo-Pacific [17,72,73].

4. Discussion

Prior to this study, three species of shrimp belonging to the infraorder Stenopodidea were known in the New Zealand region; the widely-distributed, Indo-Pacific stenopodid cleaner shrimp *Stenopus hispidus*, and the sponge-associated spongicolids *Spongicoloides novaezelandiae* and *Spongiocaris yaldwyni*, the former purportedly widely distributed from Madagascar to New Zealand [13] and the latter only known from the holotype and recently published gene sequences only [17,21]. These crustaceans are naturally rare, with only a total of 54 specimens collected in the region since 1962, representing the entire available material in natural history collections. However, this material examined for our study comprised seven species, three new to science (*Spongicoloides clarki*, *S. sonne* and *Spongiocaris antipodes*) and one new record for the region (*Spongicola goyi*), more than doubling the previously known New Zealand diversity. Additional records are provided for the previously known species, including the southernmost record for Stenopodidea on the subantarctic Macquarie Ridge for *S. novaezelandiae*. This species now most likely remains an endemic New Zealand species, with most previous records referred to *S. weijiaensis* or considered dubious and in need of verification. In contrast, the previously endemic *S. yaldwyni* is synonymized with *S. neocaledonensis* and its combined distribution range now includes New Zealand and New Caledonia.

These revisions are supported by DNA sequencing and the increased taxon sampling provided for these select spongicolids provides valuable data for ongoing wider phylogenetic studies. The classification within the Stenopodidea remains unsettled with our findings e.g., corroborating recent studies resolving the genera *Spongiocaris* and *Spongicoloides* as paraphyletic [10,17,35]. Morphologically, these two genera are currently well defined, based on distinct characters such as the plesiomorphic presence (*Spongiocaris*) or derived absence (*Spongicoloides*) of an exopod on the second maxilliped [3]. However, *Spongicoloides iheyaensis* is placed far removed from its congeners in the molecular tree (Figure 2), more similar to *Engystenopus palmipes* Alcock & Anderson, 1894 [74] that has a well-developed exopod on the second maxilliped [9]. These two deep-water spongicolids might share ontogenetic characteristics that could see them united in the future (J. W. Goy, pers. comm.), which would render the autapomorphies of the genus *Spongicoloides* homoplastic. These and other questions, such as the placement of the spongicolid *Globospongicola spinulatus* Komai & Saito, 2006 [8] within the Stenopodidae and the monotypic *Juxtastenopus* Goy, 2010 [9] rendering the genus *Stenopus* paraphyletic, are currently addressed by a number of researchers.

The overall pattern remains of a predominantly cryptic and tropical to temperate group of shrimp, including the shallow-water stenopodid and deep-water spongicolids [3]. Also, nearly all spongicolids retain collection notes or records of host associations with hexactinellid sponges (*Spongicoloides novaezelandiae* so far lacks this information), conforming with our current understanding of symbiotic relationships with primarily glass sponges (unusually, *Spongiocaris yaldwyni* was recorded with gorgonian octocorals by Goy [13]).

Little is known about the host selection among symbiotic deep-sea invertebrates, yet limited studies have indicated that deep-sea sponge-associated decapods might have lower host specificity compared with shallow-water symbiotic species. For instance, the palaemonid genus *Philarius* Holthuis, 1952 [75] is an obligatory associate of shallow-water coral of the genus *Acropora* Oken, 1815 [76] (Scleractinia, Acroporidae) [77–80]. Among the snapping shrimps of the *Alpheus armatus* (Rathbun, 1901) [81] species complex, each species is associated with specific species of anemones [82]. In contrast, deep-water species appear less specific in their host association, e.g., species of *Paralebbeus* Bruce & Chace, 1986 [83] (Caridea, Thoridae) are associated with sponges of two families, Euplectellidae Gray, 1867 [84] (Hexactinellida, Lyssacinosida) and Farreidae Gray, 1872 [85] (Hexactinellida, Sceptrulophora) (e.g., [83,86,87]). As the most well-known deep-sea sponge-associated crustaceans, spongicolid shrimps are reported to be in association with a relatively wide range of hexactinellid sponges [3,12,35]. In this study, we found *Spongicoloides sonne* sp. nov. could be associated with sponges belonging to *Corbitella* (Lyssacinosida, Euplectellidae) and *Farrea* (Sceptrulophora, Farreidae). Furthermore, *Spongicola goyi* in New Zealand was collected from a *Pheronema conicum* (Hexactinellida, Amphidiscosida), but was previously reported from a range of hexactinellids as well as a demosponge [12]. The low host specificity of spongicolid shrimps and other crustaceans could be attributed to the decrease of the diversity and number of suitable host sponges in deeper waters [35]. Conversely, *Spongicoloides clarki* sp. nov. and the two *Spongiocaris* species reported from New Zealand share the same host *Regadrella okinoseana*, which is one of the most common deep-sea hexactinellid sponges in this region [47]. Notably, a single host sponge has so far only been found to be physically occupied by a single species of spongicolid at the same time (authors, pers. obs.). Overall, these patterns corroborate the hypothesis that host shift may not play a key role in speciation of deep-sea sponge-associated decapods, as was recently found in certain axiidean shrimps [88].

Supplementary Materials: The following are available online at https://www.mdpi.com/article/10.3390/d13080343/s1, Table S1: CO1 K2P genetic distances, Table S2: 16S K2P genetic distances.

Author Contributions: Conceptualization, K.E.S. and Q.K.; methodology, all authors; molecular experiments and analyses, Q.K. and K.E.S.; resources, all authors; data curation, all authors; writing—original draft preparation, K.E.S. and Q.K.; writing, review and editing, all authors. All authors have read and agreed to the published version of the manuscript.

Funding: K.E.S. was funded by NIWA under Coasts and Oceans Research Programme 2 Marine Biological Resources: discovery and definition of the marine biota of New Zealand (2019–2021 SCI). Q.K. was funded by the Natural Science Foundation of China (No. 41876178), the Senior User Project of RV KEXUE (KEXUE2020GZ01), the National Key R&D Program of China (2018YFC0309804), Science & Technology Basic Resources Investigation Program of China (2017FY201404) and the China Ocean Mineral Resources Research and Development Association Program (Nos. DY135-E2-1-02 and DY135-E2-3-04).

Institutional Review Board Statement: Not applicable.

Informed Consent Statement: Not applicable.

Data Availability Statement: The data presented in this study are openly available in NCBI Gen Bank at https://www.ncbi.nlm.nih.gov/genbank/ (accessed on 10 July 2021), see Table 1 for Accession numbers.

Acknowledgments: Our gratitude goes to the staff of the NIWA Invertebrate Collection (Wellington), the National Museum New Zealand Te Papa Tongarewa (Wellington), the Tāmaki Paenga Hira Auckland Museum (Auckland) and the Muséum national d'Histoire naturelle (Paris) for access to collections and curatorial services. Thanks to sponge systematist Michelle Kelly, NIWA Auckland, for identifications of euplectellid and pheronematid host sponges. Thanks to Joe Goy (Harding University) and Laure Corbari (MNHN) for constructive comments on a draft of this manuscript and two anonymous reviewers for helpful suggestions and improvements. We acknowledge the following collection surveys: NORFANZ voyage TAN0308—a survey of the northern Tasman Sea sponsored by the National Oceans Office, CSIRO Marine Research, Australia, New Zealand Ministry of Fisheries and NIWA, New Zealand, project ZBD2002-16 (2003); TAN0803—"MacRidge 2" Macquarie Ridge voyage, an interdisciplinary New Zealand-Australian voyage which also contributed to CSIRO's Division of Marine and Atmospheric Research project "Biodiversity Voyages of Discovery" funded by the CSIRO Wealth from Oceans Flagship and the 'Seamounts' project (2008); TAN1104—the Ocean Survey 20/20 Mapping the Mineral Resources of the Kermadec Arc Project, funded by Land Information New Zealand, Institute of Geological and Nuclear Science, NIWA, and Woods Hole Oceanographic Institution (2011); TAN1206—the 'Impact of resource use on vulnerable deep-sea communities' project (CO1X0906), funded by the New Zealand Foundation for Research, Science and Technology (2012); TAN1612—the "Biodiversity of the Kermadec Islands and offshore waters of the Kermadec Ridge—a coastal, marine mammal and deep-sea survey", funded by the Marine Funding Advisory Research Group, NIWA, Ministry for the Environment, Te Papa Tongarewa, Auckland War Memorial Museum and The Pew Charitable Trust (2016); Project *PoribacNewZ* of the Institute for Chemistry and Biology of the Marine Environment (ICBM), Carl von Ossietzky University of Oldenburg, on the new German RV *Sonne* (voyage SO254), using the GEOMAR Helmholtz Centre for Ocean Research Kiel ROV *KIEL 6000* [89]. The MNHN specimens were collected during the following surveys (i) Deep-sea cruises: KARUBENTHOS 2 (10.17600/15005400); NORFOLK2 (10.17600/3100030); BIOCAL (10.17600/85002911); MUSORSTOM 8 (10.17600/94100040), BATHUS 1 (10.17600/93000350); SEAMOUNT 2 (10.17600/93000020); MIRIKY; SMIB 4 (10.17600/89004911); CALSUB (10.17600/89009911); MUSORSOTOM 4 (10.17600/85009111); BORDAU 2 (10.17600/100060); GUYANE 2014 (ii) Shallow-water biodiversity surveys: PAPUA NIUGINI, SANTO 2006, PAKAIHI I TE MOANA, KARUBENTHOS 2012. Deep-sea cruises, PIs S. Samadi, L. Corbari, B. Richer de Forges and P. Bouchet, were operated by Muséum National d'Histoire Naturelle (MNHN) and Institut de Recherche pour le Développement (IRD) as part of the research program "Tropical Deep-Sea Benthos". Biodiversity surveys were operated by MNHN under the exploration program "Our planet reviewed" (https://www.mnhn.fr/en/research-expertise/lieux/our-planet-reviewed (accessed on 13 July 2021)), PI P. Bouchet (MNHN). Funders and sponsors include the French Ministry of Foreign Affairs, the Total Foundation, Prince Albert II of Monaco Foundation, Stavros Niarchos Foundation, and Richard Lounsbery Foundation.

Conflicts of Interest: The authors declare no conflict of interest.

Abbreviations

AWMM	Tāmaki Paenga Hira Auckland War Memorial Museum, Auckland
CL	total carapace length
COI	mitochondrial cytochrome c oxidase I gene;
F	female
ov.	ovigerous
NIWA	National Institute of Water & Atmospheric Research, Wellington
NMNZ	Museum of New Zealand Te Papa Tongarewa, Wellington
M	male
Mxp1–3	maxillipeds 1–3
P1–5	pereiopods 1–5
PCL	Postrostral carapace length
TL	total body length

References

1. Bate, C.S. *Report on the Crustacea Macrura Collected by the Challenger during the Years 1873–76*; Neill and Company: Edinburgh, UK, 1888; Volume 24, pp. 1–942.
2. WoRMS Editorial Board. World Register of Marine Species. Available online: http://www.marinespecies.org (accessed on 18 May 2021).
3. Goy, J.W. Infraorder Stenopodidea Claus, 1872. In *Treatise on Zoology—Anatomy, Taxonomy, Biology—The Crustacea Volume 9, Part A*; Schram, F.R., von Vaupel Klein, J.C., Eds.; Koninklijke Brill NV: Leiden, The Netherlands, 2010.
4. Schwentner, M.; Richter, S.; Rogers, D.C.; Giribet, G. Tetraconatan phylogeny with special focus on Malacostraca and Branchiopoda: Highlighting the strength of taxon-specific matrices in phylogenomics. *Proc. R. Soc. B Biol. Sci.* **2018**, *285*, 20181524. [CrossRef]
5. Tan, M.H.; Gan, H.M.; Lee, Y.P.; Bracken-Grissom, H.; Chan, T.-Y.; Miller, A.D.; Austin, C.M. Comparative mitogenomics of the Decapoda reveals evolutionary heterogeneity in architecture and composition. *Sci. Rep.* **2019**, *9*, 10756. [CrossRef]
6. Wolfe, J.M.; Breinholt, J.W.; Crandall, K.A.; Lemmon, A.R.; Lemmon, E.M.; Timm, L.E.; Siddall, M.E.; Bracken-Grissom, H.D. A phylogenomic framework, evolutionary timeline and genomic resources for comparative studies of decapod crustaceans. *Proc. R. Soc. B Biol. Sci.* **2019**, *286*, 20190079. [CrossRef]
7. Alvarez, F.; Iliffe, T.M.; Villalobos, J.L. Macromaxillocarididae, a new family of stenopodidean shrimp from an anchialine cave in the Bahamas, with the description of *Macromaxillocaris bahamaensis*, n. gen., n. sp. *J. Crustac. Biol.* **2006**, *26*, 366–378. [CrossRef]
8. Komai, T.; Saito, T. A new genus and two new species of Spongicolidae (Crustacea, Decapoda, Stenopodidea) from the South-West Pacific. In *Mémoires du Muséum National d'Histoire Naturelle*; Editions du Muséum: Paris, France, 2006; Volume 193, pp. 265–284.
9. Goy, J.W. A review of the genus Engystenopus (Crustacea: Decapoda: Stenopodidea). *Juxtastenopus*, gen. nov., a new combination for *E. spinulatus* Holthuis, 1946, and transfer of *E. palmipes* Alcock & Anderson, 1894 to the family Spongicolidae Schram, 1986. *Zootaxa* **2010**, *2372*, 263–277. [CrossRef]
10. Bochini, G.L.; Cunha, A.M.; Terossi, M.; Almeida, A.O. A new genus and species from Brazil of the resurrected family Macromaxillocarididae Alvarez, Iliffe & Villalobos, 2006 and a worldwide list of Stenopodidea (Decapoda). *J. Crustac. Biol.* **2020**, *40*, 704–714. [CrossRef]
11. Komai, T.; de Grave, S.; Saito, T. Two new species of the stenopodidean shrimp genus *Spongiocaris* Bruce & Baba, 1973 (Crustacea: Decapoda: Spongicolidae) from the Indo-West Pacific. *Zootaxa* **2016**, *4111*, 27. [CrossRef]
12. Saito, T.; Komai, T. A review of species of the genera *Spongicola* de Haan, 1844 and *Paraspongicola* de Saint Laurent & Cleva, *Crustacea, Decapoda, Stenopodidea, Spongicolidae*). *Zoosystema* **2008**, *30*, 87–147.
13. Goy, J.W. Stenopodidean shrimps (Crustacea: Decapoda) from New Caledonian waters. *Zootaxa* **2015**, *4044*, 301–344. [CrossRef] [PubMed]
14. Saito, T.; Anker, A. Two new species and new records of *Microprosthema* Stimpson, 1860 (Crustacea: Decapoda: Stenopodidea: Spongicolidae) from the Indo-West Pacific. *Zootaxa* **2014**, *3857*, 24. [CrossRef]
15. Calado, R. *Marine Ornamental Shrimp: Biology, Aquaculture and Conservation*; Wiley–Blackwell: Hoboken, NJ, USA, 2008.
16. Saito, T.; Takeda, M. Phylogeny of the family Spongicolidae (Crustacea: Stenopodidea): Evolutionary trend from shallow-water free-living to deep-water sponge-associated habitat. *J. Mar. Biol. Assoc. UK* **2003**, *83*, 119–131. [CrossRef]
17. Chen, C.-L.; Goy, J.W.; Bracken-Grissom, H.D.; Felder, D.L.; Tsang, L.M.; Chan, T.-Y. Phylogeny of Stenopodidea (Crustacea : Decapoda) shrimps inferred from nuclear and mitochondrial genes reveals non-monophyly of the families Spongicolidae and Stenopididae and most of their composite genera. *Invertebr. Syst.* **2016**, *30*, 479–490. [CrossRef]
18. Hansen, H.J. Crustacea Malacostraca, I. *Dan. Ingolf Exped.* **1908**, *3*, 1–120.
19. Davie, P.J.F. *Crustacea: Malacostraca: Phyllocarida, Hoplocarida, Eucarida (Part 1)*; CSIRO Publishing: Melbourne, Australia, 2002; Volume 19.3A, p. 551.
20. Baba, K. A new stenopodidean shrimp (Decapoda, Natantia) from the Chatham Rise, New Zealand. *Pac. Sci.* **1979**, *33*, 311–314.
21. Bruce, A.J.; Baba, K. *Spongiocaris*, a new genus of stenopodidean shrimp from New Zealand and South African waters, with a description of two new species (Decapoda, Natantia, Stenopodidea). *Crustaceana* **1973**, *25*, 153–170. [CrossRef]
22. Yaldwyn, J.C. Records of, and observations on, the coral shrimp genus *Stenopus* in Australia, New Zealand and the south-west Pacific. *Aust. Zool.* **1968**, *14*, 277–289.
23. Olivier, A.G. Suite de l'Introduction à l'Histoire Naturelle des Insectes. Palèmon. In *Encyclopédie Méthodique. Histoire Naturelle. Insectes, Volume 8*; Olivier, A.G., Ed.; H. Agasse, Imprimeur-Libraire: Paris, France, 1811; pp. 656–670.
24. Charting Around New Zealand (CANZ) Group. New Zealand Region Bathymetry, 1:4 000 000, 2nd Edition. In *NIWA Miscellaneous Chart Series No. 85*; NIWA: Auckland, New Zealand, 2008.
25. Folmer, O.; Black, M.; Hoeh, W.; Lutz, R.; Vrijenhoek, R. DNA primers for amplification of mitochondrial cytochrome c oxidase subunit I from diverse metazoan invertebrates. *Mol. Mar. Biol. Biotechnol.* **1994**, *3*, 294–299.
26. Palumbi, S.R.; Benzie, J. Large mitochondrial DNA differences between morphologically similar Penaeid shrimp. *Mol. Mar. Biol. Biotechnol.* **1991**, *1*, 27–34. [PubMed]
27. Katoh, K.; Rozewicki, J.; Yamada, K.D. MAFFT online service: Multiple sequence alignment, interactive sequence choice and visualization. *Brief. Bioinform.* **2019**, *20*, 1160–1166. [CrossRef]
28. Tamura, K.; Stecher, G.; Peterson, D.; Filipski, A.; Kumar, S. MEGA6: Molecular evolutionary genetics analysis version 6.0. *Mol. Biol. Evol.* **2013**, *30*, 2725–2729. [CrossRef] [PubMed]

29. Kalyaanamoorthy, S.; Minh, B.Q.; Wong, T.K.F.; von Haeseler, A.; Jermiin, L.S. ModelFinder: Fast model selection for accurate phylogenetic estimates. *Nat. Methods* **2017**, *14*, 587–589. [CrossRef]

30. Nguyen, L.T.; Schmidt, H.A.; von Haeseler, A.; Minh, B.Q. IQ-TREE: A fast and effective stochastic algorithm for estimating maximum likelihood phylogenies. *Mol. Biol. Evol.* **2015**, *32*, 268–274. [CrossRef]

31. Anisimova, M.; Gil, M.; Dufayard, J.F.; Dessimoz, C.; Gascuel, O. Survey of branch support methods demonstrates accuracy, power, and robustness of fast likelihood-based approximation schemes. *Syst. Biol.* **2011**, *60*, 685–699. [CrossRef]

32. Guindon, S.; Dufayard, J.F.; Lefort, V.; Anisimova, M.; Hordijk, W.; Gascuel, O. New Algorithms and Methods to Estimate Maximum-Likelihood Phylogenies: Assessing the Performance of PhyML 3.0. *Syst. Biol.* **2010**, *59*, 307–321. [CrossRef]

33. Hoang, D.T.; Chernomor, O.; von Haeseler, A.; Minh, B.Q.; Vinh, L.S. UFBoot2: Improving the ultrafast bootstrap approximation. *Mol. Biol. Evol.* **2018**, *35*, 518–522. [CrossRef] [PubMed]

34. Rambaut, A. FigTree 1.4.3. Available online: http://tree.bio.ed.ac.uk/software/figtree (accessed on 5 November 2020).

35. Kou, Q.; Gong, L.; Li, X. A new species of the deep-sea spongicolid genus *Spongicoloides* (Crustacea, Decapoda, Stenopodidea) and a new species of the glass sponge genus *Corbitella* (Hexactinellida, Lyssacinosida, Euplectellidae) from a seamount near the Mariana Trench, with a novel commensal relationship between the two genera. *Deep Sea Res. Part I* **2018**, *135*, 88–107. [CrossRef]

36. Xu, P.; Zhou, Y.; Wang, C. A new species of deep-sea sponge-associated shrimp from the North-West Pacific (Decapoda, Stenopodidea, Spongicolidae). *ZooKeys* **2017**, *685*. [CrossRef] [PubMed]

37. Sun, S.; Sha, Z.; Wang, Y. Complete mitochondrial genome of the first deep-sea spongicolid shrimp *Spongiocaris panglao* (Decapoda: Stenopodidea): Novel gene arrangement and the phylogenetic position and origin of Stenopodidea. *Gene* **2018**, *676*, 123–138. [CrossRef] [PubMed]

38. Mantelatto, F.L.; Terossi, M.; Negri, M.; Buranelli, R.C.; Robles, R.; Magalhaes, T.; Tamburus, A.F.; Rossi, N.; Miyazaki, M.J. DNA sequence database as a tool to identify decapod crustaceans on the São Paulo coastline. *Mitochondrial DNA Part A* **2018**, *29*, 805–815. [CrossRef] [PubMed]

39. Latreille, P.A. *Histoire Naturelle, Générale et Particulière des Crustacés et des Insectes. Ouvrage Faisant Suite à L'histoire Naturelle Générale et Particulière, Composée par Leclerc de Buffon, et Rédigée par C.S. Sonnini, Membre de Plusieurs Sociétés Savantes*; L'Imprimerie de F. Dufart: Paris, France, 1802; Volume 3, p. 476.

40. Schram, F.R. *Crustacea*; Oxford University Press: New York, NY, USA, 1986.

41. De Haan, W. Crustacea. In *Fauna Japonica Sive Descriptio Animalium, Quae in Itinere per Japoniam, Jussu et Auspiciis Superiorum, Qui Summum in India Batava Imperium Tenent, Suspecto, Annis 1823–1830 Collegit, Notis, Observationibus et Adumbrationibus Illustravit*; von Siebold, P.F., Ed.; Lugduni-Batavorum: Leiden, The Netherlands, 1833; pp. 1–243.

42. De Grave, S.; Fransen, C.H.J.M. Carideorum catalogus: The recent species of the dendrobranchiate, stenopodidean, procarididean and caridean shrimps (Crustacea: Decapoda). *Zool. Meded. Leiden* **2011**, *89*, 195–589.

43. Lévi, C.; Lévi, P. Spongiaires Hexactinellides du Pacifique Sud-Ouest (Nouvelle-Calédonie). *Bull. Mus. Natl. D'histoire Nat.* **1982**, *4*, 283–317.

44. Schulze, F.E. Über den Bau und das System der Hexactinelliden. *Abh. Königlichen Akad. Wiss. Berl. Phys. Math. Cl.* **1886**, 1–97.

45. Alcock, A. *A Descriptive Catalogue of the Indian Deep-Sea Crustacea Decapoda Macrura and Anomala, in the Indian Museum. Being a Revised Account of the Deep-Sea Species Collected by the Royal Indian Marine Survey Ship Investigator*; Trustees of the Indian Museum: Calcutta, India, 1901; p. 286.

46. Ijima, I. Notice of New Hexactinellida from Sagami Bay. *Zool. Anz.* **1896**, *19*, 249–254.

47. Reiswig, H.M.; Kelly, M. The marine fauna of New Zealand: Euplectellid glass sponges (Hexactinellida, Lyssacinosida, Euplectellidae). *NIWA Biodivers. Mem.* **2018**, *130*, 1–170.

48. Saito, T.; Tsuchida, S.; Yamamoto, T. *Spongicoloides iheyaensis*, a new species of deep-sea sponge-associated shrimp from the Iheya Ridge, Ryukyu Islands, Southern Japan (Decapoda: Stenopodidea: Spongicolidae). *J. Crustac. Biol.* **2006**, *26*, 224–233. [CrossRef]

49. Goy, J.W. *Spongicoloides galapagensis*, a new shrimp representing the first record of the genus from the Pacific Ocean (Crustacea: Decapoda: Stenopodidea). *Proc. Biol. Soc. Wash.* **1980**, *93*, 760–770.

50. Baba, K. *Spongicoloides hawaiiensis*, a new species of shrimp (Decapoda: Stenopodidea) from the Hawaiian Islands. *J. Crustac. Biol.* **1983**, *3*, 477–481. [CrossRef]

51. Burukosvsky, R.N. A description of the shrimp *Spongicoloides tabachnicki* (Decapoda, Spongicolidae) from the glass sponge *Euplectella jovis*. *Zool. Zhurnal* **2009**, *88*, 498–503. (In Russian)

52. Saito, T.; Okuno, J.; Chan, T. A new species of *Stenopus* (Crustacea: Decapoda: Stenopodidae) from the Indo-west Pacific, with a redefinition of the genus. *Raffles Bull. Zool. Suppl.* **2009**, *20*, 109–120.

53. Ahyong, S.T. Decapod crustacea of the Kermadec Biodiscovery Expedition 2011. In *Kermadec Biodiscovery Expedition 2011*; Trnski, T., Schlumpf, A., Eds.; Auckland Museum: Auckland, New Zealand, 2015; Volume 20, pp. 406–442.

54. Schnabel, K.E. The Marine Fauna of New Zealand. Squat lobsters (Crustacea: Decapoda: Chirostyloidea). *NIWA Biodivers. Mem.* **2020**, *132*, 1–351.

55. Claus, C. *Grundzüge der Zoologie. Zum Gebrauche an Universitäten und Höheren Lehranstalten sowie zum Selbststudium (Zweite Vermehrte Auflage)*; N.G. Elwert'sche Universitäts-Buchhandlung: Marburg/Leipzig, Germany, 1872; p. 1170.

56. Latreille, P.A. Salicoques, Carides, Latr. *Nouv. Dict. D'histoire Nat.* **1819**, *30*, 68–73.

57. Holthuis, L.B. Biological results of the Snellius expedition. XIV. The Stenopodidae, Nephropsidae, Scyllaridae and Palinuridae. *Temminckia* **1946**, *7*, 1–178.

58. Doak, W.T. *Beneath New Zealand Seas*; Reed: Wellington, New Zealand, 1971; p. 112.

59. Yaldwyn, J.C. Shrimps and prawns. *N. Z. Nat. Herit.* **1974**, *38*, 1041–1046.

60. De Saint Laurent, M.; Cleva, R. Crustacés Décapodes: Stenopodidea. In *Résultats des Campagnes MUSORSTOM: I. Philippines (18–28 Mars 1976). Série A, Zoologie*; Forest, J., Ed.; Mémoires du Muséum National d'Histoire Naturelle: Paris, France, 1981; Volume 91, pp. 151–188.

61. Goy, J.W. A new species of *Stenopus* from Australia, with a redescription of *Stenopus cyanoscelis* (Crustacea: Decapoda: Stenopodidae). *J. Nat. Hist. Lond.* **1992**, *26*, 79–102. [CrossRef]

62. Poore, G.C.B. *Marine Decapod Crustacea of Southern Australia. A Guide to Identification (with Chapter on Stomatopoda by Shane Ahyong)*; CSIRO Publishing: Melbourne, Australia, 2004; p. 574.

63. Yaldwyn, J.C.; Webber, R. Annotated checklist of New Zealand Decapoda (Arthropoda: Crustacea). *Tuhinga* **2011**, *22*, 171–272.

64. Webber, W.R.; Fenwick, G.D.; Bradford-Grieve, J.M.; Eagar, S.H.; Buckeridge, J.S.; Poore, G.C.B.; Dawson, E.W.; Watling, L.; Jones, B.; Wells, J.B.J.; et al. Subphylum Crustacea—Shrimps, crabs, lobsters, barnacles, slaters, and kin. In *New Zealand Inventory of Biodiversity Volume 2. Chaetognatha, Ecdysozoa, Ichnofossils*; Gordon, D.P., Ed.; Canterbury University Press: Christchurch, New Zealand, 2010; Volume 2, pp. 98–232.

65. Williams, A.B. *Shrimps, Lobsters, and Crabs of the Atlantic Coast of the Eastern United States, Maine to Florida*; Smithsonian Institution Press: Wasington, DC, USA, 1984; p. 550.

66. Manning, R.B.; Chace, F.A.J. Decapod and Stomatopod Crustacea from Ascension Island, South Atlantic Ocean. *Smithson. Contrib. Zool.* **1990**, *503*, 1–91. [CrossRef]

67. Goy, J.W. *Microprosthema emmiltum*, new species, and other records of stenopodidean shrimps from the Eastern Pacific (Crustacea: Decapoda). *Proc. Biol. Soc. Wash.* **1987**, *100*, 717–725.

68. Risso, A. *Histoire Naturelle des Principales Productions de l'Europe Méridionale et Particulièrement de Celles des Environs de Nice et des Alpes Maritimes*; F.G. Levrault: Paris, France, 1826. [CrossRef]

69. De Man, J.G. Bericht ueber die von Herrn Dr. J. Brock im indischen Archipel gesammelten Decapoden und Stomatopoden. *Arch. Naturges* **1888**, *53*, 215–600. [CrossRef]

70. Goy, J.W. Diagnosis of three new *Stenopus* species. In *Armoured Knights of the Sea*; Debelius, H., Ed.; Alfred Kernen Verlag: Essen, Germany, 1984; pp. 116–117.

71. Wainwright, B.J.; Arlyza, I.S.; Karl, S.A. Population genetics of the banded coral shrimp, *Stenopus hispidus* (Olivier, 1811), in the Indonesian archipelago. *J. Exp. Mar. Biol. Ecol.* **2020**, *525*, 151325. [CrossRef]

72. Dudoit, A.A.; Iacchei, M.; Coleman, R.R.; Gaither, M.R.; Browne, W.E.; Bowen, B.W.; Toonen, R.J. The little shrimp that could: Phylogeography of the circumtropical *Stenopus hispidus* (Crustacea: Decapoda), reveals divergent Atlantic and Pacific lineages. *PeerJ* **2018**, *6*, e4409. [CrossRef]

73. Bracken-Grissom, H.D.; Felder, D.L.; Vollmer, N.L.; Martin, J.W.; Crandall, K.A. Phylogenetics links monster larva to deep-sea shrimp. *Ecol. Evol.* **2012**, *2*, 2367–2373. [CrossRef] [PubMed]

74. Alcock, A.; Anderson, A.R.S. Natural history notes from H.M. Royal Indian Marine Survey Steamer "Investigator", commander C.F. Oldham, R.N., commanding.—Series II, No. 14. An account of a recent collection of deep-sea Crustacea from the Bay of Bengal and Laccadive Sea. *J. Asiat. Soc. Bengal Nat. Hist.* **1894**, *63*, 141–185.

75. Holthuis, L.B. The Decapoda of the Siboga Expedition. Part XI. The Palaemonidae collected by the Siboga and Snellius Expeditions, with remarks on other Species. II. Subfamily Pontoniinae. *Siboga Exped.* **1952**, *39*, 1–253.

76. Oken, L. *Lehrbuch der Naturgeschichte. Dritter Theil: Zoologie. Erste Abtheilung: Fleischlose Thiere*; C.H. Reclam & Jena: A. Schmid: Leipzig, Germany, 1815; p. 842.

77. Bruce, A.J. Notes on some Indo–West Pacific Pontoniinae. XL. The rediscovery of *Periclimenes lifuensis* Borradaile, 1898 (Decapoda, Pontoniine) and the establishment of its systematic position. *Crustaceana* **1982**, *42*, 158–173. [CrossRef]

78. Marin, I.; Anker, A. A partial revision of the *Philarius gerlachei* (Nobili, 1905) species complex, with description of four new species from the Indo–West Pacific. *Zootaxa* **2011**, *2781*, 1–28. [CrossRef]

79. Mitsuhashi, M. A new species of the genus *Philarius* (Crustacea: Decapoda: Palaemonidae) from Ryukyu Islands, Japan. *Zootaxa* **2012**, *3481*, 82–88. [CrossRef]

80. Kou, Q.; Li, X. A new palaemonid shrimp of the "*Philarius gerlachei* (Nobili, 1905) species complex" (Crustacea: Decapoda: Palaemonidae) from Hainan Island, South China Sea. *Raffles Bull. Zool.* **2016**, *64*, 269–277.

81. Rathbun, M.J. Investigations of the Aquatic Resources and Fisheries of Porto Rico by the United States Fish Commission Steamer Fish Hawk in 1899. The Brachyura and Macrura of Porto Rico. *Bull. U. S. Fish Comm.* **1901**, *20*, 1–127.

82. Hurt, C.; Silliman, K.; Anker, A.; Knowlton, N. Ecological speciation in anemone-associated snapping shrimps (*Alpheus armatus* species complex). *Mol. Ecol.* **2013**, *22*, 4532–4548. [CrossRef]

83. Bruce, A.J.; Chace, F.A.J. *Paralebbeus zotheculatus*, n. gen, n sp., a new hippolytid shrimp from the Australian northwest shelf. *Proc. Biol. Soc. Wash.* **1986**, *99*, 237–247.

84. Gray, J.E. Notes on the Arrangement of Sponges, with the Descriptions of some New Genera. *Proc. Zool. Soc. Lond.* **1867**, *1867*, 492–558.

85. Gray, J.E. Notes on the Classification of the Sponges. *Ann. Mag. Nat. Hist.* **1872**, *9*, 442–461. [CrossRef]

86. Chace, F.A.J. The caridean shrimps (Crustacea: Decapoda) of the Albatross Philippine expedition, 1907–1910, Part 7: Families Atyidae, Eugonatonotidae, Rhynchocinetidae, Bathypalaemonellidae, Processidae and Hippolytidae. *Smithson. Contrib. Zool.* **1997**, *587*, 1–106.
87. Xu, P.; Liu, F.; Ding, Z.; Wang, C. A new species of the thorid genus *Paralebbeus* Bruce & Chace, 1986 (Crustacea: Decapoda: Caridea) from the deep sea of the Northwestern Pacific Ocean. *Zootaxa* **2016**, *4085*, 119–126. [CrossRef] [PubMed]
88. Kou, Q.; Xu, P.; Poore, G.C.B.; Li, X.; Wang, C. A New Species of the Deep-Sea Sponge-Associated Genus *Eiconaxius* (Crustacea: Decapoda: Axiidae), With New Insights Into the Distribution, Speciation, and Mitogenomic Phylogeny of Axiidean Shrimps. *Front. Mar. Sci.* **2020**, *7*, 469. [CrossRef]
89. Schupp, P.J.; Rohde, S.; Versluis, D.; Petersen, L.-E.; Clemens, T.; Conrad, K.P.; Mills, S.; Kelly, M. Section 7.14. Investigations on the biodiversity of benthic sponge and invertebrate communities and their associated microbiome. In *Functional Diversity of Bacterial Communities and the Metabolome in the Water Column, Sediment and in Sponges in the Southwest Pacific around New Zealand RV SONNE SO254 Cruise Report/Fahrtbericht*; Simon, M., Ed.; University of Oldenburg: Oldenburg, Germany, 2017; pp. 57–59.

Article

A Mysterious World Revealed: Larval-Adult Matching of Deep-Sea Shrimps from the Gulf of Mexico

Carlos Varela * and Heather Bracken-Grissom

Institute of Environment, Department of Biological Sciences, Florida International University, North Miami, FL 33181, USA; hbracken@fiu.edu
* Correspondence: cvare015@fiu.edu

Abstract: The identification of deep-sea (>200 m) pelagic larvae is extremely challenging due to the morphological diversity across ontogeny and duration of larval phases. Within Decapoda, developmental stages often differ conspicuously from their adult form, representing a bizarre and mysterious world still left to be discovered. The difficulties with sampling and rearing deep-sea larvae, combined with the lack of taxonomic expertise, argues for the use of molecular methods to aid in identification. Here, we use DNA barcoding combined with morphological methods, to match larval stages with their adult counterpart from the northern Gulf of Mexico and adjacent waters. For DNA barcoding, we targeted the mitochondrial ribosomal large subunit 16S (16S) and the protein coding cytochrome oxidase subunit 1 (COI). These data were combined with previous sequences to generate phylogenetic trees that were used to identify 12 unknown larval and two juvenile species from the infraorder Caridea and the suborder Dendrobranchiata. Once identified, we provide taxonomic descriptions and illustrations alongside the current state of knowledge for all families. For many groups, larval descriptions are missing or non-existent, so this study represents a first step of many to advance deep-sea larval diversity.

Keywords: DNA barcoding; Gulf of Mexico; Caridea; Dendrobranchiata; Decapoda; larval-adult matching; life history

Citation: Varela, C.; Bracken-Grissom, H. A Mysterious World Revealed: Larval-Adult Matching of Deep-Sea Shrimps from the Gulf of Mexico. *Diversity* **2021**, *13*, 457. https://doi.org/10.3390/d13100457

Academic Editors: Michael Wink, Patricia Briones-Fourzán and Michel E. Hendrickx

Received: 16 August 2021
Accepted: 15 September 2021
Published: 23 September 2021

1. Introduction

In order to understand the evolution, distribution and ecology of marine organisms, as well as their impact on community and ecosystem processes, it is important to study their life history and developmental biology [1–3]. Decapod crustaceans, including shrimps, lobsters and crabs and are well-known due to their economic importance in the food, aquarium and pharmaceutical industries [4,5]. However, much less is known about their often-complex life histories. Decapods have numerous reproductive strategies, and those with sexual reproduction produce eggs which are either deposited directly in the bottom of the sea floor, remain attached to the parents, or are released as free moving organisms into the pelagic environment [6]. Many species progress through a series of larval stages (i.e., nauplius, mysis, zoea, phyllosoma), often representing bizarre forms unidentifiable from their adult counterpart [7] (Figures 1 and 2). The duration of the larval stages varies between and within taxonomic groups, sometimes lasting several months before settling as juveniles or benthic adults [8–11]. Due to the morphological disparity across ontogeny and duration of larval phases, the identification of planktonic decapod larvae, especially those in the deep sea (>200 m), is extremely challenging.

Descriptions of decapod larval stages are limited, with most of the preexisting literature focused on shallow-water species of economic interest because of their food and/or ornamental value [12–14]. For example, in the Gulf of Mexico, larvae stages are known from the shrimp family Penaeidae [15–17], the crab families Menippidae (stone crabs) and Portunidae (swimming crabs) [18–21] and the spiny lobster family Palinuridae [22]. In the last decade, additional papers have been published for decapod larval stages in the Gulf of Mexico [23–25], however more studies are needed.

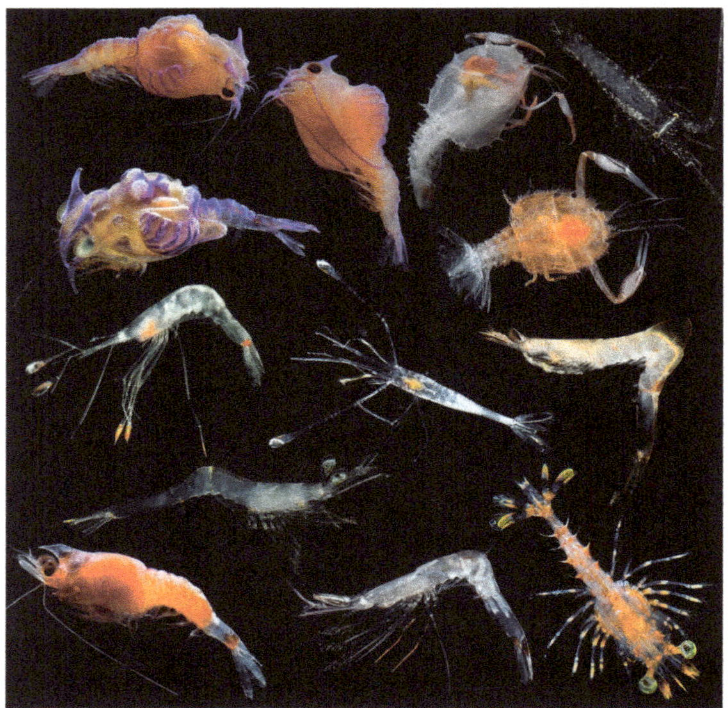

Figure 1. Examples of shrimp and lobster developmental stages collected on deep-pelagic research cruises in the northern Gulf of Mexico. ©DantéFenolio DEEPEND | RESTORE.

Figure 2. Examples of crab developmental stages collected on deep-pelagic research cruises in the northern Gulf of Mexico. ©DantéFenolio DEEPEND | RESTORE.

Our knowledge of pelagic or benthic deep-sea decapod larvae is inadequate or even non-existent and is further complicated by the technological demands and expense of sampling in deep oceanic waters. Extensive knowledge of taxonomy is required to achieve reliable larval identifications, and because this requires specialized training and years of practice, most researchers have difficulty recognizing larval stages in a plankton sample [26,27], especially those in the deep sea [28]. Those that have been identified come from larval-rearing experiments of females, and because males and females differ dramatically in larval morphology, several have been incorrectly identified [26,29]. Another factor that complicates identification is that the literature can be very old and difficult to access [7,29], however adequate library resources can aleviate this problem. Due to the abovementioned reasons, illustrated guides (based on external morphological characters that can be observed under a stereomicroscope) are necessary to aid future investigations and identifications, especially for those with limited taxonomic training.

Morphological descriptions can be given alongside molecular methods (DNA barcoding) to fully characterize and document larval-adult linkages. DNA barcoding is a molecular method for fast and accurate species identification and can be particularly useful in early life stages that differ conspicuously from their adult form [30,31]. Although rearing experiments have facilitated the taxonomic identification of larvae from plankton samples, most are difficult (or impossible) to breed and maintain in the laboratory. Molecular approaches, such as DNA barcoding, can be an excellent alternative or complementary method for larval identifications [32–35]. This method does require a reliable database of adult barcodes that are linked to vouchered museum specimens in zoological collections. When these adult datasets are available, larvae can be targeted from similar localities (or a species distributional range) and matched back to adults using DNA barcoding genes (ex. 16S and COI) and phylogenetic trees. A very recent barcoding study on adult deep-pelagic crustaceans was conducted in the Gulf of Mexico and adjacent waters [35], and we plan to use this dataset (alongside previously published datasets) to match unknown larvae collected on research expeditions into the northern Gulf of Mexico and adjacent waters over the past 5 years.

Adult-larval linkages are critical because they can enhance our basic biological understanding of the species under study. First, documenting and describing larval stages allows for the correct identification of a species during development. The correct identification of a species is arguably the most important first step to any scientific investigation. Secondly, larval-adult linkages have allowed for the description of complex life cycles and distributional ranges for many species [36–38]. An example is the deep-sea shrimp, *Cerataspis monstrosus* Gray, 1828, which can be found in the abyssal plains (up to 5000 m in the Gulf of Mexico) but has a larval form (*Cerataspis*-"monster" larvae) found in the mesopelagic (~500 m) [39]. Lastly, the correct identification and distribution of larvae is critical to understanding the food web dynamics in the Gulf of Mexico, as crustacean larvae are often the main food source for small and large migratory fishes, cephalopods and some marine mammals [40–43]. Overall, these adult-larval linkages do not only allow for advancements in taxonomy and systematics, but also provide fundamental information for studies in ecology and evolution.

In this paper we will use a molecular technique, namely DNA barcoding, to match early-life stages with their adult counterpart in an effort to better understand the life history and distribution of deep-sea (~200–1500 m) decapod crustaceans from the northern Gulf of Mexico and adjacent waters. We provide larval-adult matching for 14 species (12 larval, 2 juvenile) based on DNA barcoding and phylogenetic methods. For each species, detailed morphological illustrations and taxonomic descriptions of diagnostic characters are provided. Of the 14 species in this paper, only four have some previous larval knowledge: *Heterocarpus ensifer*, of which only the early four zoeal stages are known [44,45], *Plesionika edwardsii*, of which the seven first zoea stages are known [46], *Funchalia villosa*, of which some taxonomic data on its postlarva is known [47] and *Cerataspis monstrosus* of which some of the mysis stages are known [48]. We hope this research can guide future studies and aid in the identification of deep-sea crustacean larvae from the Gulf of Mexico.

2. Materials and Methods

2.1. Sample Collection

All material used in this study was collected during eight research expeditions totaling ~126 days at sea (Supplementary Table S1). Six of the eight research cruises were in the Gulf of Mexico on the R/V Point Sur as part of the Deep Pelagic Nekton Dynamics of the Gulf of Mexico (DEEPEND) consortium (http://www.deependconsortium.org, accessed on 14 September 2021). The other two cruises were in the Florida Straits on the R/V Walton Smith as part of a National Science Foundation grant to study bioluminescence and vision in the deep sea. During the DEEPEND cruises, every collection site was sampled during the day (entire water column from the surface to 1500 m depth, sampled at noon) and at night (surface to 1500 m depth, sampled at midnight). Sampling occurred during the wet (August) and dry (May) seasons from 2015 to 2016 and one during the dry (May) season from 2017–2018. Gulf of Mexico samples were collected with a Multiple Opening/Closing Net and Environmental Sensing System (MOC-10) composed of six 3 mm mesh nets, allowing for collected specimens to be assigned to a depth bin (0–200 m, 200–600 m, 600–1000 m, 1000–1200 m, and 1200–1500 m; the sixth net sampled from 0 to 1500 m). Samples from all nets and depths were included as part of this study. More details on DEEPEND net sampling and methods can be found in [49]. Florida Straits samples were collected with a 9 m^2. Tucker trawl fitted with a cod-end capable of closure at depth (for details see [50]), allowing for discrete depth sampling. All sampling was conducted in the midwater, from 0–800 m.

The contents of each net were placed in a large tray and crustacean larvae were sorted and preserved as whole-specimens, either in 80% EtOH or an RNA-stabilizing buffer (RNAlater) and stored at −20 °C onboard the vessel. Upon returning samples to the lab, all batch-stored individuals were transferred to the Florida International Crustacean Collection (FICC). All individuals selected for DNA barcoding were then given a unique voucher ID in the FICC database, including all relevant collection metadata. Metadata included collection date, time (day or night), collection locality and GPS coordinates, and depth. The unique voucher number ensured that the resulting DNA barcode matches to one and only one individual. Total genomic DNA was extracted from muscle tissue of the abdomen or the 3rd to 5th pleopod. Tissue collected from each vouchered specimen was stored in 80% EtOH at −20 °C and voucher specimens were preserved in 80% EtOH and deposited in the FICC.

We adopt the terminology of [51] for Dendrobranchiata and [52] for Caridea, to standardize the different life stages. The number of specimens examined per stage (N) is referred in each description. Measurements taken were Carapace length (CL), measured from the tip of rostrum to the posterior margin of the carapace and Total length (TL), corresponding to the distance from the tip of the rostrum to the posterior end of telson.

2.2. Molecular Analyses

2.2.1. DNA Extraction, PCR and Sequencing

Total genomic DNA (gDNA) was extracted from muscle tissue of the abdomen or the 3rd to 5th pleopod using DNeasy® Blood and Tissue Kits (Qiagen, Valencia, CA, USA). When the tissue did not completely digest, 10 µL of 10% DTT and an additional 10 µL Proteinase K were added, and samples were incubated until complete digestion was achieved. Visualization of total genomic DNA was performed using 2% agarose gels, run at 100 V for 90 min, and the DNA concentration was measured using a dsDNA HS Assay kit on the Qubit 2.0 Fluorometer (Invitrogen, Life Technologies, Carlsbad, CA, USA).

Two partial mitochondrial genes were selected due to their informativeness in decapod barcoding studies. These included the partial 16S large ribosomal subunit and cytochrome oxidase I (COI) gene, totalling ~550 basepairs (bps) and ~600 bps, respectively. All primers included M13 tails as a universal tag (Invitrogen, Carlsbad, CA, USA) (Table 1).

Table 1. The targeted genes, primer sequences and annealing temperatures used in this study.

Targeted Gene	Forward Primer	Reverse Primer	Anneal Temperature
16S	5′-TGCCTGTTTATCAAAAACAT-3′ 5′-CGCCTGTTTATCAAAAACAT-3′ [53]	5′-AGATAGAAACCAACCTGG-3′ [54]	45 °C
	5′-CGCCTGTTTAACAAAAACAT-3′ [55]	5′-CCGGTCTGAACTCAGATCACGT-3′ [55]	45 °C
COI	5′-GGTCAACAAATCACAAAGATATTG-3′ [56]	5′-TAAACTTCAGGGTGACCAAAAAATCA-′3 [56]	40 °C
	5′-YCAYAARGAYATTGG-3′ [35]	5′-GGRTGNCCRAARAAYCA-3′ [35]	45 °C

Polymerase chain reaction (PCR) using a thermal cycler (Pro-Flex PCR System) was used to amplify the 16S and COI gene regions. Thermal profiles were as follows: initial denaturing for 2–5 min at 94 °C; annealing for 35–40 cycles: 30–45 s at 94/95 °C, 30 s at 38–50 °C (depending on the taxon and primers used; see Table 1), 1 min at 72 °C; final extension 2–3 min at 72 °C. Both forward and reverse strands were amplified, and all PCR products were sent to GENEWIZ (South Plainfield, NJ, USA) for sequencing. Consensus sequences were generated within Geneious 9.1.7 (Biomatters Ltd., Newark, NJ, USA) and primer regions and non-readable segments at the beginning of the sequences were manually removed prior to multiple sequence alignment. To check for pseudogenes, all six possible reading frames for the COI gene were translated to ensure stop codons were not present. On several occasions, several individuals of the same species were included to help identify contamination. All obtained sequences were deposited in the GenBank database (Supplementary Table S1).

2.2.2. Phylogenetic Tree Construction

Newly generated larval sequences were aligned with a subset of data generated in [35] alongside other sequences from previously published studies (Supplementary Table S1) to help identify the unknow larvae. The Multiple Sequence Alignment Tool (MAFFT) with the E-INS-i algorithm [57] was used to align the DNA sequences. ModelFinder [58] was used to determine the model of evolution that best fit each gene. Maximum Likelihood (ML) analyses were conducted using IQ_TREE 2.0.4 [59] and a search for the best-scoring tree with 1000 replicates [60] was performed. Ultrafast Bootstrapping (UFBoot) was used to assess confidence in the resulting topologies. Bayesian Inference (BI) analyses were performed using parameters identified by ModelFinder and conducted in MrBayes (v.3.2.6) [61]. Both single-gene trees (16S and COI) and concatenated trees (16S + COI) were constructed for each major group using ML and BI approaches. Trees were visualized in FigTree v.1.4.2 and topologies were compared across all phylogenies for congruence. All support values (UFBoot and posterior probabilities) are listed on the corresponding branch. High support is indicated by values >95.

3. Results

3.1. Larval-Adult Identification using DNA-Barcoding

Phylogenetic trees were constructed to help in identification and evolutionary relationships should not be inferred based on these findings. In total, 28 larval individuals were included in this study. Our DNA barcoding efforts resulted in a total of 25 de novo 16S sequences and nine de novo COI sequences from these larvae. Using a subset of the dataset generated from [35] and previous studies, in combination with these newly generated larval sequences (Supplementary Table S1), the final tree (16S + COI) included 51 total species from the infraorder Caridea and suborder Dendrobranchiata (Figure 3).

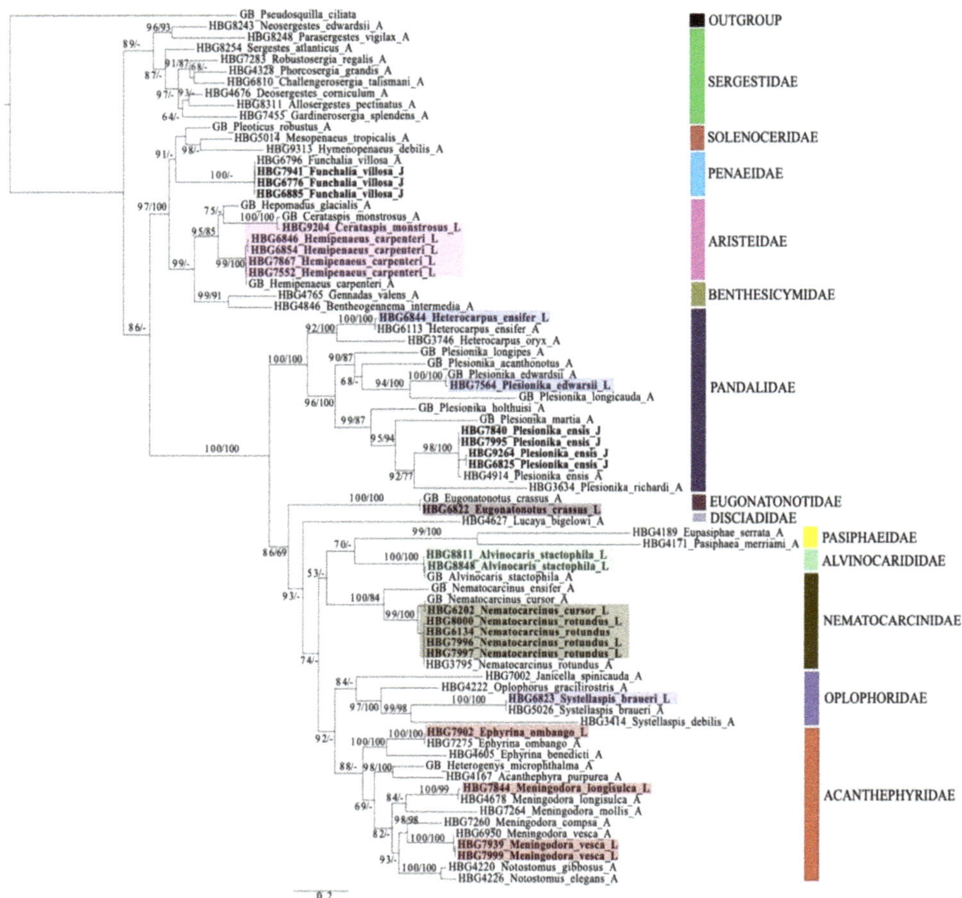

Figure 3. Maximum likelihood (ML) phylogeny of 51 barcoded individuals from the infraorder Caridea and suborder Dendrobranchiata based on the mitochondrial genes, 16S and COI. The number along the branches represent ultrafast bootstrap support (UFboot) values and Bayesian posterior probabilities (pp), respectively. UFBoot and pp values >95 indicate strong support. Voucher numbers (HBG#) represent specimens in the Florida International Crustacean Collection (FICC) and GB represents GenBank sequences. Family names are listed along the vertical bars. A = adult representative, J = juvenile representative and L = larval representative. Highlighted individuals represent the larvae matched with their adult counterpart.

Using this phylogeny, we were able to successfully match 14 larval and juvenile species (=16 developmental stages) with their adult counterparts. From the infraorder Caridea, the larvae represented six families, eight genera and eleven species. From the suborder Dendrobranchiata, the larvae represented two families, three genera and three species. The families of larval carideans identified included Acanthephyridae Spence Bate, 1888, Alvinocarididae Christoffersen, 1986, Eugonatonotidae Chace, 1937, Nematocarcinidae Smith, 1884, Pandalidae Haworth, 1825, and Oplophoridae Dana, 1852. The families of larval dendrobranchiates included Penaeidae and Aristeidae. Overall, the 14 larval and juvenile species that were successfully matched to their adult counterpart include *Alvinocaris stactophila* Williams, 1988, *Eugonatonotus crassus* (A. Milne-Edwards, 1881), *Systellaspis debilis* (A. Milne-Edwards, 1881), *Nematocarcinus cursor* A. Milne-Edwards, 1881,

N. *rotundus* Crosnier and Forest, 1973 *Plesionika edwarsii* (J.F. Brandt in von Middendorf, 1851), *P. ensis* (A. Milne-Edwards, 1881), *Heterocarpus ensifer* A. Milne-Edwards, 1881, *Meningodora vesca* (Smith, 1886), *M. longisulca* Kikuchi, 1985 and *Ephyrina ombango* Crosnier and Forest, 1973 from Caridea and *Funchalia villosa* Bouvier, 1905, *Hemipenaeus carpenteri* Wood-Mason in Wood-Mason and Alcock, 1891 and *Cerataspis monstrosus* Gray, 1868 from Dendrobranchiata. Single-gene trees for 16S and COI genes are provided as Supplementary Materials (Supplementary Figures S1 and S2).

3.2. Larval Morphology

Acanthephyridae Spence Bate, 1888
Meningodora Smith, 1882
Meningodora longisulca Kikuchi, 1985
(Figure 4)

Material examined: Gulf of Mexico: HBG 7844, R/V Point Sur, DP05-09May17-MOC10-B175N-095-N3, 28. 95125 and −87.91466, 09 May 2018, 6–1451 m, MOCNESS plankton net, L. Timm, coll.

Zoea. Size: 8 mm (Carapace length); 26 mm (Total length). N = 1.

Carapace (Figure 4A). Rostrum straight, reaching the end of the cornea, unarmed; epigastric spine present; eyes pedunculate.

Pleon (Figure 4A) with 6 somites, no spines or setae. Pleopods 1–4 missing in the specimen, pleopod 5 without setae.

Antennule (Figure 4B). Peduncle 3-segmented, article 1 the longest, slender, with 23 plumose setae; article 2 with 8 plumose setae and article 3 with 9 plumose setae and two flagella distally.

Antenna (Figure 4C). Protopod 3-segmented with a flagellum; exopod flattened with 73 plumose setae.

Mandible (Figure 4D) without mandibular palp; incisor with 7 terminal teeth.

Maxillule (Figure 4E). Coxal endite with 5 simple setae; basial endite with 15 (10 simple setae plus 5 conical setae) and protopod with one simple setae.

Maxilla (Figure 4F). Coxal endite with 6 simple setae; basial endite bilobed with 3 + 4 simple setae; endopod with 2 (1 + 1) simple setae; scaphognathite (damage in the specimen) margin with 26 plumose setae.

First maxilliped (Figure 4G). Coxa with 7 simple setae; basis with 28 simple setae; endopod unsegmented with 3 (2 + 1) plumose setae; exopod unsegmented with 35 plumose setae.

Second maxilliped (Figure 4H). Coxa with one simple setae; basis with 3 simple setae; endopod 5-segmented with 0, 1, 0, 1, 1 simple setae; exopod missing in the specimen.

Third Maxilliped (Figure 4I). Coxa and basis without setae; endopod 4-segmented with 0, 0, 0, 12, simple setae; exopod missing in the specimen.

First to fifth Pereopods missing in the specimen.

Uropod (Figure 4J). Endopod well developed with 53 plumose setae; exopod, slightly wider than endopod, with 80 plumose setae.

Telson (Figure 4K) elongate, subtriangular, armed with two pairs of dorsolateral spines close to the posterior margin. Posterior margin with a pointed projection, armed with two principal spines in each corner.

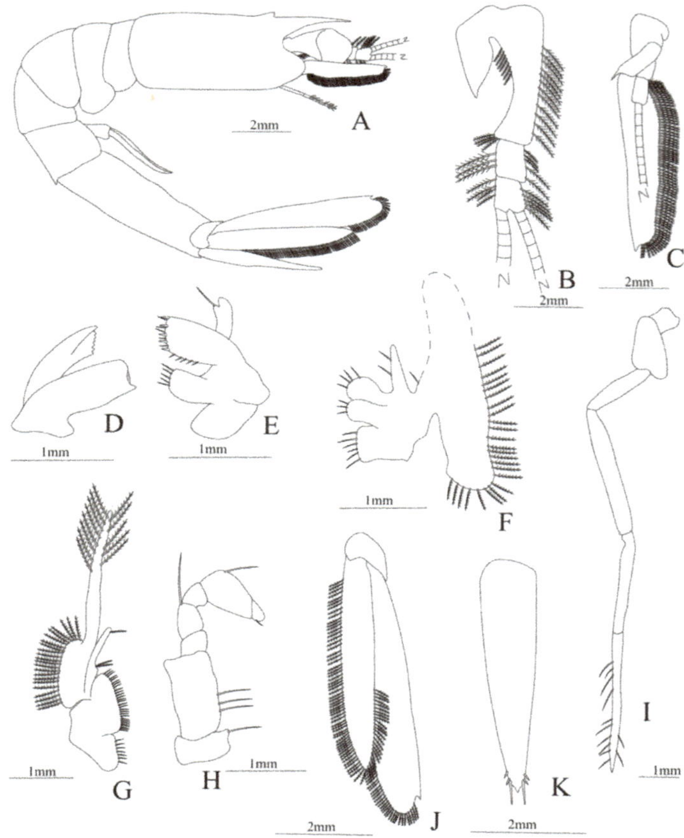

Figure 4. *Meningodora longisulca*: (**A**) lateral view; (**B**) antennule; (**C**) antenna, (**D**) mandible; (**E**) maxillule; (**F**) maxilla; (**G**) first maxilliped; (**H**) second maxilliped; (**I**) third maxilliped; (**J**) uropods; (**K**) telson.

Meningodora vesca (Smith, 1886)
(Figures 5 and 6)

Material examined: Gulf of Mexico, HBG 7939, R/V Point Sur, DP05-08May17-MOC10-B003D-092-N4, 27. 9271 and −87.0178, 8 May 2017, 600–400 m, MOCNESS plankton net, L. Timm, coll. Gulf of Mexico, HBG 7999, R/V Point Sur, DP05-03May17-MOC10-B065N-087-N3, 28.53128 and −88.0236, 3 May 2017, 1000–600 m, MOCNESS plankton net, L. Timm, coll.

Decapodite. Size. 14 mm (Carapace length); 43 mm (Total length). N = 2.

Carapace (Figure 5A). Rostrum slightly beyond the cornea and armed with 8 dorsal and one ventral spines; strong branchiostegal spine; eyes pedunculate.

Pleon (Figure 5A) with 6 somites, no spines or setae. Pleopods 1–2 well developed, pleopods 3–5 missing in the specimen.

Antennule (Figure 5B). Peduncle 3-segmented, article 1 the longest, slender, with 12–16 plumose setae; article 2 with 5–6 plumose setae and article 3, subequal in size with article 2, with 11–15 plumose setae and two flagella distally.

Antenna (Figure 5C). Protopod 3-segmented (flagellum missing in the specimen); exopod flattened with 59–74 plumose setae.

Mandible (Figure 5D). Mandibular palp 3-segmented, armed with 2, 4, 3 simple setae; incisor with 7 terminal teeth.

Maxillule (Figure 5E). Coxal endite with 38 serrulated setae; basial endite with 16 conical setae and a subterminal simple seta; protopod unarmed.

Maxilla (Figure 5F). Coxal endite with 21 plumose setae; basial endite bilobed with 16 +19 serrulated setae; exopod with 5 plumose setae; scaphognathite (damage in the specimen) margin with 102 plumose setae.

First maxilliped (Figure 5G). Coxa with 2 plus 5 plumose setae; basis with 42–46 serrulated setae; endopod with 7 (2 + 3 + 2) plumose setae; exopod with 36–38 plumose setae.

Second maxilliped (Figure 5H). Coxa without setae; basis with 4–6 simple setae; endopod 5-segmented with 5–11 simple, 0–5, 3–5 simple, 4–12 simple, 9–11 plumose setae; exopod unsegmented and armed with 12–16 plumose setae.

Third maxilliped (Figure 5I). Coxa without setae; basis with 3 simple setae; endopod 3-segmented with 33 simple, 10 simple, 21 (7 simple + 14 plumose) setae; exopod unsegmented and armed with 15 plumose setae.

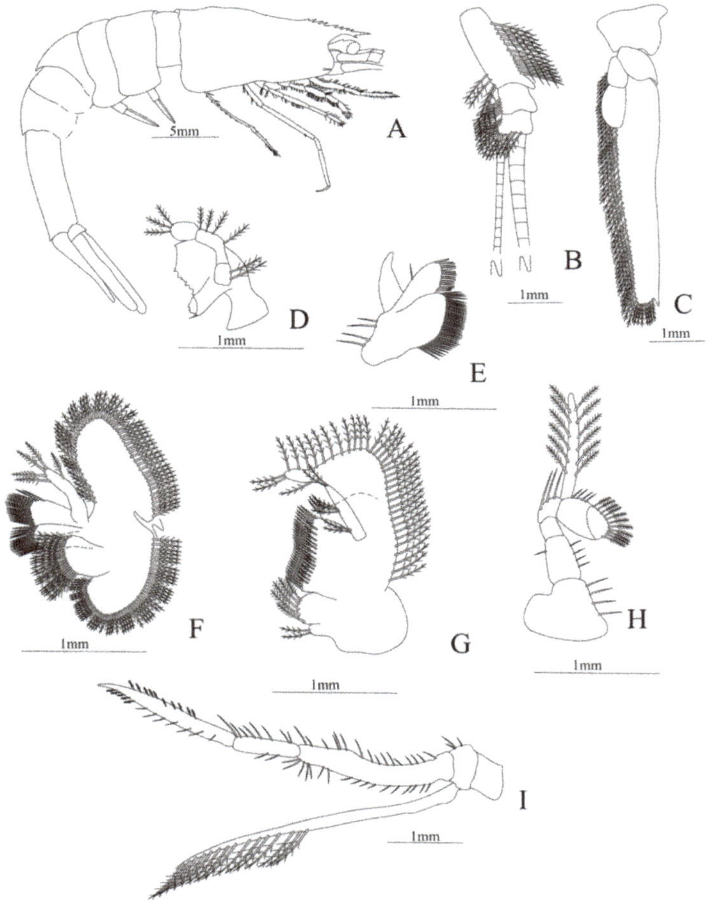

Figure 5. *Meningodora vesca*: (**A**) lateral view; (**B**) antennule; (**C**) antenna; (**D**) mandible; (**E**) maxillule; (**F**) maxilla; (**G**) first maxilliped; (**H**) second maxilliped; (**I**) third maxilliped.

First pereopod (Figure 6A). Coxa with 7–9 simple setae; basis with 4 simple setae; endopod 5-segmented with 10 (5 plumose plus 5 simple), 14–29 simple, 7–13 plumose, 7–10 simple, 2–4 simple setae; exopod unsegmented and unarmed.

Second pereopod (Figure 6B). Coxa with 4 simple setae. Basis with 3 simple setae; endopod 5-segmented with 6, 12, 2, 10, 3 simple setae; exopod unsegmented with 5 simple setae.

Third pereopod (Figure 6C). Coxa with 3 simple setae. Basis with 5 simple setae; endopod 5-segmented with 4 (3 spines plus one simple seta), one spine, 0, 0, 0 setae; exopod unsegmented and unarmed.

Fourth pereopod missing in the specimen.

Fifth pereopod (Figure 6D). Coxa and basis with one simple seta each one; endopod 4-segmented with 7 (3 spines plus 4 simple setae), 4 spines, 2 simple setae, 19 (8 simple setae plus 11 plumose setae).

Uropod (Figure 6E). Endopod well developed with 53–65 plumose setae; exopod, slightly wider than endopod, with 80–82 plumose setae.

Telson (Figure 6F) Damaged in the specimen. Elongate, subtriangular, armed with 3 pairs of dorsolateral spines. Posterior margin with a pointed projection.

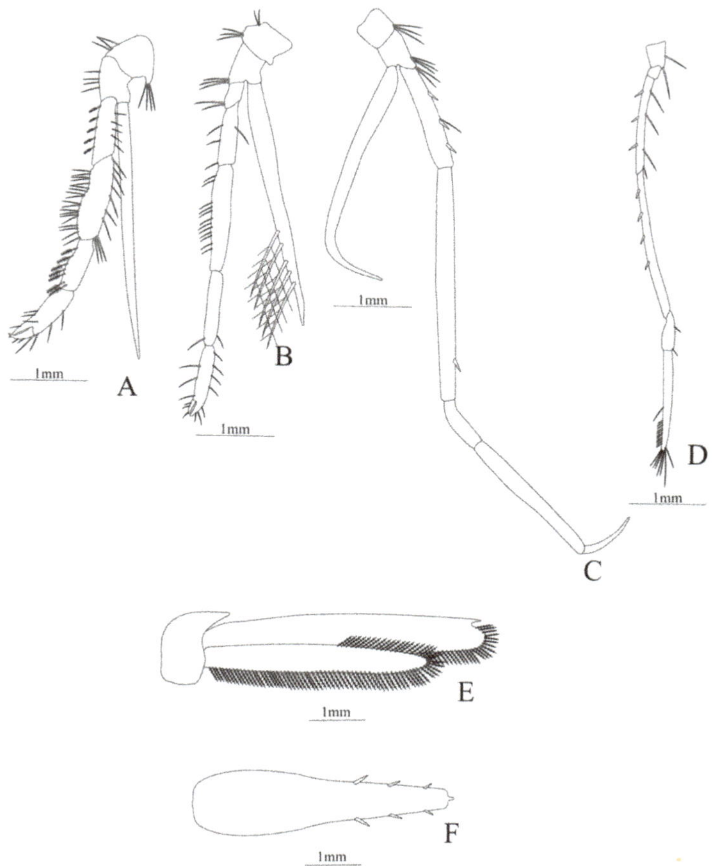

Figure 6. *Meningodora vesca*: (**A**) first Pereopod; (**B**) second Pereopod; (**C**) third Pereopod; (**D**) fifth Pereopod; (**E**) uropods; (**F**) telson.

Ephyrina Smith, 1885
Ephyrina ombango Crosnier and Forest, 1973
(Figure 7)

Material examined: Gulf of Mexico: HBG7902, R/V Point Sur, DP05-01May17-MOC10-B081D-084-N3, 28.5116, −87.0153, 1 May 2017, 1000–600 m, MOCNESS plankton net, L. Timm, coll.

Zoea. Size. 4 mm (Carapace length); 16 mm (Total length). N = 1.

Carapace (Figure 7A). Rostrum small, not reach the cornea, unarmed; anteroventral margin bearing small pterygostomian spine; eyes pedunculate.

Pleon (Figure 7A) with 6 somites, no spines or setae. Pleopods 1–3 missing in the specimen, pleopods 4–5 without setae.

Antennule (Figure 7B). Peduncle 3-segmented, article 1 the longest, slender, with 5 simple setae; article 2 also with 3 simple setae and article 3 with two flagella distally.

Antenna (Figure 7C). Protopod 2-segmented (flagellum missing in the specimen); exopod flattened with 46 plumose setae.

Mandible (Figure 7D,E). Mandibular palp 3-segmented, with 4, 1, 8 plumose setae; right incisor with 6 teeth and left incisor with 8 teeth.

Maxillule (Figure 7F). Coxal endite with 24 (10 plumose plus 14 serrulated) setae; basial endite with 18 conical serrulated setae and a subterminal simple setae; protopod with 4 simple setae.

Maxilla (Figure 7G). Coxal endite with 33 plumose setae; basial endite bilobed with 12 + 25 plumose setae; endopod with 5 (1 + 1 + 1 + 2) plumose setae; scaphognathite margin with 88 plumose setae.

First maxilliped (Figure 7H). Coxa with 16 plumose setae; basis with 42 plumose setae; endopod unsegmented with 1, 1, 1, 3, plumose setae; exopod unsegmented with 42 simple setae.

Second maxilliped (Figure 7I). Coxa with 4 plumose setae; basis with 12 plumose setae; endopod 5-segmented with 8, 1, 7, 11, 0 plumose setae, except in the article 4 where all the setae were serrulated; exopod unsegmented, armed distally with 2 plumose setae.

Third maxilliped (Figure 7J). (Damaged in the specimen) Coxa without setae; basis with 4 simple setae; endopod 4-segmented with 14, 23, 20, 7, plumose setae.

First to Fifth Pereopod missing in the specimen.

Uropod (Figure 7K) with rami subequal. Endopod (Damaged in the specimen) with 85 plumose setae; exopod, slightly wider than endopod, with 75 plumose setae.

Telson (Figure 7L) elongate, subtriangular, armed with 8 pairs of dorsolateral spines. Posterior margin armed with a terminal spine.

Alvinocarididae Christoffersen, 1986
Alvinocaris Williams and Chace, 1982
Alvinocaris stactophila Williams, 1988
(Figures 8 and 9)

Material examined: Gulf of Mexico: HBG 8811, R/V Point Sur, DP06-20Jul18-MOC10-B001D-101-N0, 28. 95125 and −87.91466, 29.01879 and −88.02719, 20 July 2018, 6–1451 m, MOCNESS plankton net, L. Timm coll; Gulf of Mexico: HBG 8848, R/V Point Sur, DP06-24Jul18-MOC10-B251N-106-N3, 28. 540167, −88.47116 and 28.5122, −88.6337, 24 July 2018, 602–1001 m, MOCNESS plankton net, L. Timm, coll.

Decapodite. Size. 7 mm (Carapace length); 19 mm (Total length). N = 2.

Carapace (Figure 8A). Rostrum straight, armed dorsally with 11–12 spines, longer than antennular peduncle; antennal spine small; anteroventral margin bearing small pterygostomian spine; eyes pedunculate.

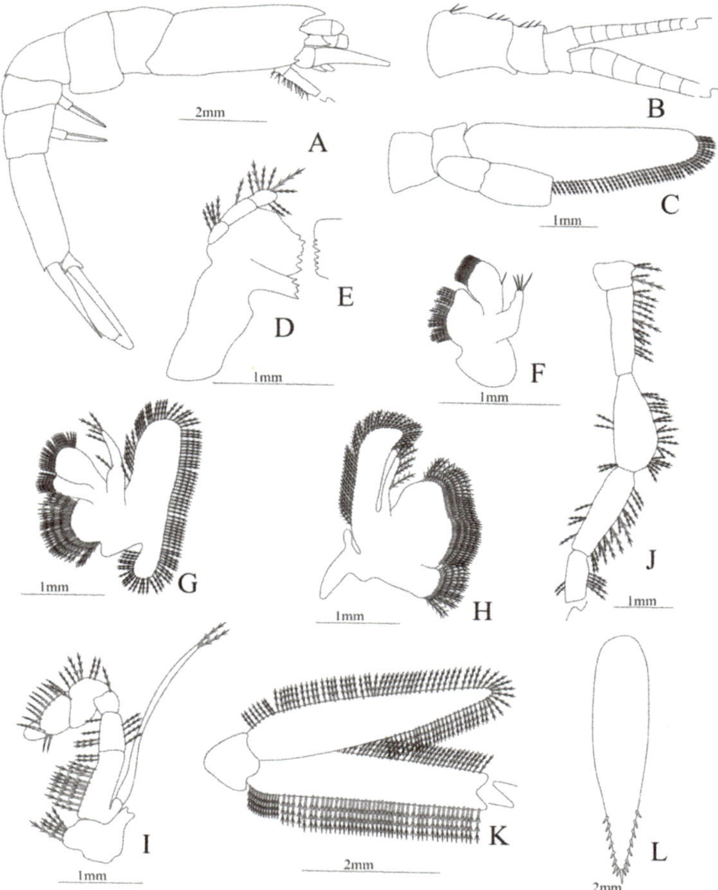

Figure 7. *Ephyrina ombango*: (**A**) lateral view; (**B**) antennule; (**C**) antenna; (**D**) left mandible; (**E**) right mandible (cutting edge); (**F**) maxillule; (**G**) maxilla; (**H**) first maxilliped; (**I**) second maxilliped; (**J**) third maxilliped; (**K**) uropods; (**L**) telson.

Pleon (Figure 8A) with 6 somites, no spines or setae. Pleopods 1–3 missing in the specimen, pleopods 4–5 well developed.

Antennule (Figure 8B). Peduncle 3-segmented, article 1 the longest, slender, article 2 also with plumose setae in both margins and article 3, the smallest, with two flagella distally. Flagella short, almost same size.

Antenna (Figure 8C). Protopod 3-segmented with a flagellum; exopod flattened with 63–65 plumose setae, endopod unarmed and unsegmented.

Mandible (Figure 8D). Mandibular palp 2-segmented, article 1 unarmed, article 2 with 4 simple setae; incisor with 5 terminal teeth.

Maxillule (Figure 8E). Coxal endite with 13 simple setae; basial endite with 11 simple setae and protopod with 6 setae (1 + 1 + 1 +2).

Maxilla (Figure 8F). Coxal endite with 21–22 simple setae; basial endite bilobed with 13 + 10 simple setae; endopod with 8 (3 + 1 + 2 + 2) plumose setae; scaphognathite margin with 116–120 plumose setae and 18–20 simple terminal long setae.

First maxilliped (Figure 8G). Coxa with 7–13 simple setae; basis with 28–29 plumose setae; endopod unsegmented with 1, 2, 1, 1, 2 (1 outer plus 1 terminal) plumose setae; exopod unsegmented with 27–31 simple setae.

Second maxilliped (Figure 8H). Coxa without setae; basis with 2 simple setae; endopod 5-segmented with 6–11, 1–3, 0–2, 0, 1–3 simple setae; exopod unsegmented, armed distally with 2–4 plumose natatory setae.

Third maxilliped (Figure 8I). Coxa without setae; basis with 4–5 simple setae; endopod 5-segmented with 2, 2, 5, 7, 2 simple setae; exopod unsegmented, armed distally with 6 plumose natatory setae.

First pereopod (Figure 8J). Coxa and basis without setae; endopod 5-segmented with 2, 2, 1, 6, 0 simple setae; exopod unsegmented, armed distally with 2–6 plumose natatory setae.

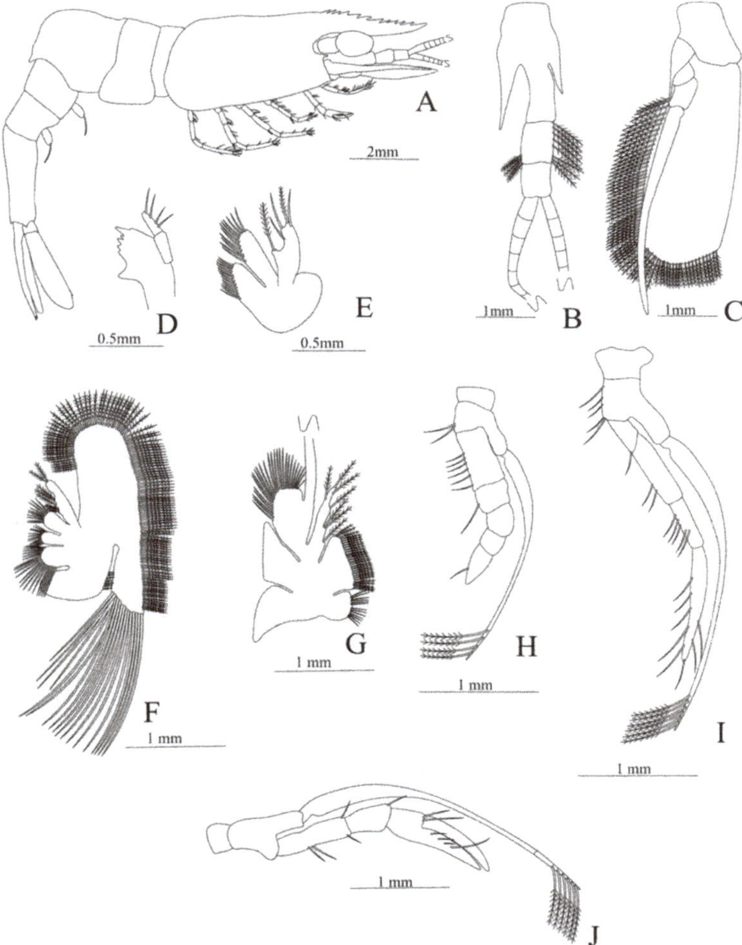

Figure 8. *Alvinocaris stactophila*: (**A**) lateral view; (**B**) antennule; (**C**) antenna; (**D**) mandible; (**E**) maxillule; (**F**) maxilla; (**G**) first maxilliped; (**H**) second maxilliped; (**I**) third maxilliped; (**J**) first pereopod.

Second pereopod (Figure 9A). Coxa without setae. Basis with 2 simple setae; endopod 5-segmented with 2, 6, 0, 7, 3 simple setae; exopod unsegmented, armed distally with 2–6 plumose natatory setae.

Third pereopod (Figure 9B). Coxa without setae. Basis with 2 simple setae; endopod 5-segmented with 3, 5, 1, 7, 0 simple setae; exopod unsegmented, armed distally with 1–8 long, plumose natatory setae.

Fourth pereopod (Figure 9C). Coxa without setae. Basis with 3 simple setae; endopod 5-segmented with 4, 5, 1, 6, 0 simple setae; exopod unsegmented, armed distally with 2–6 long, plumose natatory setae.

Fifth pereopod (Figure 9D). Coxa and basis unarmed; endopod 5-segmented with 4, 1, 1, 8, 0 simple setae.

Uropod (Figure 9E) with rami subequal. Endopod well developed with 54–58 plumose setae; exopod, slightly wider than endopod, with 64–68 plumose setae.

Telson (Figure 9F) elongate, subrectangular, armed with 4 pairs of dorsolateral spines. Posterior margin convex, armed with 2 principal spines in each corner and 6 small spines on distal margin between.

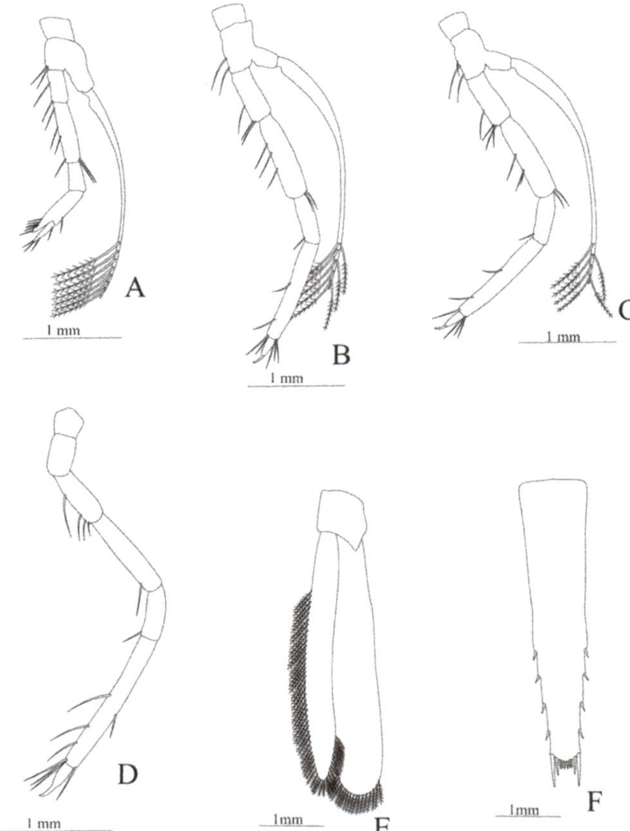

Figure 9. *Alvinocaris stactophila*: (**A**) second pereopod; (**B**) third pereopod; (**C**) fourth pereopod; (**D**) fifth pereopod; (**E**) uropods; (**F**) telson.

Eugonatonotidae Chace, 1937
Eugonatonotus Schmitt, 1926
Eugonatonotus crassus (A. Milne Edwards, 1881)
(Figures 10 and 11)

Material examined: Gulf of Mexico: HBG 6822, R/V Point Sur, DP04-08Aug16, MOC10-SE1N-063-N0, from 26.9878 and −87.9494 to 27.0591 and −88.0856, 8 August 2016, 1504.9-N/A m, MOCNESS plankton net, H. Bracken-Grissom, coll.

Zoea. Size. 6 mm (Carapace length); 19 mm (Total length). N = 1.

Carapace (Figure 10A). Rostrum short and unarmed; eyes pedunculate.

Pleon (Figure 10A) with 6 somites, no spines or setae. Pleopods 1–3 missing in the specimen, pleopods 4–5 without setae.

Antennule (Figure 10B). Peduncle 3-segmented, article 1 the longest, slender, with 17 plumose setae; article 2 with 6 plumose setae and article 3, subequal in size with article 2, with 6 plumose setae and two flagella distally, flagella short, almost same size.

Antenna (Figure 10C). Protopod 3-segmented with a flagellum; exopod flattened with 35 plumose setae.

Mandible (Figure 10D). Mandibular palp 3-segmented, article 1 and 2 unarmed, article 3 with 4 simple setae; incisor with 7 terminal teeth.

Maxillule (Figure 10E). Coxal endite with 6 simple setae; basial endite with 6 simple setae and protopod with 12 setae (2 + 2 + 12).

Maxilla (Figure 10F). (Damaged in the specimen) Coxa l without setae; basial endite with 16 simple setae; scaphognathite margin with 57 plumose setae.

First maxilliped (Figure 10G). Coxa with 4simple setae plus one plumose set; basis with 14 plumose setae; endopod 4-segmented with 7, 4, 7, 3, 2 simple setae, except the last article that bear 2 plumose and one simple setae; exopod unsegmented with 26 simple setae

Second maxilliped (Figure 10H). Coxa without setae; basis with 11 simple setae and 2 plumose setae; endopod 4-segmented with 6 simple, 13 simple, 2 simple, 10 (9 simple plus one plumose) setae; exopod unsegmented and unarmed.

Third maxilliped (Figure 10I). Coxa without setae; basis with 7 simple setae; endopod 5-segmented with 6, 8, 4, 17, 5 simple setae; exopod unsegmented and unarmed.

First pereopod (Figure 10J). Coxa and basis without setae; endopod 5-segmented with 2, 2, 1, 6, 0 simple setae; exopod unsegmented, armed distally with 6 plumose natatory setae.

Second pereopod (Figure 11A). Coxa without setae. Basis with one simple setae; endopod 5-segmented with 5, 8, 5, 16, 1 simple setae; exopod unsegmented and unarmed.

Third pereopod (Figure 11B). Coxa without setae. Basis with one simple setae; endopod 5-segmented with 6, 5, 1, 10, 1 simple setae; exopod unsegmented, armed distally with 15 simple setae.

Fourth pereopod (Figure 11C). Coxa and basis without setae; endopod 5-segmented with 1, 1, 3, 10, 4 simple setae; exopod unsegmented and unarmed.

Fifth pereopod (Figure 11D). Coxa without setae; basis with 3 simple setae and one plumose setae; endopod 5-segmented with 8, 11, 3, 10, 0 simple setae.

Uropod (Figure 11E) with rami subequal. Endopod well developed with 54 plumose setae; exopod, slightly wider than endopod, with 68 plumose setae.

Telson (Figure 11D) elongate, subtriangular. Posterior margin, armed with 2 principal spines in each corner and 6 small spines.

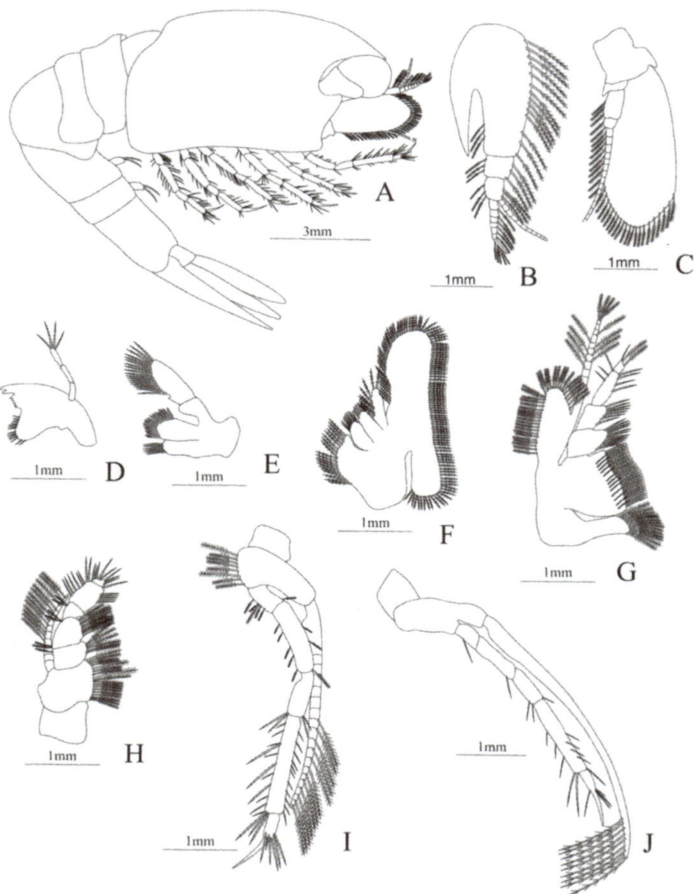

Figure 10. *Eugonatonotus crassus*: (**A**) lateral view; (**B**) antennule; (**C**) antenna; (**D**) mandible; (**E**) maxillule; (**F**) maxilla; (**G**) first maxilliped; (**H**) second maxilliped; (**I**) third maxilliped; (**J**) first pereopod.

Nematocarcinidae Smith, 1884
Nematocarcinus A. Milne-Edwards, 1881
Nematocarcinus cursor A. Milne-Edwards, 1881
(Figures 12 and 13)

Material examined: Florida Straits: HBG 6202, R/V Walton Smith, BLV01-19Jul16-STNB-D005, from 25.421423 and −79.648933 to 25.405617 and −79. 661217, 19 July 2016, 700–500 m, Trawl plankton net, H. Bracken-Grissom, coll.

Zoea. Size. 7 mm (Carapace length); 21 mm (Total length). N = 1.

Carapace (Figure 12A). Rostrum shorter than the cornea, armed dorsally with 5 spines, epigastric spine present; eyes pedunculate; pterygostomial spine present.

Pleon (Figure 12A) with 6 somites, no spines or setae. Pleopod 4 missing in the specimen, pleopods 1–2 and 4–5 without setae.

Antennule (Figure 12B). Peduncle 3-segmented, article 1 the longest, slender, with four pointed projections and with 16 plumose setae; article 2 with one plumose setae and article 3, subequal in size with article 2, with 8 plumose setae and two flagella distally, flagella almost same size.

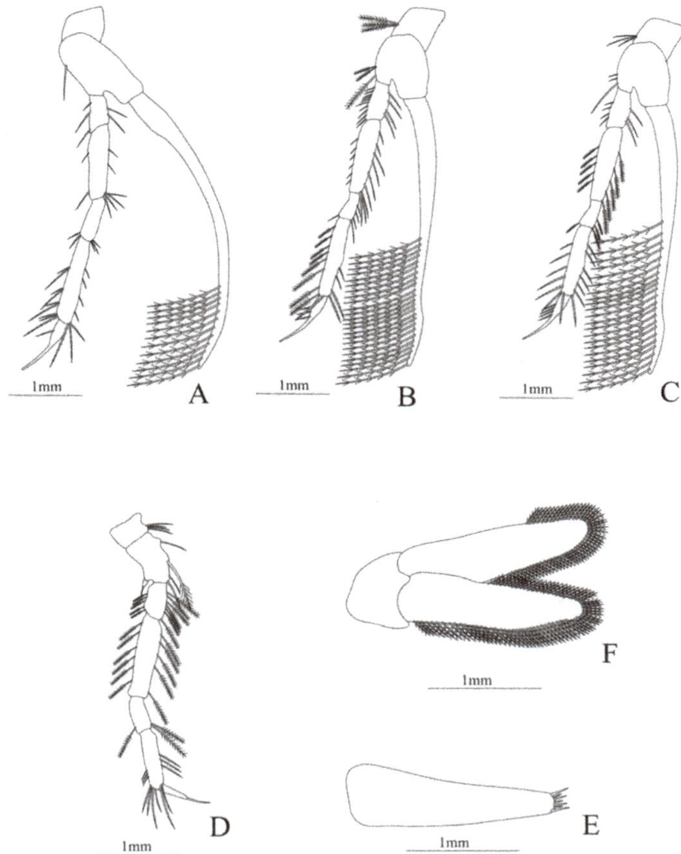

Figure 11. *Eugonatonotus crassus*: (**A**) second pereopod; (**B**) third pereopod; (**C**) fourth pereopod; (**D**) first pereopod; (**E**) uropods; (**F**) telson.

Antenna (Figure 12C). Protopod 3-segmented, segment 1 unarmed, segment 2 with two plumose setae, segment 3 with a flagellum; exopod flattened with 66 plumose setae.

Mandible (Figure 12D,E). Mandibular palp absent; left and right incisor with 3 terminal teeth.

Maxillule (Figure 12F). Coxal endite with 26 conical serrulated setae; basial endite with 11 simple setae and 13 conical serrulated setae; protopod with two articles, article 1 with two serrulated setae and article 2 with 6 serrulated setae.

Maxilla (Figure 12G). Coxa with 31 plumose setae; basial endite bilobed with 10 and 16 serrated setae respectively; scaphognathite margin with 127 plumose setae.

First maxilliped (Figure 12H). Coxa with 18 plumose setae; basis with 13 plumose and 17 serrulated setae; endopod 4-segmented with 6, 2, 2, 3, plumose setae, except the last segment that bear serrulated setae; exopod with 10 plumose setae.

Second maxilliped (Figure 12I). Coxa with 3 plumose setae; basis with 9 plumose setae; endopod 5-segmented with 5 plumose, 2 plumose, 1 plumose, 8 (5 plumose plus 3 serrulated), 5 serrulated setae; exopod unsegmented and unarmed.

Third maxilliped (Figure 12J). Coxa with 8 plumose setae; basis with 5 plumose setae; endopod 5-segmented with 4, 3, 3, plumose setae, one serrulated setae; last article subdivided in three small articles with 3, 2 and 2 serrulated setae; exopod armed with 10 plumose setae.

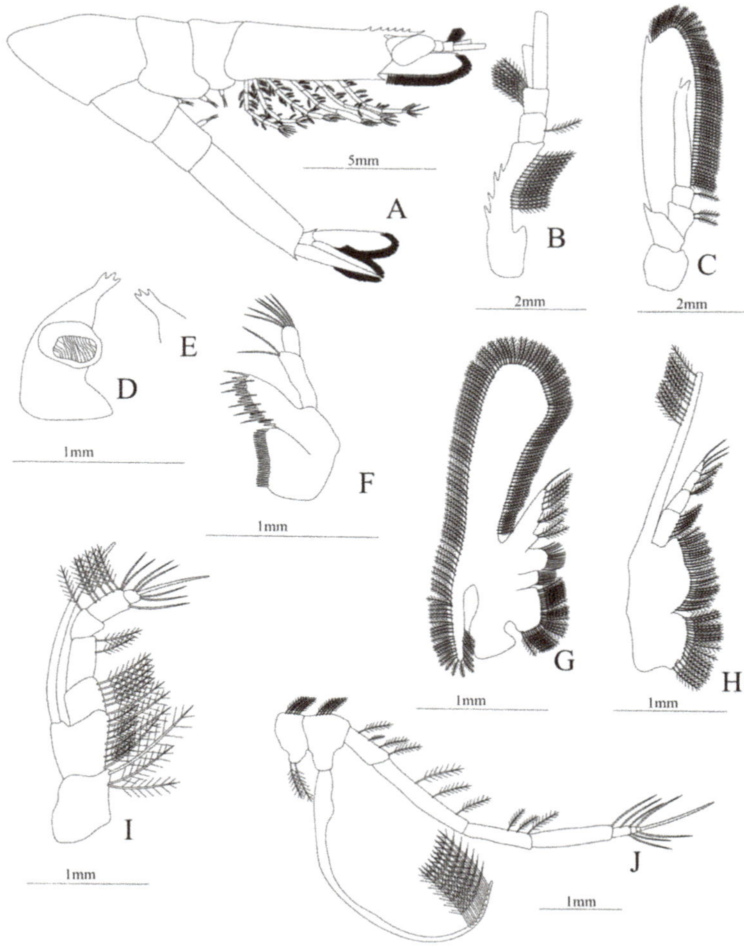

Figure 12. *Nematocarcinus cursor*: (**A**) lateral view; (**B**) antennule; (**C**) antenna; (**D**) left mandible; (**E**) right mandible (cutting edge); (**F**) maxillule; (**G**) maxilla; (**H**) first maxilliped; (**I**) second maxilliped; (**J**) third maxilliped.

First pereopod (Figure 13A). Coxa with 2 plumose setae, basis with 3 plumose setae; endopod 5-segmented with 5, 3, 2, 3, 4 plumose setae, except the last two segments that have serrulated setae; exopod, with 15 plumose setae.

Second pereopod (Figure 13B). Coxa with 2 plumose setae. Basis with 3 plumose setae; endopod 5-segmented with 3, 3, 3, 3, 3 plumose setae, except the last two segments that have serrulate setae; exopod with 9 plumose setae.

Third pereopod (Figure 13C). Coxa with 4 plumose setae, basis with one plumose setae; endopod 5-segmented with 6, 5, 5, 2, 3 plumose setae, except the last two segments that have serrulated setae; exopod with 13 plumose setae.

Four pereopod (Figure 13D). Coxa with 3 plumose setae, basis without setae; endopod 5-segmented with 6, 10, 3, 4, 3 plumose setae, except the last two segment that have serrulated setae; exopod with 7 plumose setae.

Fifth pereopod (Figure 13E). Coxa without setae; basis with 5 plumose setae; endopod 5-segmented with 2, 3, 7, 4, 3 plumose setae setae, except the last two segments that have serrulated setae.

Uropods (Figure 13F). Endopod well developed with 72 plumose setae, slightly wider than exopod; exopod, with 76 plumose setae.

Telson (Figure 13G) elongate, subtriangular. Lateral margin with 8 pairs of spines. Posterior margin, armed with 2 principal spines in each corner and 6 small spines.

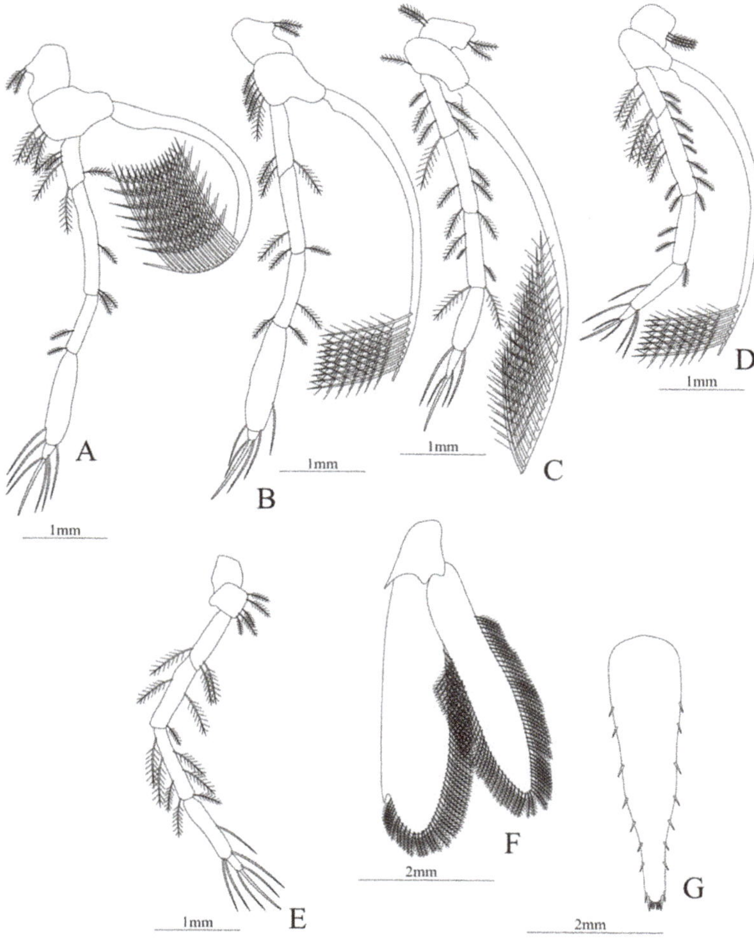

Figure 13. *Nematocarcinus cursor*: (**A**) first pereopod; (**B**) second pereopod; (**C**) third pereopod; (**D**) fourth pereopod; (**E**) fifth pereopod; (**F**) uropods; (**G**) telson.

Nematocarcinus rotundus
(Figures 14–16)

Material examined: Gulf of Mexico: HBG 7555, R/V Point Sur, DP04-11Aug16, MOC10-SW3D-068-N5, from 27.01226 and −88.4618 to 26.9255 and −88.5970, 11 August 2016, 199.8–5 m, MOCNESS plankton net, H. Bracken-Grissom, coll.

Zoea. Size. 7 mm (Carapace length); 21 mm (Total length). N = 1.

Carapace (Figure 14A). Rostrum shorter than the cornea, armed dorsally with four spines, epigastric spine present; eyes pedunculate; pterygostomial spine present.

Pleon (Figure 14A) with 6 somites, no spines or setae. Pleopods 3–4 missing in the specimen, pleopods 1, 2 and 5 without setae.

Antennule (Figure 14B). Peduncle 3-segmented, article 1 the longest, slender, with four pointed projections and with 16 plumose setae; article 2 with one plumose setae and article 3, subequal in size with article 2, with 8 plumose setae and two flagella distally, flagella almost same size.

Antenna (Figure 14C). Protopod 3-segmented, segment 1 unarmed, segment 2 with two plumose setae, segment 3 with a flagellum; exopod flattened with 66 plumose setae.

Mandible (Figure 14D,E). Mandibular palp absent; left and right incisor with 3 terminal teeth.

Maxillule (Figure 14F). Coxal endite with 28 conical serrulated setae; basial endite with 11 simple setae and 13 conical serrulated setae; protopod with two articles, article 1 with two serrulated setae and article 2 with 6 serrulated setae.

Maxilla (Figure 14G). Coxal with 31 plumose setae; basial endite bilobed with 8 and 10 serrated setae respectively; scaphognathite margin with 122 plumose setae.

First maxilliped (Figure 14H). Coxa with 18 plumose setae; basis with 13 plumose and 17 serrulated setae; endopod 4-segmented with 5, 3, 1, 2, plumose setae, except the last article that bear 2 serrulated setae; exopod with 15 plumose setae.

Second maxilliped (Figure 14I). Coxa with 3 plumose setae; basis with 9 plumose setae; endopod 5-segmented with 3 plumose, 3 plumose, 1 plumose, 5 (2 plumose plus 3 serrulated), 5 serrulated setae; exopod unsegmented and unarmed.

Third maxilliped (Figure 14J). Coxa with 8 plumose setae; basis with 5 plumose setae; endopod 5-segmented with 3, 2, 2, plumose setae, one serrulated setae; last article subdivided in three small articles with 3, 1 and 2 serrulated setae; exopod armed with 14 plumose setae.

First pereopod (Figure 15A). Coxa and basis with 3 plumose setae each one; endopod 5-segmented with 5, 3, 4, 3, 4 plumose setae, except the last two segments that have serrulated setae; exopod, with 15 plumose setae.

Second pereopod (Figure 15B). Coxa with 2 plumose setae. Basis with 3 plumose setae; endopod 5-segmented with 4, 5, 0, 2, 3 plumose setae, except the last two segments that have serrulate setae; exopod damage in the specimen.

Third pereopod (Figure 15C). Coxa with 2 plumose setae, basis with 3 plumose setae; endopod 5-segmented with 6, 7, 4, 2, 3 plumose setae, except the last two segments that have serrulated setae; exopod with 18 plumose setae.

Four pereopod (Figure 15D). Coxa and basis with one plumose seta each; endopod 5-segmented with 7, 7, 5, 4 (one plumose and 3 serrulated), 3 plumose setae, except the last segment that have serrulated setae; exopod with 8 plumose setae.

Fifth pereopod (Figure 15E). Coxa without setae; basis with 5 plumose setae; endopod 5-segmented with 3, 4, 4, plumose setae, 4 (one plumose and 3 serrulated), 3 serrulated setae; exopod with 8 plumose setae.

Uropods (Figure 15F). Endopod well developed with 72 plumose setae, slightly wider than exopod; exopod, with 76 plumose setae.

Telson (Figure 15G). (Damaged in the specimen) elongate, subtriangular. Lateral margin with 7 pairs of spines. Posterior margin damage in the specimen.

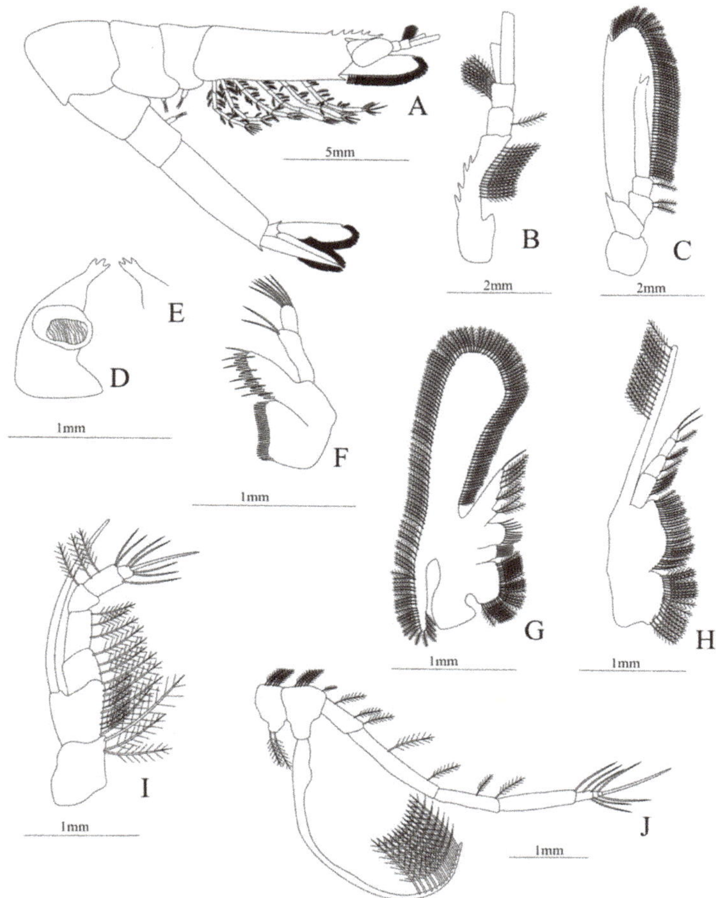

Figure 14. *Nematocarcinus rotundus*: (**A**) lateral view; (**B**) antennule; (**C**) antenna; (**D**) left mandible; (**E**) right mandible (cutting edge); (**F**) maxillule; (**G**) maxilla; (**H**) first maxilliped; (**I**) second maxilliped; (**J**) third maxilliped.

Material examined: Gulf of Mexico: HBG 6134, R/V Point Sur, DP03-06May16-MOC10-B079N-045-N3, 27. 4613 and −86.8992, 27.5005 and −86.9771; 6 May 2016, 601.4–996.1 m. MOCNESS plankton net, L. Timm, coll. Gulf of Mexico: HBG 7996, R/V Point Sur, (DP05-06May17-MOC10-B287N-089-N3), 28.1179 and −87.3899; 6 May 2017, 1000–600 m, MOCNESS plankton net, L. Timm, coll. Gulf of Mexico: HBG 7997, R/V Point Sur, (DP05-06May17-MOC10-B287N-089-N3), 28. 1179 and −87.3899, 6 May 2017, 1000–600 m, MOCNESS plankton net, L. Timm, coll. Gulf of Mexico: HBG 8000, R/V Point Sur, DP05-03May17-MOC10-B065N-087-N3, 28. 5312 and −88.0236, 5 May 2017, 1000–600 m, MOCNESS plankton net, L. Timm, coll.

Decapodite. Size: 8 mm (Carapace length); 26 mm (Total length). N = 4.

Carapace (Figure 16A). Rostrum straight, armed with 7–12 dorsal spines, sligthly longer than antennular peduncle; eyes pedunculate.

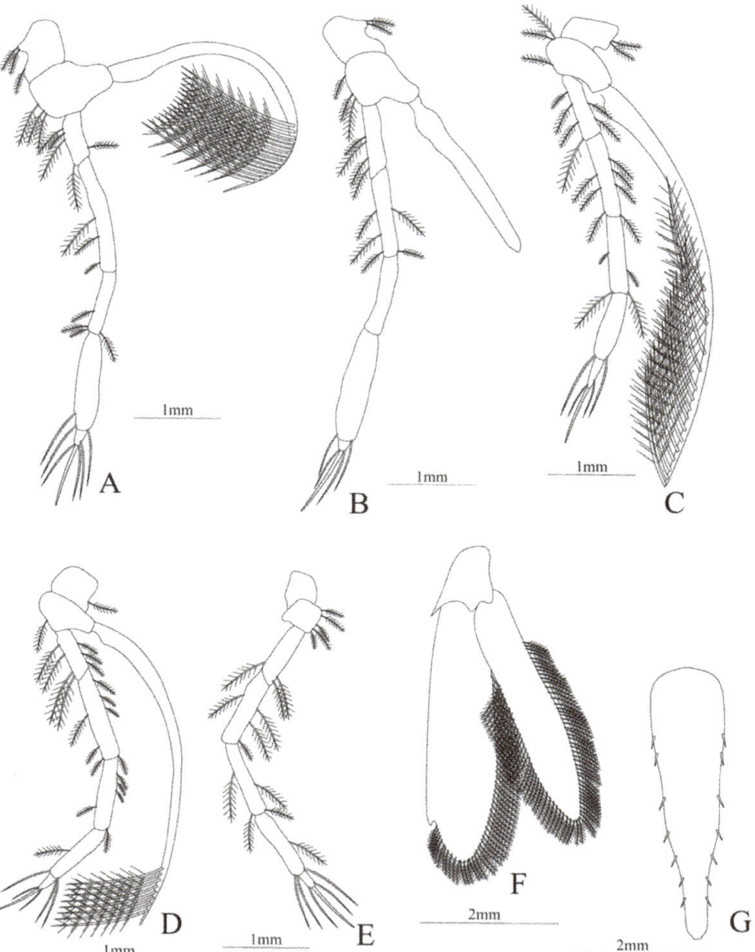

Figure 15. *Nematocarcinus rotundus*: (**A**) first pereopod; (**B**) second pereopod; (**C**) third pereopod; (**D**) fourth pereopod; (**E**) fifth pereopod; (**F**) uropods; (**G**) telson.

Pleon (Figure 16A) with 6 somites, no spines or setae. Pleopods well developed.

Antennule (Figure 16B). Peduncle 3-segmented, article 1 the longest, slender, with 15–24 plumose setae; article 2 with 15–17 plumose setae and article 3, subequal in size with article 2, with 8–16 plumose setae and two flagella distally, flagella almost same size.

Antenna (Figure 16C). Protopod 3-segmented, segment 1 unarmed, segment 2 with two plumose setae, segment 3 with a flagellum; exopod flattened with 66–83 plumose setae.

Mandible (Figure 16D). Mandibular palp 3-segmented, with 1, 8, 13 simple setae; incisor with 7 terminal teeth.

Maxillule (Figure 16E). Coxal endite with 8 serrulated setae; basial endite with 15 conical setae; protopod with 3 plumose setae.

Maxilla (Figure 16F). Coxal endite with 36 plumose setae; basial endite bilobed with 23 (12 plumose plus 11 conical) + 36 plumose setae; endopod with 6 plumose setae; scaphognathite margin with 149 plumose setae.

First maxilliped (Figure 16G). Coxa without setae; basis with 47 (10 conical plus 10 plumose plus 27 serrulated) setae; endopod unsegmented with 21 plumose setae; exopod unsegmented with 21 simple setae

Second maxilliped (Figure 16H). Coxa without setae; basis with 8 simple setae; endopod 5-segmented with 11, 6, 2, all plumose, 25 (5 simple plus 20 serrulated setae), 11 serrulated; exopod unsegmented and unarmed.

Third maxilliped (Figure 16I). Coxa without setae; basis with 3 simple setae; endopod 4-segmented with 16, 20, 9 all simple, 29 serrulated setae; exopod missing in the specimen.

First to fifth Pereopods missing in the specimens.

Uropods (Figure 16J). Endopod well developed with 81–96 plumose setae, slightly wider than exopod; exopod, with 72–75 plumose setae.

Telson (Figure 16K) elongate, subtriangular. Lateral margin with 5 pairs of spines. Posterior margin, armed with 2 principal spines in each corner and 2 distal spines.

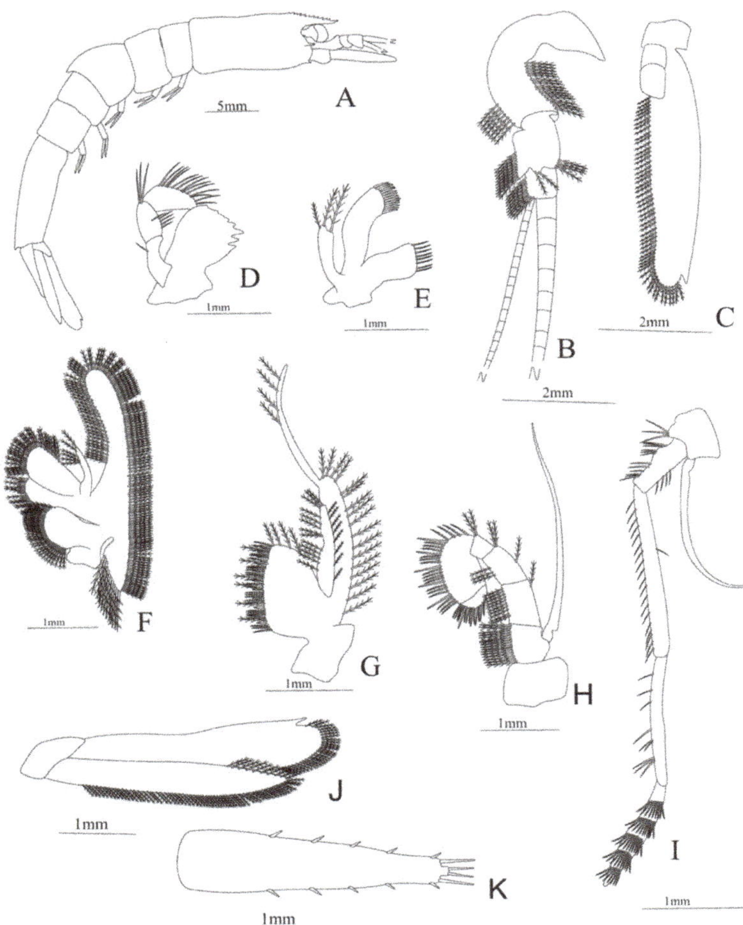

Figure 16. *Nematocarcinus rotundus*: (**A**) lateral view; (**B**) antennule; (**C**) antenna; (**D**) mandible; (**E**) maxillule; (**F**) maxilla; (**G**) first maxilliped; (**H**) second maxilliped; (**I**) third maxilliped; (**J**) uropods; (**K**) telson.

Oplophoridae Dana, 1852
Systellaspis Spence Bate, 1888
Systellaspis braueri (Balss, 1914)
(Figures 17 and 18)

Material examined: Gulf of Mexico: HBG6823, R/V Point Sur, DP04-08Aug16-MOC10-SE1N-063-N0, from 26.9878, −87.9494 to 27.0591, −88.0856, 8 August 2016, 1504-NA m, MOCNESS plankton net, H. Bracken-Grissom, coll.

Decapodite. Size. 8 mm (Carapace length); 26 mm (Total length). N = 1.

Carapace (Figure 17A). Rostrum straight, armed dorsally with 9 spines and ventrally with one small spine, same length of the eye; antennal spine small, anteroventral margin bearing one small spine and a pterygostomian spine; eyes pedunculate.

Pleon (Figure 17A) with 6 somites, no spines or setae. Pleopods 1–2 missing in the specimen, pleopods 3–5 well developed.

Antennule (Figure 17B). Peduncle 3-segmented, article 1 the longest armed with 5 simple setae, article 2 also with 3 simple setae and article 3 the smallest, with one simple setae and two flagella distally, flagella subequal in size.

Antenna (Figure 17C). Protopod 3-segmented, flagellum missing in the specimen; exopod flattened with 52 plumose setae and a pointed process distally.

Mandible (Figure 17D). Mandibular palp 3-segmented, article 1 armed with 3 simple setae, article 2 with 2 lateral simple setae and article 3 with 6 simple setae plus 3 plumose setae, right incisor with 9 teeth.

Maxillule (Figure 17E). Coxal endite with 19 plumose setae; basial endite with 18 conical serrulate setae plus 2 plumose setae and protopod with one plumose subterminal seta

Maxilla (Figure 17F). Coxal endite with 10 plumose setae; basial endite bilobed with 11 + 19 (17 plumose plus 2 simple) setae; endopod with 3 plumose setae; scaphognathite margin with 124 plumose setae.

First maxilliped (Figure 17G). Coxa with 8 plumose setae; basis with 28 plumose setae; endopod unsegmented with 12 plumose setae; exopod unsegmented, armed with 14 plumose setae.

Second maxilliped (Figure 17H). Coxa without setae; basis with 6 plumose setae; endopod 5-segmented with 18, 8, 2, plumose setae plus 23, 12 serrulate setae; exopod unsegmented, armed distally with 8 plumose natatory setae.

Third maxilliped (Figure 18A). Coxa with 3 plumose setae; basis with 6 plumose setae, endopod 3-segmented with 40 (22 inner setae, 3 of them serrulate setae, all the others plumose + 18 outer plumose setae), 9 serrulate setae and 23 serrulate setae; exopod unsegmented, armed distally with 7 plumose natatory setae.

First pereopod (Figure 18B). Coxa with 3 and basis with 6 plumose setae; endopod 5-segmented with 10 plumose setae, 18 plumose setae, 5, 11, 1 serrulate setae; exopod unsegmented and unarmed.

Second pereopod (Figure 18C). Coxa with 8 plumose setae, basis with 4 plumose setae; endopod 5-segmented with 17 plumose setae, 15 plumose setae and 4, 7, 1 serrulate setae.

Third pereopod missing in the specimen.

Fourth pereopod (Figure 18D). Coxa with 9 simple setae, basis with 4 simple setae; endopod 5-segmented with 10 (5 spines + 5 simple setae), 12 (4 spines + 8 simple setae), 1 simple setae, 5 spines, 0, 0; exopod unsegmented and unarmed.

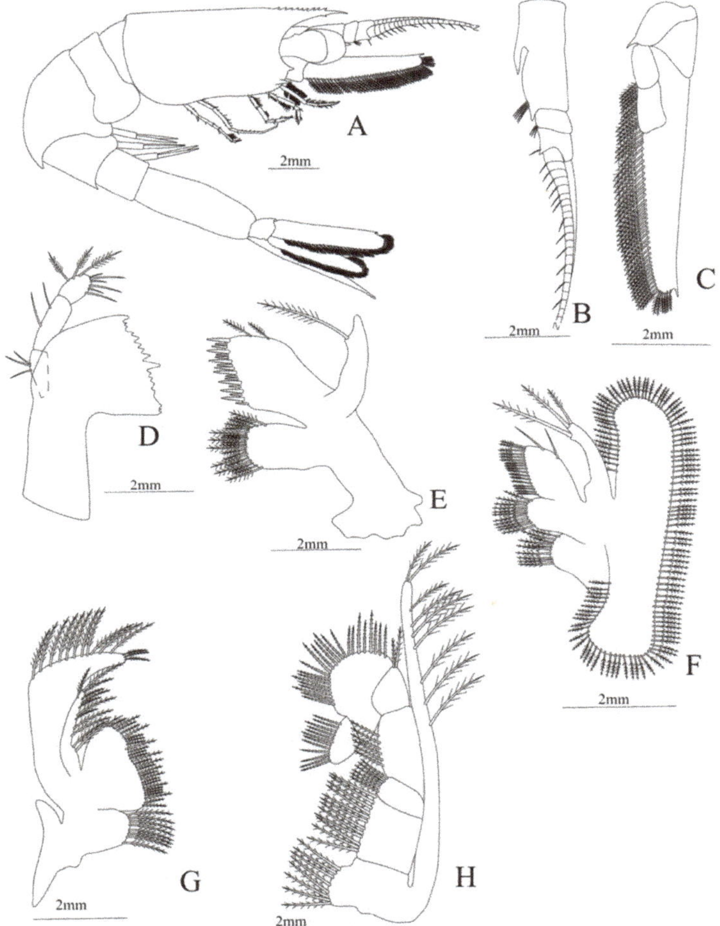

Figure 17. *Systellaspis braueri*: (**A**) lateral view; (**B**) antennule; (**C**) antenna; (**D**) mandible; (**E**) maxillule; (**F**) maxilla; (**G**) first maxilliped; (**H**) second maxilliped.

Fifth pereopod (Figure 18E). Coxa and basis without setae; endopod 5-segmented with 5 (2 spine + 4 simple setae), 2, 3, 12, 8 simple setae; exopod unsegmented and unarmed.

Uropod (Figure 18F). Endopod well developed with 54 plumose setae; exopod with 42 plumose setae

Telson (Figure 18G) elongate, subtriangular, with 11 pairs of lateral spines, 1 pair of large mobile spines and 10 pairs of spines on the distal part near the tip of the telson; one small spine on the distal margin.

Material examined: Gulf of Mexico: HBG6844, R/V Point Sur, DP04-17Aug16-MOC10-B252N-080-N5, from 28.5272, −87.4972 to 28. 3842, −87.4866, 17 August 2016, 199.5–5 m, MOCNESS plankton net, L. Timm, coll.

Zoea. Size. 22 mm (Carapace length); 36 mm (Total length). N = 1.

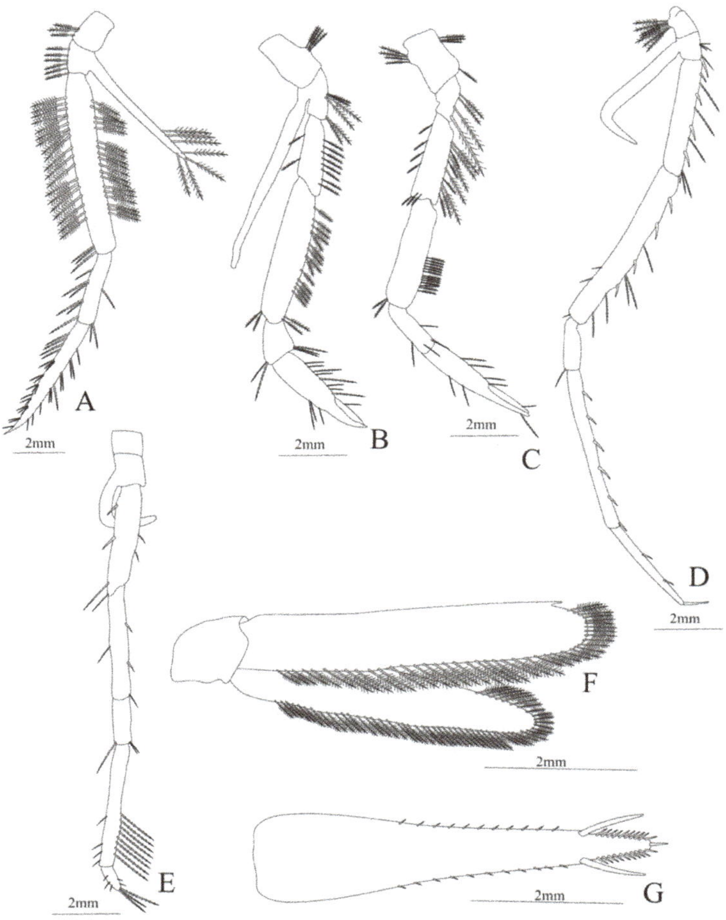

Figure 18. *Systellaspis braueri*: (**A**) third maxilliped; (**B**) first pereopod; (**C**) second pereopod; (**D**) fourth pereopod; (**E**) fifth pereopod; (**F**) telson; (**G**) uropods.

Pandalidae Haworth, 1825
Heterocarpus ensifer A. Milne-Edwards, 1881
(Figures 19 and 20)

Carapace (Figure 19A). Rostrum large armed dorsally with 21 spines and 9 ventral spines, one spine near the posterior margin of the carapace, suborbital spine strong.

Pleon (Figure 19A) with a pointed projection on segments 3 and 4. Other segments without spines or setae. Pleopods 1–4 missing in the specimen, pleopod 5 without setae.

Antennule (Figure 19B). Peduncle 3-segmented, article 1 the longest, slender, with 9 plumose setae in both margins, article 2 with 2 plumose setae and article 3, the smallest, with 3 plumose setae and with two flagella distally.

Antenna (Figure 19C). Protopod 3-segmented, article 1 and 2 unarmed, article 3 with 5 small spines and a flagellum; exopod flattened, subtriangular, with a slender and pointed projection on its distal region and 13 pointed projections on the superior margin and 64 plumose setae in the inferior margin.

Mandible (Figure 19D, E) without palp, right mandible with 6 teeth and left mandible with 4 teeth.

Maxillule (Figure 19F). Coxal endite with 19 conical serrulated setae; basial endite with 12 conical serrulated setae; protopod with 4 plumose setae.

Maxilla (Figure 19G). Coxal endite bilobed with 17 plumose plus 2 serrated and one plumose setae; basial endite bilobed with 10 plus 12 plumose setae; endopod with 8 (2 + 2 + 1 + 1 + 2) plumose setae, segmentation not well defined; scaphognathite margin with 143 plumose setae.

First maxilliped (Figure 19H). Coxa with 7 plumose setae; basis with 23 plumose setae; exopod with 50 plumose setae; endopod 4-segmented, armed with 22 setae, five of them plumose all the others simple.

Second maxilliped (Figure 19I). Coxa with one plumose seta; basis with 10 plumose plus 4 serrulated setae; endopod 5-segmented with 4, 3, 2, 4, 8 plumose setae, except the first and the last articles which have one serrated seta each; exopod armed distally with 17 plumose setae.

Third Maxilliped (Figure 19J). Coxa with 3 simple setae; basis with 9 simple setae; endopod 4-segmented with 13, 9, 21, 2 simple setae; exopod armed distally with 6 plumose setae.

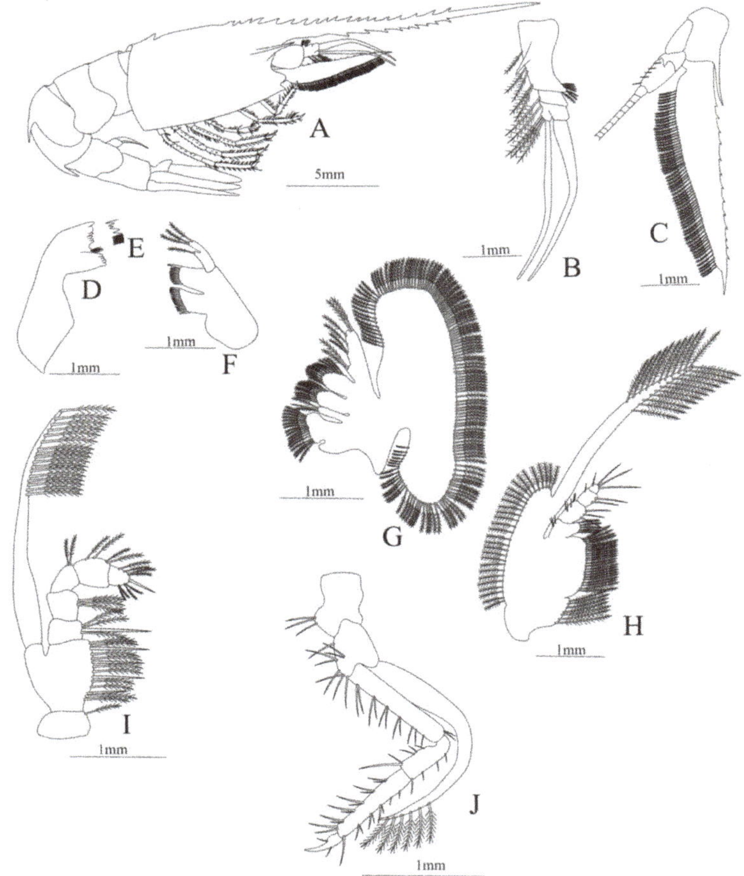

Figure 19. *Heterocarpus ensifer*: (**A**) lateral view; (**B**) antennule; (**C**) antenna; (**D**) right mandible; (**E**) left mandible (cutting edge); (**F**) maxillule; (**G**) maxilla; (**H**) first maxilliped; (**I**) second maxilliped; (**J**) third maxilliped.

First pereopod (Figure 20A). Coxa without setae; Basis with 5 simple setae; endopod 5-segmented with 5, 8, 14, 27, 4 simple setae; exopod armed distally with 10 plumose setae.

Second pereopod (Figure 20B). Coxa without setae; basis with 5 simple setae; endopod 5-segmented with 10, 8, 8, 7, 3 simple setae; exopod armed distally with 6 plumose setae.

Third pereopod (Figure 20C). Coxa without setae; basis with 4 setae; endopod 5-segmented with 5, 21, 7, 23, 5 simple setae; exopod armed distally with 6 plumose setae.

Fourth pereopod (Figure 20D). Coxa without setae; basis with 2 simple setae; endopod 5-segmented with 9 (6 simple setae plus 3 spines), 17 (10 simple setae plus 7 spines), 7, 27, 5 simple setae; exopod armed distally with 7 plumose setae.

Fifth pereopod (Figure 20E). Coxa with one simple setae; Basis with 6 simple setae; endopod 5-segmented with 11 (3 spines plus 8 simple setae), 14, 7, 34, 8 simple setae.

Uropod (Figure 20F). Endopod and exopod well developed, exopod with 84 plumose setae and endopod with 90 plumose setae.

Telson (Figure 20G) enlarged, subtriangular, with 4 pairs of lateral spines and posterior margin bearing row of 5 diminute spines and one pairs of spines on outer margin.

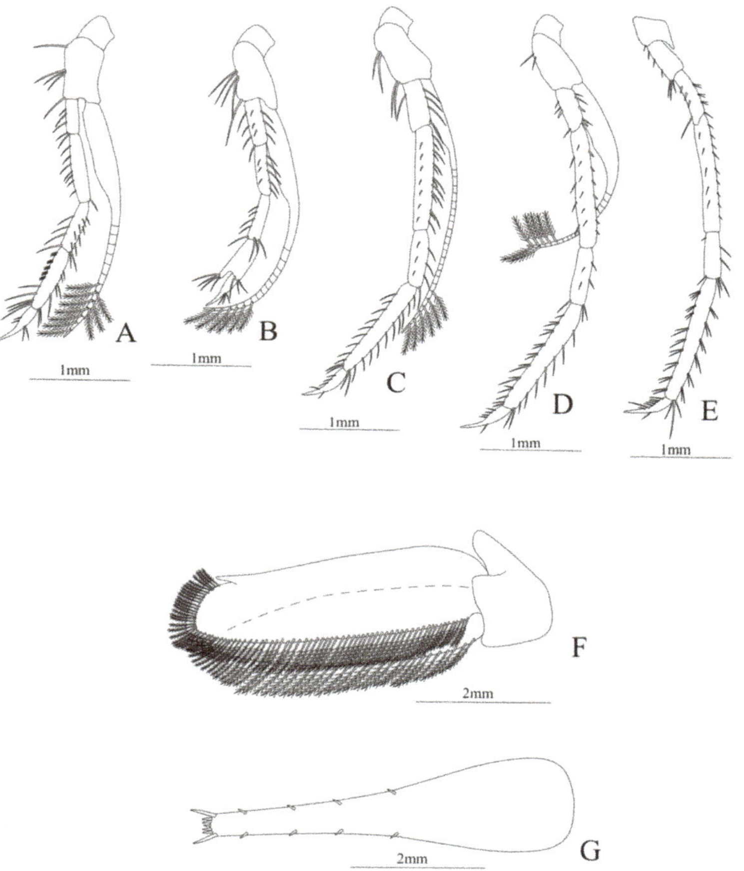

Figure 20. *Heterocarpus ensifer*: (**A**) first pereopod; (**B**) second pereopod; (**C**) third pereopod; (**D**) fourth pereopod; (**E**) fifth pereopod; (**F**) uropods; (**G**) telson.

Plesionika Spence Bate, 1888
Plesionika edwardsii (J.F. Brandt in von Middendorf, 1851)
(Figures 21 and 22)

Material examined: Gulf of Mexico: HBG 7584, R/V Point Sur, DP04-09Aug16-MOC10-SE3N-065-N5, from 26.9997, −86.9912 to 26.9903, −87.1491; 9 August 2016, 199.2–5 m, MOCNESS plankton net, H. Bracken-Grissom, coll.

Decapodite. Size: 15 mm (Carapace length); 58 mm (Total length). N = 1.

Carapace (Figure 21A). Rostrum long and unarmed, slender, longer than carapace; antennal spine small; anteroventral margin bearing 1 strong pterygostomian spine; eyes pedunculate.

Pleon (Figure 21A) with 6 somites, no spines or setae. Pleopods 1–4 missing in the specimen, pleopod 4 well developed.

Antennule (Figure 21B). Peduncle 3-segmented, article 1, the longest, armed with 27 (15 outer plus 12 inner) plumose setae and one spine, article 2 with 9 (6 outer plus 3 inner) plumose setae and article 3 with 5 outer plumose setae and two flagella distally.

Antenna (Figure 21C). Protopod 3-segmented with a flagellum; exopod flattened with 71 plumose setae and a pointed process distally.

Mandible. Palp absent; right and left slightly asymmetrical, right incisor with 3 terminal teeth (Figure 21D); left incisor with 4 teeth (Figure 21E).

Maxillule (Figure 21F). Coxal endite with 12 conical serrate setae; basial endite with 7 conical serrate setae and 4 simple setae; endopod unsegmented, with 1 + 3 serrated setae.

Maxilla (Figure 21G). Coxal endite bilobed with 12 plumose plus 3 simple setae; basial endite bilobed with 4 + 7 simple setae; endopod unsegmented with 6 (2 + 2 + 2) simple setae; scaphognathite margin with 120 plumose setae.

First maxilliped (Figure 21H). Coxa with 3 large plumose plus 3 simple setae; basis with 12 plumose setae; endopod 4-segmented with 6 (5 simple plus one conical serrate) 3 (2 simple plus one conical serrate), 2 (one simple pus one conical serrate), 3 simple setae; endopod armed with 21 plumose setae and exopod armed distally with 12 plumose setae.

Second maxilliped (Figure 21I). Coxa with one plumose seta; basis with 11 (4 simple plus 4 plumose plus 3 conical serrated) setae; endopod 5-segmented with 3 (one conical serrated plus 2 simple), 2 simple, 1 simple, 7 simple, 8 (5 conical serrated and 3 simple) setae; exopod unarmed.

Third maxilliped (Figure 21J). Coxa without setae; basis with 5 simple setae; endopod 5-segmented with 2 simple, 19 (13 simple setae plus 6 spines), 11 simple, 12 simple, 0 setae; exopod armed with 9 plumose setae.

First pereopod (Figure 22A). Coxa and basis unarmed; endopod 5-segmented with 5, 14 (7 spines plus 7 simple setae), 12 (4 spines plus 8 simple setae), 9 simple, 0 setae; exopod unarmed.

Second pereopod (Figure 22B). Coxa unarmed, basis with 2 simple setae; endopod 5-segmented with 4 spines, 11 (6 spines plus 5 simple setae), 19 (6 spines plus 13 simple setae), 2 simple setae, 0 setae; exopod unarmed.

Third pereopod (Figure 22C). Basis armed with 2 simple setae; endopod 5-segmented with 6 simple setae, 18 spines, 5 spines, 19 (9 spines plus 10 setae), 0 setae; exopod unarmed.

Fourth pereopod (Figure 22D). Coxa and basis unarmed; endopod 5-segmented with 3 spines, 19 (9 spines plus 10 simple setae), 5 (4 spines plus one simple seta), 9 (7 spines plus 2 simple setae), 0 simple setae; exopod unarmed.

Fifth pereopod (Figure 22E). Coxa unarmed, basis with 2 simple setae; endopod 5-segmented with 4 simple setae, 21 (10 spines plus 11 simple setae), 8 (5 spines plus 3 simple setae), 15 (9 spines plus 6 simple setae), 0 simple setae; exopod absent.

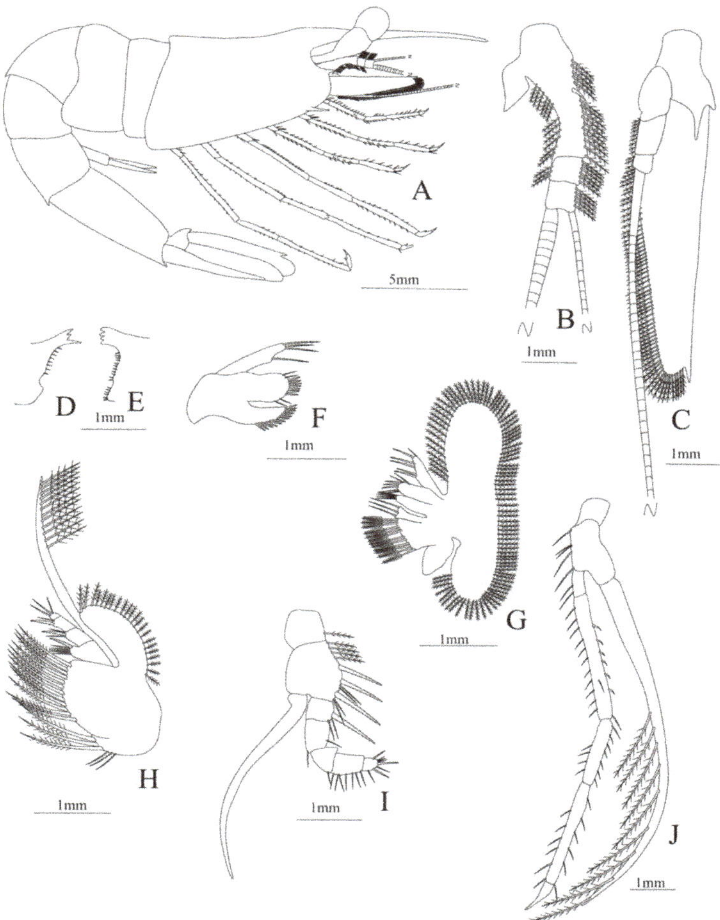

Figure 21. *Plesionika edwardsii*: (**A**) lateral view; (**B**) antennule; (**C**) antenna; (**D**) left mandible (cutting edge); (**E**) rigth mandible (cutting edge); (**F**) maxillule; (**G**) maxilla; (**H**) first maxilliped; (**I**) second maxilliped; (**J**) third maxilliped.

Uropods (Figure 22F). Endopod well developed with 96 plumose setae; exopod, with 84 plumose setae

Telson (Figure 22G) elongate, subtriangular, with three pairs of lateral spines; distally with one central large spine and 3 pairs of small spines and one spine on each corner.

Plesionika ensis (A. Milne-Edwards, 1881)
(Figures 23 and 24)

Material examined: Gulf of Mexico: HBG6825, R/V Point Sur, DP04-07Aug16-MOC10-SW4N-061-N0, 26.8887, −89.0389, and 26.9936, −88.9987, 7 August 2016, 1500.8-NA m, MOCNESS plankton net, H. Bracken-Grissom, coll. Gulf of Mexico: HBG7845, R/V Point Sur, DP05-10May17-MOC10-B175D-096-N2, 28.9922 and −87.4786, 29.0336 and −87.6491, 10 May 2017, 1199–995 m, MOCNESS plankton net, L. Timm, coll. Gulf of Mexico: HBG7995, R/V Point Sur, DP05-06May17-MOC10-B287N-089-N3, 28.1179 and −87.3899, 28.0467 and −87.5559, 6 May 2017, 1000–600 m, MOCNESS plankton net, L. Timm, coll. Gulf of Mexico:

HBG9264, R/V Point Sur, DP06-20Jul18-MOC10-B175N-102-N0, 29.0045 and −87.4658, 20 July 2018, 600 m, MOCNESS plankton net, H. Bracken-Grissom, coll.

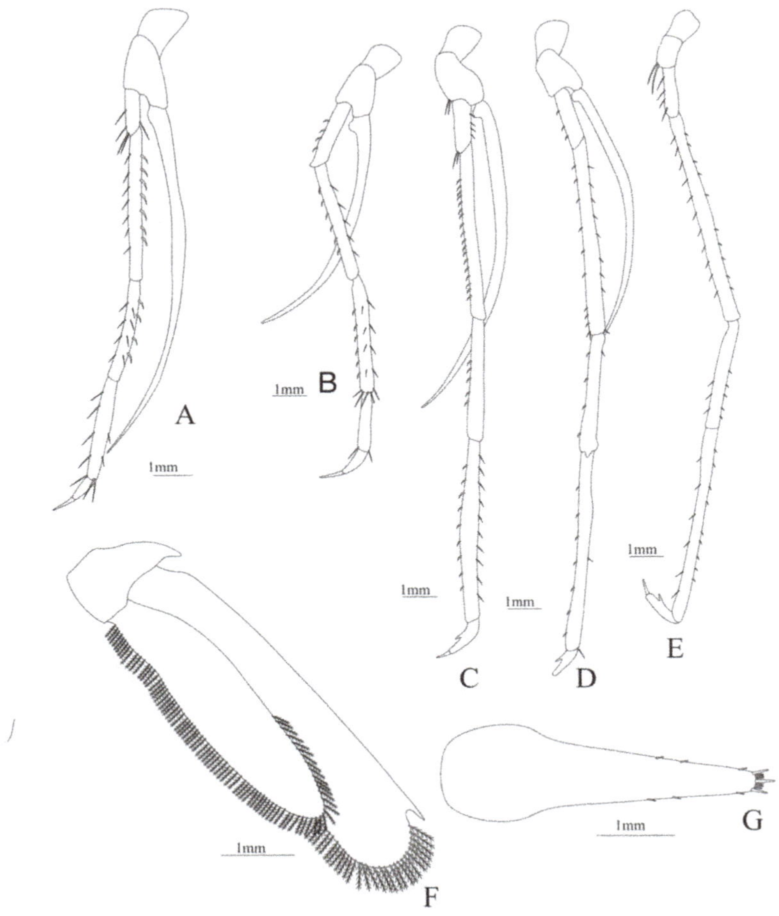

Figure 22. *Plesionika edwardsii*: (**A**) first pereopod; (**B**) second pereopod; (**C**) third pereopod; (**D**) fourth pereopod; (**E**) fifth pereopod; (**F**) telson; (**G**) Uropods.

Juvenile. Size. 12 mm (Carapace length); 36 mm (Total length). N = 4.

Carapace (Figure 23A). Rostrum long, slender, with 3 basal spines, slightly curved upwards and longer than antennular peduncle; antennal spine present; eyes pedunculate.

Pleon (Figure 23A) with 6 somites, no spines or setae. Pleopods 3–4 missing in the specimen, pleopods 1–2 and 5 well developed.

Antennule (Figure 23B). Peduncle 3-segmented, article 1 with 16–18 plumose setae, article 2 with 9 plumose setae and article 3 with two flagella.

Antenna (Figure 23C). Protopod 3-segmented; article 1 with two sharp projections, article 2 with 4 simple setae and article 3 with 5 simple setae. exopod flattened with 63–66 plumose setae and a pointed process distally.

Mandible (Figure 23D). Palp 3-segmented, article 1 unarmed, article 2 with 3 simple setae and article 3 with 16 simple setae, right incisor with 5 terminal teeth.

Maxillule (Figure 23E). Coxal endite with 10–12 simple setae plus 10–18 serrulate setae; basial endite with 15–18 simple setae plus 10–12 conical setae; endopod unsegmented, with 6 simple setae plus one plumose seta; exopod absent.

Maxilla (Figure 23F). Coxal endite with 12–16 plumose setae; basial endite bilobed both armed with 28–30 and 28–32 serrulated setae respectively; endopod unsegmented with 4 (1 + 1 + 2) plumose setae; scaphognathite margin with 89–93 plumose setae.

First maxilliped (Figure 23H). Coxa with 15–17 serrulate setae; basis endite with 43–52 serrulate setae; endopod with 28–32 plumose setae; exopod unsegmented, armed distally with 10–13 plumose setae.

Second maxilliped (Figure 23G). Coxa with 4 serrulated setae; basis with 14 serrulated setae; endopod 5-segmented with 1 plumose seta, 6 plumose setae and 4–5, 11–20, 5–10 serrulated setae; exopod armed with 8–10 plumose setae.

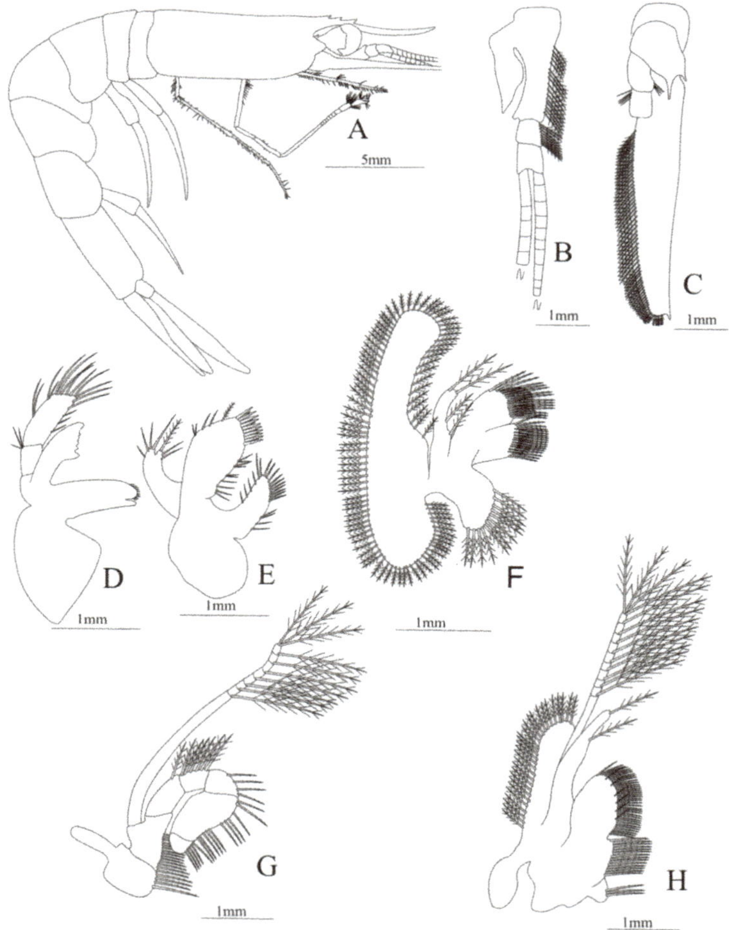

Figure 23. *Plesionika ensis*: (**A**) lateral view; (**B**) antennule; (**C**) antenna; (**D**) mandible; (**E**) maxillule; (**F**) maxilla; (**G**) first maxilliped; (**H**) second maxilliped.

Third maxilliped (Figure 24A). Coxa without setae; basis with 7 simple setae; endopod 3-segmented with 24, 13, 12, simple setae; exopod unsegmented, armed distally 16 simple setae.

First pereopod missing in the specimen.

Second pereopod (Figure 24B). Coxa and basis without setae; endopod 5-segmented with 14, 0, 7, 0 (with 8 divisions), 24, 6 simple setae.

Third and fourth pereopods missing in the specimen.

Fifth pereopod (Figure 24C). Coxa without setae, basis with 10 simple setae, endopod 5-segmented with 13, 26, 26, 26, 3 simple setae.

Uropod (Figure 24D). Endopod well developed with 67–76 plumose setae; exopod, with 92–97 plumose setae.

Telson (Figure 24E) elongate, subtriangular, with 3 pairs of lateral spines and 2 pairs of distal spines.

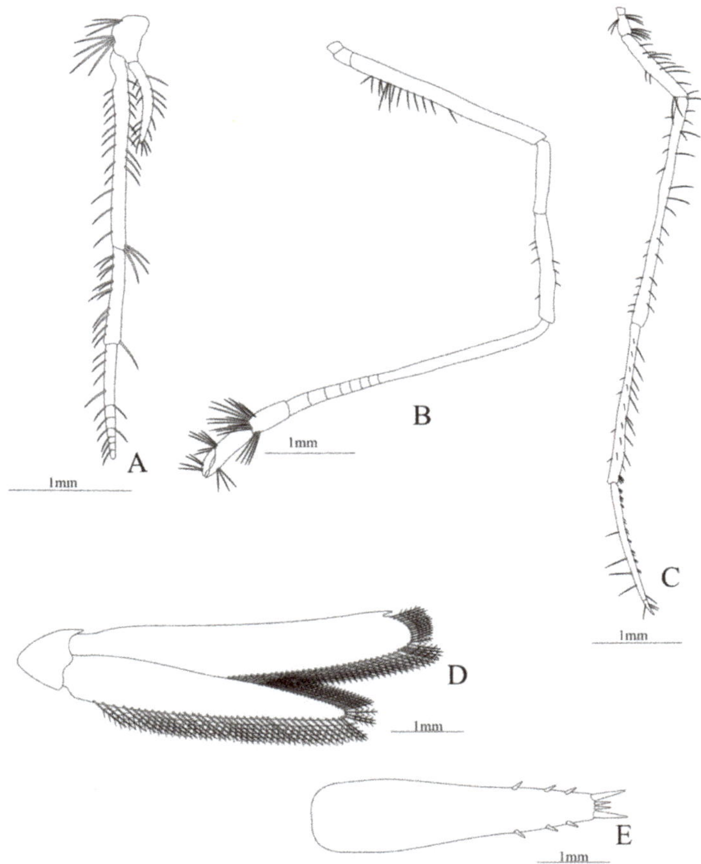

Figure 24. *Plesionika ensis*: (**A**) third maxilliped; (**B**) second pereopod; (**C**) fifth pereopod; (**D**) Uropods; (**E**) telson.

Aristeidae Wood-Mason in Wood-Mason and Alcock, 1891

Hemipenaeus Spence Bate, 1881

Hemipenaeus carpenteri Wood-Mason in Wood-Mason and Alcock, 1891

(Figures 25 and 26)

Material examined: Gulf of Mexico: HBG 6846, R/V Point Sur, DP04-09Aug16-MOC10-SE3N-065-N3, 26.9997, −86.9912 and 26.9909, −87.1491, 9 August 2016, 1000.5–3 m, MOC-NESS plankton net, H. Bracken-Grissom, coll.

Mysis. Size. 6 mm (Carapace length); 16 mm (Total length). N = 1.

Carapace (Figure 25A) with two lateral swollen process near the posterior margin, rostrum long, extend until the end of the article 1 of the antennule, slightly curved; anteroventral margin bearing 1 strong pterygostomial spine and 1 postorbital spine; eyes pedunculate.

Pleon (Figure 25A) with 6 somites, no spines or setae. Pleopods 1–5 without setae.

Antennule (Figure 25B). Peduncle 3-segmented, article 1 the longest, slender, with 21 plumose setae in both margins, article 2 with 11 plumose setae in both margins and article 3, the smallest with 5 plumose setae and two flagella distally. Flagella short, same size, inner 5-segmented and outer 6-segmented with plumose setae.

Antenna (Figure 25C). Protopod 3-segmented with a flagellum; exopod with 66 plumose setae.

Mandible (Figure 25D). Palp 3-segmented, article 1 unarmed, article 2 with 5 simple setae and article 3 with 10 simple setae.

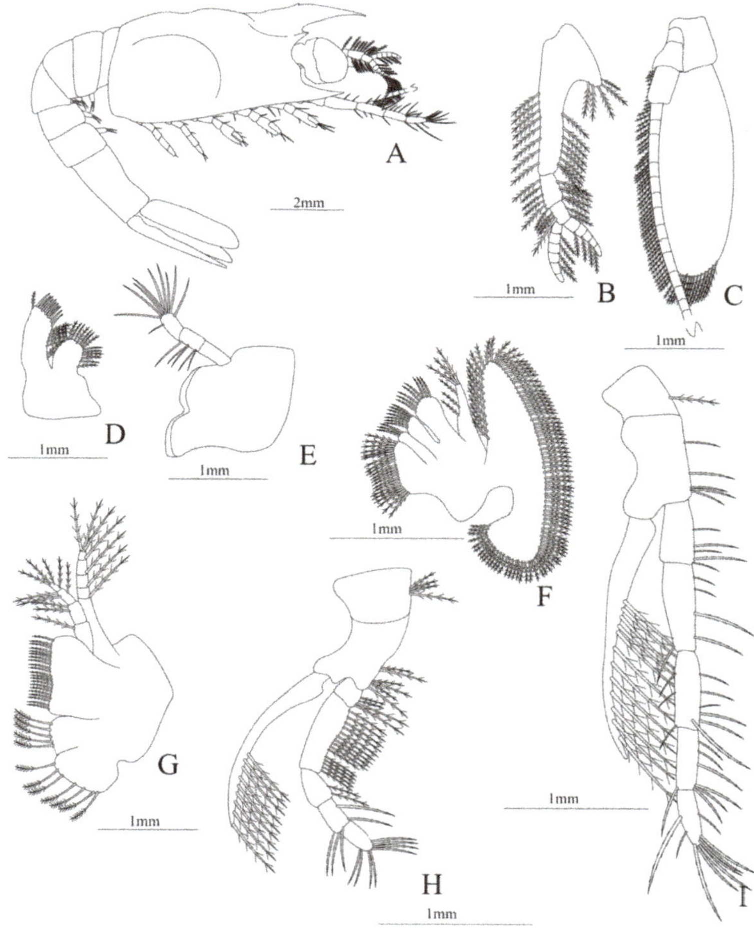

Figure 25. *Hemipenaeus carpenteri*: (**A**) lateral view; (**B**) antennule; (**C**) antenna; (**D**) maxillule; (**E**) mandible; (**F**) maxilla; (**G**) first maxilliped; (**H**) second maxilliped; (**I**) third maxilliped.

Maxillule (Figure 25E). Coxal endite with 15 (10 serrated plus 5 plumose) setae; basial endite with 11 conical setae and one plumose subdistal setae.

Maxilla (Figure 25F). Coxal endite bilobed with 21 (15 plumose plus 6 serrated) setae; basial endite bilobed with 15 (7 plus 8 serrated) setae; endopod with 6 (1 + 1 + 1 + 3) plumose setae, segmentation not well defined; scaphognathite margin with 89 plumose setae.

First maxilliped (Figure 25G). Coxa with two endites and 12 (5 + 7) plumose setae; basis with 21 serrated setae; endopod 4-segmented with 1, 2, 4, 3 plumose setae; exopod unsegmented, armed with 7 plumose setae.

Second maxilliped (Figure 25H). Coxa with 4 plumose setae; basis with 6 plumose setae; endopod 5-segmented with 4, 15, 2, 5, 9 plumose setae, except the last two articles which have serrated setae; exopod unsegmented, armed distally with 9 long plumose natatory setae.

Third maxilliped (Figure 25I). Coxa with 1 plumose seta; basis with 5 serrated setae; endopod 5-segmented with 5, 5, 9, 7, 9 serrated setae; exopod unsegmented, armed distally with 12 long plumose natatory setae.

First pereopod (Figure 26A). Basis with 2 simple setae; endopod 5-segmented with 1, 1, 1, 2, 3 setae; exopod armed distally with 10 plumose setae.

Second pereopod (Figure 26B). Basis unarmed; endopod 5-segmented with 2, 1, 1, 3, 2 setae; exopod armed distally with 14 plumose setae.

Third pereopod (Figure 26C). Basis unarmed; endopod 5-segmented with 0, 1, 1, 1, 4 (2 inner + 2 terminal) setae; exopod armed distally with 11 plumose setae.

Fourth pereopod (Figure 26D). Basis unarmed; endopod 5-segmented with 0, 1, 1, 0, 2 setae; exopod armed distally with 9 plumose setae.

Fifth pereopod (Figure 26E). Basis unarmed; endopod 5-segmented with 0, 1, 0, 0, 1 setae; exopod armed distally with 12 plumose setae.

Uropod (Figure 26F). Endopod and exopod well developed, both missing setae.

Telson (Figure 26G) enlarged, subrectangular, with two pairs of lateral spines and posterior margin bearing row of 4 pairs of minute spinules and 2 pairs of spines on outer margin.

Hemipenaeus carpenteri Wood-Mason in Wood-Mason and Alcock, 1891
(Figures 27 and 28)

Material examined: Gulf of Mexico: HBG 6854, R/V Point Sur DP04-08Aug16-MOC10-SE1N-063-N5, 26. 9878, −87.9494, and 27.0591, −88.0856, 8 August 2016, 202.7–5 m, MOCNESS plankton net, H. Bracken-Grissom, coll. Gulf of Mexico: HBG 7552, R/V Point Sur DP04-11Aug16-MOC10-SW3D-068-N5, 27. 0122, −88.4618, and 26.9255, −88.5970, 11 August 2016, 199.8–5 m, MOCNESS plankton net, H. Bracken-Grissom, coll. Gulf of Mexico: HBG 7867, R/V Point Sur DP05-11May17-MOC10-B175D-098-N0, 26. 9690, −87.4396, 11 May 2017, 1500–0 m, MOCNESS plankton net, L. Timm, coll.

Mysis. Size: 9 mm (Carapace length); 21 mm (Total length). N = 3.

Carapace (Figure 27A) with two lateral swollen process near the posterior margin, rostrum long, extend until the end of the article 1 of the antennule; orbital spine as a projected bump; antennal spine is a small bump; anteroventral margin bearing 1 strong and curved pterygostomial spine; eyes pedunculate.

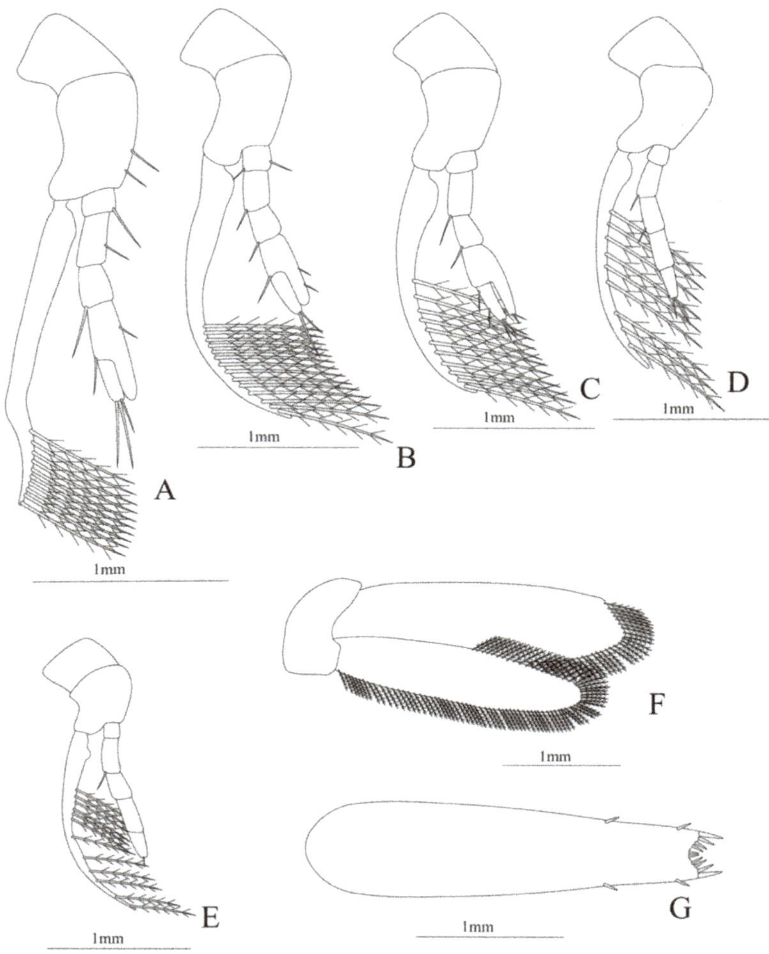

Figure 26. *Hemipenaeus carpenteri*: (**A**) first pereopod; (**B**) second pereopod; (**C**) third pereopod; (**D**) fourth pereopod; (**E**) fifth pereopod; (**F**) uropods; (**G**) telson.

Pleon (Figure 27A) with 6 somites, no spines or setae. Pleopods without setae.

Antennule (Figure 27B). Peduncle 3-segmented, article 1 the longest, slender, with 3 simple and 9–12 plumose setae, article 2 also with 6 plumose setae in the outer margins and article 3, the smallest with 3 lateral simple setae and two distal flagella, outer flagella unarmed and inner flagella with 4 lateral simple setae and 2 distal setae.

Antenna (Figure 27C). Protopod 3-segmented with a flagellum; exopod with 62–69 plumose setae.

Mandible (Figure 27D). Palp 2-segmented, article 1 with 7–10 plumose setae and article 2 with 13–15 plumose setae (7 lateral plus 6 terminal).

Maxillule (Figure 27E). Coxal endite with 7 curved conical spines and 1 subterminal simple setae; basial endite with 11 plumose setae.

Maxilla (Figure 27F). Coxal endite bilobed with 6 + 8 simple setae; basial endite bilobed with 6 + 8 plumose setae; endopod with 5 (2 + 1 + 2) plumose setae, segmentation not well defined; scaphognathite margin with 89–92 plumose setae.

First maxilliped (Figure 27G). Coxa with 8–10 plumose setae; basis with 14–18 plumose setae in the margin and 10–12 simple setae; endopod unsegmented with 11 (4 + 2 + 1 + 1 + 3) simple setae; exopod unsegmented, armed with 8 plumose setae.

Second maxilliped (Figure 27H). Coxa without setae; basis with 5–8 simple setae; endopod 5-segmented with 5–6, 5–7, 5, 7–12, 8–9 serrulated setae; exopod unsegmented, armed distally with 7–9 plumose setae.

Third maxilliped (Figure 27I). Coxa without setae; basis with 4 simple setae; endopod 5-segmented with 5, 3, 4, 6, 8, all simple setae; exopod unsegmented armed distally with 5–7 plumose setae.

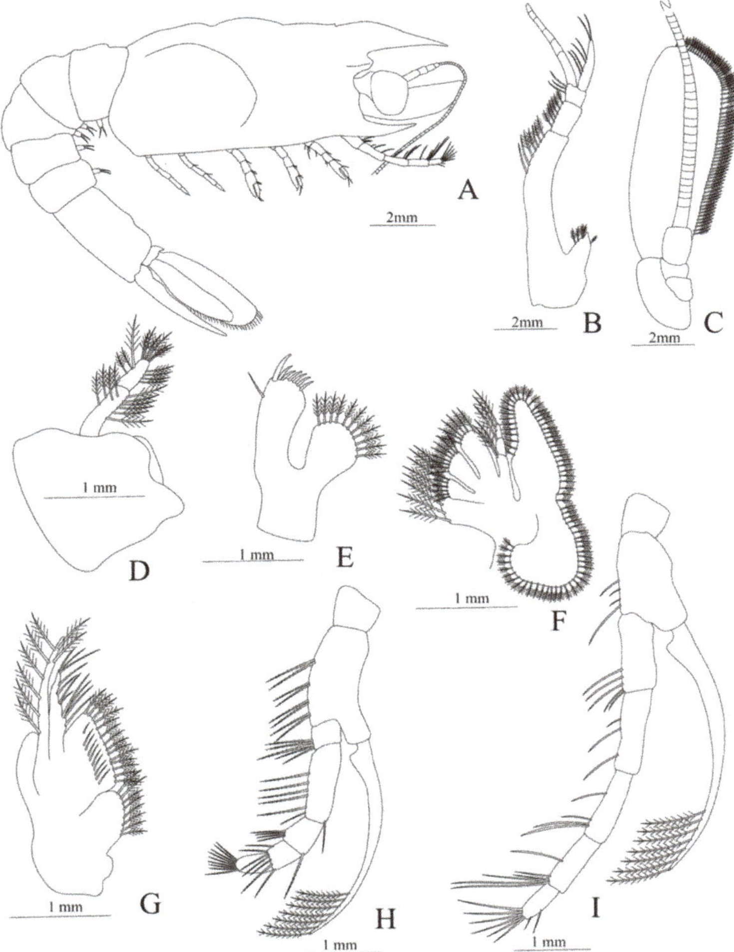

Figure 27. *Hemipenaeus carpenteri*: (**A**) lateral view; (**B**) antennule; (**C**) antenna; (**D**) mandible; (**E**) maxillule; (**F**) maxilla; (**G**) first maxilliped; (**H**) second maxilliped; (**I**) third maxilliped.

First pereopod (Figure 28A). Coxa and basis without setae; endopod 5-segmented with 0, 0, 2, 3, 2 setae; exopod unsegmented, armed with 7–10 plumose natatory setae.

Second pereopod (Figure 28B). Coxa without setae, basis with 2 simple setae; endopod 5-segmented with 3, 2, 3, 1, 4 simple setae; exopod unsegmented, armed with 7–9 plumose natatory setae.

Third pereopod (Figure 28C). Coxa and basis without setae; endopod 5-segmented with 0, 1, 1, 3, 3 simple setae; exopod unsegmented, armed with 9–12 long, plumose natatory setae.

Fourth pereopod (Figure 28B). Coxa and basis without setae; endopod 5-segmented with 0, 1, 1, 0, 1 simple seta; exopod unsegmented, armed with 11–12 long plumose natatory setae.

Fifth pereopod (Figure 28A). Coxa and basis unarmed; endopod 5-segmented with 0, 0, 0, 0, 1 simple setae; exopod unsegmented armed with 10–12 long plumose natatory setae.

Uropod (Figure 28F). Endopod well developed with 80–85 plumose setae; exopod with 60–63 plumose setae.

Telson (Figure 28G) elongate, subtriangular, with 3 pairs of lateral spines and 5 pairs of distal spines.

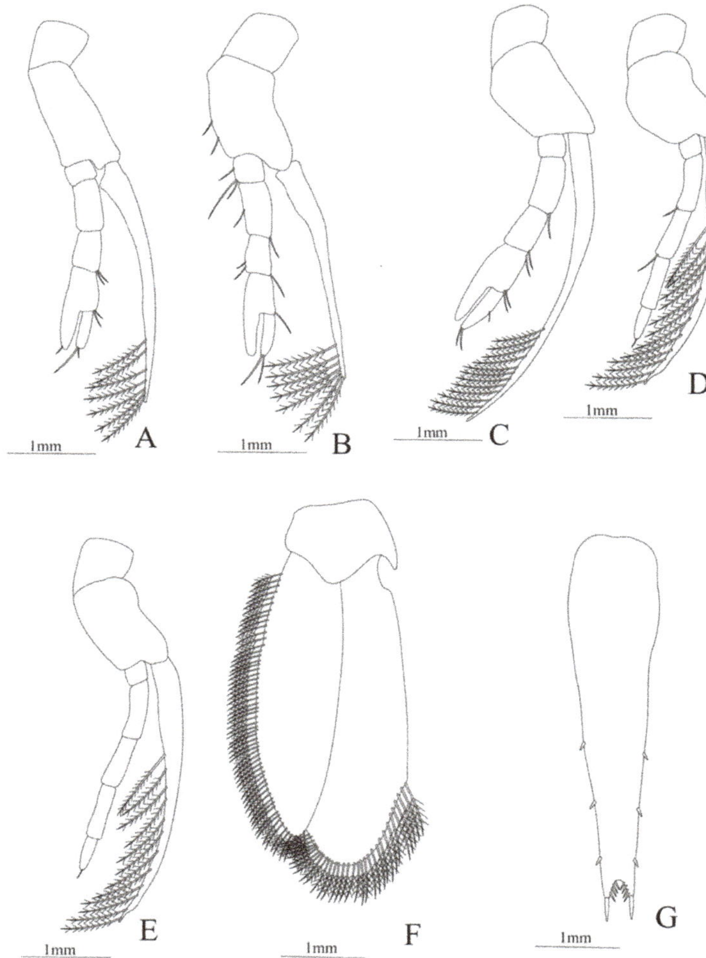

Figure 28. *Hemipenaeus carpenteri*: (**A**) first pereopod; (**B**) second pereopod; (**C**) third pereopod; (**D**) fourth pereopod; (**E**) fifth pereopod; (**F**) uropods; (**G**) telson.

Cerataspis monstrosus (Gray, 1828)
(Figures 29 and 30)

Material examined: Gulf of Mexico: HBG 9204, R/V Point Sur, DP06-24Jul18-MOC10-B251N-106-N1, 28.5401, −88.4711 and 28.5122, −88.6337, 24 July 2018, 1201–1475 m, MOCNESS plankton net, H. Bracken-Grissom, coll.

Mysis. Size. 6 mm (Carapace length); 20 mm (Total length). N = 1.

Carapace (Figure 29A) with two small lateral swollen process near the posterior margin, rostrum long, extend until the end of the article 1 of the antennule, slightly curved; anteroventral margin bearing one small pterygostomian spine; eyes pedunculate.

Pleon (Figure 29A) with 6 somites, small spine on dorsal third somite. Pleopods 4–5 missing in the specimen, pleopods 1–3 without setae.

Antennule (Figure 29B). Peduncle 3-segmented, article 1 the longest, slender, with 35 plumose setae in both margins, article 2 with 18 plumose setae in both margins and article 3, the smallest with 6 plumose setae and two flagella distally.

Antenna (Figure 29C). Protopod 2-segmented with a flagellum; exopod with 86 plumose setae and a pointed process distally.

Mandible (Figure 29D). Palp 4-segmented, articles 1- 3 unarmed, article 4 with 7 simple setae.

Maxillule (Figure 29E). Coxal endite with 13 conical setae; basial endite with 15 conical setae, protopod with two simple setae.

Maxilla (Figure 29F). (Damaged in the specimen). Coxal endite and the bilobed basial endite bilobed unarmed; endopod with 5 (1 + 2 + 2) simple setae, segmentation not well defined; scaphognathite margin with 38 plumose setae.

First maxilliped (Figure 29G). (Damaged in the specimen). Coxa and basis unarmed; endopod unsegmented with 17 plumose setae; exopod 4 segmented with 0, 2,14, 8 plumose setae.

Second maxilliped (Figure 29H). (Damaged in the specimen). Coxa and basis unarmed; endopod 4-segmented with 3, 1, 1, 2 simple setae; exopod unsegmented and unarmed.

Third maxilliped missing in the specimen.

First pereopod (Figure 29I). Coxa and basis unarmed; endopod 5-segmented with 2, 1, 0, 0, 0 setae; exopod unsegmented and unarmed.

Second pereopod (Figure 30A). Coxa unarmed; Basis with 3 setae; endopod 5-segmented with 3, 0, 4, 0, 0 setae; exopod unsegmented and unarmed.

Third pereopod (Figure 30B). Coxa and basis unarmed; endopod 5-segmented with 2, 0, 0, 0, 0 setae; exopod unsegmented and unarmed.

Fourth pereopod (Figure 30C). (Damaged in the specimen) Coxa and basis unarmed; endopod 5-segmented with 3, 0, 1, 0, 0 setae; exopod unsegmented and unarmed.

Fifth pereopod (Figure 30D). (Damaged in the specimen). Coxa and basis unarmed; endopod 5-segmented with 2, 5, 0, 0, 0 setae; exopod unsegmented and unarmed.

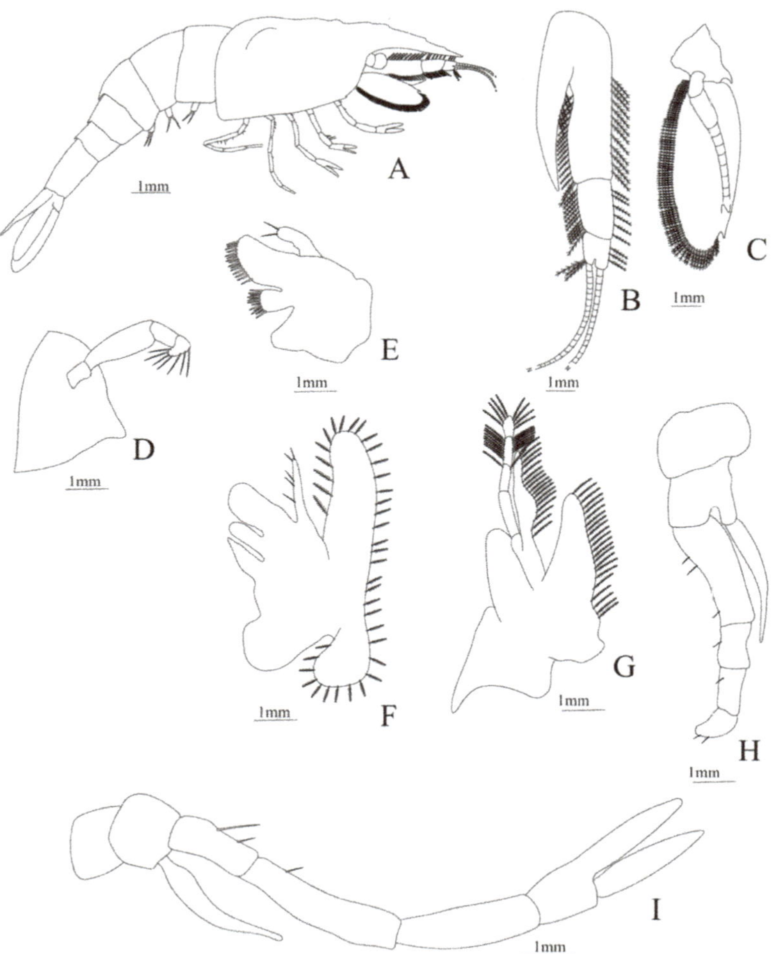

Figure 29. *Cerataspis monstrosus*: (**A**) lateral view; (**B**) antennule; (**C**) antenna; (**D**) mandible; (**E**) maxillule; (**F**) maxilla; (**G**) first maxilliped; (**H**) second maxilliped; (**I**) first pereopod.

Uropod (Figure 30E). Endopod well developed with 96 plumose setae; exopod with 120 plumose setae.

Telson (Figure 30F). (Damaged in the specimen) Subrectangular, distal margin bearing row of 13 min spines and 3 pairs of spines on lateral margin, small simple setae between the lateral spines.

Family Penaeidae Rafinesque, 1815
Genus *Funchalia* J. Y. Johnson, 1868
Funchalia villosa (Bouvier, 1905)
(Figures 31 and 32)

Material examined: Gulf of Mexico: HBG 6776, R/V Point Sur, DP04-06Aug16-MOC10-SW6N-059-N4, from 26.9936, −89.9941 to 27.0451, −90.0844, 6 August 2016, 601–4 m, MOCNESS plankton net, H. Bracken-Grissom, coll. Gulf of Mexico: HBG 6885, R/V Point Sur, DP04-06Aug16-MOC10-SW6D-058-N0, from 26.9942, −89, 9938 to 27.0611, −90.0923, 6 August 2017, 1510.6-NA m, MOCNESS plankton net, H. Bracken-Grissom, coll. Gulf of

Mexico: HBG 7941, R/V Point Sur, DP05-08May17-MOC10-B081N-083-N0, from 28.5187, −87, 9897, 8 May 2017, 1500–0 m, MOCNESS plankton net, L. Timm, coll.
Juvenile. Size. 11 mm (Carapace length); 32 mm (Total length). N = 3.

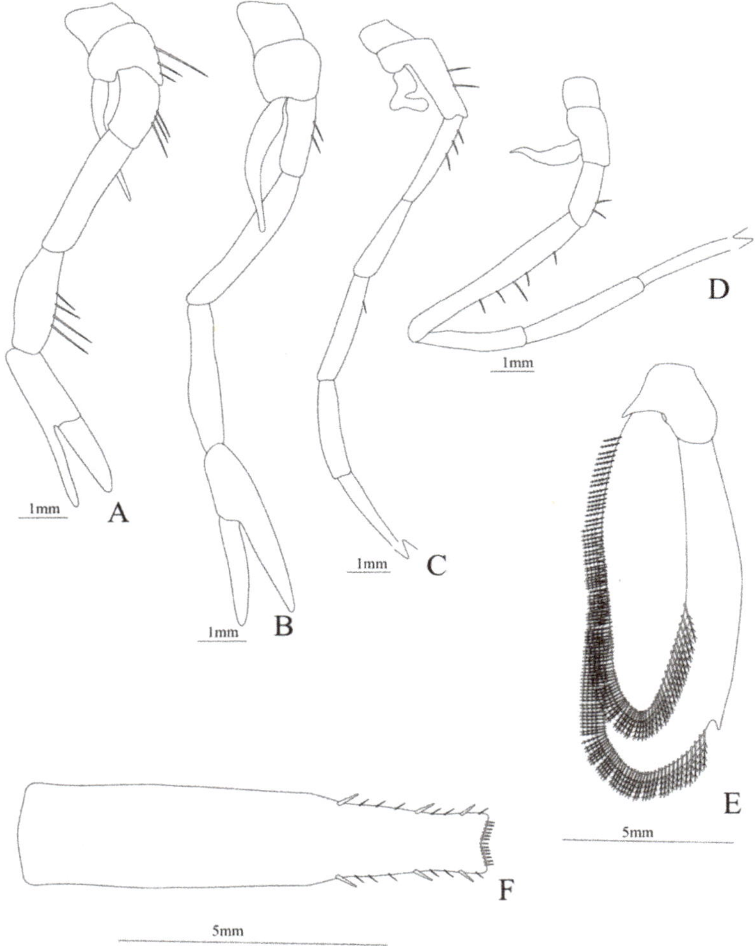

Figure 30. *Cerataspis monstrosus*: (**A**) second pereopod; (**B**) third pereopod; (**C**) fourth pereopod; (**D**) fifth pereopod; (**E**) uropods; (**F**) telson.

Carapace (Figure 31A) with rostrum short, armed with 5–7 dorsal spines, epigastric tooth present.

Pleon (Figure 31A) with 6 somites, without spines or setae. Pleopods 2 and 4 missing in the specimen, pleopods 1, 3 and 5 well development.

Antennule (Figure 31B). Peduncle 3-segmented, article 1 the longest, slender, with 28 simple plus 6 plumose setae, article 2 with 24 simple setae and article 3, the smallest with 10 simple setae and two flagella distally.

Antenna (Figure 31C). Protopod 3-segmented with a flagellum; exopod with 30–48 plumose setae.

Mandible (Figure 31D). Palp 2-segmented, articles 1armed with 3–8 simple setae and article 2 with 18–44 plumose setae.

Maxillule (Figure 31E): Coxal endite with 26–43 (12–22 serrulated plus 14–21 conical serrulated) setae; basial endite with 18 plumose setae setae.

Maxilla (Figure 31F). Coxal endite with one simple setae, basial endite bilobed with 6–12 + 8–16 simple setae; endopod with one simple setae, segmentation not well defined; scaphognathite margin with 65–126 plumose setae.

First maxilliped (Figure 31G). Coxa with 6 simple setae, basis with 14–26 simple setae; endopod unsegmented with 5 simple setae; exopod with 11–19 simple setae.

Second maxilliped (Figure 31H). Coxa without setae, basis with 5–7 simple setae; endopod 4-segmented with 11–18, 0–3, 12–22 serrated, 6–16 serrated setae; exopod unsegmented and unarmed.

Third maxilliped (Figure 31I). Coxa and basis without setae, endopod 5-segmented with 7–10, 3–5, 11–16, 11–21, 9–21 simple setae; exopod with 8–34 setae.

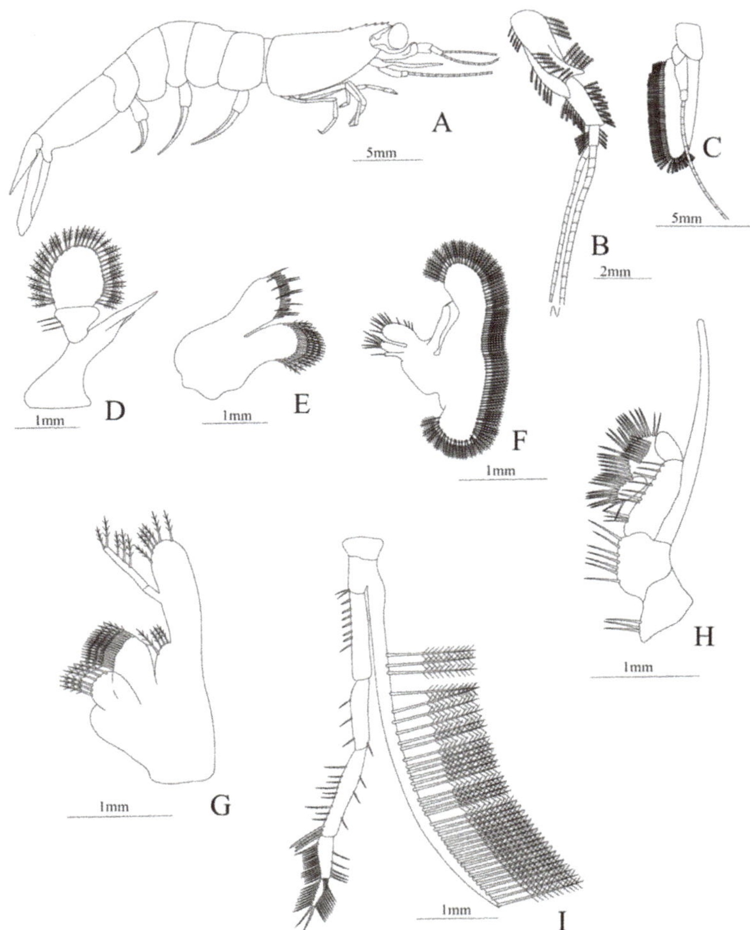

Figure 31. *Funchalia villosa*: (**A**) lateral view; (**B**) antennule; (**C**) antenna; (**D**) mandible; (**E**) maxillule; (**F**) maxilla; (**G**) first maxilliped; (**H**) second maxilliped; (**I**) third maxilliped.

First pereopod (Figure 32A). Coxa and basis with 2 setae; endopod 5-segmented with 4–5, 4–8, 7–15, 6–11, 3–7 setae.

Second pereopod (Figure 32B). Coxa and basis with 2 simple setae; endopod 5-segmented with 3–6, 9–20 (3–9 spines plus 6–11 simple), 8–21, 6–8, 4–5 simple setae.

Third pereopod (Figure 32C). Coxa with 2 simple setae, basis without setae; endopod 5-segmented with 4–14, 10–16, 7–14, 7–9, 1–7 simple setae.

Fourth pereopod (Figure 32D). Coxa with 2 simple setae, basis with one seta; endopod 5-segmented with 6–15, 16–39, 8–10, 12–21, 0 simple setae.

Fifth pereopod (Figure 32E). Coxa with 3–6 simple setae, basis with 2–4 setae; endopod 5-segmented with 5–14, 10–16, 3–13, 3–9, 0 simple setae.

Uropod (Figure 32F). Endopod well developed with 30–126 plumose setae; exopod with 54–143 plumose setae.

Telson (Figure 32G) enlarged, subtriangular, distal margin with a pointed projection, 3 pairs of spines near the distal margin, lateral margins with small simple setae.

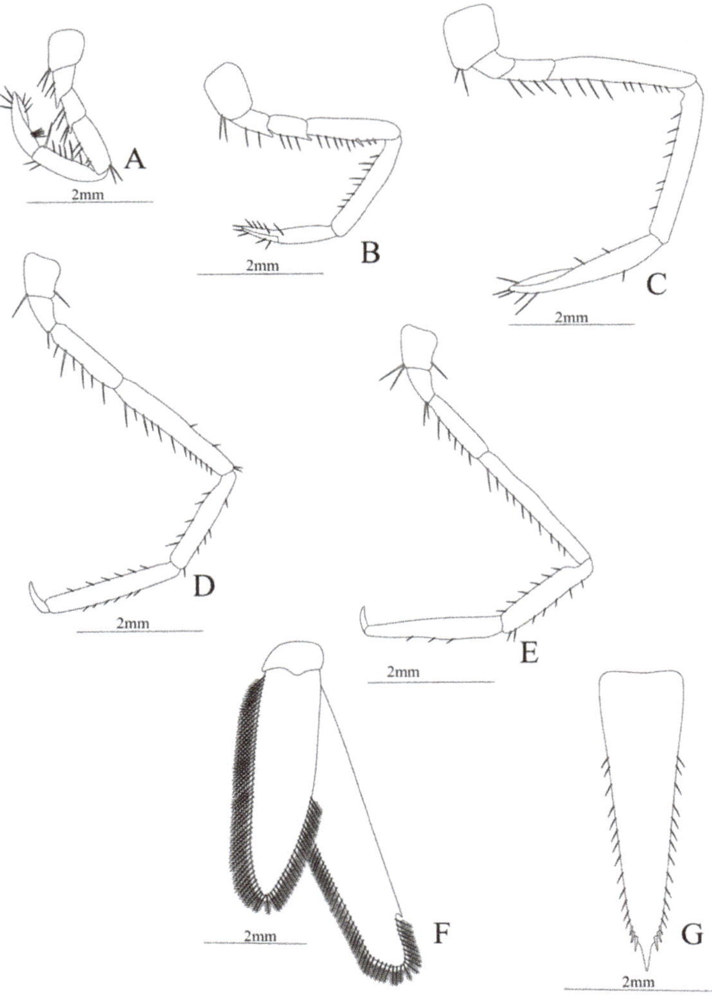

Figure 32. *Funchalia villosa*: (**A**) first pereopod; (**B**) second pereopod; (**C**) third pereopod; (**D**) fourth pereopod; (**E**) fifth pereopod; (**F**) uropods; (**G**) telson.

4. Discussion

Here, we use DNA barcoding to successfully match 16 developmental stages and 14 larval and juvenile species with their adult counterpart. In the Section 3 we provide the phylogenetic evidence for the larval-adult identification accompanied by taxonomic descriptions and illustrations. Below, we summarize our main findings with a brief description of the current state of knowledge for deep-sea larval biology across each group. For many of these deep-sea shrimp species and some families, larval descriptions are scarce or non-existent.

It is important to note that many of these species likely have multiple larval stages and much more work is needed to fully describe the life history. Developmental plasticity in the number of larval stages is common for shrimps and several factors, including temperature, salinity and available food, can influence this variability [62–67]. These factors affect the molting cycle and can produce morphological differences across larvae stages [68]. Even at the population level, the same species can have a different number of larval stages and variation in the morphology (ex. the armature of thoracopods and pereopods [6,69].

4.1. Suborder Dendrobranchiata

4.1.1. Family Aristeidae

The family Aristeidae contains nine genera, of which only 6 are present in the Gulf of Mexico [39,70]. The species in this family predominantly occupy deep-sea benthic habitats, although there are species that inhabit the meso- and bathypelagic zones, where they play an important role in the oceanic food web [71,72]. For almost 180 years, the larval stages of some genera within this family including *Plesiopenaeus* (=*Cerataspis*) and *Aristaeomorpha* Wood-Mason, 1891 were called *"Cerataspis"*. These "cerataspis-like" individuals were so morphologically distinct and bizarre they were considered a valid genus and their affinity to other groups was unknown [73]. However, in 2012, Bracken-Grissom et al., used molecular techniques to unravel the mystery surrounding one larval form called *Cerataspis montrosus*, identifying the adult counterpart to be *Plesiopenaeus armatus* within the family Aristeidae. Larval stages of these deep-sea shrimp are frequently found in the stomach contents of fish and collected in nekton nets in shallow water and deep-sea waters. In the Gulf of Mexico, the mysis stage of *Cerataspis montrosus* Gray, 1828 is the only record from this family [74] and the larval stages of *Aristeus antennatus* and *Aristeomorpha foliacea* have been previously recorded [75–78].

In the present study, two mysis stages of *Hemipenaeus carpenteri* and an additional mysis stage of *Cerataspis monstrosus* are illustrated. Identifications were made using sequences obtained by [74]. In the case of the two zoea stages of *H. carpenteri*, we have found that both stages morphologically resemble the mysis II and mysis III stages described by [48] for *Cerataspis monstrosus*. This finding verifies that it is typical for multiple species within the family Aristeidae to present these bizarre "cerataspis-like" pelagic larval stages. In the case of the zoea *Cerataspis monstrosus*, our material appears to be an undescribed mysis stage and could be a more advanced developmental stage that the ones described by [48] due to the reduction of the exopods in the 1–5 pereopods. Nevertheless, more material is needed to confirm this result.

4.1.2. Family Penaeidae

The family Penaeidae consists of 27 genera, of which only eight are present in the Gulf of Mexico [39,70]. Many species within this family are considered valuable resources for fisheries and aquaculture, both in tropical and subtropical regions [79,80]. Many of the larval stages of species of commercial interest are known, such as the genera *Penaeus* Fabricius, 1798, *Metapenaeopsis* Bouvier, 1905, *Rimapenaeus* Perez-Farfante and Kensley, 1997 and *Trachypenaeus* Perez-Farfante, 1972, nevertheless, there are still problems in the identification of these larval stages [25,81–83].

In this paper, we have a juvenile *Funchalia villosa* which was identified using sequences obtained by [35]. This species is pelagic, and it is known to perform diel vertical migrations, descending to 2608 m deep during the day and migrating to shallow water of around 50 m deep at night [84,85]. Our material does not present exopods on the pereopods and resembles an adult specimen according to [86].

4.2. Infraorder Caridea

4.2.1. Family Acanthephyridae

The family Acanthephyridae consists of seven genera, with six genera present in the Gulf of Mexico: *Acanthephyra, Heterogenys, Hymenodora, Ephyrina, Meningodora* and *Notostomus* [39,70]. This family inhabits only deep waters and meso-bathypelagic habitats, and many perform daily vertical migrations [87–89]. Past studies examining the larval biology of this family within the Gulf of Mexico is lacking, however some work does exist for species of *Acanthephyra* [90–93]. Egg size across the family varies drastically and much work is still to be done [67]. Past studies have divided the family into two major groups based on developmental characertistics. Group one consists of the genera *Ephyrina* and *Hymenodora* which have large lipid-filled eggs and five or fewer zoeal stages, whereas group two includes the genera *Acanthephyra, Meningodora* and *Notostomus* which have small eggs and nine or more planktotrophic stages [67,90].

In our study we identified one zoea stage of *Meningodora longisulca* and *Ephyrina ombango,* and one decapodite stage of *M. vesca* using sequences from [35]. In all cases, these are the first descriptions and illustrations of developmental stages belonging to these pelagic species. For *Meningodora longisulca,* the zoea is half the reported size for the adult and differs in several morphological characteristics. These include a zoea with (1) an unarmed rostrum in contrast to an armed adult rostrum with 7–10 dorsal spines, (2) the cornea wider than the peduncle in contrast to the adult cornea slightly narrowed than the eyestalk, and (3) underdeveloped mouthparts. For *Meningodora vesca,* the decapodite has characters very similar to those of the adult [87,94]. For *Ephyrina ombango,* the zoea differs from the adult in the shape of the rostrum. This includes the zoea possessing a blunt projection compared to the adult rostrum directed slightly anterodorsally [87–89].

4.2.2. Family Alvinocarididae

The family Alvinocarididae consists of 9 genera, but only the genus *Alvinocaris* is present in the Gulf of Mexico [39,70]. The members of this family are understudied and inhabit deep-sea cold seeps and hydrothermal vents areas around the world, with depths that vary from 250 to 4500 m [95–97]. This family is fraugth with taxonomic problems because larval stages have been erroneously described as new genera or species [98,99].

Across all alvinocaridids, only the morphology of the first zoea of four species is known including *Alvinocaris muricola* Williams, 1988, *Mirocaris fortunata* (Martin and Christiansen, 1995), *Nautilocaris saintlaurentae* Komai and Segonzac, 2004 and *Rimicaris exoculata* Williams and Rona, 1986 [98]. In this study we found a decapodite stage of *Alvinocaris stactophila,* and to identify this material, we used the sequences obtained by [99]. Our material is close to the adult size range; however, it still differs in some characteristics. This includes the shape of the decapodite carapace which is longer than wide, and the adult carapace is almost as long as wide. Differences also exist in mouthparts including the armature of the maxillipeds 2 and 3 lacking setae, which is a larval characteristic of this family. However, the remaining mouthparts such as the maxillula, maxilla and maxilliped 1 present an armature similar to that described for the adult [93,99]. As reported for several other species of alvinocaridids, the larval stages of *Alvinocaris stactophila* are pelagic [100,101]. This was confirmed with our material since the decapodite was captured using a MOCNESS trawl at depths of 600–1000 m. The adult of this species is benthic, inhabiting cold seeps at a depth of 534 m, making this a new depth record for this species. It is still unknown how the pelagic larval forms locate cold and hydrothermal seeps as they are presumably located 10 s to 100 s of meters from these ecosystems.

4.2.3. Family Eugonatonotidae

The family Eugonatonotidae consists of only one genus, *Eugonatonus* Schmitt, 1926, which is present in the Gulf of Mexico [39,70]. The collection of this deep-sea species has been considered rare or very unusual [102,103]. The lack of knowledge surrounding the larval stages of the species has led to the description of *Galatheacaris abyssalis* and the creation of the family Galatheacarididae (=Eugonatonotidae) and the superfamily Galatheacaridoidea (=Nematocarcinoidea) [104]. This mistake was later corrected by [105] which found the new discovery to be a larval stage of *Eugonatonotus chacei* Chan and Wu, 1991.

Our material contains a zoea stage of *Eugonatonotus crassus*, which was identified using the sequences of [106]. De Grave et al., [105], states that this genus of benthic shrimp possibly has several planktonic zoeal stages. This is the first time that illustrations for the zoeal stage of *Eugonatonotus crassus* have been documented.

4.2.4. Family Nematocarcinidae

The family Nematocarcinidae consists of five genera, of which only two are present in the Gulf of Mexico [39,70]. The members of this family represent a wide bathyal distribution and can be found associated with the benthic community [87,88,107]. Illustrations of larval stages have only been recorded for very few species within the genus *Nematocarcinus* [65].

In the present study, the zoea and decapodite of *N. rotundus* and the zoea of *N. cursor* are illustrated. To identify this material, we used the sequences of [108] and sequences obtained from adult specimen material found in the Florida International Crustacean Collection (FICC) that were identified using [88,107]. It appears that both zoeal stages of *N. rotundus* and *N. cursor* are advanced based on size [65]. As for the decapodite of *N. rotundus*, the specimen shows characters similar to those of the adult. These include a short rostrum (with dorsal teeth) that does not exceed the article 2 of the antennule and a telson that does not exceed the uropods. This is the first time that illustrations of these developmental stages have been recorded for *N. cursor* and *N. rotundus*.

4.2.5. Family Oplophoridae

The family Oplophoridae consists of three genera, all of which are present in the Gulf of Mexico [39,70]. The members of this family, like those of the family Acanthephyridae, inhabit deep waters in meso-bathypelagic habitats and perform daily vertical migrations [87,89]. For this family, larval stage illustrations have only been reported for two species, *Oplophorus spinosus* and *Systellaspis debilis* [67,90,91].

In this paper, information on the decapodite stage of *Systellaspis braueri* is provided for the first time and identifications were given using the sequences obtained by [35]. The complete larval development of *S. debilis* has four zoeal stages and one decapodite stage, which suggests that the species of this genus are lecithotrophic and have a short larval development with few stages. Lecithotrophy is considered an adaptation to the deep-sea environment where they live [67].

4.2.6. Family Pandalidae

The family Pandalidae consists of 19 genera, of which only three, *Heterocarpus*, *Pantomus* and *Plesionika*, are present in the Gulf of Mexico [39,70]. The representatives of this family are distributed world-wide, and many species inhabit deep waters [109]. In addition, due to their size, some species are of commercial interest [110–113]. The number of zoeal stages varies greatly among species within the family Pandalidae, where the complete life cycle of these species has been studied. For example, in the genus *Pandalopsis* (=*Pandalus*), the life cycle is completed in only 3–5 zoeal stages, while in the genus *Pandalus* Leach, 1814 (in Leach, 1813–1815), depending on the species, the life cycle is completed in 2–7 zoeal stages [114]. It is also known that species within the genus *Plesionika* have at least 7 to 8 zoeal stages [115].

In the present study, the juvenile stage of *Plesionika ensis*, decapodite stage of *P. edwarsii* and a zoea stage of *Heterocarpus ensifer* are presented. All material was identified using sequences obtained by [35,116]. Although the complete larval development of species belonging to the genus *Plesionika* are still unknown [114,115], past studies have reported the larval stages from seven species. This includes the following: the first zoeal stages for *Plesionika acanthonotus* (Smith, 1882), *P. crosnieri* Chan and Yu, 1991, *P. ortmanni* Doflein, 1902 and *P. semilaevis* Bate, 1888; the first to the seventh zoeal stages for *P. edwardsii* (Brandt, 1851); the first to the eighth zoeal stages for *P. grandis* Doflein, 1902; and the first five zoeal and the decapodite stages for *P. narval* (J. C. Fabricius, 1787) [10,46,116,117]. In the material presented here, the zoeal stages of the species in the genus *Plesionika* have the dorsal connection between carapace and abdomen at an almost 180° angle, an eye peduncle narrowed at base, antennular peduncles strongly concave, a well-developed rostrum since the first stage and with dorsal spines in later stages, supraorbital spines present, and pereiopod 5 without an exopod [114]. The decapodite stages have a carapace with anterior and posterior dorsomedial tubercles, supraorbital spines present, a mandible without palp, the first four pereopods with exopods, and a carpus of pereiopod two not multi-articulated [114]. The material of *Plesionika ensis* represents a juvenile specimen, in which, the pereopods show reduction of the exopods, the mandibular palp is developed and the carpus of pereopod two is subdivided. Our material of *P. edwarsii* seems to be a decapodite stage due to the absence of the mandibular palp, the non-segmentation of the carpus in pereopod two, and the reduction in the pereopodal exopods. The reduction of exopods in the pereopods has also been recorded for the decapodite state of *Plesionika narval* [114].

Regarding *Heterocarpus ensifer*, only the first four zoeal stages of this species have been recorded [44]. Our material appears to be a more advanced zoea stage, presenting characters common to the zoea of the family Pandalidae, such as dorsal connection between carapace and abdomen at an almost 180-degree angle, the eye peduncle narrowed at base, well-developed rostrum, and supraorbital spines present. However, our material lacks a mandibular palp, subdivision of the carpus of pereopod two and has exopods on pereopods 1–4. These findings support our hypothesis that our *Heterocarpus* material is from a more advanced zoea stage. Our material represents the first illustrations of a juvenile of *P. ensis*, a decapodite stage of *P. edwardsii* and a zoea of *H. ensifer*.

5. Conclusions

This study represents the benefits of using DNA barcoding to help advance the field of larval biodiversity. More specifically, these methods can be used as a complementary approach alongside taxonomy to assist in species identification. This is especially useful for species where the larval morphology differs significantly from the adult and those that are difficult to rear in the laboratory [118,119]. Together, molecular and morphological methods hold great promise in the conservation of marine biodiversity [120] and should be used to reveal the unseen, bizarre and mysterious world that exists in the deep sea.

Supplementary Materials: The following are available online at https://www.mdpi.com/article/10.3390/d13100457/s1, Table S1. Taxonomy, voucher catalog numbers, localities and GenBank (GB) accession numbers for gene sequences used in the study. An "N/A" = not available. Gulf of Mexico = GOM), Florida Straits = FL Straits, and Mediterranean = Mediterr, Figure S1. Maximum Likelihood (ML) phylogeny of 70 barcoded individuals from the suborder Dendrobranchiata and infraorder Caridea based on the mitochondrial 16S gene, Figure S2. Maximum Likelihood (ML) phylogeny of 43 barcoded individuals from the suborder Dendrobranchiata and infraorder Caridea based on the mitochondrial COI gene.

Author Contributions: C.V. and H.B.-G. have contributed to the conceptualization, methodology and writing of this work; C.V. contributed to the generation and analysis of the data and development of the figures; H.B.-G. contributed to securing funding for the work. Both authors have read and agreed to the published version of the manuscript.

Funding: This research was made possible by grants from The Gulf of Mexico Research Initiative (GOMRI), the National Science Foundation (NSF) Division of Environmental Biology Grant (# 1556059 awarded to H.B.-G.), the National Oceanic and Atmospheric Administration's (NOAA) RESTORE Science Program (# NA19NOS4510193 to Florida International University and Nova Southeastern University) and the NOAA Ocean Exploration Research (NOAA-OER) Grant (#NA170AR0110208; awarded to S. Johnsen).

Institutional Review Board Statement: Not applicable.

Data Availability Statement: Data are publicly available through the Gulf of Mexico Research Initiative Information and Data Cooperative (GRIIDC) at https://data.gulfresearchinitiative.org. https://doi.org/10.7266/N70P0X3T and https://doi.org/10.7266/n7-1xs7-4n30.

Acknowledgments: The authors thank all members of the Deep-Pelagic Nekton Dynamics of the Gulf of Mexico (DEEPEND) research consortium and the NSF Bioluminescence and Vision team for their assistance during the cruises. A special thanks to Danté Fenolio for the captivating images of crustacean larvae and plate assembly, the CRUSTOMICS lab for helping sort and curate samples, and the crew of the R/V *Point Sur* and University of Miami R/V *Walton Smith* for their assistance in collecting these specimens. We would also like to extend gratitude to PhD Candidate Carlos Varela's PhD committee: Ligia Collado-Vides, Elizabeth Anderson, Jose M. Eirin-Lopez and DeEtta Mills. This is contribution #289 from the Coastlines and Oceans Division in the Institute of Environment at Florida International University.

Conflicts of Interest: The authors declare no conflict of interest. The funders had no role in the design of the study; in the collection, analyses, or interpretation of data; in the writing of the manuscript, or in the decision to publish the results.

References

1. Heppell, S.S.; Caswell, H.; Crowder, L.B. Life histories and elasticity patterns: Perturbation analysis for species with minimal demographic data. *Ecology* **2000**, *81*, 654–665. [CrossRef]
2. McGill, B.J.; Enquist, B.J.; Weiher, E.; Westoby, M. Rebuilding community ecology from functional traits. *Trends Ecol. Evol.* **2006**, *21*, 178–185. [CrossRef] [PubMed]
3. Williams, E.H.; Bunkley-Williams, L. Life Cycle and Life History Strategies of Parasitic Crustacea. In *Parasitic Crustacea*; Zoological Monographs; Smit, N., Bruce, N., Hadfield, K., Eds.; Springer: Cham, Switzerland, 2019; Volume 3, pp. 179–266.
4. Bondad-Reantaso, M.G.; Subasinghe, R.P.; Josupeit, H.; Cai, J.; Zhou, X. The role of crustacean fisheries and aquaculture in global food security: Past, present and future. *J. Invertebr. Pathol.* **2012**, *110*, 158–165. [CrossRef] [PubMed]
5. Wolfe, J.; Breinholt, J.; Crandall, K.; Lemmon, A.; Lemmon, E.; Timm, L.E.; Siddall, M.; Bracken-Grissom, H. A phylogenomic framework, evolutionary timeline, and genomic resources for comparative studies of decapod crustaceans. *Proc. R. Soc. B* **2019**, *286*, 20190079. [CrossRef] [PubMed]
6. Anger, K. *The Biology of Decapod Crustacean Larvae*; Crustacean Issues; A.A. Balkema Publishers: Amsterdam, The Netherlands, 2001; Volume 14, 419p.
7. Martin, J.W.; Olesen, J.; Hoeg, J.T. (Eds.) Introduction. In *Atlas of Crustacean Larvae*; John Hopkins University Press: Baltimore, MD, USA, 2014; pp. 1–7.
8. Matsuda, H.; Takenouchi, T.; Goldstein, J.S. The Complete Larval Development of the Pronghorn Spiny Lobster *Panulirus penicillatus* (Decapoda: Palinuridae) in Culture. *J. Crustacean Biol.* **2006**, *26*, 579–600. [CrossRef]
9. Goldstein, J.S.; Matsuda, H.; Takenouchi, T.; Butler, M.J. IV The Complete Development of Larval Caribbean Spiny Lobster *Panulirus argus* (Latreille, 1804) in Culture. *J. Crustacean Biol.* **2008**, *28*, 306–327. [CrossRef]
10. Jiang, G.C.; Chan, T.Y.; Shih, T.W. Larval development to the first eighth zoeal stages in the deep-sea caridean shrimp *Plesionika grandis* Doflein, 1902 (Crustacea, Decapoda, Pandalidae). *Zookeys* **2017**, *719*, 23–44. [CrossRef]
11. Sanvicente-Añorve, L.; Zavala-Hidalgo, J.; Allende-Arandía, E.; Hermoso-Salazar, M. Larval dispersal in three coral reef decapod species: Influence of larval duration on the metapopulation structure. *PLoS ONE* **2018**, *13*, e0193457. [CrossRef]
12. Harvey, A.W.; Martin, J.W.; Wetzer, R. Phylum Arthropods: Crustacea. In *Atlas of Marine Invertebrate Larvae*; Young, C.M., Sewell, M.A., Rice, M.E., Eds.; Academic Press: San Diego, CA, USA, 2002; pp. 337–370.
13. Torres, A.P.; dos Santos, A.; Alemany, F.; Massuti, E. Larval stages of crustacean species of interest for conservation and fishing exploitation in the western Mediterranean. *Sci. Mar.* **2013**, *77*, 149–160. [CrossRef]
14. Epifanio, C.E. Early Life History of the Blue Crab *Callinectes sapidus*: A Review. *J. Shellfish Res.* **2019**, *38*, 1–22.
15. Cook, H.L. A generic key to the protozoean, mysis and post-larval stages of the littoral penaeidae of the northwestern Gulf of Mexico. *Fish Bull* **1966**, *65*, 437–447.
16. Pérez-Farfante, I. *Diagnostic Characters of Juveniles of the Shrimps* Penaeus aztecus aztecus, P. duorarum duorarum, *and* P. brasiliensis *(Crustacea; Decapoda; Penaeidae)*; Special Scientific Report; United States Fisheries Wildlife Service: Washington, DC, USA, 1970; Volume 559, pp. 1–26.
17. Subrahmanyam, C.B. Descriptions of Shrimp Larvae (Family Penaeidae) Off the Mississippi Coast. *Gulf Res. Rep.* **1971**, *3*, 241–258. [CrossRef]

18. Porter, H.J. Zoeal stages of the stone crab, *Menippe mercenaria* Say. *Chesap. Sci.* **1960**, *1*, 168–177. [CrossRef]
19. Martin, J.W. Notes and bibliography on the larvae of xanthid crabs, with a key to the known xanthid zoeas of the western Atlantic and Gulf of Mexico. *Bull. Mar. Sci.* **1984**, *34*, 220–239.
20. Martin, J.W.; Truesdale, F.M.; Felder, D.L. Megalopa stage of the Gulf stone crab *Menippe adina* Williams and Felder, 1986 with a comparison in the megalopae in the genus Menippe. *Fish Bull.* **1988**, *86*, 289–297.
21. Stuck, K.; Perry, H.; Graham, D.; Heard, R.W. Morphological Characteristics of Early Life History Stages of the Blue Crab, *Callinectes sapidus* Rathbun, from the Northern Gulf of Mexico with a Comparison of Studies from the Atlantic Seaboard. *Gulf Caribb. Res.* **2009**, *21*, 37–55. [CrossRef]
22. Manzanilla-Dominguez, H.; Gasca, R. Distribution and Abundance of Phyllosoma Larvae (Decapoda, Palinuridae) in the Southern Gulf of Mexico and the Western Caribbean Sea. *Crustaceana* **2004**, *77*, 75–93.
23. Ditty, J.G.; Salas, J.A. Misidentification of mysis stages of *Xiphopenaeus kroyeri* (Heller, 1862) and *Rimapenaeus* Perez—Farfante and Kensley, 1997 (Decapoda: Penaeidae) in the Western Atlantic. *J. Crustacean Biol.* **2012**, *32*, 931–939. [CrossRef]
24. Cházaro-Olvera, S.; Winfield Aguilar, I.; Ortiz Touzet, M.; Vázquez-López, H.; Horta-Puga, G.J. Morphology of the Zoeae Larvae of Brachyura (Crustacea, Decapoda) in Veracruz, Southwestern Gulf of Mexico. *Am. J. Life Sci.* **2013**, *1*, 238–242. [CrossRef]
25. Criales, M.; Varela, C. Morphological Features of Taxonomical Value for the Identification of Three Western Atlantic Penaeid Shrimp Genera (Decapoda: Penaeidae). *Gulf Caribb. Res.* **2018**, *29*, 34–41. [CrossRef]
26. González-Gordillo, J.I.; dos Santos, A.; Rodríguez, A. Checklist and annotated bibliography of decapod Crustacea larvae from the Southwestern European coast (Gibraltar Strait area). *Sci. Mar.* **2001**, *65*, 275–305. [CrossRef]
27. Dos Santos, A.; Gonzalez-Gordillo, J.A. An illustrated key for the identification of zoeal stages of Pleocyemata larvae (Crustacea, Decapoda) from the southwestern European coast. *J. Mar. Biol. Assoc. UK* **2004**, *84*, 205–227. [CrossRef]
28. Vecchione, M. Importance of assessing taxonomic adequacy in determining fishing effects on marine biodiversity. *ICES J. Mar. Sci.* **2000**, *57*, 677–681. [CrossRef]
29. Vela, M.J.; González-Gordillo, J.I. Larval descriptions of the family Porcellanidae: A worldwide annotated compilation of the literature (Crustacea, Decapoda). *ZooKeys* **2016**, *564*, 47–70.
30. Savolainen, V.; Cowan, R.S.; Vogler, A.P.; Roderick, G.K.; Lane, R. Towards writing the encyclopedia of life: An introduction to DNA barcoding. *Philos. Trans. R. Soc. Lond.* **2005**, *B360*, 1805–1811. [CrossRef]
31. Hebert, P.D.N.; Ratnasingham, S.; de Waard, J.R. Barcoding animal life: Cytochrome c oxidase subunit 1 divergences among closely related species. *Proc. R. Soc. Lond.* **2003**, *B270*, S96–S99. [CrossRef]
32. Mantelatto, F.L.; Carvalho, F.L.; Simões, S.M.; Negri, M.; Souza-Carvalho, E.A.; Terossi, M. New primers for amplification of cytochrome c oxidase subunit I barcode region designed for species of Decapoda (Crustacea). *Nauplius* **2016**, *24*, e2016030. [CrossRef]
33. Mantelatto, F.L.; Terossi, M.; Negri, M.; Buranelli, R.C.; Robles, R.; Magalhães, T.; Tamburus, A.F.; Rossi, N.; Miyazaki, M.J. DNA sequence database as a tool to identify decapod crustaceans on the São Paulo coastline. *Mitochondrial DNA Part A DNA Mapp. Seq. Anal.* **2018**, *29*, 805–815. [CrossRef]
34. Terossi, M.; Almeida, A.O.; Buranelli, R.C.; Castilho, A.L.; Costa, R.C.; Zara, F.J.; Mantelatto, F.L. Checklist of decapods (Crustacea) from the coast of the São Paulo state (Brazil) supported by integrative molecular and morphological data: I. Infraorder Caridea: Families Hippolytidae, Lysmatidae, Ogyrididae, Processidae and Thoridae. *Zootaxa* **2018**, *4370*, 76–94. [CrossRef] [PubMed]
35. Varela, C.; Golightly, C.; Timm, L.E.; Wilkins, B.; Frank, T.; Fenolio, D.; Collins, S.B.; Bracken-Grissom, H.D. DNA barcoding enhances large-scale biodiversity initiatives for deep-pelagic crustaceans within the Gulf of Mexico and adjacent waters. *J. Crustacean Biol.* **2021**, *41*, ruab005. [CrossRef]
36. Gimenez, L. Phenotypic links in complex life cycles: Conclusions from studies with decapod crustaceans. *Integr. Comp. Biol.* **2006**, *46*, 615–622. [CrossRef]
37. Gattolliat, J.L.; Monaghan, M.T. DNA-based association of adults and larvae in Baetidae (Ephemeroptera) with the description of a new genus *Adnoptilum* in Madagascar. *J. N. Am. Benthol. Soc.* **2010**, *29*, 1042–1057. [CrossRef]
38. Walsh, H.J.; Richardson, D.E.; Marancik, K.E.; Hare, J.A. Long-Term Changes in the Distributions of Larval and Adult Fish in the Northeast U.S. Shelf Ecosystem. *PLoS ONE* **2015**, *10*, e0137382. [CrossRef]
39. Felder, D.L.; Álvarez, F.; Goy, J.W.; Lemaitre, R. Decapoda (Crustacea) of the Gulf of Mexico, with comments on the Amphionidacea. In *Gulf of Mexico Origin, Water and Biota*; Felder, D.L., Camp, D.K., Eds.; Texas A&M University Press: College Station, TX, USA, 2009; Volume 1, pp. 1019–1104.
40. Villanueva, R. Decapod crab zoeae as food for rearing cephalopod paralarvae. *Aquaculture* **1994**, *128*, 143–152. [CrossRef]
41. Dunham, J.S.; Duffus, D.A. Diet of graywhales (*Eschrichtius robustus*) in Clayoquot Sound, British Columbia, Canada. *Mar. Mammal Sci.* **2002**, *18*, 419–437. [CrossRef]
42. Villanueva, R.; Perricone, V.; Fiorito, G. Cephalopods as Predators: A Short Journey among Behavioral Flexibilities, Adaptions, and Feeding Habits. *Front. Physiol.* **2017**, *8*, 598. [CrossRef]
43. Acevedo, J.; Aguayo-Lobo, A.; González, A.; Haro, D.; Olave, C.; Quezada, F.; Martínez, F.; Garthe, S.; Cáceres, B. Occurrence of Sei Whales (*Balaenoptera borealis*) in the Magellan Strait from 2004–2015, Chile. *Aquat. Mamm.* **2017**, *43*, 63–72. [CrossRef]
44. Landeira, J.M.; Lozano-Soldevilla, F.; Almansa, E.; Gonzalez-Gordillo, J.I. Early larval morphology of the armed nylon shrimp *Heterocarpus ensifer ensifer* A. Milne-Edwards, 1881 (Decapoda, Caridea, Pandalidae) from laboratory culture. *Zootaxa* **2010**, *2427*, 1–14. [CrossRef]

45. Jiang, G.C.; Chan, T.Y.; Shih, T.W. Morphology of the first zoeal stage of three deep-water pandalid shrimps, *Heterocarpus abulbus* Yang, Chan & Chu, 2010, *H. hayashii* Crosnier, 1988 and *H. sibogae* De Man, 1917 (Crustacea: Decapoda: Caridea). *Zootaxa* **2014**, *3768*, 428–436.
46. Landeira, J.M.; Lozano-Soldevilla, F.; González-Gordillo, J.I. Morphology of first seven larval stages of the striped soldier shrimp, *Plesionika edwardsii* (Brandt, 1851) (Crustacea: Decapoda: Pandalidae) from laboratory reared material. *Zootaxa* **2009**, *1986*, 51–66. [CrossRef]
47. Lindley, J.A.; Hernandez, F.; Scatllar, J.; Docoito, J. *Funchalia* sp. (Crustacea: Penaeidae) associated with *Pyrosoma atlanticum* (Thaliacea: Pyrosomidae) of the Canary Islands. *J. Mar. Biol. Assoc. UK* **2001**, *81*, 173–174. [CrossRef]
48. Heegard, P. *Larvae of Decapod Crustacea: The Oceanic Penaeids Solenocera-Cerataspis-Cerataspides*; Dana Report No. 67; Host Norway: Copenhagen, Denmark, 1966; pp. 1–147.
49. Cook, A.B.; Bernard, A.M.; Boswell, K.M.; Bracken-Grissom, H.; D'Elia, M.; de Rada, S.; English, D.; Eytan, R.I.; Frank, T.; Hu, C.; et al. A Multidisciplinary approach to investigate deep-pelagic eco- system dynamics in the Gulf of Mexico following Deepwater Horizon. *Front. Mar. Sci.* **2020**, *7*, 548880. [CrossRef]
50. Frank, T.; Widder, E. Comparative study of the spectral sensitivities of mesopelagic crustaceans. *J. Comp. Physiol. A* **1999**, *185*, 255–265. [CrossRef]
51. Martin, J.W.; Criales, M.M.; dos Santos, A. Dendrobranchiata. In *Atlas of Crustacean Larvae*; Martin, J.W., Olesen, J., Hoeg, J.T., Eds.; John Hopkins University Press: Baltimore, MD, USA, 2014; pp. 235–242.
52. Guerao, G.; Cuesta, J.A. Caridea. In *Atlas of Crustacean Larvae*; Martin, J.W., Olesen, J., Hoeg, J.T., Eds.; John Hopkins University Press: Baltimore, MD, USA, 2014; pp. 250–255.
53. Palumbi, S.; Martin, A.; Romano, S.; McMillan, W.; Stice, L.; Grabowski, G. *The Simple Fool's Guide to PCR*, version 2.0; University of Hawaii: Honolulu, HI, USA, 2002.
54. Crandall, K.A.; Fitzpatrick, J.F. Crayfish Molecular Systematics: Using a Combination of Procedures to Estimate Phylogeny. *Syst. Biol.* **1996**, *45*, 1–26. [CrossRef]
55. Simon, C.; Frati, F.; Beckenbach, A.; Crespi, B.; Liu, H.; Flook, P. Evolution, weighting and phylogenetic utility of mitochondrial gene sequences and a compilation of conserved polymerase chain reaction primers. *Ann. Entomol. Soc. Am.* **1994**, *87*, 651–701. [CrossRef]
56. Folmer, O.; Black, M.; Hoeh, W.; Lutz, R.; Vrijenhoek, R. DNA primers for amplification of mitochondrial Cytochrome C oxidase subunit I from diverse metazoan invertebrates. *Mol. Mar. Biol. Biotechnol.* **1994**, *3*, 294–299.
57. Katoh, K.; Misawa, K.; Kuma, K.; Miyata, T. MAFFT: A novel method for rapid multiple sequence alignment based on fast Fourier transform. *Nucleic Acids Res.* **2002**, *30*, 3059–3066. [CrossRef]
58. Kalyaanamoorthy, S.; Minh, B.Q.; Wong, T.K.F.; von Haeseler, A.; Jermiin, L.S. ModelFinder: Fast model selection for accurate phylogenetics estimates. *Nat. Methods* **2017**, *14*, 587–589. [CrossRef]
59. Nguyen, L.T.; von Schmidt, H.A.; Haeseler, A.; Minh, B.Q. IQ-TREE: A fast and effective stochastic algorithm for estimating maximum-likelihood phylogenies. *Mol. Biol. Evol.* **2015**, *32*, 268–274. [CrossRef]
60. Minh, B.Q.; Nguyen, M.A.T.; von Haeseler, A. Ultrafast approximation for phylogenetic bootstrap. *Mol. Biol. Evol.* **2013**, *30*, 1188–1195. [CrossRef]
61. Huelsenbeck, J.P.; Ronquist, F. MRBAYES: Bayesian inference of phylogenetic trees. *Bioinformatics* **2001**, *17*, 745–755. [CrossRef] [PubMed]
62. Thatje, S.; Schnack-Schiel, S.; Arntz, W.E. Developmental trade- offs in Subantarctic meroplankton communities and the enigma of low decapod diversity in high southern latitudes. *Mar. Ecol. Prog. Ser.* **2003**, *260*, 195–207. [CrossRef]
63. Thatje, S.; Lovrich, G.A.; Torres, G.; Hagen, W.; Anger, K. Changes in biomass, lipid, fatty acid and elemental composition during abbreviated larval development of the Subantarctic shrimp *Campylonotus vagans*. *J. Exp. Mar. Biol. Ecol.* **2004**, *301*, 159–174. [CrossRef]
64. Thatje, S.; Lovrich, G.A.; Anger, K. Egg production, hatching rates, and abbreviated larval development of *Campylonotus vagans* Bate, 1888 (Crustacea: Decapoda: Caridea) in Subant- arctic waters. *J. Exp. Mar. Biol. Ecol.* **2004**, *301*, 15–27. [CrossRef]
65. Thatje, S.; Bacardit, R.; Arntz, W.E. Larvae of the deep-sea Nematocarcinidae (Crustacea: Decapoda: Caridea) from the Southern Ocean. *Polar Biol.* **2005**, *28*, 290–302. [CrossRef]
66. Oliphant, A.; Hauton, C.; Thatje, S. The implications of temperature-mediated plasticity in larval instar number for development within a marine invertebrate, the shrimp *Palaemonetes varians*. *PLoS ONE* **2013**, *8*, e75785. [CrossRef]
67. Bartilotti, C.; dos Santos, A. The secret life of deep sea shrimps: Ecological and evolutionary clues from the larval description of Systellaspis debilis (Caridea: Oplophoridae). *PeerJ* **2019**, *7*, e7334. [CrossRef]
68. Wehrtmann, I.; Albornoz, L. Larvae of *Nauticaris magellanica* (Decapoda: Caridea: Hippolytidae) reared in the laboratory differ morphologically from those in nature. *J. Marine Biol. Assoc. UK* **2003**, *83*, 949–957. [CrossRef]
69. Thatje, S.; Bacardit, R. Morphological variability in larval stages of *Nauticaris magellanica* (A. Milne Edwards, 1891) (Decapoda: Caridea: Hippolytidae) from South American waters. *Bull. Mar. Sci.* **2000**, *66*, 375–398.
70. WoRMS Editorial Board World Register of Marine Species. 2021. Available online: https://www.marinespecies.org (accessed on 13 August 2021).

71. Pérez Farfante, I.; Kensley, B. *Penaeoid and sergestoid shrimps and prawns of the world. Keys and Diagnoses for the Families and Genera*; Mémoires du Muséum National d'Histoire Naturelle; Muséum National d'Histoire Naturelle: Paris, France, 1997; Volume 175, pp. 1–233.

72. Tavares, C.R.; Serejo, C.S. Taxonomy of Aristeidae (Dendrobranchiata: Penaeoidea) from the central coast of Brazil, collected by the Revizee program, between 19° and 22° S. *Zootaxa* **2007**, *1585*, 1–44. [CrossRef]

73. Franks, J.S.; Russell, A.D. First record of *Cerataspis*, a larval oceanic penaeoid crustacean, from the Gulf of Mexico. *Gulf Caribb. Res.* **2008**, *20*, 87–89. [CrossRef]

74. Bracken-Grissom, H.D.; Felder, D.L.; Vollmer, N.L.; Martin, J.W.; Crandall, K.A. Phylogenetics links monster larva to deep-sea shrimp. *Ecol. Evol.* **2012**, *2*, 2367–2373. [CrossRef] [PubMed]

75. Heldt, J.H. Contribution al etude de la biologie des crevettes peneides *Aristeomorpha foliacea* (Risso) et *Aristeus antennatus* (Risso) (formes larvaires). *Bull. Société Des. Sci. Nat. Tunis.* **1955**, *VIII*, 1–29.

76. Carbonell, A.; dos Santos, A.; Alemany, F.; Vélez-Belchi, P. Larvae of the red shrimp *Aristeus antennatus* (Decapoda: Dendrobranchiata: Aristeidae) in the Balearic Sea: New occurrences fifty years later. *Mar. Biodivers. Rec.* **2010**, *3*, e103. [CrossRef]

77. Carreton, M.; Company, J.B.; Planella, L.; Heras, S.; García-Marín, J.L.; Agulló, M.; Clavel-Henry, M.; Rotllant, G.; dos Santos, A.; Roldán, M.I. Morphological identification and molecular confirmation of the deep sea blue and red shrimp *Aristeus antennatus* larvae. *PeerJ* **2019**, *7*, e6063. [CrossRef] [PubMed]

78. Carreton, M.; dos Santos, A.; de Sousa, L.F.; Rotllant, G.; Compan, J.B. Morphological description of the first protozoeal stage of the deep-sea shrimps *Aristeus antennatus* and *Gennadas elegans*, with a key. *Sci. Rep.* **2020**, *10*, 11178. [CrossRef] [PubMed]

79. Calo-Mata, P.; Pascoal, A.; Fernández-No, I.; Böhme, K.; Gallardo, J.M.; Barros-Velázquez, J. Evaluation of a novel 16S rRNA/tRNA^{Val} mitochondrial marker for the identification and phylogenetic analysis of shrimp species belonging to the superfamily Penaeoidea. *Anal. Biochem.* **2009**, *391*, 127–134. [CrossRef]

80. Gusmão, J.; Lazoski, C.; Solé-Cava, A. A new species of *Penaeus* (Crustacea: Penaeidae) revealed by allozyme and cytochrome oxidase I analyses. *Mar. Biol.* **2000**, *137*, 435–446. [CrossRef]

81. Ditty, J.G. Young of *Litopenaeus setiferus, Farfantepenaeus aztecus* and *F. duorarum* (Decapoda: Penaeidae): A re-assessment of characters for species discrimination and their variability. *J. Crustacean Biol.* **2011**, *31*, 458–467. [CrossRef]

82. Ditty, J.G. Sternal Spines in Penaeid Postlarvae (Decapoda: Penaeidae): Life-Phase-Specific and Systematically Significant? *J. Crustacean Biol.* **2014**, *34*, 618–628. [CrossRef]

83. Ditty, J.G.; Alvarado-Bremer, J.R. Species discrimination of postlarvae and early juvenile brown shrimp (*Farfantepenaeus aztecus*) and pink shrimp (*F. duorarum*) (Decapoda: Penaeidae): Coupling molecular genetics and comparative morphology to identify early life stages. *J. Crustacean Biol.* **2011**, *31*, 126–137. [CrossRef]

84. Crosnier, A. Crevettes pénéides d'eau profonde récoltés dans l'océan Indien lors des campagnes Benthedi, Safari I et II, MD 32/Réunion. *Bulletin du Muséum National d'Histoire Naturelle Sér. 4 Section A* **1985**, *9*, 695–726.

85. González, J.A.; Santana, J.I. The family Penaeidae from the Canary Islands (Norteastern Atlantic), with first record of *Penaeus kerathurus*. *Boletim Museu de Historia Natural do Funchal* **2014**, *LXIV*, 29–34.

86. De Freitas, A.J. The Penaeoidea of southeast Africa. II-The Family Penaeidae (excluding genus *Penaeus*). *Investig. Rep. Oceanogr. Res. Inst.* **1984**, *58*, 1–104.

87. Chace, F.A., Jr. Plankton of the Bermuda Oceanographic Expeditions. IX. The bathypelagic caridean Crustacea. *Zoologica* **1940**, *25*, 117–209.

88. Crosnier, A.; Forest, J. Les crevettes profondes de l'Atlantique Oriental Tropical. *Faune Trop.* **1973**, *19*, 1–409.

89. Alves, F.A., Jr.; dos Santos Silva, E.; Araujo, M.S.L.C.; Cardoso, I.; Bertrand, A.; Souza-Filho, J.F. Taxonomy of deep-sea shrimps of the Superfamily Oplophoroidea Dana 1852 (Decapoda: Caridea) from Southwestern Atlantic. *Zootaxa* **2019**, *4613*, 401–442. [CrossRef]

90. Gurney, R.; Lebour, M.V. On the larvae of certain Crustacea Macrura, mainly from Bermuda. *J. Linn. Soc. Zool.* **1941**, *41*, 89–181. [CrossRef]

91. Gurney, R. Larvae of decapod crustacea. *Ray Soc. Lond.* **1942**, *129*, 1–306.

92. Foxton, P. Observations on the Early Development and Hatching of the Eggs of *Acanthephyra purpurea* A. Milne-Edwards. *Crustaceana* **1964**, *6*, 235–237. [CrossRef]

93. Fernandes, L.D.A.; De Souza, M.F.; Bonecker, S.L.C. Morphology of Oplophorid and Bresiliid larvae (Crustacea, Decapoda) of Southwestern Atlantic plankton, Brazil. *Pan-Am. J. Aquat. Sci.* **2007**, *2*, 199–230.

94. Cardoso, I.A. New record of *Meningodora vesca* (Smith, 1887) (Caridea, Oplophoridae) to the Southwestern Atlantic. *Nauplius* **2006**, *14*, 1–7.

95. Williams, A.B. New marine decapod crustaceans from waters influenced by hydrothermal discharge, brine, and hydrocarbon seepage. *Fish Bull.* **1988**, *86*, 263–287.

96. Martin, J.W.; Haney, T. Decapod crustaceans from hydrothermal vents and cold seeps: A review through 2005. *Zool. J. Linn. Soc.* **2005**, *145*, 445–522. [CrossRef]

97. Vereshchaka, A.L.; Kulagin, D.N.; Lunina, A.A. Phylogeny and New Classification of Hydrothermal Vent and Seep Shrimps of the Family Alvinocarididae (Decapoda). *PLoS ONE* **2015**, *10*, e0129975. [CrossRef]

98. Hernández-Ávila, I.; Cambon-Bonavita, M.A.; Pradillon, F. Morphology of First Zoeal Stage of Four Genera of Alvinocaridid Shrimps from Hydrothermal Vents and Cold Seeps: Implications for Ecology, Larval Biology and Phylogeny. *PLoS ONE* **2015**, *10*, e0144657. [CrossRef] [PubMed]

99. Shank, T.M.; Black, M.B.; Halanych, K.M.; Lutz, R.A.; Vrijenhoek, R.C. Miocene radiation of deep-sea hydrothermal vent shrimp (Caridea: Bresiliidae): Evidence from mitochondrial cytochrome oxidase subunit I. *Mol. Phylogenetics Evol.* **1999**, *13*, 244–254. [CrossRef] [PubMed]

100. Vereshchaka, A.L. A new genus and species of caridean shrimp (Crustacea: Decapoda: Alvinocarididae) from Nort Atlantic hydrothermal vents. *J. Mar. Biol. Assoc. UK* **1996**, *76*, 951–961. [CrossRef]

101. Herring, P.; Gaten, E.; Shelton, P. Are vent shrimps blinded by science? *Nature* **1999**, *398*, 116. [CrossRef]

102. Chan, T.Y.; Yu, H.P. An uncommon deep-sea shrimp *Eugonatonotus crassus* (A. Milne Edwards, 1881) (Crustacea: Decapoda: Eugonatonotidae) from Taiwan. *Bull. Inst. Zool. Acad. Sin.* **1988**, *7*, 259–263.

103. Chan, T.Y.; Yu, H.P. *Eugonatonotus chacei* sp. nov., second species of the genus (Crustacea, Decapoda, Eugonatonotidae). *Bulletin du Museum National d'Histoire Naturelle* **1991**, *13*, 143–152.

104. Vereshchaka, A.L. New family and superfamily for a deep-sea caridean shrimp from the Galathea collections. *J. Crustacean Biol.* **1997**, *17*, 361–373. [CrossRef]

105. De Grave, S.; Chu, K.H.; Chan, T.Y. On the systematic position of *Galatheacaris abyssalis* (Decapoda: Galatheacaridoidea). *J. Crustacean Biol.* **2010**, *30*, 521–527. [CrossRef]

106. Aznar-Cormano, L.; Brisset, J.; Chan, T.Y.; Corbari, L.; Puillandre, N.; Utge, J.; Zbinden, M.; Zuccon, D.; Samadi, S. An improved taxonomic sampling is a necessary but not sufficient condition for resolving inter-families relationships in Caridean decapods. *Genetica* **2015**, *143*, 195–205. [CrossRef]

107. Burukovsky, R.N. Taxonomy of *Nematocarcinus* (Decapoda, Nematocarcinidae). Description of *Nematocarcinus* from waters of the American continent. *Zool. Zhurnal* **2001**, *80*, 1429–1443.

108. Wong, J.M.; Pérez-Moreno, J.L.; Chan, T.Y.; Frank, T.M.; Bracken-Grissom, H.D. Phylogenetic and transcriptomic analyses reveal the evolution of bioluminescence and light detection in marine deep-sea shrimps of the family Oplophoridae (Crustacea: Decapoda). *Mol. Phylogenetics Evol.* **2015**, *83*, 278–292. [CrossRef] [PubMed]

109. González, J.A.; Santana, J.I. Shrimps of the family Pandalidae (Crustacea, Decapoda) off the Canary Islands, Eastern Central Atlantic. *S. Afr. J. Mar. Sci.* **1996**, *17*, 173–182. [CrossRef]

110. Holthuis, L.B. FAO species catalogue. I. Shrimps and prawns of the world. An annotated catalogue of species of interest to fisheries. *FAO Fish. Synop.* **1980**, *125*, 271.

111. Holthuis, L.B. Crevelles. In *Guide FAO d'Identification des Especes pour les Besoins de la Peche. (Revision I). Mediterranee et Mer Noire. Zone de Pêche 37. Vegetaux et Invertebres*; Fischer, W., Bauchot, M.L., Schneider, M., Eds.; FAO: Rome, Italy, 1987; Volume 1, pp. 189–292.

112. Chilari, A.; Thessalou-Legaki, M.; Petrakis, G. Population structure and reproduction of the deep-water shrimp *Plesionika martia* (Decapoda: Pandalidae) from the eastern Ionian Sea. *J. Crustacean Biol.* **2005**, *25*, 233–241. [CrossRef]

113. Chakraborty, R.D.; Chan, T.Y.; Maheswarudu, G.; Kuberan, G.; Purushothaman, P.; Chang, S.C.; Jomon, S. On *Plesionika quasigrandis* Chace, 1985 (Decapoda, Caridea, Pandalidae) from southwestern India. *Crustaceana* **2015**, *88*, 923–930. [CrossRef]

114. Landeira, J.M.; Chan, T.Y.; Aguilar-Soto, N.; Jiang, G.C.; Yang, C.H. Description of the decapodid stage of *Plesionika narval* (Fabricius, 1787) (Decapoda: Caridea: Pandalidae) identified by DNA barcoding. *J. Crustacean Biol.* **2014**, *34*, 377–387. [CrossRef]

115. Landeira, J.M.; Gonzalez-Gordillo, J.I. Description of first five larval stages of *Plesionika narval* (Fabricius, 1787) (Crustacea, Decapoda, Pandalidae) obtained under laboratory conditions. *Zootaxa* **2009**, *2206*, 45–61. [CrossRef]

116. Matzen da Silva, J.; Creer, S.; dos Santos, A.; Costa, A.C.; Cunha, M.R.; Costa, F.O.; Carvalho, G.R. Systematic and Evolutionary Insights Derived from mtDNA COI Barcode Diversity in the Decapoda (Crustacea: Malacostraca). *PLoS ONE* **2011**, *6*, e19449. [CrossRef] [PubMed]

117. Jiang, G.C.; Landeira, J.M.; Shih, T.W.; Chan, T.Y. First zoeal stage of *Plesionika crosnieri* Chan & Yu, 1991, *P. ortmanni* Doflein, 1902, and *P. semilaevis* Bate, 1888, with remarks on the early larvae of *Plesionika* Bate, 1888 (Crustacea, Decapoda). *Zootaxa* **2018**, *4532*, 385–395.

118. DeSalle, R. Species discovery versus species identification in DNA barcoding efforts: Response to Rubinoff. *Conserv. Biol.* **2006**, *20*, 154–1547. [CrossRef] [PubMed]

119. Landschoff, J.; Gouws, G. DNA barcoding as a tool to facilitate the taxonomy of hermit crabs (Decapoda: Anomura: Paguroidea). *J. Crustacean Biol.* **2018**, *38*, 780–793. [CrossRef]

120. Trivedi, S.; Aloufi, A.A.; Ansari, A.A.; Ghosh, S.K. Role of DNA barcoding in marine biodiversity assessment and conservation: An update. *Saudi J. Biol. Sci.* **2016**, *23*, 161–171. [CrossRef]

Article

Diversity and Distribution of Mid- to Late-Stage Phyllosomata of Spiny and Slipper Lobsters (Decapoda: Achelata) in the Mexican Caribbean

Rubén Muñoz de Cote-Hernández [†], Patricia Briones-Fourzán *, Cecilia Barradas-Ortiz, Fernando Negrete-Soto and Enrique Lozano-Álvarez

Unidad Académica de Sistemas Arrecifales, Instituto de Ciencias del Mar y Limnología, Universidad Nacional Autónoma de México, Puerto Morelos 77580, Mexico; rub.hhe.n@gmail.com (R.M.d.C.-H.); barradas@cmarl.unam.mx (C.B.-O.); fnegrete@cmarl.unam.mx (F.N.-S.); elozano@cmarl.unam.mx (E.L.-Á.)
* Correspondence: briones@cmarl.unam.mx
† Current address: Grupo Interdisciplinario de Gestión Ambiental, A.C., Ciudad de México 04909, Mexico.

Abstract: Achelata (Palinuridae and Scyllaridae) have a flat, transparent, long-lived planktonic larva called phyllosoma, which comprises multiple stages and has a duration from a few weeks (some scyllarids) to >20 months (some palinurids). The larval development of many Achelata occurs in oceanic waters, where conventional plankton nets usually collect the early- to mid-stages but not the later stages, which remain poorly known. We examined the diversity and distribution of mid- and late-stage phyllosomata in the oceanic waters of the Mexican Caribbean, where the swift Yucatan Current is the dominant feature. The plankton samples were collected at night with a large mid-water trawl in autumn 2012 (55 stations) and spring 2013 (34 stations). In total, we obtained 2599 mid- and late-stage phyllosomata (1742 in autumn, 857 in spring) of five palinurids (*Panulirus argus, Panulirus guttatus, Panulirus laevicauda, Palinurellus gundlachi, Justitia longimana*) and three scyllarids (*Parribacus antarcticus, Scyllarides aequinoctialis, Scyllarus chacei*). Overall, the mid-stages were ~2.5 times as abundant as the late stages. The palinurids far outnumbered the scyllarids, and *P. argus* dominated over all the other species, followed at a distance by *P. guttatus*. The densities of all the species were generally low, with no clear spatial pattern, and the phyllosomata assemblage composition greatly overlapped between seasons. These results suggest the extensive mixing of the organisms entrained in the strong Yucatan Current, which clearly favors the advection of the phyllosomata in this region despite the presence of some local sub-mesoscale features that may favor short-term retention.

Keywords: decapods; spiny lobsters; slipper lobsters; phyllosoma; Caribbean Sea; Yucatan Current

Citation: Muñoz de Cote-Hernández, R.; Briones-Fourzán, P.; Barradas-Ortiz, C.; Negrete-Soto, F.; Lozano-Álvarez, E. Diversity and Distribution of Mid- to Late-Stage Phyllosomata of Spiny and Slipper Lobsters (Decapoda: Achelata) in the Mexican Caribbean. *Diversity* **2021**, *13*, 485. https://doi.org/10.3390/d13100485

Academic Editor: Bert W. Hoeksema

Received: 31 August 2021
Accepted: 1 October 2021
Published: 5 October 2021

Publisher's Note: MDPI stays neutral with regard to jurisdictional claims in published maps and institutional affiliations.

1. Introduction

The decapod crustacean infraorder Achelata comprises two families: Palinuridae (spiny lobsters) and Scyllaridae (slipper lobsters) [1,2]. Palinurids are characterized by their long and spiny second antennae, whereas the second antennae of scyllarids are modified as a hinged series of five flat plates. Both families share a distinctive type of planktonic, transparent larva called "phyllosoma" (plural: phyllosomata), which differs broadly in morphology from the benthic juveniles and adults. The term phyllosoma ("leaf body" in Greek) refers to the extremely flattened body of this larva. The duration of the larval phase varies with species but may encompass 4 to 22 months in Palinuridae and 30 days to 9 months in Scyllaridae [3,4].

The adaptations of phyllosomata to a long life in oceanic waters include a virtually transparent body that helps to avoid predation and the development of long and narrow pereopods with numerous setae and exopods that are used for flotation and to swim well enough to catch and retain prey [3,5]. The final stage metamorphoses into a nektonic postlarva, known as puerulus in Palinuridae and nisto in Scyllaridae, which swims back to the coastal benthic habitats where it settles [3,4]. Based on their size and degree of development, phyllosomata are grouped into different stages, each of which can encompass one or more instars [6]. The number of stages depends on the degree of morphological differentiation and the duration of the larval phase. The early stages can be more similar between species, but, as they grow, morphological changes are useful for identification purposes [3,7].

In the wider Caribbean region, adult populations of 13 species of Achelata occur (five Palinuridae and eight Scyllaridae) [8,9]. Of these species, the complete larval series is known for six: *Panulirus argus* [10,11], *Justitia longimana* [12], *Scyllarus americanus* [13], *Scyllarus depressus* [14], *Scyllarus chacei* [15], and *Scyllarus planorbis* [16]. The incomplete larval series is known for *Panulirus guttatus* [17,18], *Panulirus laevicauda* [19], *Palinurellus gundlachi* [20], *Scyllarides aequinoctialis* [21], *Scyllarides nodifer*, and *Parribacus antarcticus* [22,23]. There are no data regarding the larval development of *Bathyarctus faxoni*. One palinurid, *P. argus*, is a major fisheries resource throughout the region [24], whereas *P. guttatus*, *P. laevicauda*, *S. aequinoctialis*, *P. antarcticus*, and *S. nodifer* constitute minor fishing resources of varying importance in different countries [8,9].

The distribution of the Achelata species is determined by factors such as their larval duration, the survival rate of subsequent larval stages, their behavior and capacity to swim vertically, as well as the influence of local and regional oceanic processes [3,6]. The longer the larval duration, the greater the potential for dispersal via ocean currents [25]. However, despite limited swimming abilities, phyllosomata can control their vertical position in the water column, which may modulate the range of dispersion. Ontogenetic behavioral changes of phyllosomata may result in the stratification of different larval stages [26–28]. Regardless, phyllosomata spend most of their planktonic life in reasonably well-illuminated regions, as indicated by their eye structure [29], and they can only partially control their dispersion, which is mainly determined by the ocean currents [3,6,30]. These currents influence the connectivity and flow of larvae among different areas, although short- or long-term larval retention may occur in mesoscale processes [31,32].

Larval retention and dispersion dynamics may impact the distribution and abundance of phyllosomata [33–35], with important implications for the regional larval biodiversity. Processes of postlarval recruitment to coastal habitats may underlie fluctuations in the abundance of adult populations, and understanding these processes requires extensive information on the distribution of larvae [3,36]. In the Mexican Caribbean, studies addressing the diversity and distribution of phyllosomata have mainly obtained information on the early- to mid-stages, with little knowledge about the late stages [37–41]. The aim of the present study was to investigate the diversity and distribution of the mid- to late-stage phyllosomata of spiny and slipper lobsters along the Mexican Caribbean in two contrasting seasons in order to increase insight into the larval dynamics and connectivity of Achelata in the wider Caribbean region.

2. Materials and Methods

2.1. Study Area

The study area is in the NW Caribbean Sea, off the coast of the state of Quintana Roo (Mexico), from the shelf brake to approximately 100 km offshore. Along this coast, the continental shelf is very narrow (<10 km in general, but <3 km in many parts), and depth increases rapidly a short distance from the shore [42]; therefore, our samplings were conducted almost exclusively in oceanic waters (Figure 1). In the NW Caribbean Sea, the Cayman Current flows from the Caribbean basin in an E-W direction and arrives almost perpendicular to the coast close to the southern sector of our study area. There, it veers north and becomes the Yucatan Current, which flows parallel to the coast and reaches its maximum velocities (up to 3 m s^{-1}) in the Yucatan Channel. Upon entering the Gulf of Mexico, it turns into the Loop Current [43]. The velocity of the Yucatan Current is modulated by the passage of cyclonic and anticyclonic eddies and by the latitude of impingement of the Cayman Current [42,44,45]. Moreover, a coastal counter-current and a persistent

sub-mesoscale eddy south of Cozumel Island have been repeatedly reported [42,46–48].
The dynamic complexity of this current system has important biological implications,
with some areas favoring larval retention and others favoring larval advection [42,48] (see
Section 3).

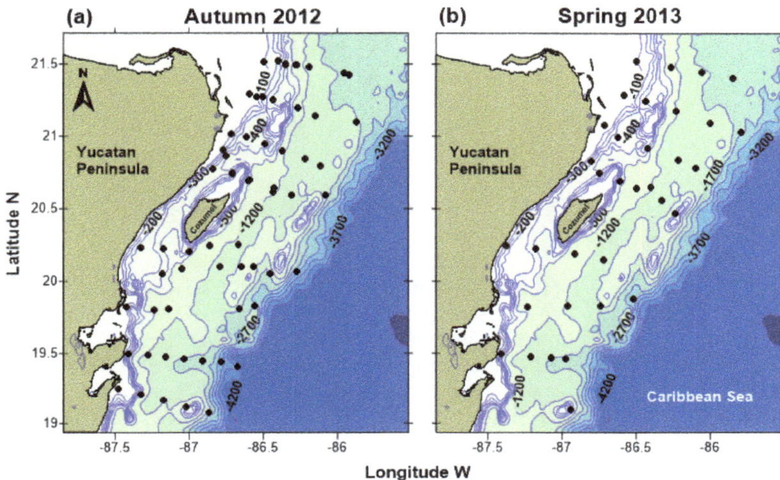

Figure 1. Study area showing location of plankton sampling stations (black dots) during (**a**) the
autumn 2012 cruise and (**b**) the spring 2013 cruise.

2.2. Larval Sampling and Identification

Plankton samples were obtained during a research project addressing the location of
metamorphosis zones of *P. argus* and *P. guttatus* and the nutritional condition of *P. argus*
during the phyllosoma-puerulus-juvenile transition [48–51]. Sampling was conducted in
numerous stations located along several transects perpendicular to the shelf break over
oceanic waters during two oceanographic cruises in UNAM's R/V Justo Sierra, "Metamorfosis 1", conducted in autumn 2012 (13–24 November; 55 stations) and "Metamorfosis 2",
conducted in spring 2013 (4–14 April; 38 stations). Hereafter, these cruises will be referred
to as the "autumn cruise" and the "spring cruise", respectively (Figure 1).

Samples were obtained with two different nets used simultaneously: a larger mid-
water Tucker trawl (effective mouth area: 9 m^2; length: 12 m; mesh size: 10 mm) primarily
designed to collect late-stage phyllosomata, and a smaller neuston net (mouth area: 1.5 m^2;
length: 3 m; mesh size: 3 mm) primarily designed to collect swimming postlarvae of spiny
lobsters [48,49]. The Tucker trawl was towed from stern at depths of 5–15 m and was
fitted with a Sea-Bird SBE39 data logger to record time, depth, and temperature during
the tow. The neuston net was towed from starboard approximately mid-ship within the
uppermost meter of the water column. Tows were conducted against the prevailing current
for 30–35 min at an average speed of 1.3 m s^{-1}. Samples were taken exclusively during
the hours of darkness (i.e., between 1 h after sunset and 1 h before sunrise) to increase the
chances of catching phyllosomata, which tend to concentrate in the top 25 m of the water
column in dark nights [27,52–54].

Surface geostrophic current fields during each cruise were derived from satellite-
obtained sea surface dynamic altimetry data obtained from the Archiving, Validation, and
Interpretation of Satellite Oceanographic data (AVISO). Results on hydrographic analyses
and geostrophic flow velocities obtained from CTD (Conductivity, Temperature and Depth)
profiles carried out during the day along the same transects in which larval sampling took
place during the night were reported by Briones-Fourzán et al. [48].

Upon retrieving the nets, the fresh plankton were immediately examined for phyllosomata, which were preserved in 80% ethanol. In the laboratory, the larvae were identified and staged using specialized literature for palinurids: *P. argus* [10,11], *P. guttatus* [17], *P. larvicauda* [19], *J. longimana* [12], and *P. gundlachi* [20], and for scyllarids: *S. americanus* [13], *S. depressus* [14], *S. chacei* [15], *S. planorbis* [16], *S. nodifer* and *S. aequinoctialis* [15,21], and *P. antarcticus* [22,23,55,56]. To aid in phyllosomata identification and staging, the following measurements were taken under a stereoscopic microscope: total body length (BL), from the anterior margin of the cephalic shield between the eyestalks to the posterior end of the pleon; cephalic length (CL), from the anterior to the posterior margin of the cephalic shield; and cephalic width (CW), measured at the widest part of the cephalic shield [11,18]. To distinguish between similar stages of *P. argus*, *P. guttatus*, and *P. laevicauda*, we also used several ratios between these and other measurements (e.g., distance between mouth parts, see [17]). To distinguish between similar scyllarid species, we followed the keys provided by Robertson [15].

Earlier papers reported 11 larval stages for *P. argus* (e.g., [10,53]); however, Goldstein et al. [11] obtained the complete larval development of *P. argus* in culture, recognizing only 10 stages prior to metamorphosis. Therefore, we followed Goldstein et al. [11] to stage phyllosomata of *P. argus*. In the case of *P. guttatus*, Goldstein et al. [18] also obtained stages I–IX (which they considered the subfinal stage) of *P. guttatus* in culture, but their article was not available at the time our study was conducted. Instead, we followed Baisre and Alfonso [17], who suggested 11 stages for this species.

2.3. Data Analyses

Because phyllosomata are highly diluted, individual stages of a given species may be scarce or completely absent in some sampling stations; therefore, a common procedure to analyze their distribution consists of grouping them into three categories: early-stage, mid-stage, and late-stage phyllosomata [30,41,52,57,58]. We used the following criteria for this categorization: early stages: appendages not well developed and with no exopods, not all pereopods present, antennules longer than or equal in length to antennae; mid-stages: exopods without setae in almost all appendages, antennae may exceed the length of antennules, abdomen starting to segment or already segmented, pleopods and uropods beginning to develop; late stages: appendages well developed with exopods and setae present, all pereopods present and well developed, antennae exceed length of antennules, abdomen well segmented, pleopods and uropods completely developed, gills (when present) may appear as papillae, unilobed buds, or completely developed bilobed buds. For *P. gundlachi*, we also considered the degree of extension of the cephalic shield over the thorax (see [20]). These categories do not necessarily include the same stages for all species (Table 1).

Table 1. Categorization of individual phyllosomata stages of eight Achelata species into early stages, mid stages, and late stages.

Species	Early Stages	Mid Stages	Late Stages
Panulirus argus	I–III	IV–VIII	IX–X
Panulirus guttatus	I–III	IV–VIII	IX–XI
Panulirus laevicauda	I–III	IV–VIII	IX–XI
Palinurellus gundlachi	I–V	VI–VIII	IX–XII
Justitia longimana	I–IV	V–VIII	IX–X
Parribacus antarcticus	I–V	VI–VIII	IX–XI
Scyllarides aequinoctialis	I–V	VI–IX	X–XI
Scyllarus chacei	I–III	IV–V	VI–VII

To examine the spatial distribution of larvae, the number of mid-and late-stage larvae of each species was standardized to 1000 m^3 of filtered water. Densities were estimated only for samples taken with the mid-water Tucker trawl, which caught 94% of all phyllosomata.

The filtered volume (V) was estimated with the equation V = D × A, where D is distance travelled, derived from the ship's speed (m s^{-1}) during the tow × duration of the tow (s), and A = net mouth effective area (9 m^2).

For each species, the mean and standard deviation of all body dimensions were estimated by stage [59]. Only for *P. argus*, which was by far the dominant species (see Section 3) with multiple larvae of each stage in both seasons, the mean BL of each mid- and late-stage was compared between seasons with a non-parametric Mann–Whitney test.

To characterize the biodiversity of phyllosomata along our study area, we estimated five ecological indices for each sampling station during each cruise: species richness (*S*) and abundance (*N*), as well as Shannon–Wiener's diversity (*H'*), Pielou's evenness (*J'*), and Simpson's dominance (λ). The latter three combine measures of richness and abundance, hence providing greater ability to discriminate samples than *S* or *N* alone [58]. Each ecological index was compared between seasons (cruises) with a Student's *t* test after checking for normality and homogeneity of variances.

The potential effect of season (cruise) on the assemblage composition of phyllosomata was examined using multivariate analyses. For these analyses, we considered as separate groups the mid- and late-stage phyllosomata of each species based on differences in age, vertical movement range, diet, and trophic position with increasing size [60–63]. Results were visualized with a non-metric multidimensional scaling (NMDS) ordination on the square-root transformed abundance data using the Bray–Curtis similarity measure [64]. The significance of the observed differences was further tested with a one-way analysis of similarity (ANOSIM). This test provides an R-value indicative of the degree of difference between samples. R-values close to 0 are indicative of little difference, while values close to 1 are indicative of a large difference in sample composition [65]. We then did a similarity percentage analysis (SIMPER, [64]) to identify those groups of phyllosomata responsible for the observed similarities between seasons. The software PRIMER 6 v6.1.9 [66] and PAST v4.05 [67] were used for these analyses.

3. Results

3.1. Surface Geostrophic Current Fields

The conditions of the altimetry-derived current fields remained fairly constant throughout each cruise; therefore, we chose the central date of each cruise to visualize the current field during that cruise. During the autumn cruise, an anticyclonic eddy was located south of Cuba and the velocity of the geostrophic currents off the Yucatan Peninsula increased northwards (Figure 2a). During the spring cruise, there was a large anticyclonic/cyclonic eddy system also south of Cuba and to the east of our study area (Figure 2b). In this cruise, differences in the sea surface height were greater than during the autumn cruise, resulting in stronger geostrophic currents. During both cruises, the geostrophic calculations from the CTD survey revealed the presence of a coastal counter-current and a submesoscale eddy south of Cozumel [48], but these features were close to the coast where altimetry measurements are not reliable. Therefore, they were not registered on the altimetry-deduced flow field. However, it is important to mention their existence because of their potential influence on temporal larval retention [48].

3.2. Species Abundance and Stage Composition

In total, we collected 2599 phyllosomata of both Achelata families, of which 2563 (98.6%) were identified to species. The 36 remaining larvae (1.4%) were too damaged to be identified with certainty.

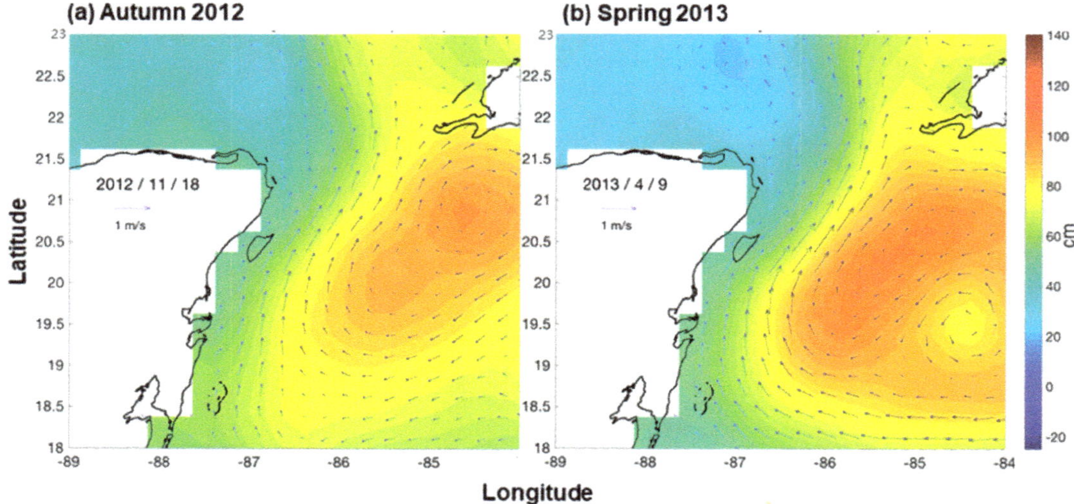

Figure 2. Surface geostrophic current field as derived from satellite-obtained sea surface dynamic altimetry data obtained from the Archiving, Validation, and Interpretation of Satellite Oceanographic data (AVISO) for (**a**) 18 November 2012 and (**b**) 9 April 2013.

In the autumn cruise, 53 of the 55 sampled stations (96.4%) yielded 1742 phyllosomata, 1715 (98.5%) of which were identified to species. Five species of Palinuridae accounted for 94.1% of these larvae, and three species of Scyllaridae for the remaining 4.9%. The most abundant species was *Panulirus argus* (61.2% of all the larvae), followed by *P. guttatus* (29%), *Parribacus antarcticus* (3%), *Palinurellus gundlachi* (2.7%), *Justitia longimana* (2.2%), *Scyllarides aequinoctialis* (1.9%), *Panulirus laevicauda* and *Scyllarus chacei* (0.9% of all the autumn larvae each) (Figure 2).

In the spring cruise, 29 of the 34 sampled stations (85.3%) yielded 857 phyllosomata, of which 848 (99%) were identified to species. Palinurids accounted for 93.6% of all the spring larvae, distributed in the five species; the rest (6.4%) corresponded to two scyllarid species. Again, the most abundant species by far was *P. argus* (78.0% of all the larvae), followed at a distance by *P. guttatus* (8.1%), *P. gundlachi* (7.2%), *Scyllarus chacei* (3.8%), *Scyllarides aequinoctialis* (2.6%), *P. laevicauda* (0.2%), and *J. longimana* (0.1%) (Figure 3).

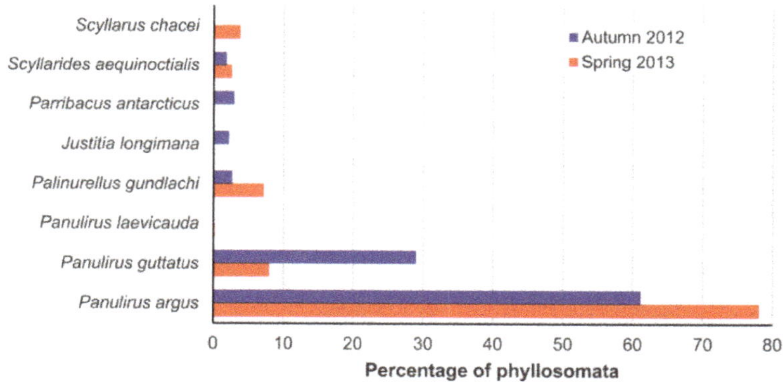

Figure 3. Percentage of phyllosomata (all stages) per species collected in the autumn 2012 cruise (N = 1715) and the spring 2013 cruise (N = 849).

As expected, the vast majority of phyllosomata were mid- and late stages (Table 2), although a few early-stage larvae of *P. gundlachi*, *P. antarcticus*, and *S. aequinoctialis* were collected in either or both cruises (Table 2). In *P. argus*, *P. guttatus*, *S. aequinoctialis*, and *J. longimana*, mid-stage larvae outnumbered late-stage larvae in both cruises. This was also the case for *P. antarcticus* in the autumn cruise, the only one in which this species occurred (Table 2). Only one final phyllosoma of *S. chacei* (stage VII) and one final phyllosoma of *P. laevicauda* (stage XI) were collected in the autumn cruise. In the spring cruise, late-stage larvae outnumbered mid-stage larvae for *P. gundlachi*, whereas all the *S. chacei* larvae were late-stage phyllosomata. In this cruise, two phyllosomata X of *P. laevicauda* and one phyllosoma V of *J. longimana* were collected (Table 2).

Table 2. Number of phyllosomata of each species categorized into early-stage, mid-stage, and late-stage larvae by cruise (N = total number of larvae).

Species	Autumn 2012				Spring 2013			
	N	Early	Mid	Late	N	Early	Mid	Late
Panulirus argus	1049	0	763	286	662	0	476	153
Panulirus guttatus	497	0	392	105	68	0	56	11
Panulirus laevicauda	1	0	0	1	2	0	0	2
Palinurellus gundlachi	46	1	9	36	61	14	12	35
Justitia longimana	38	0	25	13	1	0	1	0
Parribacus antarcticus	51	1	46	4	0	0	0	0
Scyllarides aequinoctialis	32	1	23	8	22	1	20	1
Scyllarus chacei	1	0	0	1	32	0	0	32

The mean and standard deviation of each measurement by stage are provided for all the palinurids (Table 3) and scyllarids (Table 4) collected. Within the palinurids, the species with the largest larvae by stage were *P. guttatus* and *J. longimana*, and with the smallest larvae by stage *P. gundlachi* (Table 3). Within the scyllarids, the largest larvae by stage were those of *S. aequinoctialis*, and the smallest, by far, those of *S. chacei* (Table 4). In *P. argus*, significant differences in size between the seasons occurred in stages V ($U = 99$, $n_1 = 38$, $n_2 = 16$, $p < 0.001$), VII ($U = 19,594$, $n_1 = 233$, $n_2 = 191$, $p = 0.034$), VIII ($U = 6997$, $n_1 = 214$, $n_2 = 15$, $p < 0.001$), and X ($U = 263$, $n_1 = 46$, $n_2 = 18$, $p = 0.024$) but not in stages VI ($U = 18,992$, $n_1 = 241$, $n_2 = 162$, $p = 0.645$) or IX ($U = 4679$, $n_1 = 136$, $n_2 = 73$, $p = 0.495$).

Table 3. Number of mid- and late stage palinurid phyllosomata by species and stage (N), and morphometric measurements: BL (total body length), CL (cephalic length), CW (cephalic width). All measurements in mm. SD: standard deviation.

Species and Stage	N	BL (Range)	BL (Mean ± SD)	CL (Mean ± SD)	CW (Mean ± SD)
Panulirus argus					
Stage V	54	4.3–8.1	6.0 ± 0.9	4.3 ± 0.8	2.3 ± 0.4
Stage VI	403	5.7–12.8	9.3 ± 1.4	6.8 ± 1.1	3.7 ± 0.6
Stage VII	424	9.2–17.0	13.3 ± 1.4	9.6 ± 1.1	5.4 ± 0.6
Stage VIII	319	13.4–22.6	18.4 ± 1.8	12.4 ± 1.2	7.2 ± 0.7
Stage IX	209	18.0–23.0	20.1 ± 0.9	13.3 ± 1.1	7.8 ± 0.6
Stage X	64	22.0–30.0	25.0 ± 1.3	15.1 ± 1.1	8.5 ± 0.7

Table 3. *Cont.*

Species and Stage	N	BL (Range)	BL (Mean ± SD)	CL (Mean ± SD)	CW (Mean ± SD)
Panulirus guttatus					
Stage V	28	6.9–11.0	9.2 ± 1.1	6.7 ± 0.8	3.6 ± 0.6
Stage VI	234	10.0–16.4	12.6 ± 1.4	9.4 ± 1.4	5.3 ± 0.8
Stage VII	123	11.5–21.0	16.7 ± 1.9	12.5 ± 1.8	7.3 ± 1.4
Stage VIII	64	14.5–28.5	21.7 ± 2.1	16.0 ± 1.5	9.6 ± 1.2
Stage IX	43	19.5–32.1	26.9 ± 2.7	19.1 ± 1.9	11.6 ± 1.6
Stage X	10	28.0–33.0	30.6 ± 1.4	21.0 ± 1.0	13.0 ± 0.5
Stage XI	4	37.5–42.5	39.8 ± 2.1	24.8 ± 0.9	14.9 ± 0.8
Palinurellus gundlachi					
Stage V	11	5.0–6.5	5.9 ± 0.6	3.7 ± 0.9	3.8 ± 1.0
Stage VI	5	6.3–7.0	6.7 ± 0.3	5.0 ± 0.4	4.4 ± 0.4
Stage VII	3	7.4–8.5	8.0 ± 0.6	5.3 ± 0.6	5.8 ± 0.2
Stage VIII	9	8.5–11.2	9.6 ± 0.8	7.0 ± 0.7	6.3 ± 1.2
Stage IX	8	9.0–10.7	10.0 ± 0.6	6.6 ± 1.2	6.8 ± 0.4
Stage X	18	11.0–13.5	11.7 ± 0.7	8.0 ± 0.7	8.1 ± 1.2
Stage XI	14	11.3–16.0	13.0 ± 1.6	8.9 ± 1.1	8.3 ± 0.8
Stage XII	33	12.2–22.0	15.7 ± 1.1	10.4 ± 1.2	9.4 ± 0.7
Justitia longimana					
Stage V	4	8.0–10.0	8.8 ± 0.9	6.8 ± 0.6	4.1 ± 0.6
Stage VI	8	11.3–16.1	13.5 ± 1.6	10.6 ± 1.2	7.0 ± 1.2
Stage VII	5	17.0–20.0	18.1 ± 1.2	13.9 ± 1.0	10.1 ± 1.4
Stage VIII	6	20.0–23.8	21.0 ± 1.4	14.8 ± 2.3	11.8 ± 0.7
Stage IX	12	24.0–28.0	26.0 ± 1.3	18.3 ± 1.0	14.4 ± 0.8
Stage X	1		35.5	22.6	16.6
Panulirus laevicauda					
Stage IX	1		14.8	11.5	7.0
Stage X	1		16.5	12.5	8.0
Stage XI	1		22.5	17.0	10.5

Table 4. Number of mid- and late stage scyllarid phyllosomata by species and stage (N), and morphometric measurements: BL (total body length), CL (cephalic length), CW (cephalic width). All measurements in mm. SD: standard deviation.

Species and Stage	N	BL (Range)	BL (mean ± SD)	CL (mean ± SD)	CW (mean ± SD)
Parribacus antarcticus					
Stage V	1		6.0	4.8	2.7
Stage VI	24	8.5–15.0	11.0 ± 1.7	8.9 ± 1.6	5.8 ± 1.1
Stage VII	12	13.0–17.5	15.3 ± 1.3	12.5 ± 1.0	9.3 ± 3.2
Stage VIII	9	18.8–28.3	21.3 ± 2.8	17.4 ± 2.1	13.0 ± 3.0
Stage IX	2	22.9–25.1	24.0 ± 1.6	20.0 ± 1.3	14.9 ± 1.2
Stage X	2	33.0–36.1	34.6 ± 2.2	28.3 ± 1.8	22.3 ± 1.5
Scyllarides aequinoctialis					
Stage IV	2	4.0–4.4	4.2 ± 0.3	3.4 ± 0.6	1.9 ± 0.2
Stage VI	7	6.3–9.8	7.8 ± 1.1	6.0 ± 1.1	3.8 ± 0.7
Stage VII	9	7.3–10.3	9.3 ± 0.8	7.7 ± 1.8	4.5 ± 0.6
Stage VIII	13	10.3–18.4	14.4 ± 2.5	11.1 ± 2.0	7.6 ± 1.6
Stage IX	13	18.0–26.0	23.3 ± 2.7	18.6 ± 2.0	14.0 ± 2.1
Stage X	7	27.0–33.0	29.8 ± 2.3	22.2 ± 1.5	17.1 ± 1.3
Stage XI	2	38.6–41.2	39.9 ± 1.8	27.4 ± 1.5	20.1 ± 0.8
Stage XII	1		42.0	28.0	22.0
Scyllarus chacei					
Stage VI	2	5.0–7.3	6.2 ± 1.6	4.2 ± 1.1	4.7 ± 0.7
Stage VII	31	9.2–11.4	10.5 ± 0.5	5.9 ± 0.5	7.1 ± 1.3

3.3. Phyllosomata Density and Spatial Distribution

The densities of the mid- to late-stage phyllosomata (all the species combined) were low across the study area. The values per sampling station ranged from 0 to 4.1 ind 1000 m^{-3} in the autumn 2012 cruise (Figure 4a), and from 0 to 9.7 ind 1000 m^{-3} in the spring 2013 cruise (Figure 4b). In general, greater densities of phyllosomata occurred south of Cozumel Island in the autumn cruise and north of Cozumel in the spring.

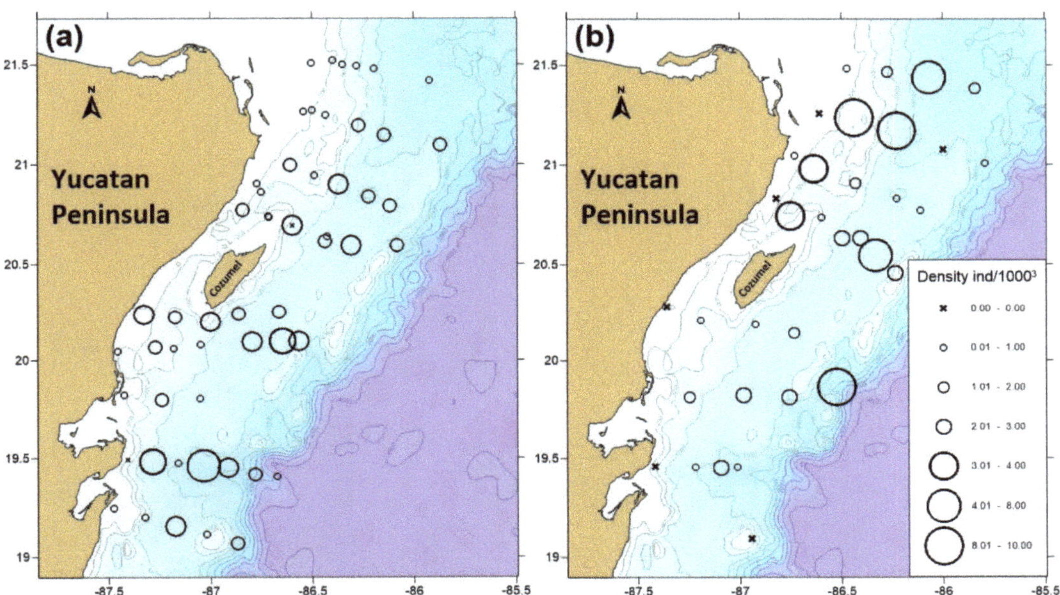

Figure 4. Density (number of larvae per 1000 m^3 of water) of phyllosomata (all species and stages combined) by sampling station. (**a**) autumn cruise; (**b**) spring cruise. The scale is the same for both panels.

The most frequent larvae were the mid- and late-stage phyllosomata of *P. argus*, occurring in 89% and 87% of the 55 sampling stations in autumn, respectively, and in 70% and 61% of the 34 stations in spring, respectively. In general, the densities of the mid-stage larvae were lower in the autumn (range: 0–2.6 ind 1000 m^{-3}), with greater concentrations south of Cozumel (Figure 5a), than in the spring (range 0.1–7.0 ind 1000 m^{-3}), when higher values mostly occurred close to the Yucatan Channel (Figure 5b). The densities of the late stages of *P. argus* were even lower (0.1–1.2 ind 1000 m^{-3} in autumn and 0.1–2.7 ind 1000 m^{-3} in the spring), with the highest values recorded in two stations in the southern zone, very close to the coast, in the autumn cruise (Figure 5c) and in one southern station far from the coast and another station close to the Yucatan Channel in the spring cruise (Figure 5d).

The mid-stage phyllosomata of *P. guttatus* occurred in 80% of the autumn cruise stations and 51.5% of the spring cruise stations. The maximum densities were 1.02 and 0.91 ind 1000 m^{-3}, respectively, but there were more stations with 0.5–1 ind 1000 m^{-3} in the autumn cruise (Figure 6a,c). The late stages of *P. guttatus* were collected in 63.3% of the autumn cruise stations but in only 27.3% of the spring cruise stations. The maximum densities were 0.38 and 0.15 ind 1000 m^{-3}, respectively. Therefore, densities >0.2 ind 1000 m^{-3} occurred only in seven stations of the autumn cruise, distributed throughout the study area (Figure 6b,d).

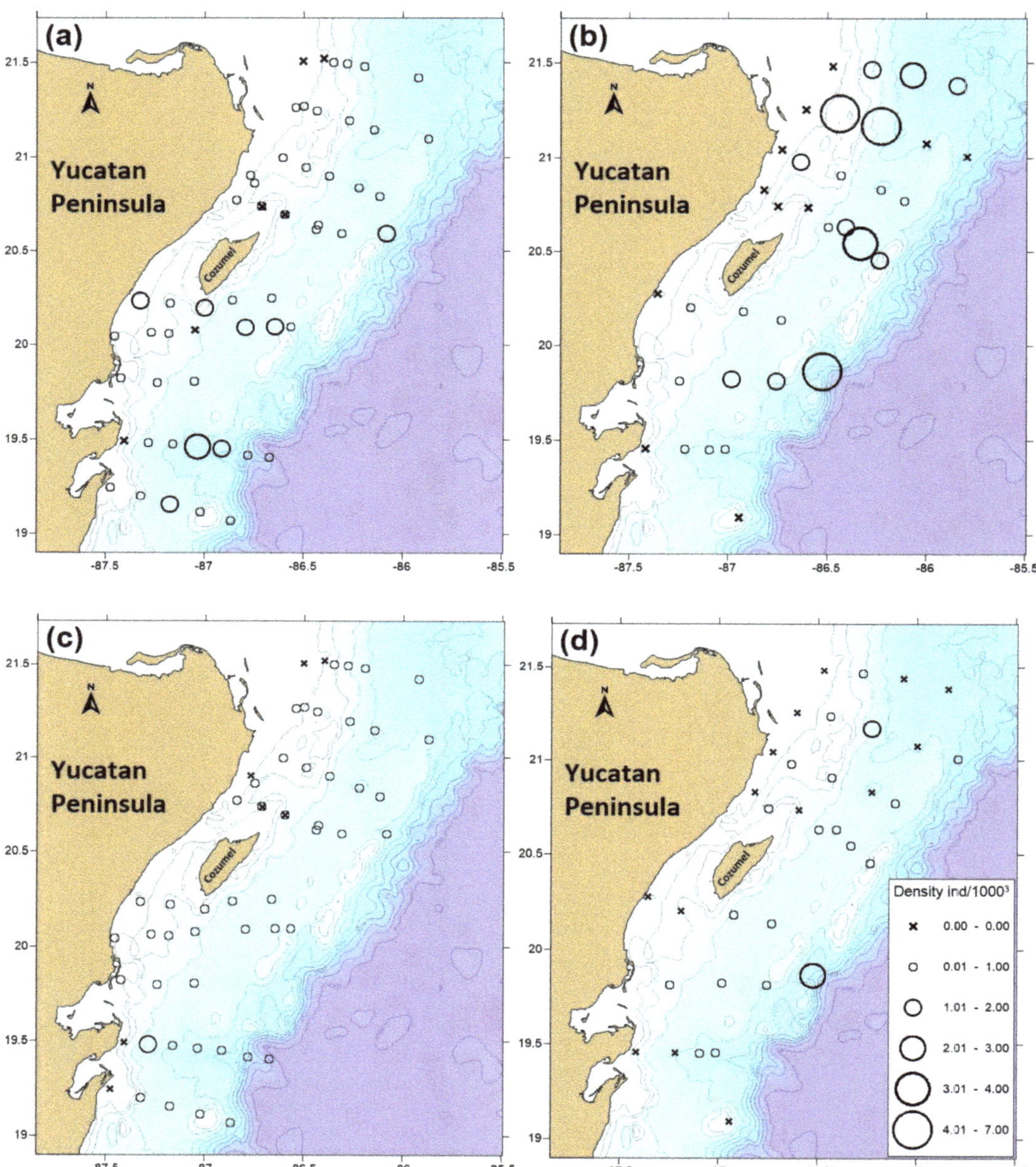

Figure 5. Density (number of larvae per 1000 m^3 of water) of phyllosomata of *Panulirus argus* by sampling station. (**a**) mid-stage phyllosomata, autumn cruise; (**b**) mid-stage phyllosomata, spring cruise; (**c**) late-stage phyllosomata, autumn cruise; (**d**) late-stage phyllosomata, spring cruise. The scale is the same for all panels.

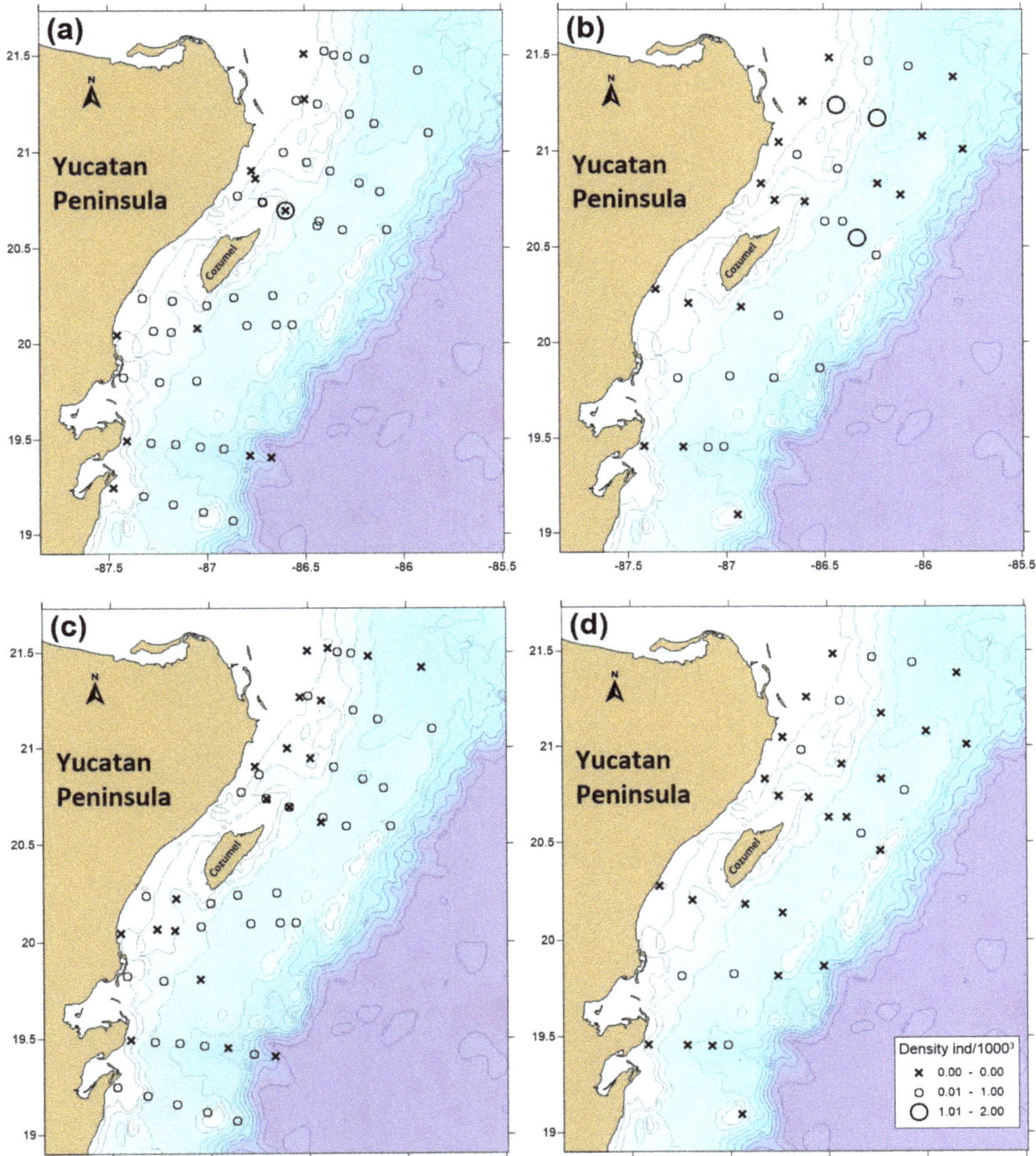

Figure 6. Density (number of larvae per 1000 m^3) of water of phyllosomata of *Panulirus guttatus* by sampling station. (**a**) mid-stage phyllosomata, autumn cruise; (**b**) mid-stage phyllosomata, spring cruise; (**c**) late-stage phyllosomata, autumn cruise; (**d**) late-stage phyllosomata, spring cruise. The scale is the same for all panels.

Palinurellus gundlachi was the third most frequent species. In the autumn cruise, the mid- and late-stage phyllosomata of *P. gundlachi* appeared in 80% and 34.5% of the stations, respectively, with densities similar to those of *P. guttatus* (Figure 7a,b). The maximum values were 1.02 and 0.36 ind 1000 m^{-3}, respectively. In the spring cruise, the mid- and late-stage larvae occurred in 23.5% and 38.2% of the stations, respectively, with higher densities for late stages (1.18 ind 1000 m^{-3}) than for mid-stages (0.47 ind 1000 m^{-3}) (Figure 7c,d). In the spring cruise as well, early stages appeared in 25.5% of the stations but at rather low densities (maximum 0.3 ind 1000 m^{-3}) (not shown). In the case of *S. aequinoctialis*, the mid-stage larvae occurred in 21.8% of the autumn stations at maximum densities of 0.23 ind 1000 m^{-3}, and in 30.3% of the spring stations at densities of up to 0.47 ind 1000 m^{-3} (Figure 8a,b). In contrast, the late stages were present only in the autumn (in 12.7% of the stations) at a maximum density of 0.1 ind 1000 m^{-3} (Figure 8c). The late stages of *S. chacei* only occurred in the spring cruise, appearing in 39.4% of the sampled stations at densities of up to 2.12 ind 1000 m^{-3} (Figure 8d).

Two species, *J. longimana* and *P. antarcticus*, virtually only occurred in the autumn cruise. The mid- and late-stage larvae of *J. longimana* occurred in 36.7 and 18.2% of the stations at maximum densities of 0.15 and 0.22 ind 1000 m^{-3}, respectively (Figure 9a,b). As for *P. antarcticus*, the mid- and late-stage larvae occurred in 34.5% and only 7.3% of the sampling stations, respectively, at maximum densities of 0.19 and 0.07 ind 1000 m^{-3}, respectively (Figure 10a,b).

3.4. Phyllosomata Biodiversity and Assemblage Composition

The biodiversity and assemblage composition analysis of the phyllosomata included 14 groups: the mid- and late-stage phyllosomata of *P. argus*, *P. guttatus*, *P. gundlachi*, *J. longimana*, *P. antarcticus*, and *S. aequinoctialis*, and the late-stage larvae of *S. chacei* and *P. laevicauda*. In all of the ecological indices, the variances between the cruises were homogeneous (Levene's tests, all of the p's > 1.0) and the data were normally distributed (Shapiro–Wilk tests, all of the p's > 0.1). Therefore, t-tests were used to compare all of the indices between the cruises (Table 5). Neither S nor N varied with the cruise. Prior to estimating the three compound indices, we removed the data from six sampling stations in the autumn cruise and three in the spring cruise that yielded only one phyllosoma. All the compound indices differed significantly between the cruises. H' and J' were higher in the autumn, whereas λ was higher in the spring (Table 5).

The NMDS 2D-plot showed great overlap in the assemblage composition of the phyllosomata from both cruises. This was confirmed by the results of an ANOSIM (R = 0.143). (Figure 11). A SIMPER (Appendix A) revealed that, within the autumn cruise, the assemblage composition exhibited an average similarity among the samples of 45.9%, mainly due to the contribution of the mid-stage larvae of *P. argus* and *P. guttatus* (36.1% and 24.6%, respectively), followed by the late-stage larvae of the same species (21.7% and 9.3%, respectively). Within the spring cruise, the average similarity among the samples was 44.4%, with the mid-stage larvae of *P. argus* emerging again as the main contributor (51.7%), followed by the late-stage *P. argus* (18.6%) and mid-stage *P. guttatus* (9.4%). In the autumn cruise, only four larval groups accounted for 90% of the observed similarity, whereas, in the spring cruise, six groups accounted for this percentage. Between the seasons, the phyllosomata assemblage had a mean dissimilarity of 59.7%, with eight groups accounting for nearly 90% of this dissimilarity. The mid- and late-stage larvae of *P. argus* were the main contributors to this dissimilarity between the cruises (25.3% and 15.7%, respectively), followed by the mid- and late-stage larvae of *P. guttatus* (15.6 and 8.6%, respectively), and the late-stage larvae of *S. chacei* (7.8%) and *P. gundlachi* (7.1%) (Appendix A).

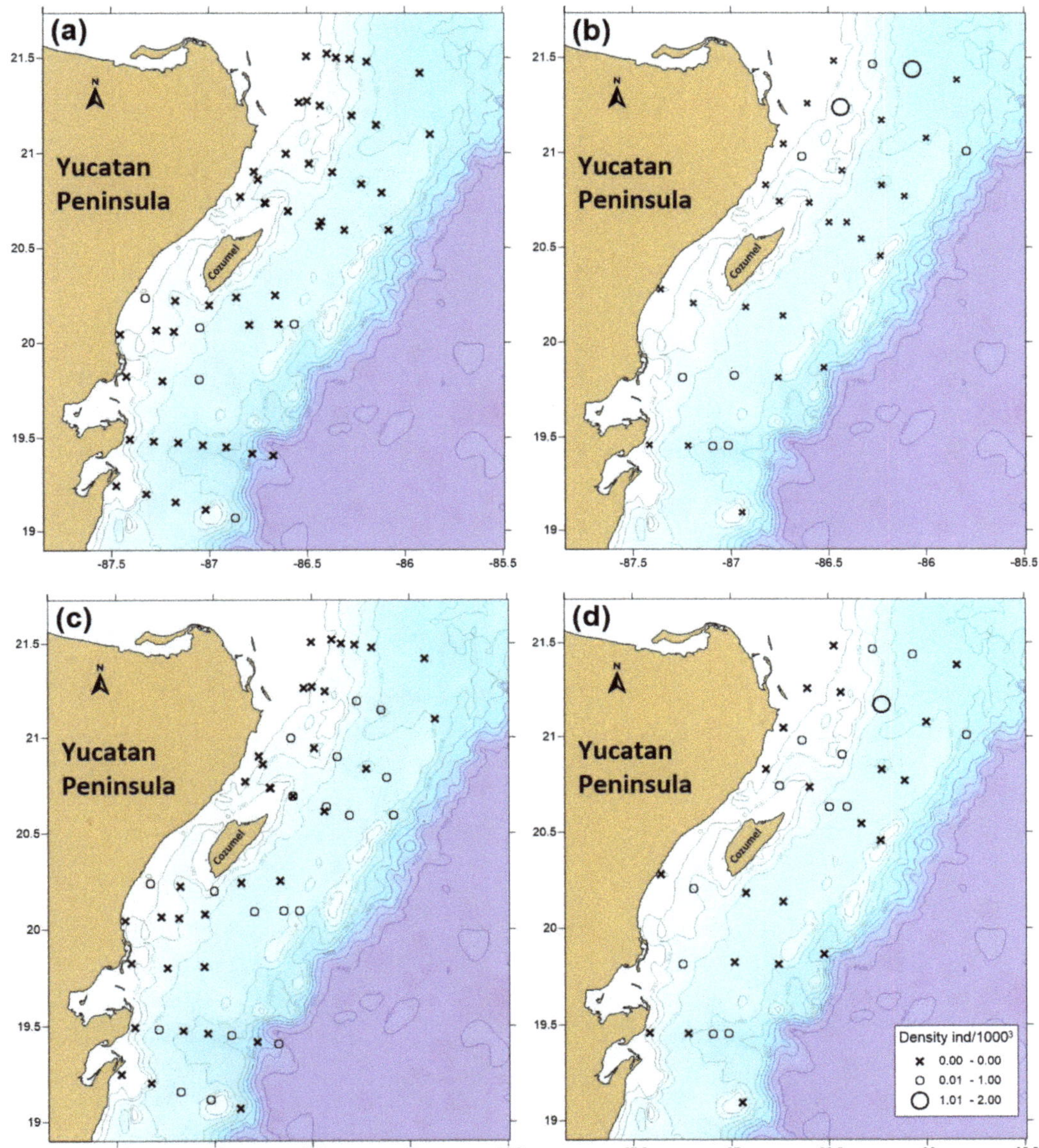

Figure 7. Density (number of larvae per 1000 m^3 of water) of phyllosomata of *Palinurellus gundlachi* by sampling station. (**a**) mid-stage phyllosomata, autumn cruise; (**b**) mid-stage phyllosomata, spring cruise; (**c**) late-stage phyllosomata, autumn cruise; (**d**) late-stage phyllosomata, spring cruise. The scale is the same for all panels.

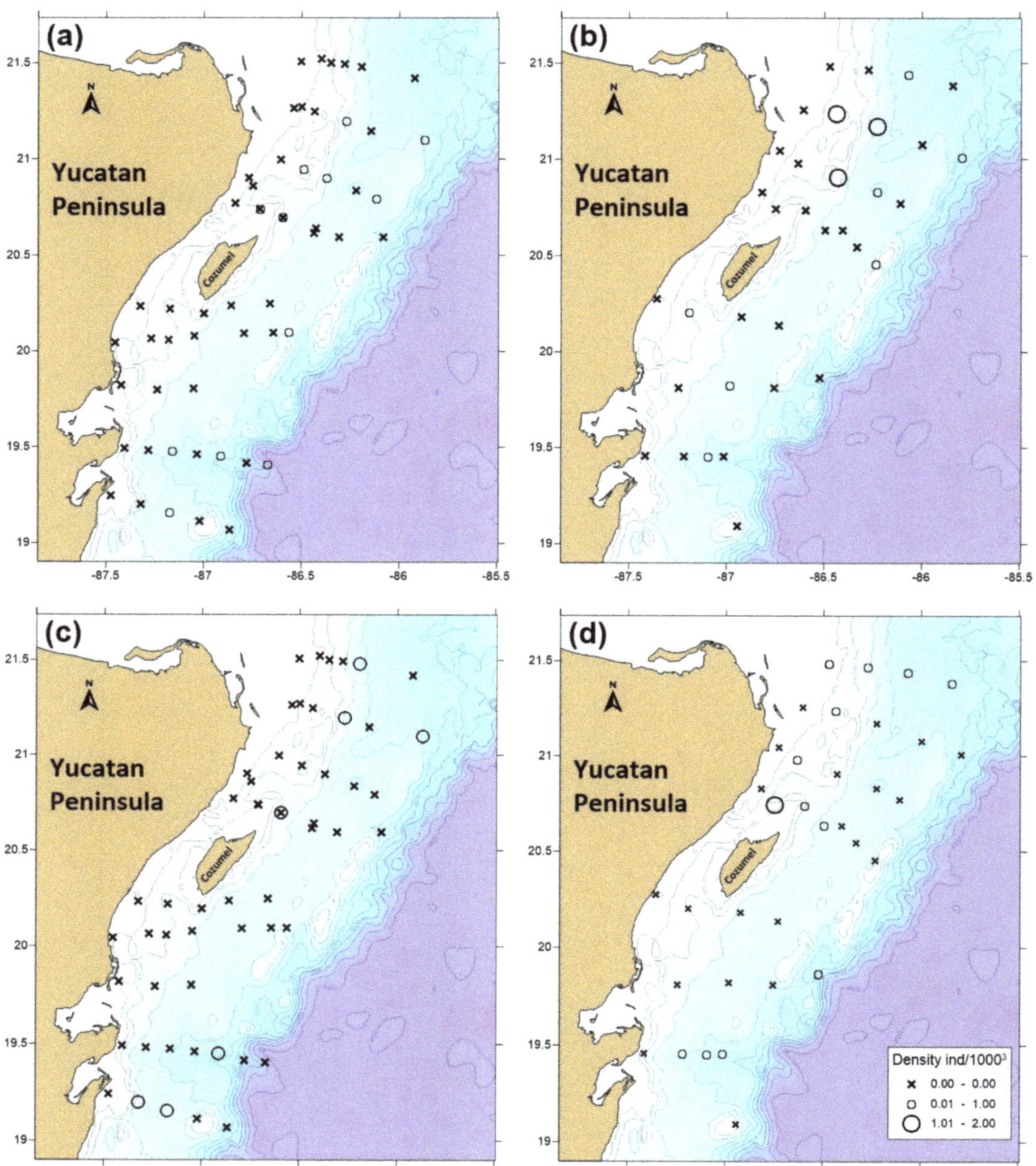

Figure 8. Density (number of larvae per 1000 m^3 of water) of phyllosomata by sampling station. *Scyllarides aequinoctialis*: (**a**) mid-stage phyllosomata, autumn cruise; (**b**) mid-stage phyllosomata, spring cruise; (**c**) late-stage phyllosomata, autumn cruise. *Scyllarus chacei*: (**d**) late-stage phyllosomata, spring cruise. The scale is the same for all panels.

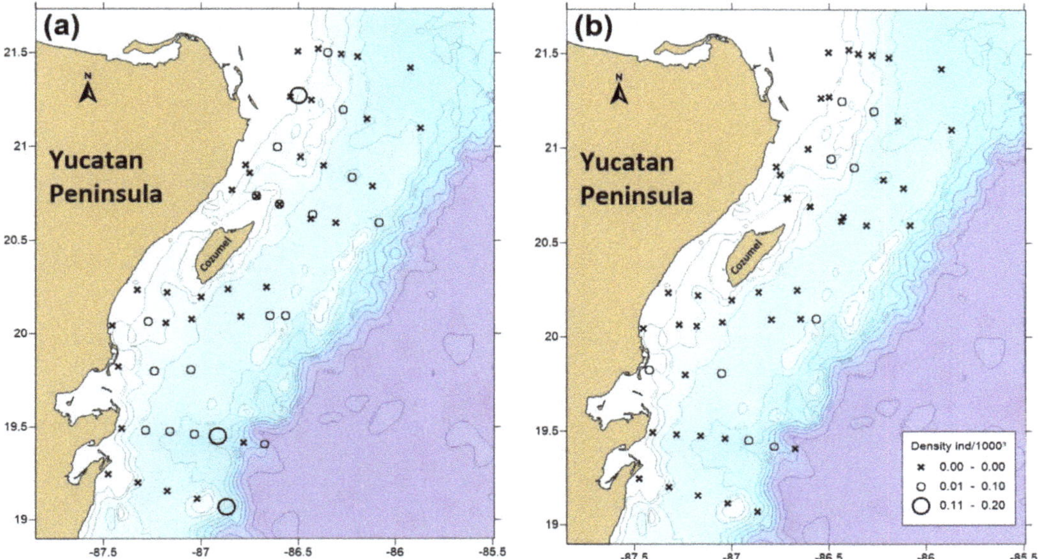

Figure 9. Density (number of larvae per 1000 m^3 of water) of (**a**) mid-stage phyllosomata and (**b**) late-stage phyllosomata of *Justitia longimana* by sampling station during the autumn 2012 cruise. The scale is the same for both panels.

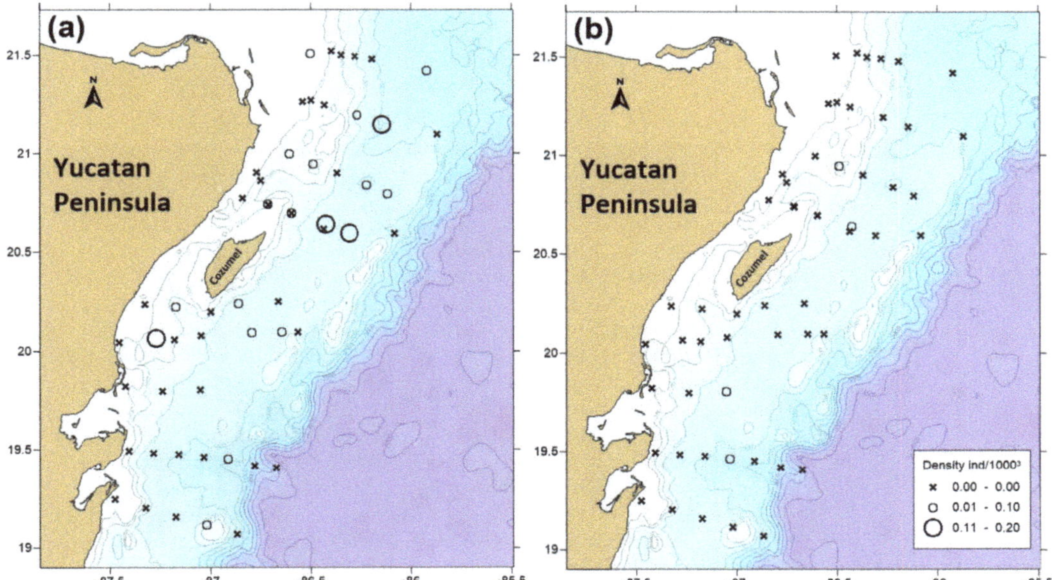

Figure 10. Density (number of larvae per 1000 m^3 of water) of (**a**) mid-stage phyllosomata and (**b**) late-stage phyllosomata of *Parribacus antarcticus* by sampling station during the autumn 2012 cruise. The scale is the same for both panels.

Table 5. Ecological indices for phyllosomata by cruise. Mean ± standard deviation of species richness (*S*), number of individuals (*N*), Shannon–Wiener's diversity (*H'*), Pielou's evenness (*J'*), and Simpson's dominance (λ), and results of Student's *t*-tests comparing each index between the autumn 2012 and the spring 2013 cruises.

Ecological Index	Autumn 2012	Spring 2013	df	*t*	*p*
S	4.9 ± 0.29	4.3 ± 0.38	85	1.276	0.2056
N	29.0 ± 3.28	28.5 ± 4.97	85	0.082	0.9348
H'	1.2396 ± 0.0425	1.0023 ± 0.0625	76	3.15	0.0050
J'	0.7380 ± 0.0159	0.6989 ± 0.0258	76	2.888	0.0023
λ	0.3203 ± 0.0163	0.4472 ± 0.0256	76	4.303	<0.0001

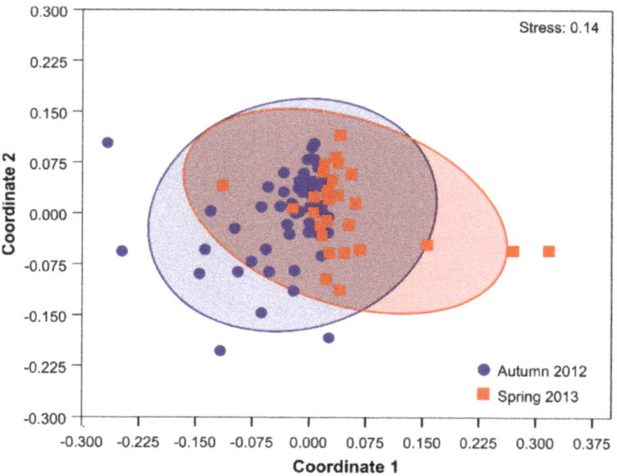

Figure 11. Non-metric multi-dimensional scaling ordinations (NMDS) for the phyllosomata assemblage structure in plankton samples obtained with the Tucker trawl in the autumn cruise (blue dots) and the spring cruise (red dots). The ordinations show the 95% concentration ellipses.

4. Discussion

This study is the first targeting the diversity and spatial distribution of mid- to late-stage phyllosomata of Achelata in two different seasons (autumn 2012 and spring 2013) in the oceanic waters of the western Caribbean Sea. Our sampling scheme focused on relatively large larvae, using nets of large dimensions and mesh sizes towed at high velocities [48]. Moreover, by sampling exclusively during the night in the top 20 m of the water column, we were able to obtain large numbers of phyllosomata, which rise near the surface at night and descend deeper in the water column during the day [27,28,52].

Of the 13 species of adult Achelata (five palinurids and eight scyllarids) present in the area, we obtained the phyllosomata of eight: all five palinurids (*Panulirus argus, P. guttatus, P. laevicauda, Justitia longimana, Palinurellus gundlachi*) but only three scyllarids (*Scyllarides aequinoctialis, Parribacus antarcticus, Scyllarus chacei*). Previous studies [37–41] had already reported on the phyllosomata of these species, alone or in different combinations, along the Mexican Caribbean, although those studies collected far more early- than mid-stage phyllosomata and very few late-stage larvae.

The density distributions of all the species confirm that mid- to late-stage phyllosomata, but particularly the latter, are highly diluted [68]. The mortality throughout the larval phase has been estimated at 97.3–99.6% for several palinurid species (*Panulirus interruptus*: [69], *Jasus lalandii*: [70], *J. edwardsii*: [71]); therefore, the abundance of late-stage phyllosomata would be expected to be lower than that of early- and mid-stage phyllosomata [68,72].

4.1. Palinurids

In both our cruises, the palinurid larvae far outnumbered the scyllarid larvae, which is a common pattern in offshore oceanic waters (e.g., [41,54,73]). The overwhelming dominance of *P. argus* phyllosomata over those of all the other species is consistent with other studies throughout the wider Caribbean region [37,38,40,41,74,75]. Benthic populations of *P. argus* occupy many different habitats—from shallow reef lagoons to deep reefs and rocky habitats down to ~70 m in depth [76]—and sustain important fisheries [24]. Although subtropical populations of *P. argus* exhibit seasonal reproduction [77], tropical populations reproduce year-round, with females spawning up to two to four times per year depending on their size [77–80]. These features of *P. argus*, in conjunction with a relatively short larval duration (for a palinurid) of 5–7 months [11], may explain the dominance of the phyllosomata of this species in both our cruises. In our study area, the recruitment of *P. argus* postlarvae to coastal habitats also occurs year-round [33], and the mean size of the postlarvae has been found to vary through time [48,81]. In our cruises, the mean size of *P. argus* phyllosomata varied with the season in some stages but not in other stages. These differences may be related to variations in the individual larval histories, including temperature regimes and variability in food availability throughout the larval development [51,63,82].

The second most abundant species in our samples was *P. guttatus*, particularly in the autumn cruise. Canto-García et al. [41] reported *P. guttatus* as the second most abundant species in their winter cruise but as the third most abundant in their spring cruise. These authors used DNA barcoding to discern between the early stages of *P. argus* and *P. guttatus*, which are morphologically very similar [15,41]. However, the size and morphological differences between *P. argus* and *P. guttatus* become more evident in the more advanced stages. Baisre and Alfonso [17] described phyllosomata VI to X (which they considered the subfinal stage) of *P. guttatus* from plankton samples. We found several final phyllosomata of *P. guttatus* (i.e., with bilobed gill buds in pereopods 1 to 4) and assigned them to stage XI following Baisre and Alfonso [17]. More recently, Goldstein et al. [18] successfully reared larvae of *P. guttatus* up to stage IX (which they considered the subfinal stage), but this paper was not available when our study was conducted.

Numerous individuals of all the mid- and late stages of *P. guttatus* were obtained in our autumn cruise but not in our spring cruise, when stages V and VIII were altogether absent and stages IX to XI very scarce. The year-round reproduction of *P. guttatus* is common in most of its geographic range [83,84], with individual females spawning three to four times per year [85]. However, unlike *P. argus*, *P. guttatus* is a rather small lobster restricted to the coral reef habitat throughout its benthic life and a minor fishing resource in only a few locations [76,86,87]. Moreover, its average larval duration, estimated in culture at 410 days (~13.6 months) [18], is among the longest within the genus *Panulirus*. According to Goldstein et al. [18], such a long larval duration may be a strategy to avoid sub-optimal conditions or settlement sites before attaining larval competency. However, our results suggest that a potential trade-off of this strategy could be a wider temporal variation in larval abundance, especially of the late stages.

The other three palinurids present in our samples were far less abundant, and data on their biology are also very scarce. *Palinurellus gundlachi* is a small, highly cryptic coral-dwelling lobster that occurs down to about 35 m in depth [9], whereas *Justitia longimana* occurs from 40 m to over 100 m in depth [88]. Larval stages II–X of *J. longimana* were described from plankton samples, but the larval duration has not been determined, although it would appear to be protracted [12]. Larval stages I–XII of *P. gundlachi*, with an estimated duration of 10 months, were also described from plankton samples [20]. Interestingly, we obtained more late- than mid-stage phyllosomata of *P. gundlachi* in both our cruises, and even several early-stage larvae in the spring cruise. However, there is no information on the reproductive activities of these two species with which to confront our findings. *Panulirus laevicauda*, an abundant species in Brazil [9], rarely occurs in the wider Caribbean region [89]. We only obtained three phyllosomata of *P. laevicauda*, two stage X and one that we ascribed to stage IX, and only four larvae of this species (in stages VI, VIII, X, and XI) had been previously obtained in the Caribbean Sea [19,38].

4.2. Scyllarids

The family Scyllaridae has four subfamilies, of which three (Arctidinae, Ibacinae, and Scyllarinae) occur in the Atlantic [90]. Arctidinae (e.g., *Scyllarides*) and Ibacinae (e.g., *Parribacus*) are large slipper lobsters associated with complex coral reef habitats [8]. Both have long larval phases that develop in oceanic waters, much like palinurids. The complete larval development of *S. aequinoctialis* and *P. antarcticus* (comprising 11 and 12 stages, respectively) was described from a combination of wild and cultured larvae and was estimated at 8–9 months and 9 months, respectively [21–23]. We obtained more mid- than late-stage phyllosomata of *S. aequinoctialis* in both cruises, especially in the autumn. Interestingly, *P. antarcticus* was the third most abundant species in our autumn cruise, with mostly mid-stage larvae, but was completely absent in our samples of the spring cruise. Unfortunately, almost nothing is known about the reproductive dynamics of these two scyllarids [91].

Scyllarinae (e.g., *Scyllarus* and *Bathyarctus*) are small slipper lobsters with relatively short larval phases that tend to develop in coastal and neritic waters [4]. Of the five species of Scyllarinae occurring in the western Caribbean, we only obtained the phyllosomata of one: *Scyllarus chacei*, mostly during the spring cruise and only the late stages. Early-stage larvae tentatively ascribed to this species had been previously reported in our study area but over shallower depths [39]. The larval duration of *S. chacei* has been estimated at only 44 days [15], suggesting an adaptation to enhance larval retention close to the parental populations [3]. The adults of *S. chacei* have a similar bathymetric distribution as those of *S. depressus* and *S. planorbis* (~11–300 m, [8,92,93], whereas the adults of *B. faxoni* occur at even greater depths (229–457 m). Yet, we did not obtain the larvae of *S. depressus* or *S. planorbis* in our cruises (there is no knowledge on the larvae of *B. faxoni*). Whether this result reflects interspecific differences in the reproductive periods or in the horizontal or vertical distribution patterns of the larvae remains to be studied. Larvae of *S. americanus*, which were also absent in our samples, are very common in the Gulf of Mexico, with only one report in the Mexican Caribbean [37], reflecting the distribution of adult populations of this species from Massachusetts to the Gulf of Mexico [8].

4.3. Phyllosomata Assemblages and Oceanographic Features

Panulirus argus larvae overwhelmingly dominated the catch, but, by considering the mid- and late stages of each species as separate groups to analyze diversity indices and assemblage composition, interesting patterns emerged. For example, the higher values of diversity and evenness in the autumn and dominance in the spring reflect the proportionally greater abundance of mid-stage phyllosomata of *P. argus* in the spring cruise relative to the autumn cruise, the lower abundance of mid-stage phyllosomata of *P. guttatus* in the spring cruise relative to the autumn cruise, and the presence of both mid- and late stages of *J. longimana* and *P. antarcticus* in the autumn cruise but their absence in the spring cruise.

In contrast, the assemblage composition of phyllosomata did not vary much with the season. A few sampling stations had particularly different compositions (e.g., stations with few larvae but each of a different group vs. stations with many larvae of several different groups), but, in general, there was great overlap between the cruises. This result, in conjunction with the lack of a clear spatial pattern in the density distribution of these and the rest of the larval groups, suggests that there is extensive mixing of organisms upon becoming entrained in the strong Yucatan Current, which, during our cruises, had an average velocity of 1 m s^{-1} (see Figure 2) and was coherent down to ~200 m in depth [48]. Oceanographic features play a significant role in the distribution of phyllosomata [68], and the Yucatan Current is clearly the main factor influencing the dispersion of these larvae in this region. However, although not evident in Figure 2, a coastal counter-current along the southern half of our study area and a persistent cyclonic sub-mesoscale eddy south of Cozumel Island were detected in both our cruises [48], and evidence of their existence was also found in previous studies [42,46,47]. These features may aid in the local, short-term retention of phyllosomata, as has been found for *P. argus* in the Florida Keys, USA [94] and for *J. edwardsii* in New Zealand [95,96], but probably cannot offset the highly advective environment of the Yucatan Current.

Around the Kuroshio Current in the northwest Pacific region, late stages of *Panulirus japonicus* have been reported at low densities of 0.1 to 0.5 ind/1000 m^3 [97,98], much like the ones reported in the present study. Both the Kuroshio Current and the Yucatan Current are strong western boundary currents that increase the potential for the dispersion of phyllosomata through advection. Field studies [41,45] and biophysical modelling [28,33,34] suggest that the potential for the dispersion of larvae is greater in the northern half of the Mexican Caribbean area than in the Central America area, which has a more retentive oceanographic environment. Thus, for *P. argus*, the levels of genetic connectivity were higher in Central America than in the Mexican Caribbean [99]. Further north, in Florida, genetic studies and biophysical models [32,35,46] have suggested that the local populations are highly dependent on the larval supply from more southern and southeastern populations.

The long larval duration of Achelata lobsters makes them particularly vulnerable to environmental variations and the effects of climate change because environmental conditions during the larval period affect the growth, survival, and dispersion of phyllosomata [68]. In the ocean, climate change involves increases in sea temperature and the intensity of ultraviolet radiation, as well as changes in circulation patterns, which may affect phyllosomata directly or indirectly [100,101]. For example, in the temperate East Australia Current, the phyllosomata of three tropical palinurid species are being transported by warm water eddies into temperate areas well outside the geographic range of the adult populations, providing them the opportunity to establish in such regions if environmental conditions become conducive to settlement [102]. Therefore, more field studies are needed to increase insight into the diversity and distribution of phyllosomata and their interaction with oceanographic processes that may influence the connectivity of lobster populations, and to understand how they might react to future environmental changes.

Author Contributions: Conceptualization, R.M.d.C.-H., P.B.-F. and E.L.-Á.; Data curation, R.M.d.C.-H., P.B.-F. and C.B.-O.; Formal analysis, R.M.d.C.-H. and P.B.-F.; Funding acquisition, P.B.-F.; Investigation, R.M.d.C.-H., P.B.-F., C.B.-O., F.N.-S. and E.L.-Á.; Methodology, R.M.d.C.-H., P.B.-F., C.B.-O., F.N.-S. and E.L.-Á.; Project administration, P.B.-F. and F.N.-S.; Resources, P.B.-F., F.N.-S. and E.L.-Á.; Supervision, P.B.-F. and E.L.-Á.; Validation, R.M.d.C.-H., P.B.-F. and C.B.-O.; Visualization, R.M.d.C.-H., P.B.-F. and E.L.-Á.; Writing—original draft, P.B.-F.; Writing—review & editing, R.M.d.C.-H., P.B.-F., C.B.-O., F.N.-S. and E.L.-Á. All authors have read and agreed to the published version of the manuscript.

Funding: This research was funded by Consejo Nacional de Ciencia y Tecnología, México (CONACYT) through the provision of a grant (number CB-101200) to P.B.-F., and an M.Sc. scholarship to R.M.d.C.-H.

Institutional Review Board Statement: Not applicable.

Data Availability Statement: The data presented in this study are openly available at https://hdl.handle.net/20.500.12201/11335 (accessed on 13 September 2021).

Acknowledgments: We thank Captain L. Ríos-Mora and the crew of the R/V Justo Sierra, as well as E. Escalante-Mancera, C. A. Coronado-Méndez, A. F. Espinosa-Magaña, L. Carrillo, I. Segura-García, J. P. Huchin-Mian, R. Candia-Zulbarán, R. Martínez-Calderón, N. Herrera-Salvatierra, C. Flores-Cabrera, J. M. Ojeda-Cota, A. Almazán-Becerril, S. Escobar-Morales, S. Martínez-Gomez, and M. Ortiz-Matamoros for their assistance during the oceanographic cruises. We also thank J. Candela for producing Figure 2.

Conflicts of Interest: The authors declare no conflict of interest. The funders had no role in the design of the study; in the collection, analyses, or interpretation of data; in the writing of the manuscript, or in the decision to publish the results.

Appendix A

Table A1. Similarity measures within and between seasons (cruises). Analysis of similarity percentage (SIMPER) for phyllosomata assemblages within the autumn cruise and the spring cruise, and of dissimilarity percentage between cruises. Av.Abund: average abundance; Av.Sim: average similarity; Sim/SD: similarity/standard deviation; Contrib%: contribution in %; Cum%: cumulative contribution in %; Av.Diss: average dissimilarity; Diss/SD, dissimilarity/standard deviation.

Autumn Cruise: Average Similarity: 45.94%					
Species	**Av.Abund**	**Av.Sim**	**Sim/SD**	**Contrib%**	**Cum.%**
Panulirus argus mid-stages	3.0	16.6	1.4	36.1	36.1
Panulirus guttatus mid-stages	2.1	11.3	1.2	24.6	60.7
Panulirus argus late stages	1.8	10.0	1.3	21.7	82.5
P. guttatus late stages	1.0	4.3	0.7	9.3	91.8

Spring Cruise: Average Similarity: 44.36%					
Species	**Av.Abund**	**Av.Sim**	**Sim/SD**	**Contrib%**	**Cum.%**
Panulirus argus mid-stages	3.49	22.95	1.43	51.7	51.7
Panulirus argus late stages	1.74	8.26	0.86	18.6	70.4
Panulirus guttatus mid-stages	1.05	4.17	0.71	9.4	79.8
Scyllarus chacei late stages	0.7	3.35	0.39	7.6	87.3
Palinurellus gundlachi late stages	0.7	2.44	0.48	5.5	92.8

Autumn and Spring Cruises: Average Dissimilarity = 59.68%					
Species	**Autumn Av.Abund**	**Spring Av.Abund**	**Av.Diss**	**Diss/SD**	**Contrib%**	**Cum.%**
Panulirus argus mid-stages	3.00	3.49	14.5	1.2	24.3	24.3
Panulirus argus late stages	1.83	1.74	9.4	1.2	15.7	40.0
Panulirus guttatus mid-stages	2.12	1.05	9.3	1.2	15.6	55.6
Panulirus guttatus late stages	0.98	0.35	5.2	0.9	8.6	64.3
Scyllarus chacei late stages	0.02	0.70	4.6	0.6	7.8	72.0
Palinurellus gundlachi late stages	0.42	0.70	4.2	0.9	7.1	79.1
Scyllarides aequinoctialis mid-stages	0.29	0.52	3.6	0.7	6.1	85.2
Parribacus antarcticus mid-stages	0.52	0.00	2.7	0.6	4.5	89.7
Palinurellus gundlachi mid-stages	0.11	0.34	1.9	0.6	3.2	92.9

References

1. Chan, T.Y. Annotated checklist of the world's marine lobsters (Crustacea: Decapoda: Astacidea, Glypheidea, Achelata, Polychelida). *Raffles Bull. Zool.* **2010**, *23*, 153–181.
2. Chan, T.Y. Updated checklist of the world's marine lobsters. In *Lobsters: Biology, Fisheries and Aquaculture*; Radhakrishnan, E.V., Phillips, B.F., Achamveetil, G., Eds.; Springer: Singapore, 2019; pp. 35–64.
3. Booth, J.D.; Webber, W.R.; Sekiguchi, H.; Coutures, E. Diverse larval recruitment strategies within the Scyllaridae. *N. Z. J. Mar. Freshw. Res.* **2005**, *39*, 581–592. [CrossRef]
4. Sekiguchi, H.; Booth, J.D.; Webber, W.R. Early life histories of slipper lobsters. In *The Biology and Fisheries of the Slipper Lobster*; Lavalli, K.L., Spanier, E., Eds.; CRC Press: Boca Raton, FL, USA, 2007; pp. 69–86.
5. McWilliam, P.S.; Phillips, B.F. Metamorphosis of the final phyllosoma and secondary lecithotrophy in the puerulus of *Panulirus cygnus* George: A review. *Mar. Freshw. Res.* **1997**, *48*, 783–790. [CrossRef]
6. Booth, J.D.; Phillips, B.F. Early life history of spiny lobster. *Crustaceana* **1994**, *66*, 271–294.
7. Phillips, B.F.; McWilliam, P.S. Spiny lobster development: Where does successful metamorphosis to the puerulus occur? A review. *Rev. Fish Biol. Fish.* **2009**, *19*, 193–215. [CrossRef]
8. Lyons, W.G. Scyllarid lobsters (Crustacea, Decapoda). *Mem. Hourglass Cruises* **1970**, *1*, 1–74.
9. Holthuis, L.B. *Marine Lobsters of the World*; FAO Species Catalogue No. 13; FAO: Rome, Italy, 1991.
10. Lewis, J.B. The phyllosoma larvae of the spiny lobster *Panulirus argus*. *Bull. Mar. Sci. Gulf Carib.* **1951**, *1*, 89–103.
11. Goldstein, J.S.; Matsuda, H.; Takenouchi, T.; Butler, M.J., IV. The complete development of larval Caribbean spiny lobster *Panulirus argus* (Latreille, 1804) in culture. *J. Crust. Biol.* **2008**, *28*, 306–327. [CrossRef]
12. Robertson, P.B. Phyllosoma larvae of a palinurid lobster, *Justitia longimana* (H. Milne Edwards), from the western Atlantic. *Bull. Mar. Sci.* **1969**, *19*, 922–944.
13. Robertson, P.B. The complete larval development of the sand lobster, *Scyllarus americanus* (Smith) (Decapoda, Scyllaridae) in the laboratory, with notes on larvae from the plankton. *Bull. Mar. Sci.* **1968**, *18*, 294–342.

14. Robertson, P.B. The larvae and postlarva of the scyllarid lobster *Scyllarus depressus* (Smith). *Bull. Mar. Sci.* **1971**, *21*, 841–865.

15. Robertson, P.B. The Larval Development of Some Western Atlantic Lobsters of the Family Scyllaridae. Ph.D. Thesis, University of Miami, Coral Gables, FL, USA, 1968.

16. Robertson, P.B. Larval development of the scyllarid lobster *Scyllarus planorbis* (Holthuis) reared in the laboratory. *Bull. Mar. Sci.* **1979**, *29*, 320–328.

17. Baisre, J.A.; Alfonso, I. Later stage larvae of *Panulirus guttatus* (Latreille, 1804) (Decapoda, Palinuridae) with notes on the identification of phyllosomata of *Panulirus* in the Caribbean Sea. *Crustaceana* **1994**, *66*, 32–44. [CrossRef]

18. Goldstein, J.S.; Matsuda, H.; Matthews, T.R.; Fumihiko, A.; Yamakawa, T. Development in culture of larval spotted spiny lobster *Panulirus guttatus* (Latreille, 1804) (Decapoda: Achelata: Palinuridae). *J. Crust. Biol.* **2019**, *39*, 574–581. [CrossRef]

19. Baisre, J.A.; Ruiz de Quevedo, M.E. Two phyllosome larvae of *Panulirus laevicauda* (Latreille, 1817) (Decapoda, Palinuridae) from the Caribbean Sea with a discussion about larval groups within the genus. *Crustaceana* **1982**, *43*, 147–153. [CrossRef]

20. Sims, H.W., Jr. The phyllosoma larvae of the spiny lobster *Palinurellus gundlachi* von Martens (Decapoda, Palinuridae). *Crustaceana* **1966**, *11*, 205–215. [CrossRef]

21. Robertson, P.B. The early larval development of the scyllarid lobster *Scyllarides aequinoctialis* (Lund) in the laboratory, with a revision of the larval characters of the genus. *Deep Sea Res.* **1969**, *16*, 557–586. [CrossRef]

22. Sims, H.W., Jr. The phyllosoma larvae of *Parribacus* (Decapoda, Palinuridae). *Quart. J. Florida Acad. Sci.* **1965**, *28*, 142–172.

23. Johnson, M.W. The phyllosoma larvae of slipper lobsters from the Hawaiian Islands and adjacent areas (Decapoda, Scyllaridae). *Crustaceana* **1971**, *20*, 77–103. [CrossRef]

24. Ehrhardt, N.M.; Puga, R.; Butler, M.J., IV. Implications of the ecosystem approach to fisheries management in large ecosystems: The case of the Caribbean spiny lobster. In *Towards Marine Ecosystem-Based Management in the Wider Caribbean*; Fanning, L., Mahon, R., McConney, P., Eds.; Amsterdam University Press: Amsterdam, The Netherlands, 2011; pp. 157–175.

25. Shanks, A.L. Pelagic larval duration and dispersal distance revisited. *Biol. Bull.* **2009**, *216*, 373–385. [CrossRef]

26. Minami, H.; Inoue, N.; Sekiguchi, H. Vertical distribution of phyllosoma larvae of palinurid and scyllarid lobsters in the western North Pacific. *J. Oceanogr.* **2001**, *57*, 743–748. [CrossRef]

27. Bradford, R.W.; Bruce, B.D.; Chiswell, S.M.; Booth, J.D.; Jeffs, A.; Wotherspoon, S. Vertical distribution and diurnal migration patterns of *Jasus edwardsii* phyllosomas off the east coast of the North Island, New Zealand. *N. Z. J. Mar. Freshw. Res.* **2005**, *39*, 593–604. [CrossRef]

28. Butler, M.J., IV; Paris, C.B.; Goldstein, J.S.; Matsuda, H.; Cowen, R.K. Behavior constrains the dispersal of long-lived spiny lobster larvae. *Mar. Ecol. Prog. Ser.* **2011**, *422*, 223–237. [CrossRef]

29. Mishra, M.; Jeffs, A.; Meyer-Rochow, V.B. Eye structure of the phyllosoma larva of the rock lobster *Jasus edwardsii* (Hutton, 1875): How does it differ from that of the adult? *Invertebr. Reprod. Dev.* **2006**, *49*, 213–222. [CrossRef]

30. Funes-Rodríguez, R.; Ruíz-Chavarría, J.A.; González-Armas, R.; Durazo, R.; Guzmán-del Proó, S.A. Influence of hydrographic conditions on the distribution of spiny lobster larvae off the west coast of Baja California. *Trans. Am. Fish. Soc.* **2015**, *144*, 1192–1205. [CrossRef]

31. Snyder, E.R.; Paris, C.B.; Vaz, C.A. How much do marine connectivity fluctuations matter? *Am. Nat.* **2014**, *184*, 523–530. [CrossRef] [PubMed]

32. Segura-García, I.; Garavelli, L.; Tringali, M.; Matthews, T.; Chérubin, L.M.; Hunt, J.; Box, S.J. Reconstruction of larval origins based on genetic relatedness and biophysical modeling. *Sci. Rep.* **2019**, *9*, 7100. [CrossRef]

33. Briones-Fourzán, P.; Candela, J.; Lozano-Álvarez, E. Postlarval settlement of the spiny lobster *Panulirus argus* along the Caribbean coast of Mexico: Patterns, influence of physical factors, and possible sources of origin. *Limnol. Oceanogr.* **2008**, *53*, 970–985. [CrossRef]

34. Kough, A.S.; Paris, C.B.; Butler, M.J., IV. Larval connectivity and the international management of fisheries. *PLoS ONE* **2013**, *8*, e64970. [CrossRef]

35. Lara-Hernández, J.A.; Zavala-Hidalgo, J.; Sanvicente-Añorve, L.; Briones-Fourzán, P. Connectivity and larval dispersal pathways of *Panulirus argus* in the Gulf of Mexico: A numerical study. *J. Sea Res.* **2019**, *155*, 101814. [CrossRef]

36. Saunders, M.I.; Thompson, P.A.; Jeffs, A.G.; Säwström, C.; Sachlikidis, N.; Beckley, L.E.; Waite, A.M. Fussy feeders: Phyllosoma larvae of the Western Rock lobster (*Panulirus cygnus*) demonstrate prey preference. *PLoS ONE* **2012**, *7*, e36580. [CrossRef]

37. Olvera-Limas, R.M.; Ordóñez-Alcalá, L. Distribución, abundancia relativa y desarrollo larvario de langostas *Panulirus argus* y *Scyllarus americanus* en la ZEE del Golfo de México y mar Caribe. *Cienc. Pesq.* **1988**, *6*, 7–31.

38. Manzanilla-Domínguez, H.; Gasca, R. Distribution and abundance of phyllosoma larvae (Decapoda, Palinuridae) in the southern Gulf of Mexico and the western Caribbean Sea. *Crustaceana* **2004**, *77*, 75–94.

39. Briones-Fourzán, P.; Lozano-Álvarez, E.; Monroy-Velázquez, L.V. Langostas Palinuroideas. In *Camarones, Langostas y Cangrejos de la Costa Este de México*; Hernández-Aguilera, J.L., Ruiz-Nuño, J.A., Toral-Almazán, R.E., Arenas-Fuentes, V., Eds.; CONABIO: Mexico City, Mexico, 2005; Volume 1, pp. 207–235.

40. Manzanilla-Domínguez, H.; Gasca, R.; Suárez-Morales, E. Notes on the distribution of phyllosoma larvae in an oceanic atoll-like reef system in the western Caribbean. *Crustaceana* **2005**, *78*, 505–512.

41. Canto-García, A.; Goldstein, J.; Sosa-Cordero, E.; Carrillo, L. Distribution and abundance of *Panulirus* spp. phyllosomas off the Mexican Caribbean coast. *Bull. Mar. Sci.* **2016**, *92*, 207–227. [CrossRef]

42. Carrillo, L.; Largier, J.L.; Johns, E.; Smith, R.; Lamkin, J. Pathways and upper hydrography in the Mesoamerican Barrier Reef System–Part 1: Circulation. *Contin. Shelf Res.* **2015**, *109*, 164–176. [CrossRef]
43. Badan, A.; Candela, J.; Sheinbaum, J.; Ochoa, J. Upper-layer circulation in the approaches to Yucatan Channel. In *Circulation in the Gulf of Mexico: Observations and Models*; Geophysical Monograph Series 161; American Geophysical Union: Washington, DC, USA, 2005; pp. 57–69.
44. Cetina, P.; Candela, J.; Sheinbaum, J.; Ochoa, J.; Badam, A. Circulation along the Mexican Caribbean coast. *J. Geophys. Res.* **2006**, *111*, C08021. [CrossRef]
45. Muhling, B.A.; Smith, R.H.; Vázquez-Yeomans, L.; Lamkin, J.T.; Johns, E.M.; Carrillo, L.; Sosa-Cordero, E.; Malca, E. Larval fish assemblages and mesoscale oceanographic structure along the Mesoamerican Barrier Reef System. *Fish. Oceanogr.* **2013**, *22*, 409–428. [CrossRef]
46. Merino, M. Aspectos de la circulación costera superficial del Caribe mexicano con base en observaciones utilizando tarjetas de deriva. *An. Inst. Cienc. del Mar y Limnol. Univ. Nal. Autón. México* **1986**, *13*, 31–46.
47. Chávez, G.; Candela, J.; Ochoa, J. Subinertial flows and transports in Cozumel Channel. *J. Geophys. Res.* **2003**, *108*, 3037. [CrossRef]
48. Briones-Fourzán, P.; Candela, J.; Carrillo, L.; Espinosa-Magaña, A.; Negrete-Soto, F.; Barradas-Ortiz, C.; Escalante-Mancera, E.; Muñoz de Cote-Hernández, R.; Martínez-Calderón, R.; Lozano-Álvarez, E. Metamorphosis of spiny lobsters (*Panulirus argus* and *Panulirus guttatus*) in the Yucatan current as inferred from the distribution of pueruli and final stage phyllosomata. *Limnol. Oceanogr.* **2021**, *66*, 3421–3438. [CrossRef]
49. Lozano-Álvarez, E.; Briones-Fourzán, P.; Huchin-Mian, J.P.; Segura-García, I.; Ek-Huchim, J.P.; Améndola-Pimenta, M.; Rodríguez-Canul, R. *Panulirus argus* virus 1 detected in oceanic postlarvae of Caribbean spiny lobster: Implications for disease dispersal. *Dis. Aquat. Org.* **2015**, *117*, 165–170. [CrossRef]
50. Espinosa-Magaña, A.; Briones-Fourzán, P.; Jeffs, A.; Lozano-Álvarez, E. Energy cost of the onshore transport of postlarvae of the Caribbean spiny lobster. *Bull. Mar. Sci.* **2018**, *94*, 801–819. [CrossRef]
51. Briones-Fourzán, P.; Espinosa-Magaña, A.F.; Lozano-Álvarez, E.; Jeffs, A. Analysis of fatty acids to examine larval and settlement biology of the Caribbean spiny lobster *Panulirus argus*. *Mar. Ecol. Prog. Ser.* **2019**, *630*, 137–148. [CrossRef]
52. Rimmer, D.W.; Phillips, B.F. Diurnal migration and vertical distribution of phyllosoma larvae of the Western rock lobster *Panulirus cygnus*. *Mar. Biol.* **1979**, *54*, 109–124. [CrossRef]
53. Alfonso, I.; Frías, M.P.; Baisre, J. Distribución, abundancia y migración vertical de la fase larval de la langosta comercial *Panulirus argus* en aguas cubanas. *Rev. Inv. Mar.* **1999**, *20*, 23–32.
54. Yeung, C.; McGowan, M.F. Differences in inshore-offshore and vertical distribution of phyllosoma larvae of *Panulirus*, *Scyllarus* and *Scyllarides* in the Florida Keys in May–June, 1989. *Bull. Mar. Sci.* **1991**, *49*, 699–714.
55. Robertson, P.B. A giant phyllosoma larva from the Caribbean Sea, with notes on smaller specimens. *Crustaceana* **1968**, *15* (Suppl. 2), 83–97.
56. Palero, F.; Guerao, G.; Hall, M.; Chan, T.Y.; Clark, P.F. The 'giant phyllosoma' are larval stages of *Parribacus antarcticus* (Decapoda: Scyllaridae). *Invertebr. Syst.* **2014**, *28*, 258–276. [CrossRef]
57. Phillips, B.F.; Rimmer, D.W.; Reid, D.D. Ecological investigations of the late-stage phyllosoma and puerulus larvae of the western rock lobster *Panulirus longipes cygnus*. *Mar. Biol.* **1978**, *45*, 347–357. [CrossRef]
58. Phillips, B.F.; Brown, P.A.; Rimmer, D.W.; Braine, S.J. Distribution and dispersal of the phyllosoma larvae of the western rock lobster *Panulirus cygnus*, in the south-eastern Indian Ocean. *Aust. J. Mar. Freshw. Res.* **1979**, *30*, 773–783. [CrossRef]
59. Mallol, S.; Mateo-Ramírez, A.; Alemany, F.; Álvarez-Berastegui, D.; Díaz, D.; López-Jurado, J.L.; Goñi, R. Abundance and distribution of scyllarid phyllosoma larvae (Decapoda: Scyllaridae) in the Balearic Sea (Western Mediterranean). *J. Crust. Biol.* **2014**, *34*, 442–452. [CrossRef]
60. Morris, E.K.; Caruso, T.; Buscot, F.; Fischer, M.; Hancock, C.; Maier, T.S.; Meiners, T.; Müller, C.; Obermaier, E.; Prati, D.; et al. Choosing and using diversity indices: Insights for ecological applications from the German Biodiversity Exploratories. *Ecol. Evol.* **2014**, *4*, 3514–3524. [CrossRef] [PubMed]
61. Phleger, C.F.; Nelson, M.M.; Mooney, B.D.; Nichols, P.D.; Ritar, A.J.; Smith, G.G.; Hart, P.R.; Jeffs, A.G. Lipids and nutrition of the southern rock lobster, *Jasus edwardsii*, from hatch to puerulus. *Mar. Freshw. Res.* **2001**, *52*, 1475–1486. [CrossRef]
62. Wang, M.; O'Rorke, R.; Waite, A.M.; Beckley, L.E.; Thompson, P.; Jeffs, A.G. Fatty acid profiles of phyllosoma larvae of western rock lobster (*Panulirus cygnus*) in cyclonic and anticyclonic eddies of the Leeuwin Current off Western Australia. *Prog. Oceanogr.* **2014**, *122*, 153–162. [CrossRef]
63. Sánchez, A.; Gasca, R.; Sosa-Cordero, E.; Camacho-Cruz, K. Stable carbon and nitrogen isotopes in *Panulirus argus* phyllosomas in the Mexican Caribbean. *Reg. Stud. Mar. Sci.* **2021**, *42*, 101617. [CrossRef]
64. Clarke, K.R. Non-parametric multivariate analyses of changes in community structure. *Aust. J. Ecol.* **1993**, *18*, 117–143. [CrossRef]
65. Clarke, K.R.; Warwick, R.M. *Change in Marine Communities: An Approach to Statistical Analysis and Anterpretation*, 2nd ed.; PRIMER-E: Plymouth, UK, 2001.
66. Clarke, K.R.; Gorley, R.N. *PRIMER v6: User/Manual Tutorial*; PRIMER-E: Plymouth, UK, 2006.
67. Hammer, Ø.; Harper, D.A.T.; Ryan, P.D. PAST: Paleontological statistics software package for education and data analysis. *Palaeontol. Electron.* **2001**, *4*, 1–9.
68. Phillips, B.F.; Booth, J.D.; Stanley, J.C.; Jeffs, A.G.; McWilliam, P. Larval and postlarval ecology. In *Lobsters: Biology, Management, Aquaculture and Fisheries*; Phillips, B.F., Ed.; Blackwell: Oxford, UK, 2006; pp. 231–235.

69. Johnson, M.W. Production and distribution of larvae of the spiny lobster *Panulirus interruptus* (Randall) with records on *P. gracilis* (Streets). *Bull. Scripps Inst. Oceanogr.* **1960**, *7*, 412–461.
70. Lazarus, B.I. The occurrence of phyllosomata off the Cape with particular reference to *Jasus lalandii*. *S. Afr. Div. Sea Fish. Invest. Rep.* **1967**, *63*, 1–38.
71. Lesser, J.H.R. Phyllosoma larvae of *Jasus edwardsii* (Hutton) (Crustacea: Decapoda: Palinuridae) and their distribution off the east coast of the North Island, New Zealand. *N. Z. J. Mar. Freshw. Res.* **1978**, *12*, 357–370. [CrossRef]
72. Inoue, N.; Minami, H.; Sekiguchi, H. Distribution of phyllosoma larvae (Crustacea: Decapoda: Palinuridae, Scyllaridae and Synaxidae) in the western north Pacific. *J. Oceanogr.* **2004**, *60*, 963–976. [CrossRef]
73. Johnson, M.J. The palinurid and scyllarid lobster larvae of the tropical eastern Pacific and their distribution as related to the prevailing hydrography. *Bull. Scripps Inst. Oceanogr.* **1971**, *19*, 1–36.
74. Sims, H.W., Jr.; Ingle, R.M. Caribbean recruitment of Florida's spiny lobster population. *Quart. J. Florida Acad. Sci.* **1966**, *29*, 207–242.
75. Yeung, C.; Jones, D.L.; Criales, M.M.; Jackson, T.L.; Richards, W.J. Influence of coastal eddies and counter-currents on the influx of spiny lobster, *Panulirus argus*, postlarvae into Florida Bay. *Mar. Freshw. Res.* **2001**, *52*, 1217–1232. [CrossRef]
76. Briones-Fourzán, P.; Lozano-Álvarez, E. Essential habitats for *Panulirus* spiny lobsters. In *Lobsters: Biology, Management, Aquaculture and Fisheries*, 2nd ed.; Phillips, B.F., Ed.; Wiley-Blackwell: Oxford, UK, 2013; pp. 186–220.
77. Bertelsen, R.D.; Matthews, T.R. Fecundity dynamics of female spiny lobster (*Panulirus argus*) in a south Florida fishery and Dry Tortugas National Park lobster sanctuary. *Mar. Freshw. Res.* **2001**, *52*, 1559–1565. [CrossRef]
78. Fonseca-Larios, M.E.; Briones-Fourzán, P. Fecundity of the spiny lobster *Panulirus argus* (Latreille, 1804) in the Caribbean coast of Mexico. *Bull. Mar. Sci.* **1998**, *63*, 21–32.
79. Castillo, A.; Lessios, H.A. Lobster fishery by the Kuna Indians in the San Blas Region of Panama (Kuna Yala). *Crustaceana* **2001**, *74*, 459–475.
80. Yallonardo, M.; Posada, J.M.; Schweizer, D.M. Current status of the spiny lobster, *Panulirus argus*, fishery in the Los Roques Archipelago National Park, Venezuela. *Mar. Freshw. Res.* **2001**, *52*, 1615–1622. [CrossRef]
81. Martínez-Calderón, R.; Lozano-Álvarez, E.; Briones-Fourzán, P. Morphometric relationships and seasonal variation in size, weight, and a condition index of post-settlement stages of the Caribbean spiny lobster. *PeerJ* **2018**, *6*, e5297. [CrossRef]
82. Matsuda, H.; Yamakawa, T. Effects of temperature on growth of the Japanese spiny lobster, *Panulirus japonicus* phyllosomas under laboratory conditions. *Mar. Freshw. Res.* **1997**, *48*, 791–796. [CrossRef]
83. Marfin, J.P. Biologie et pêche de la langouste *Panulirus guttatus* en Martinique. *Sci. Pêche* **1978**, *278*, 1–10.
84. Sharp, W.C.; Hunt, J.H.; Lyons, W.G. Life history of the spotted spiny lobster, *Panulirus guttatus*, an obligate reef-dweller. *Mar. Freshw. Res.* **1997**, *28*, 142–172. [CrossRef]
85. Briones-Fourzán, P.; Contreras-Ortiz, G. Reproduction of the spiny lobster *Panulirus guttatus* (Decapoda: Palinuridae) on the Caribbean Coast of Mexico. *J. Crust. Biol.* **1999**, *19*, 171–179. [CrossRef]
86. Losada-Tostesón, V.; Posada, J.M.; Losada, F. Size and reproductive status of fished spotted spiny lobster, *Panulirus guttatus*, in Morrocoy National Park, Venezuela: A preliminary report. *Mar. Freshw. Res.* **2001**, *52*, 1599–1603. [CrossRef]
87. Wynne, S.; Côté, I. Effects of habitat quality and fishing on Caribbean spotted spiny lobster populations. *J. Appl. Ecol.* **2007**, *44*, 488–494. [CrossRef]
88. Monod, T.; Postel, E. Notes sur une langouste brévicorne peu connue, *Justitia longimana* (H. Milne Edwards). *Crustaceana* **1968**, *14*, 178–184. [CrossRef]
89. Briones-Fourzán, P.; Barradas-Ortiz, C.; Negrete-Soto, F.; Segura-García, I.; Lozano-Álvarez, E. Occurrence of *Panulirus meripurpuratus* and *P. laevicauda* (Decapoda: Achelata: Palinuridae) in Bahía de la Ascensión, Mexico. *Lat. Am. J. Aquat. Res.* **2019**, *47*, 694–698.
90. Webber, W.R.; Booth, J.D. Taxonomy and evolution. In *The Biology and Fisheries of the Slipper Lobster*; Lavalli, K.L., Spanier, E., Eds.; CRC Press: Boca Raton, FL, USA, 2007; pp. 25–52.
91. Sharp, W.C.; Hunt, J.H.; Teehan, W.H. Observations on the ecology of *Scyllarides aequinoctialis*, *Scyllarides nodifer*, and *Parribacus antarcticus* and a description of the Florida scyllarid lobster fishery. In *The Biology and Fisheries of the Slipper Lobster*; Lavalli, K.L., Spanier, E., Eds.; CRC Press: Boca Raton, FL, USA, 2007; pp. 231–242.
92. Holthuis, L.B. Biological investigations of the deep sea. 42. A new species of shovel-nose lobster, *Scyllarus planorbis*, from the southwestern Caribbean and northern South America. *Bull. Mar. Sci.* **1969**, *19*, 149–158.
93. Navas, G.R.; Campos, N.H. Las langostas chinas (Crustacea: Decapoda: Scyllaridae) del Caribe colombiano. *Bol. Invest. Mar. Cost.* **1998**, *27*, 51–66. [CrossRef]
94. Yeung, C.; Lee, T.E. Larval transport and retention of the spiny lobster, *Panulirus argus*, in the coastal zone of the Florida Keys, USA. *Fish. Oceanogr.* **2002**, *11*, 286–309. [CrossRef]
95. Chiswell, S.M.; Booth, J.D. Rock lobster *Jasus edwardsii* larval retention by the Wairarapa Eddy off New Zealand. *Mar. Ecol. Prog. Ser.* **1999**, *183*, 227–240. [CrossRef]
96. Chiswell, S.M.; Booth, J.D. Distribution of mid- and late-stage *Jasus edwardsii* phyllosomas: Implications for larval recruitment processes. *N. Z. J. Mar. Freshw. Res.* **2005**, *39*, 1157–1170. [CrossRef]
97. Yoshimura, T.; Yamakawa, H.; Kozasa, E. Distribution of final stage phyllosoma larvae and free-swimming pueruli of *Panulirus japonicus* around the Kuroshio Current off southern Kyushu, Japan. *Mar. Biol.* **1999**, *133*, 293–306. [CrossRef]

98. Inoue, N.; Sekiguchi, H. Distribution of late-stage phyllosoma larvae of *Panulirus japonicus* in the Kuroshio Subgyre. *Mar. Freshw. Res.* **2001**, *52*, 1201–1209. [CrossRef]
99. Truelove, N.K.; Griffiths, S.; Ley-Cooper, K.; Azueta, J.; Majil, I.; Box, S.J.; Behringer, D.C.; Butler, M.J., IV; Preziosi, R.F. Genetic evidence from the spiny lobster fishery supports international cooperation among Central American marine protected areas. *Conserv. Genet.* **2015**, *16*, 347–358. [CrossRef]
100. Caputi, N.; de Lestang, S.; Frusher, S.; Wahle, R.A. The impact of climate change on exploited lobster stocks. In *Lobsters: Biology, Management, Aquaculture and Fisheries*, 2nd ed.; Phillips, B.F., Ed.; Wiley-Blackwell: Oxford, UK, 2013; pp. 84–112.
101. Phillips, B.F.; Pérez-Ramírez, M.; de Lestang, S. Lobsters in a changing climate. In *Climate Change Impact on Fisheries and Aquaculture: A Global Analysis*; Phillips, B.F., Pérez-Ramírez, M., Eds.; Wiley-Blackwell: Hoboken, NJ, USA, 2018; pp. 815–849.
102. Woodings, L.N.; Murphy, N.P.; Jeffs, A.; Suthers, I.M.; Liggins, G.W.; Strugnell, J.M. Distribution of Palinuridae and Scyllaridae phyllosoma larvae within the East Australian Current: A climate change hot spot. *Mar. Freshw. Res.* **2019**, *70*, 1020–1033. [CrossRef]

Population Structure and Seasonal Variability of two Luciferid Species (Decapoda: Sergestoidea) in the Western Gulf of Mexico

Laura Sanvicente-Añorve [1,*], **Juan Hernández-González** [2], **Elia Lemus-Santana** [3], **Margarita Hermoso-Salazar** [2] **and Marco Violante-Huerta** [3]

[1] Instituto de Ciencias del Mar y Limnología, Universidad Nacional Autónoma de México, Circuito Exterior S/N, Ciudad Universitaria, Mexico City 04510, Mexico

[2] Facultad de Ciencias, Universidad Nacional Autónoma de México, Circuito Exterior S/N, Ciudad Universitaria, Mexico City 04510, Mexico; juancahgo13@ciencias.unam.mx (J.H.-G.); margaritahermoso@ciencias.unam.mx (M.H.-S.)

[3] Posgrado en Ciencias del Mar y Limnología, Universidad Nacional Autónoma de México, Circuito Exterior S/N, Ciudad Universitaria, Mexico City 04510, Mexico; lesael@ciencias.unam.mx (E.L.-S.); marco.violante@comunidad.unam.mx (M.V.-H.)

* Correspondence: lesa@unam.mx

Citation: Sanvicente-Añorve, L.; Hernández-González, J.; Lemus-Santana, E.; Hermoso-Salazar, M.; Violante-Huerta, M. Population Structure and Seasonal Variability of two Luciferid Species (Decapoda: Sergestoidea) in the Western Gulf of Mexico. *Diversity* **2021**, *13*, 301. https://doi.org/10.3390/d13070301

Academic Editors: Viatcheslav Ivanenko and Bert W. Hoeksema

Received: 23 April 2021
Accepted: 10 June 2021
Published: 2 July 2021

Publisher's Note: MDPI stays neutral with regard to jurisdictional claims in published maps and institutional affiliations.

Abstract: The population ecology of luciferids has been scarcely studied. This study examined the distribution and population parameters of *Belzebub faxoni* and *Lucifer typus* in the western Gulf of Mexico. Samples were collected using a Bongo net at 82 sampling stations during three periods (July, January, and October–November). Abundance data of species were subjected to a regression tree analysis to determine the main factors affecting their distribution. In addition, total length data of individuals were fitted to a logistic equation to estimate the size at first maturity. Food availability was the key factor affecting the distribution of *B. faxoni*, which exhibited its highest abundance over the inner shelf. In contrast, *L. typus* avoided low salinity waters (<35.9 psu), and its highest densities were found over the slope and oceanic areas. The sex ratio in *B. faxoni* was always biased towards females, but *L. typus* showed variable sex proportions. Several hypotheses attempting to explain these disparities are discussed. Size structure in *B. faxoni* favored small individuals in summer, the reproduction peak. In contrast, the *L. typus* population was always dominated by large individuals, probably transported by currents from the main hatching sites. Size at first maturity was higher in *L. typus* (females: 8.16 mm; males: 8.63 mm) than in *B. faxoni* (females: 6.53 mm; males: 6.74 mm). Information here provided may help to better understand the population dynamics of these species.

Keywords: *Belzebub*; *Lucifer*; sex ratio; size structure; size at first maturity; population ecology

1. Introduction

Species of the family Luciferidae (Decapoda) are typical components—sometimes dominant—of plankton communities, commonly found in tropical and subtropical epipelagic waters from all oceans [1]. Luciferids are small shrimps of about 9 to 13 mm total length in the adult stage [2]. They spend their entire life cycle as plankton and have developed several morphological adaptations to remain in the pelagic environment. These animals are characterized by an extremely compressed body, a reduction of mouthparts and first and second pereopods [3,4]. The numerous setae on the thoracic and abdominal appendages and flabellate uropods enable luciferids to increase frictional resistance to sinking and remain afloat in the water.

Luciferids play a vital role in marine and estuarine food webs. They consume not only zooplankton prey of moderate size and even their own eggs [5,6], but also phytoplankton, as revealed by laboratory experiments [7]. In turn, they constitute significant food items for commercially or ecologically important fishes, shrimps, and even whale sharks [8–11]. Furthermore, luciferids and other related shrimps are used in the food manufacturing industry in the production of fermented pastes and sauces in some countries of Southeast Asia [12].

All luciferid species exhibit sexual dimorphism, evident in mature developmental stages of growth. Adult males are characterized by a sex organ called the petasma, located on the first pair of pleopods, two ventral processes on the sixth abdominal somite, and a prominent process on the telson [2–4,13]. In females, the sex organ, the thelycum, is located at the base of the third pair of pereopods. Uniquely among dendrobranchiate shrimps, luciferid females hold their eggs at the base of the third pair of pereopods and remain attached until eclosion, thus reducing offspring mortality. After eclosion, individuals have several developmental stages until maturity: nauplius, protozoea, zoea, mysis, and postlarva (juvenile) [5,13].

Luciferidae was considered a monotypic family in the superfamily Sergestoidea [14] until the phylogenetic analysis of Vereshchaka et al. [4], who recognized a second genus named *Belzebub* Vereshchaka, Olesen & Lunina, 2016, in addition to the former described genus *Lucifer* Thompson, 1829. From the seven currently accepted species [15], only *B. faxoni* (Borradaile, 1915) and *L. typus* Milne-Edwards, 1937 coexist in the western Atlantic including the Gulf of Mexico [4,16]. General distribution of *L. typus* comprises oceanic waters of tropical and temperate zones around the world, but it is not common to find the near the coast [17,18]. By contrast, *B. faxoni* is only found in the Atlantic Ocean with a more restricted distribution than *L. typus*, and it is especially abundant in the neritic and coastal areas of eastern America and western Africa [4,17].

The Gulf of Mexico is a semi-closed basin that provides goods and services for the social and economic development of Mexico, Cuba, and the United States. The gulf includes a great variety of ecosystems, harbors high biological diversity, supports important fisheries, and is a region of obligatory transit between the ports of the gulf facing the Atlantic. Despite their ecological relevance, studies targeting luciferids in the Gulf of Mexico are very scarce. Some works refer to the distribution of species [16] and others to the population characteristics of *B. faxoni* [5,19]. This study jointly addresses the population ecology of *B. faxoni* and *L. typus* through the analysis of the factors affecting their distribution and the main population parameters in the neritic and oceanic waters of the Mexican western Gulf of Mexico.

2. Materials and Methods

2.1. Study Area

The study area encompassed neritic and oceanic waters of the western Gulf of Mexico and was bounded by the 95° meridian to the east and the coastline of the Mexican state of Tamaulipas to the west (Figure 1).

Upper circulation in the Gulf of Mexico is dominated by the Loop Current, a warm ocean current that enters the gulf through the Yucatan Channel and flows out through the Florida Straits. The Loop Current sheds large anticyclonic eddies (200 to 400 km) that move westward into the western gulf and eventually dissipate near the Mexican slope [19]. The time interval between eddy detachments from the Loop Current is highly variable, ranging from 3 to 17 months, but physical causes of the shedding process are not well understood [20,21].

Over the shelves, surface circulation is affected by seasonal variations of wind stress and continental water discharges. Currents over the western shelf are mainly driven by the wind and run toward the southeast from September to March and toward the northwest from May to August [22]. The marine coastal zone in the western gulf is influenced by the runoff of the Bravo, San Fernando, Soto la Marina, and Pánuco Rivers as well as the Madre and Tamiahua coastal lagoons.

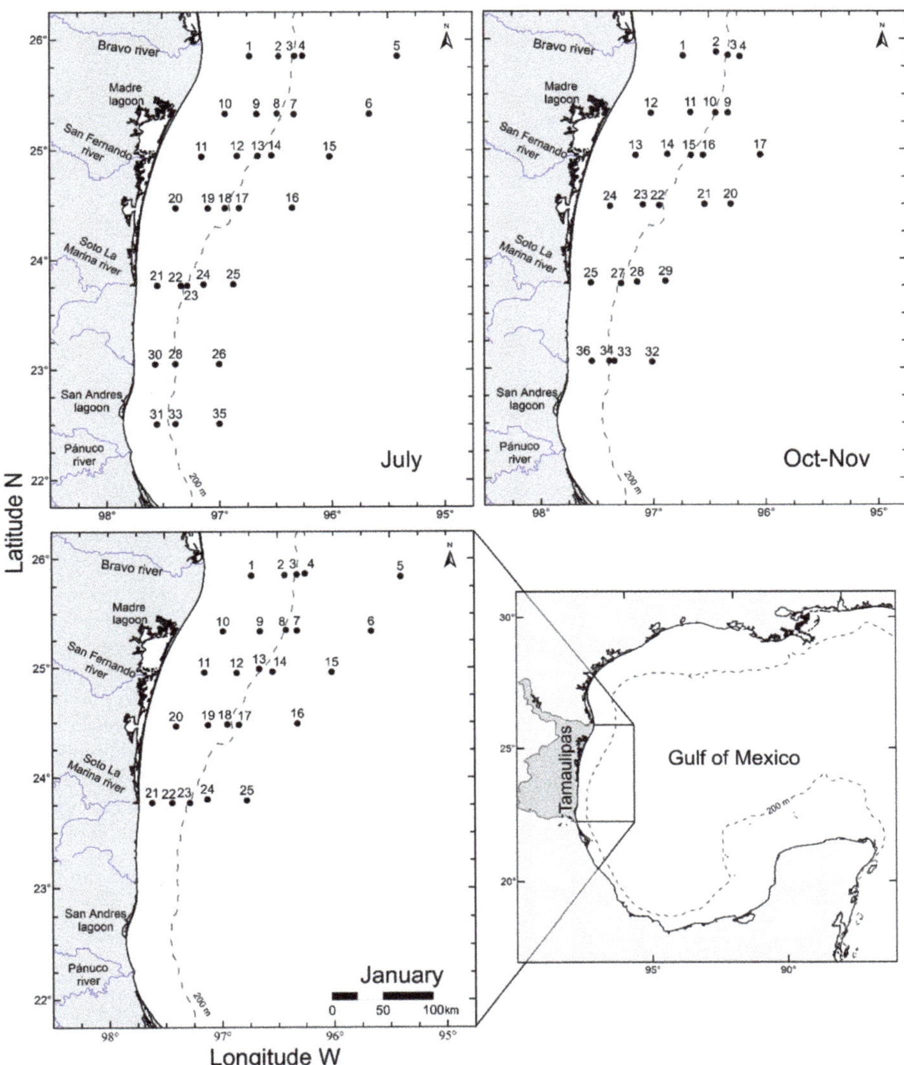

Figure 1. Study area and location of sampling stations during the three sampling periods.

2.2. Sampling and Laboratory Analysis

Samples analyzed in this study were collected within the framework of the monitoring program called "Environmental Framework of the Oceanographic Conditions in Mexico's Northwestern Exclusive Economic Zone in the Gulf of Mexico (MARZEE)". Zooplankton sampling was carried out during three oceanographic cruises named MARZEE 1 (2 to 7 July 2010), MARZEE 2 (27 to 31 January 2011), and MARZEE 3 (26 October to 2 November 2012). A total of 82 oceanographic stations located in both neritic and oceanic zones were sampled using a Bongo net (333 and 505 μm mesh) equipped with two flowmeters to estimate the volume of filtered water. Zooplankton samples were collected by oblique tows, and sampling depth varied between 15 and 200 m according to bottom depth. Samples were fixed with 4% formalin neutralized with sodium borate. At each oceanographic station, records of temperature and salinity were taken using a SBE 9Plus CTD profiler, and measurements of the chlorophyll concentration were also taken with a Wet Labs FLRTD sensor adapted to the profiler. These measurements were taken from the surface up to 1500 m depth, depending on the bottom depth.

Zooplankton biomass was estimated as wet weight (g 100 m^{-3}), using the samples obtained with the 333 µm mesh size after vacuum filtration to remove interstitial water within organisms [23]. All zooplankters were included in the measurements because samples did not contain large gelatinous organisms. Afterward, those samples were divided using a Folsom splitter, and all luciferids were sorted from the aliquots. Individuals were identified and sexed, when possible, based on previously published descriptions [3,4,18]. Each individual was measured from the base of the eye-stalk to the end of the sixth abdominal somite [24] (hereafter L). Processed individuals were quantified and standardized to 100 m^3 of filtered water (ind 100 m^{-3}).

2.3. Data Analysis

The influence of environmental variables on the distribution of species throughout the sampling period was evaluated using regression tree analysis (RTA). It is a powerful analytical method that helps identify the most important explanatory variables explaining the variability of a response variable. A tree is built by repeatedly splitting the response variable based on a single best explanatory variable, i.e., the one that minimizes the variance in the response variable. The process begins allocating data of the response variable into two mutually exclusive groups, each of them being as homogeneous as possible. Next, each of the groups is subjected to the same splitting procedure; the process continues until reaching a user-specified criterion. The result is a tree showing the mean values of the response variable at the terminal nodes [25–27]. In this study, the response variable was the abundance value of each species in each of the 82 sampling stations, and the explanatory variables were the temperature, salinity, chlorophyll concentration, bottom depth, distance to the shore, and the zooplankton biomass (as a measure of food availability). Examining the zooplankton samples, we thought that small copepods, ostracods, naked mollusks, and decapods as well as other crustacean larvae could be important prey for luciferids in the study area. These animals represented approximately 30% of the zooplankton biomass; thus, only this fraction was considered for the analysis. For salinity, temperature, and chlorophyll concentration, we took the mean integrated values in the upper 50 m, a layer in which the zooplankton were more abundant [28,29].

For each sampling season, the number of females (F) and males (M) of each species were counted, and the deviation from the 1:1 sex ratio (F:M) was evaluated using a χ^2 test at a significance level of $\alpha = 0.05$.

The non-parametric 'analysis of similarities' (ANOSIM) test was used to assess if there were differences between small and large individuals of each species among the seasons. The *Ro* statistic issued from this analysis was compared against a set of *R* random values that resulted from recomputing the statistic *n* times (9999 in this case) after randomly shuffling the sample labels [30]. A significance value of $\alpha = 0.05$ was used in this analysis

For each species and sex, we estimated the size at first maturity ($L_{0.5}$) by fitting the L to a logistic equation:

$$P_i = \frac{1}{1 + Ae^{BL_i}} \tag{1}$$

where, P_i is the cumulative frequency in percentage at the L_i value (in other words, the probability of finding a mature individual at size L_i), and *A* and *B* are parameters of the model. Particularly, the *L* value at which *P* is 0.5 represents the size at first maturity, conceived as the average size at which 50% of the individuals are mature [31].

3. Results

3.1. Environmental Conditions

High temperature values were recorded in July and October–November; the highest values (around 27 °C) were registered in neritic waters of the northern study area in July. The lowest temperatures (18 to 23 °C) were registered in January with a clear coastal–ocean gradient (Figure 2). Homogeneous salinity values (36.3 to 36.7 psu) were recorded in October–November in the entire study area; January registered the highest variability, with the lowest values (33 to 35 psu) over the inner shelf (Figure 3).

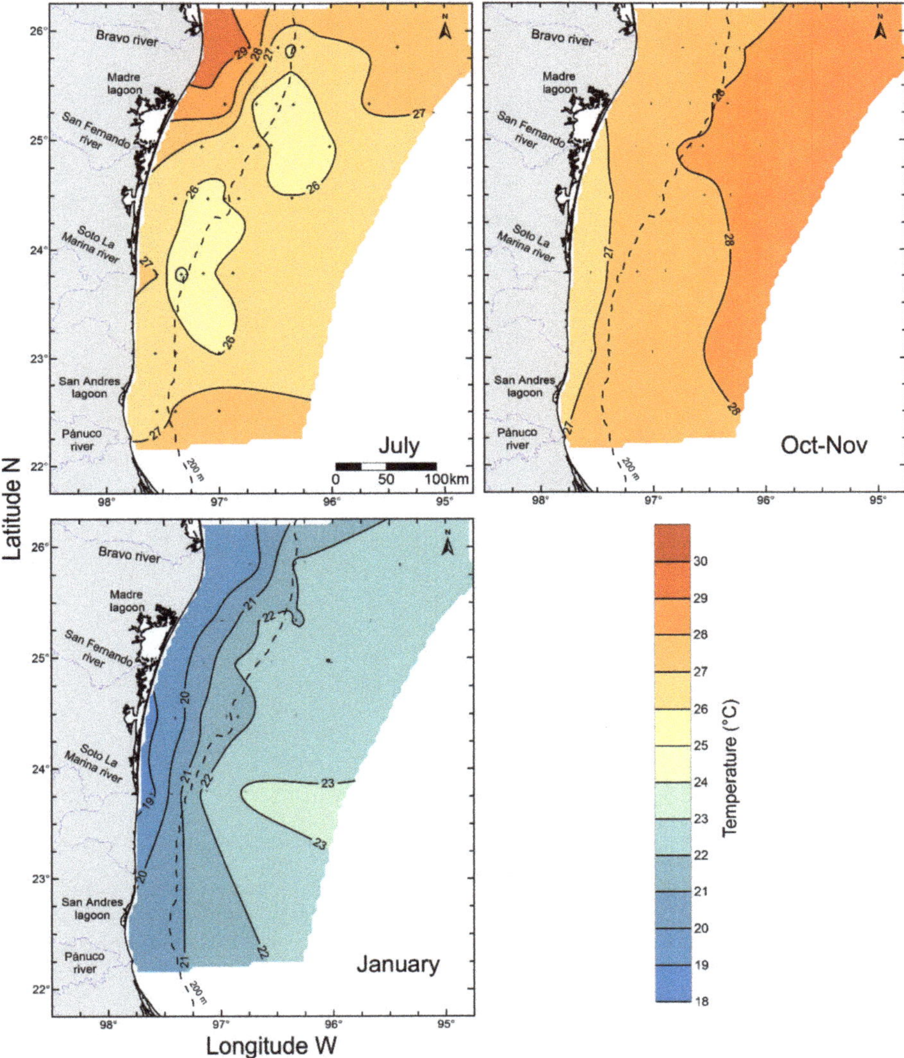

Figure 2. Mean integrated temperature (°C) in the upper 50 m layer in the western Gulf of Mexico.

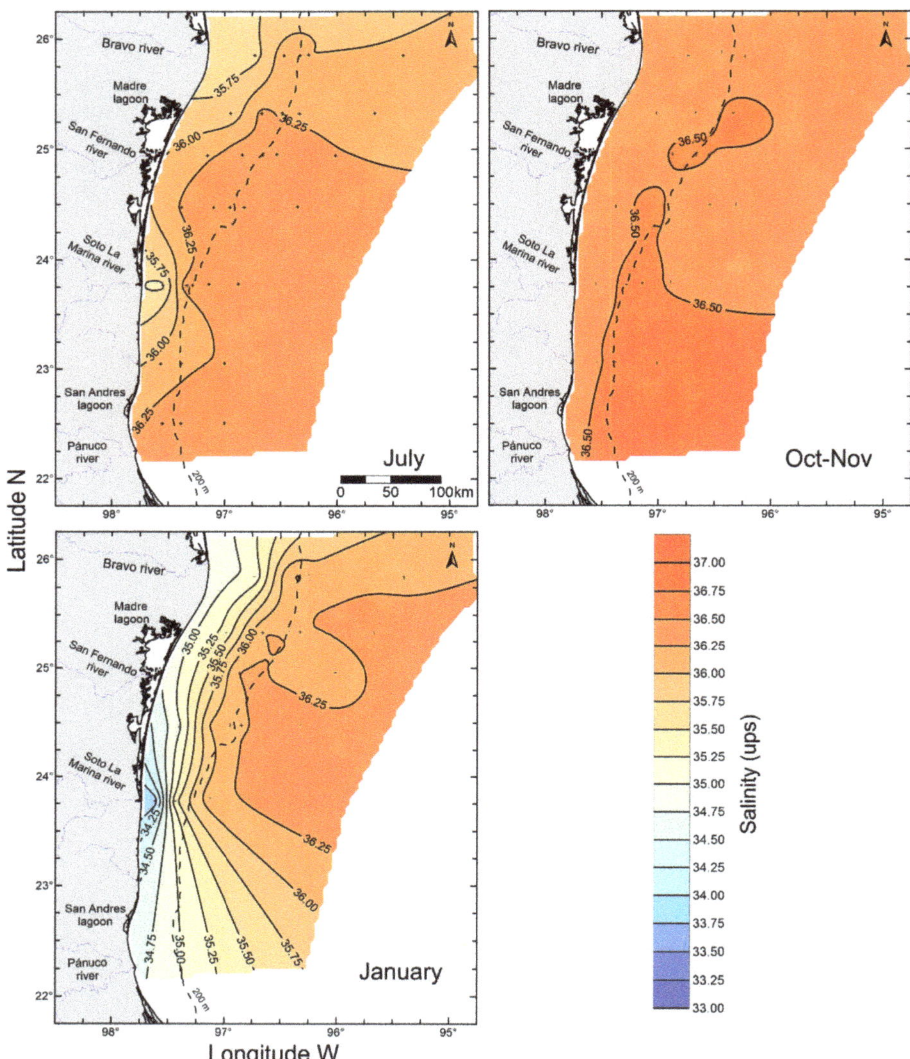

Figure 3. Mean integrated salinity (psu) in the upper 50 m layer in the western Gulf of Mexico.

Chlorophyll concentration was low (<1.5 mg m^{-3}) in all the studied seasons. The highest values were registered on the inner shelf in July and January (Figure 4). In addition, mean zooplankton biomass in the neritic zone was higher in July (9.7 $\pm$ 4.5 g 100 m^{-3}) and January (8.3 $\pm$ 4.9 g 100 m^{-3}) than in October–November (4.5 $\pm$ 3.5 g 100 m^{-3}). The highest biomass values were found at the stations close to the coast in July and January. In October–November, values were low both in the neritic zone and in the oceanic zone, with a maximum value close to the mouth of the Soto la Marina River (Figure 5).

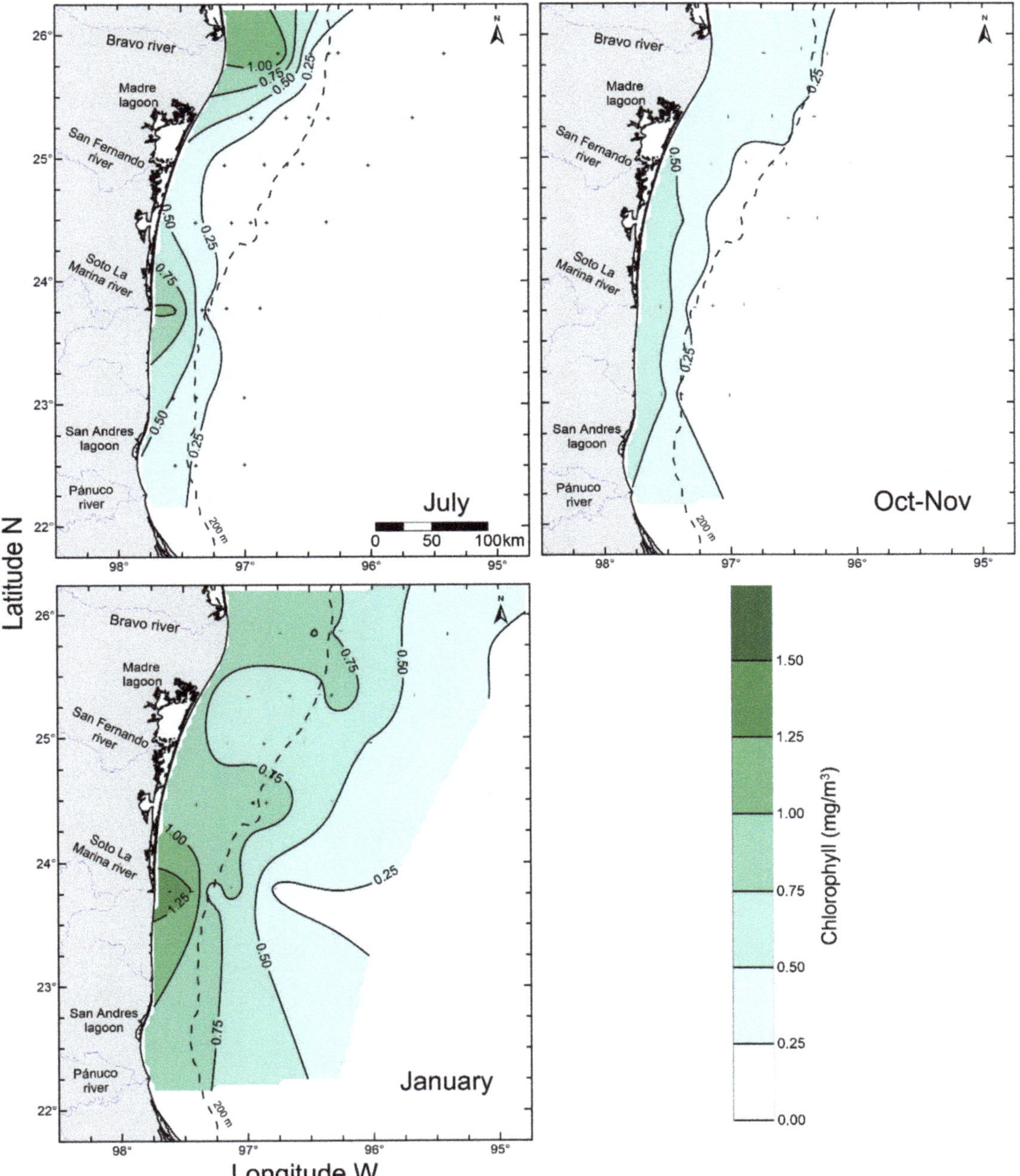

Figure 4. Mean integrated chlorophyll concentration (mg m^{-3}) in the upper 50 m layer in the western Gulf of Mexico.

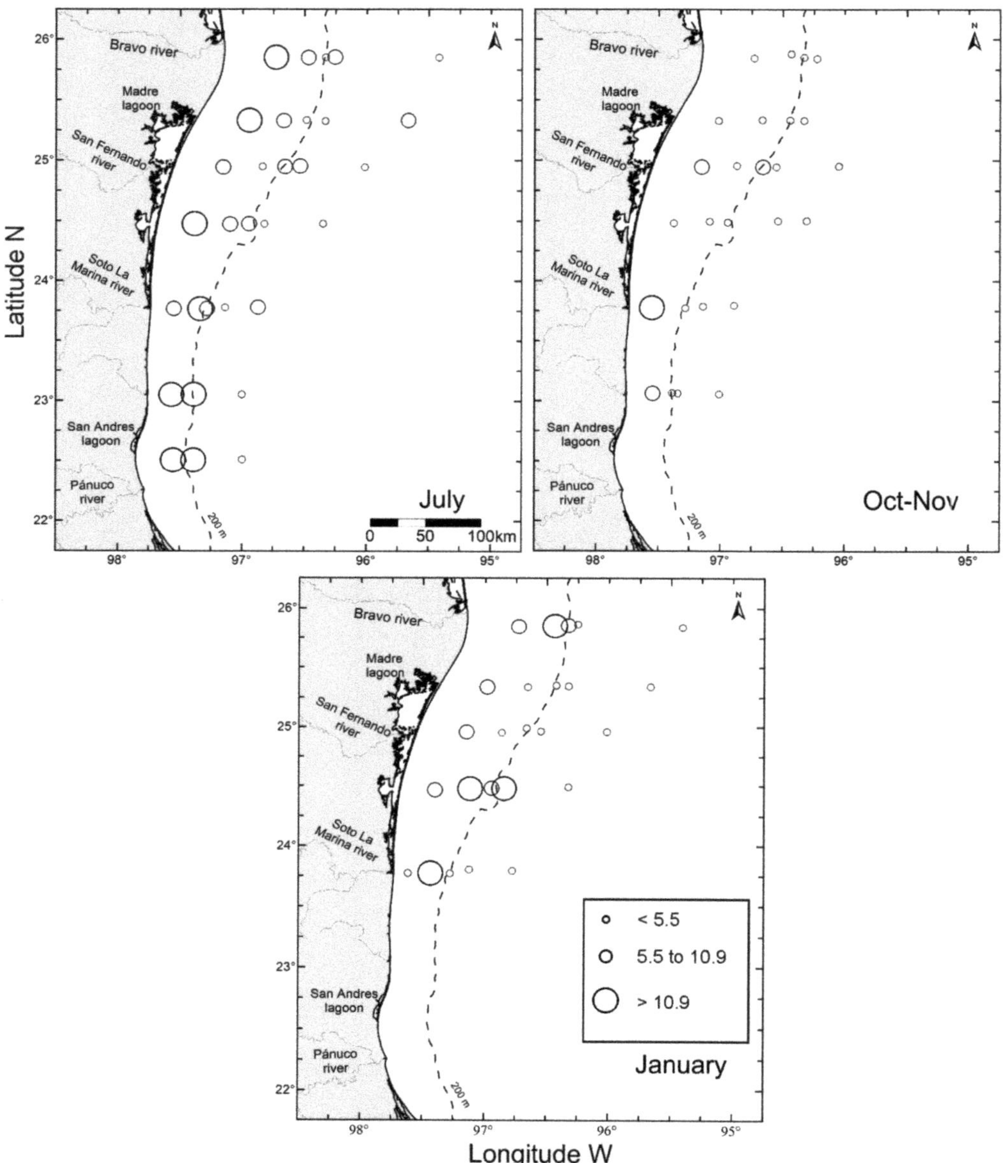

Figure 5. Distribution of zooplankton biomass (g 100 m^{-3}) values in the western Gulf of Mexico.

3.2. Distribution and Population Structure of Belzebub faxoni

The highest density values of *B. faxoni* were found in July, particularly in neritic waters, with a maximum of 4639.4 ind 100 m^{-3}. In October–November, the density of the species was less than 100 ind 100 m^{-3} except for the station close to San Andrés lagoon; in January, the species rebounded (Figure 6).

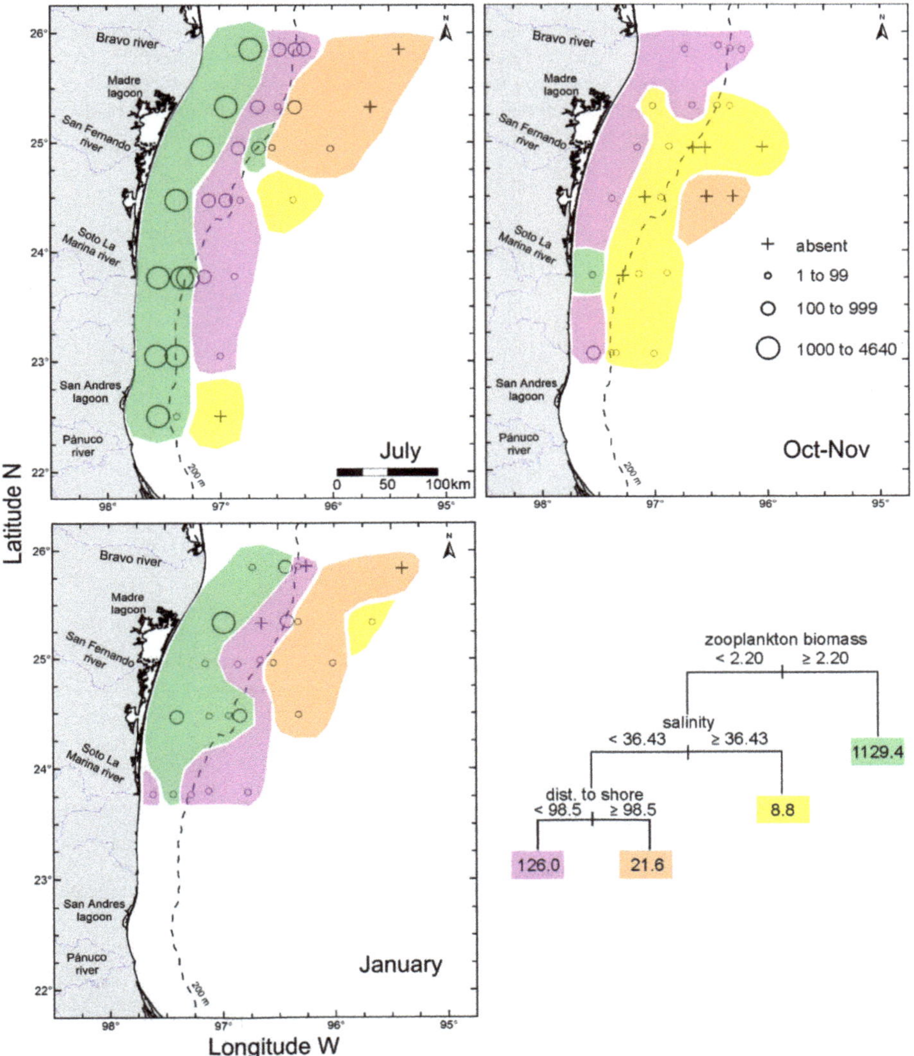

Figure 6. Abundance (ind 100 m^{-3}) of *Belzebub faxoni* in the western Gulf of Mexico and results of the regression tree analysis showing the mean abundance value of the response variable at the end of the branch.

Results of RTA indicated that the areas with the highest *B. faxoni* densities were associated with high zooplankton biomass (>2.20 g 100 m^{-3}). By contrast, low *B. faxoni* densities were registered in areas with available zooplankton biomass less than 2.20 g 100 m^{-3} and salinities higher than 36.43 psu (Figure 6).

Sex ratio (F:M) was significantly (χ^2, $p < 0.05$) skewed toward females in all the seasons: 1.8:1 in July, 1.7:1 in October–November, and 1.9:1 in January. These results showed that the number of females was almost twice the number of males in all the seasons.

Total length of *B. faxoni* individuals varied between 2.42 and 12.64 mm, and five size classes were considered to analyze the seasonal variation of the population structure (Figures 7 and 8). These classes were arbitrarily chosen in order to show a general trend in the size distribution. In July, size classes II and III registered the highest density values with a gradual decline toward classes IV and V. In the other seasons, size structure did not show a clear trend, and density values of all the size classes were low, especially in October–November. Comparing the abundance values of classes I and V among the seasons, the ANOSIM test indicated that July was different from the other months ($p < 0.05$), but January was not significantly different ($p > 0.05$) from October–November.

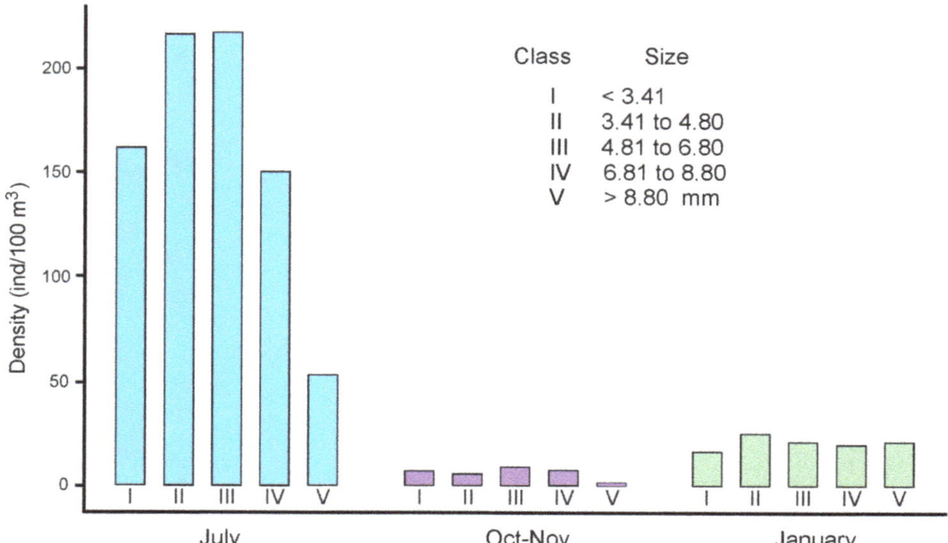

Figure 7. Size structure of the *Belzebub faxoni* population throughout three sampling seasons.

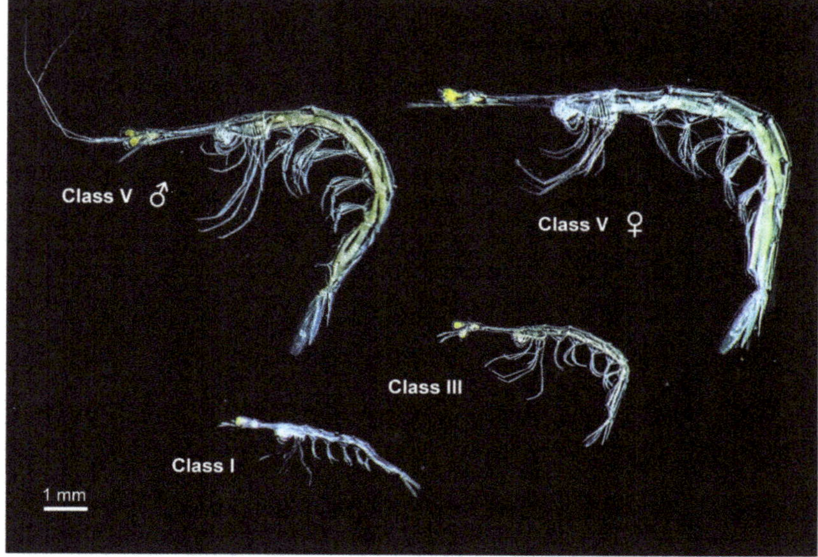

Figure 8. *Belzebub faxoni*. Habitus in lateral view from size classes I, III, and V.

Size at first maturity was 6.53 mm in females and 6.74 mm in males, as shown by fitting the L data to a logistic equation.

3.3. Distribution and Population Structure of Lucifer typus

July was the season with the highest abundance of *L. typus*, with a maximum value of 64.2 ind 100 m^{-3}; the highest densities were located in oceanic waters and over the shelf break (Figure 9). In October–November and January, the density decreased, with maximum values of 12.8 and 27.7 ind 100 m^{-3}, respectively.

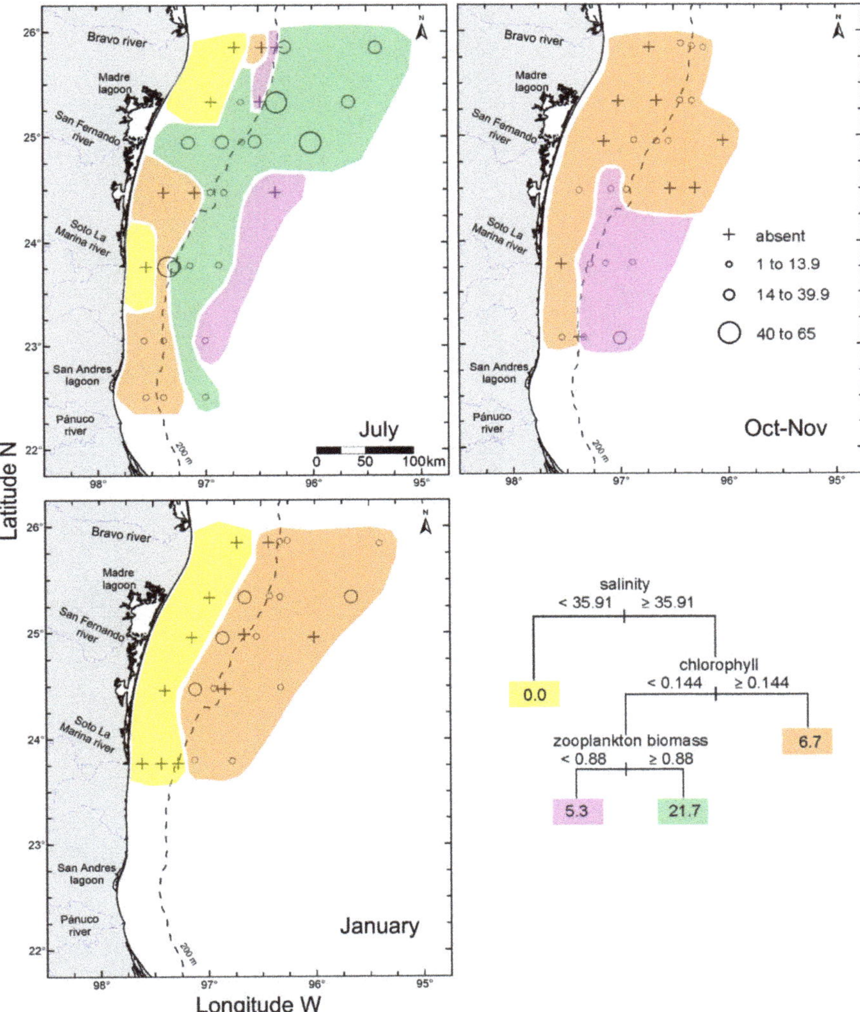

Figure 9. Abundance (ind 100 m^{-3}) of *Lucifer typus* in the western Gulf of Mexico and results of the regression tree analysis showing the mean abundance value of the response variable at the end of the branch.

Results of the RTA showed that the species was absent in areas where salinity was less than 35.91. The highest mean density value of this species was recorded in areas with salinity higher than 35.91 and available zooplankton biomass higher than 0.88 g 100 m^{-3} (Figure 9).

The sex proportion (F:M) was significantly (χ^2, $p < 0.05$) skewed towards the females (1.3:1) in July and toward the males (0.6:1) in January. In October–November, the proportion of both sexes was almost equal (0.9:1) (χ^2, $p > 0.05$).

The total lengths of individuals were between 2.33 and 11.18 mm. Size structure was similar in all the seasons, showing increasing abundance with size (Figures 10 and 11). However, the ANOSIM test indicated that there were no differences ($p > 0.05$) between densities of class I and V among the seasons.

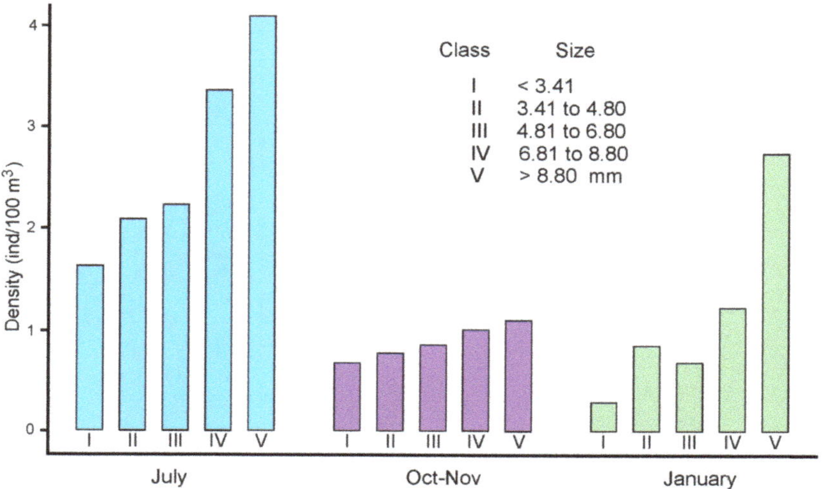

Figure 10. Size structure of the *Lucifer typus* population throughout three sampling seasons.

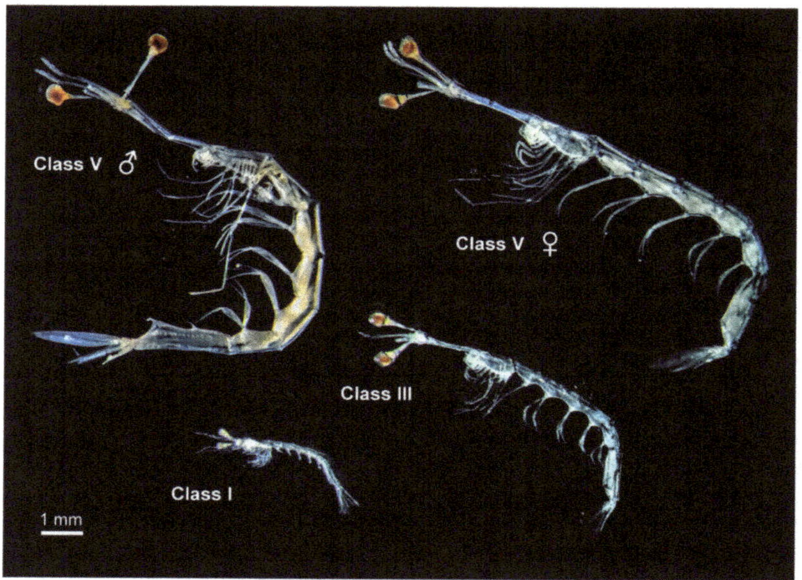

Figure 11. *Lucifer typus*. Habitus in lateral view from size classes I, III, and V.

Size at first maturity, estimated by fitting *L* data to a logistic equation, was 8.16 mm in females and 8.63 mm in males.

4. Discussion

In the western Gulf of Mexico, *Belzebub faxoni* was recorded in all the sampling periods and exhibited a marked variation in seasonal abundance with the peak in July. Similarly, in the northern Gulf, Harper [16] found that the species showed their maximum abundance in August–September and decreased in October–November. In southern Brazil, field observations indicated that the species had continuous reproductive activity throughout the year, greater during the spring–summer period [32,33]. On the other hand, *Lucifer typus* was also recorded in all the sampling periods. Density values were low throughout the year, with slightly higher abundance in July. Despite being related species, abundance values registered by *L. typus* were between two or three orders of magnitude lower than those recorded by *B. faxoni*. Along the western Atlantic, high abundance values for *B. faxoni* were also found in the coastal waters of Texas [16], the southern Gulf of Mexico [19], Venezuela [34], and northern and southern Brazil [24,35,36]. The success of *B. faxoni* in plankton communities could be attributable to a short adult lifespan associated with a rapid turnover of generations and the protection of eggs until eclosion [5].

Results of the RTA suggested that zooplankton biomass (food availability) was the main factor influencing the distribution of *B. faxoni*, which registered its greatest density values on the inner shelf (Figure 6). As stated, we think that the main potential prey for luciferids in the study area includes small copepods, ostracods, naked mollusks, and decapods as well as other crustacean larvae. However, these animals share the habitat with potential predators for luciferids; thus, the inclusion of this parameter in the analysis may also represent the risk of luciferids being preyed upon. The examination of samples showed that chaetognaths, shrimp post larvae, gelatinous organisms (medusae and siphonophores), shelled mollusks, and polychaetes could be potential predators for luciferids. However, regarding the high numbers of the species, it seemed that *B. faxoni* found suitable conditions to grow and reproduce in the study area despite the detrimental consequences of predation and other environmental factors. In situ and laboratory studies indicated that adults of *B. faxoni* were mainly carnivorous, although their feeding appendages indicated the possibility of omnivory [5]. Main prey for adults are difficult to determine because the shrimp completely masticates the food, whereas larvae can consume microalgae and small rotifers, as revealed by laboratory experiments [5,7]. The food consumption rate of this species depends on prey density [5,6], a fact that could explain the major abundance of *B. faxoni* associated with higher food availability. Concerning *L. typus*, the RTA revealed that the species avoided coastal areas where salinity was less than 35.9 psu (Figure 9). Field observations of Xu [37] showed that *L. typus* was a eurythermal but stenohaline species. In addition, Bowman and McCain [17] considered *L. typus* as an indicator of oceanic waters. In this study, the major mean abundance of this species corresponded to areas over the slope and open waters, associated with available zooplankton biomass higher than 0.88 g 100 m^{-3} (Figure 9).

The sex ratio favored the females in the *B. faxoni* population in all the sampled seasons as observed in other localities of the northern and southern Gulf of Mexico [19,38], southern Brazil [36], and the South Atlantic Gyre [39]. By contrast, the *L. typus* population showed variable sex proportions that were only skewed towards females in July when the species exhibited its highest abundance values. If reproductive costs of producing male or female progeny are the same, an equal number of males and females are expected [40]. However, in crustaceans and other marine invertebrates, deviation of this 1:1 Fisherian proportion may occur due to several factors: (i) Mechanisms of sexual determination. Sex determination can result from genetic or environmental causes. In the former, sex is determined by the genotype, while in the second, sex is not given during fertilization but in response to certain environmental factors such as temperature [41]; (ii) Sex reversal. Also called sequential hermaphroditism, it occurs when at a certain size, individuals change sex [42,43].

Sex reversal is a phenomenon commonly observed in crustaceans [43,44]; however, no studies evidence sequential hermaphroditism in luciferids or close taxa; (iii) Differential sex longevity and/or mortality. If one of the sexes has a higher mortality rate and/or lower longevity, the proportion will favor the other [42,45]. This phenomenon may occur if one of the sexes exhibits foraging or mating behavior, which represent a greater risk of being predated [46,47]; (iv) Migration. Differential migratory behavior between sexes could be the cause of sex disparity in a sample. Concerning luciferids, Marafon-Almeida et al. [39] supported differential vertical migration between the sexes, but Álvarez [48] found no differences in the migration patterns of males and females. The causes of sex disparity observed in this study are not fully understood. As in many species of crustaceans, genetic factors must play a significant role in this difference along with environmental factors and ecological relationships.

Regarding the size structure of *B. faxoni* in July, it was noteworthy that the small size classes were the most abundant except for class I (Figure 7). However, if we considered that many of the small organisms of this species escaped from the plankton nets, class I may have been the most abundant. Significantly higher abundance of class I over class V ($p < 0.05$) may indicate that the species reproduces in the study area with a peak in July. The predominance of small individuals is generally indicative of growing populations [49,50]. Townsend et al. [51] stated that the coexistence of many young individuals and few adults occurs when the birth and mortality rates are high. This means that many individuals are born, but they do not have a high probability of survival. Perhaps summer environmental conditions such as high temperatures have a positive influence on the reproductive processes of the species as has been observed for other holoplanktonic organisms [52,53]. The size structure in *L. typus* showed a gradual ascent towards larger sizes in all the seasons (Figure 10). Molles [50] indicated that in the absence of young individuals, a population could not persist. As indicated above, *L. typus* preferentially occurred in the open ocean [17,54,55]; therefore, it was likely that the studied area did not represent the optimal site for population growth. Class V was more abundant than class I, but the ANOSIM test indicated that abundance values between these classes were not significantly different among seasons ($p > 0.05$). This result could indicate that the species reproduced outside the study area and that large individuals had more time to disperse in the ocean from the main hatching sites.

Sizes at first maturity ($L_{0.5}$), estimated here for B. faxoni (females: 6.53 mm, males: 6.74 mm), are similar to those found in Ubatuba, Brazil (females: 5.97 mm, males: 6.92 mm) [36]. In another locality of southern Brazil, López [32] calculated $L_{0.5}$ in terms of the pre-buccal somite length (females: 1.19 mm; males: 1.18 mm), which in this study, corresponded to 6.08 mm for females and 6.41 mm for males. Estimations of $L_{0.5}$ for L. typus (females: 8.16 mm; males: 8.63 mm) given in this study represented the first approach for the species. Differences in the estimated values of $L_{0.5}$ among studies could be due to the model employed to calculate the parameters, the size of the sample, the environmental conditions of populations, or the genetic heritage [31,56,57]. Knowledge of size at first maturity and other population parameters have significant importance in the planning and management of luciferid resources [58].

5. Conclusions

This study provides new and valuable information on the population ecology of the two luciferid species coexisting in the western Atlantic, *Belzebub faxoni* and *Lucifer typus*. In the study area, these species were found during the three examined seasons, indicating continuous reproduction throughout the year with a peak in the summer. *Belzebub faxoni* was more abundant in the waters of the inner shelf, associated with sites of high food availability. In contrast, *L. typus* was mainly found over the slope and open waters, avoiding inshore low-salinity areas. The global distribution pattern of both species, in which *B. faxoni* mostly inhabited the neritic province and *L. typus* the oceanic one, could be the result of a long competition process causing partial resource partitioning. Population density values of *B. faxoni* were two or three orders of magnitude greater than *L. typus*. Previous studies attributed the success of *B. faxoni* in plankton communities to its short adult lifespan, a rapid turnover of generations, and the protection of eggs. Sex ratio favored the females in the *B. faxoni* population in all the periods analyzed, whereas in the *L. typus* population, the sex ratio was biased toward females in only one season. The underlying causes of the skewed sex ratios in the luciferid populations are not fully understood. The size structure of *B. faxoni*, favorable to small size classes in July, may have been indicative of a growing population; in contrast, the *L. typus* population was dominated by large individuals in all seasons, which could be the result of a long transport of these individuals from the main hatching sites. Size at first maturity for both males and females in the *B. faxoni* population agreed with previous records; information regarding this parameter for the *L. typus* population provided here represents the first approximation for the species. As an important part of the zooplankton, knowledge of the distribution and population parameters of luciferids provides important elements to better understand the ecological relationships of pelagic communities and develop future management strategies of shrimps.

Author Contributions: Conceptualization, L.S.-A.; Formal analysis, L.S.-A., J.H.-G. and M.H.-S.; Funding acquisition, L.S.-A.; Investigation, L.S.-A., J.H.-G., E.L.-S., M.H.-S. and M.V.-H.; Original draft preparation, L.S.-A.; Review and editing, L.S.-A., J.H.-G., E.L.-S., M.H.-S. and M.V.-H. All authors have read and agreed to the published version of the manuscript.

Funding: This research was funded by Instituto Nacional de Ecología y Cambio Climático and Universidad Nacional Autónoma de México.

Institutional Review Board Statement: Not applicable.

Informed Consent Statement: Not applicable.

Data Availability Statement: Data are available at: https://hdl.handle.net/20.500 (accessed on 14 May 2021).

Acknowledgments: Authors are grateful to Luis A. Soto and Alfonso V. Botello, coordinators of the MARZEE mission, for the opportunity to participate in the project. Authors also thank F. Zavala-García for his kind assistance during the development of this work. Authors appreciate the helpful comments of the reviewers to improve the manuscript. Special thanks go to P. Briones-Fourzán and M. Hendrickx, guest editors of this volume, for the invitation to participate in this interesting issue.

Conflicts of Interest: The authors declare no conflict of interest.

References

1. Omori, M. Occurrence of two species of *Lucifer* (*Dendrobranchiata: Sergestoidea: Luciferidae*) off the Pacific coast of America. *J. Crustac. Biol.* **1992**, *12*, 104–110. [CrossRef]
2. Hayashi, K.I.; Tsumura, S. Revision of Japanese *Luciferinae* (*Decapoda, Penaeidae, Sergestidae*). *Bull. Jpn. Soc. Sci. Fish* **1981**, *47*, 1437–1441. [CrossRef]
3. Pérez-Farfante, I.; Kensley, B. Penaeoid and sergestoid shrimps and prawns of the world. Keys and diagnoses for the families and genera. *Mém. Mus. Natl. Hist. Nat.* **1997**, *175*, 1–233.
4. Vereshchaka, A.L.; Olesen, J.; Lunina, A.A. A phylogeny-based revision of the family *Luciferidae* (*Crustacea: Decapoda*). *Zool. J. Linn. Soc.* **2016**, *178*, 15–32. [CrossRef]
5. Lee, W.Y.; Omori, M.; Peck, R.W. Growth, reproduction and feeding behavior of the planktonic shrimp, *Lucifer faxoni Borradaile*, off the Texas coast. *J. Plankton Res.* **1992**, *14*, 61–69. [CrossRef]
6. Vega-Pérez, L.A.; Ara, K.; Liang, T.H.; Pedreira, M.M. Feeding of the planktonic shrimp *Lucifer faxoni Borradaile*, 1915 (*Crustacea: Decapoda*) in the laboratory. *Rev. Bras. Oceanogr.* **1996**, *44*, 1–8. [CrossRef]
7. Zimmerman, S.G. The transformation of energy by *Lucifer chacei* (*Crustacea, Decapoda*). *Pac. Sci.* **1973**, *27*, 247–259.
8. Martins, A.S.; Haimovici, M.; Palacios, R. Diet and feeding of the cutlassfish *Trichiurus lepturus* in the subtropical convergence ecosystem of southern Brazil. *J. Mar. Biol. Assoc. U.K.* **2005**, *85*, 1223–1229. [CrossRef]
9. Cellamare, M.; Gómez, G. Alimentación de la sardina *Sardinella aurita* (*Clupeidae*) en el sureste de la Isla Margarita, Venezuela. *Bol. Inst. Ocean. Venez.* **2007**, *46*, 23–36.
10. Rohner, C.A.; Armstrong, A.J.; Pierce, S.J.; Prebble, C.E.M.; Cagua, E.F.; Cochran, J.E.M.; Berumen, M.L.; Richardson, A.J. Whale sharks target dense prey patches of sergestid shrimp off Tanzania. *J. Plankton Res.* **2015**, *37*, 352–362. [CrossRef]

11. Willems, T.; De Backer, A.; Kerkhove, T.; Dakriet, N.N.; De Troch, M.; Vincx, M.; Hostens, K. Trophic ecology of Atlantic seabob shrimp *Xiphopenaeus kroyeri*: Intertidal benthic microalgae support the subtidal food web off Suriname. *Estuar. Coast. Shelf Sci.* **2016**, *182*, 146–157. [CrossRef]

12. Ruddle, K. The supply of marine fish species for fermentation in southeast Asia. *Bull. Natl. Mus. Ethnol.* **1986**, *11*, 997–1036.

13. Tavares, C.; Martin, J.W. Suborder Dendrobranchiata Bate, 1888. In *Treatise on Zoology—Anatomy, Taxonomy, Biology—The Crustacea, Decapoda, Part A Eucarida: Euphausiacea, Amphionidacea, and Decapoda (Partim)*; Schram, F.R., von Vaupel Klein, J.C., Forest, J., Charmantier-Daures, M., Eds.; Brill: Leiden, The Netherlands, 2010; Volume 9, pp. 99–164.

14. De Grave, S.; Fransen, C.H.J.M. *Carideorum catalogus*: The recent species of the dendrobranchiate, stenopodidean, procarididean and caridean shrimps (*Crustacea: Decapoda*). *Zool. Meded.* **2011**, *85*, 195–588.

15. WoRMS. Luciferidae De Haan, 1849 [in De Haan, 1833–1850]. Available online: https://www.marinespecies.org/aphia.php?p=taxdetails&id=106730 (accessed on 19 April 2021).

16. Harper, D.E., Jr. Distribution of *Lucifer faxoni* (*Crustacea: Decapoda: Sergestidae*) in neritic waters off the Texas coast, with a note on the occurrence of *Lucifer typus*. *Contrib. Mar. Sci.* **1968**, *13*, 1–16.

17. Bowman, T.E.; McCain, J.C. Distribution of the planktonic shrimp, *Lucifer*, in the western north Atlantic. *Bull. Mar. Sci.* **1967**, *17*, 660–671.

18. Naomi, T.S.; Antony, G.; George, R.M.; Jasmine, S. Monograph on the planktonic shrimps of the genus *Lucifer* (family *Luciferidae*) from the Indian EEZ. *CMFRI Bull.* **2006**, *49*, 12–34.

19. Cházaro-Olvera, S.; Ortiz, M.; Winfield, I.; Pérez-Ramos, J.A.; Meiners-Mandujano, C. Distribución, densidad, proporción sexual y fecundidad de *Belzebub faxoni* (*Decapoda, Luciferidae*) en el sistema arrecifal veracruzano, SO del Golfo de México. *Rev. Biol. Mar. Oceanogr.* **2017**, *52*, 467–478. [CrossRef]

20. Hamilton, P.; Bower, A.; Furey, H.; Leben, R.; Pérez-Brunius, P. The loop current: Observations of deep eddies and topographic waves. *J. Phys. Oceanogr.* **2019**, *49*, 1463–1483. [CrossRef]

21. Sturges, W.; Leben, R. Frequency of ring separations from the loop current in the Gulf of Mexico: A revised estimate. *J. Phys. Oceanogr.* **2000**, *30*, 1814–1818. [CrossRef]

22. Zavala-Hidalgo, J.; Morey, S.L.; O'Brien, J.J. Seasonal circulation on the western shelf of the Gulf of Mexico using a high-resolution numerical model. *J. Geophys. Res. Ocean.* **2003**, *108*, 1–19. [CrossRef]

23. Zavala-García, F.; Flores-Coto, C. Medición de biomasa zooplanctónica. *An. Cent. Cienc. Mar Limnol. Univ. Nac. Autón. México* **1989**, *16*, 273–278.

24. Teodoro, S.D.S.A.; Negreiros-Fransozo, M.; Simões, S.M.; Lopes, M.; Da Costa, R.C. Population ecology of the planktonic shrimp *Lucifer faxoni Borradaile*, 1915 (*Crustacea, Sergestoidea, Luciferidae*) of the southeastern coast of Brazil. *Braz. J. Oceanogr.* **2012**, *60*, 245–253. [CrossRef]

25. Iverson, L.R.; Prasad, A.M. Predicting abundance of 80 tree species following climate change in the eastern United States. *Ecol. Monogr.* **1998**, *68*, 465–485. [CrossRef]

26. De'ath, G.; Fabricius, K.E. Classification and regression trees: A powerful yet simple technique for ecological data analysis. *Ecology* **2000**, *81*, 3178–3192. [CrossRef]

27. Morgan, J. *Classification and Regression Tree Analysis*; Technical Report, No. 1; Boston University School of Public Health: Boston, MA, USA, 2014; pp. 1–14.

28. Hopkins, T.L. The vertical distribution of zooplankton in the eastern Gulf of Mexico. *Deep Sea Res. Part A Oceanogr. Res. Pap.* **1982**, *29*, 1069–1083. [CrossRef]

29. Sanvicente-Añorve, L.; Alatorre, M.A.; Flores-Coto, C.; Alba, C. Relationships between fish larvae and siphonophores in the water column: Effect of wind-induced turbulence and thermocline depth. *ICES J. Mar. Sci.* **2007**, *64*, 878–888. [CrossRef]

30. Clarke, K.R.; Gorley, R.N.; Somerfield, P.J.; Warwick, R.M. *Change in Marine Communities: An Approach to Statistical Analysis and Interpretation*, 3rd ed.; Primer-E Ltd.: Auckland, New Zealand, 2014; pp. 1–258.

31. Roa, R.; Ernst, B.; Tapia, F. Estimation of size at sexual maturity: An evaluation of analytical and resampling procedures. *Fish Bull.* **1999**, *97*, 570–580.

32. López, M.T. Biología de *Lucifer faxoni*, Borradaile 1915, en Cananéia, Brasil (*Crustácea, Decapoda, Luciferidae*). *Bol. Inst. Ocean.* **1966**, *15*, 47–54. [CrossRef]

33. Alvarez, M.P.J. Estudo do desenvolvimento de *Lucifer faxoni* Borradaile, 1915 (*Crustacea, Decapoda, Sergestidae*) através das medidas do somito pré-bucal. *Rev. Bras. Zool.* **1988**, *5*, 371–379. [CrossRef]

34. Gómez-Gaspar, A.; Hernández-Ávila, I. Abundancia interanual del zooplancton nocturno en la costa este de Isla Margarita, Venezuela. *Bol. Inst. Ocean Venez.* **2008**, *47*, 91–102.

35. Melo, N.; Neumann-Leitão, S.; Gusmão, L.; Martins Neto, F.; Palheta, G. Distribution of the planktonic shrimp *Lucifer* (Thompson, 1829) (*Decapoda, Sergestoidea*) off the Amazon. *Braz. J. Biol.* **2014**, *74*, 45–51. [CrossRef]

36. Teodoro, S.D.S.A.; Pantaleão, J.A.F.; Negreiros-Fransozo, M.L.; Da Costa, R.C. Ecological aspects and sexual maturity of a southwestern Atlantic population of the planktonic shrimp *Lucifer faxoni* (*Decapoda: Sergestoidea*). *J. Crustac. Biol.* **2014**, *34*, 422–430. [CrossRef]

37. Xu, Z.L. Determining optimal temperature and salinity of *Lucifer* (*Dendrobranchiata: Sergestoidea: Luciferidae*) based on field data from the East China Sea. *Plankton Benthos Res.* **2010**, *5*, 136–143. [CrossRef]

38. Woodmansee, R.A. Daily vertical migration of *Lucifer*, planktonic numbers in relation to solar and tidal cycles. *Ecology* **1966**, *47*, 847–850. [CrossRef]

39. Marafon-Almeida, A.; Pereira, J.B.; Fernandes, L.F.L. Distribution of the species of *Lucifer* Thompson, 1829 in the subtropical South Atlantic between parallels 20° and 30° s. *Braz. J. Oceanogr.* **2016**, *64*, 217–226. [CrossRef]

40. Pianka, E.R. *Evolutionary Ecology*, 3rd ed.; Harper and Row: New York, NY, USA, 1983.

41. Korpelainen, H. Sex ratios and conditions required for environmental sex determination in animals. *Biol. Rev. Camb. Philos. Soc.* **1990**, *65*, 147–184. [CrossRef]

42. Wenner, A.M. Sex ratio as a function of size in marine *Crustacea*. *Am. Nat.* **1972**, *106*, 321–350. [CrossRef]

43. Gusmão, L.F.M.; McKinnon, A.D. Sex ratios, intersexuality and sex change in copepods. *J. Plankton Res.* **2009**, *31*, 1101–1117. [CrossRef]

44. Legrand, J.J.; Legrand-Hamelin, E.; Juchault, P. Sex determination in *Crustacea*. *Biol. Rev.* **1987**, *62*, 439–470. [CrossRef]

45. Kiørboe, T. Sex, sex-ratios, and the dynamics of pelagic copepod populations. *Oecologia* **2006**, *148*, 40–50. [CrossRef] [PubMed]

46. Kiørboe, T. Optimal swimming strategies in mate-searching pelagic copepods. *Oecologia* **2008**, *155*, 179–192. [CrossRef]

47. Kiørboe, T.; Ceballos, S.; Thygesen, U.H. Interrelations between senescence, life-history traits, and behavior in planktonic copepods. *Ecology* **2015**, *96*, 2225–2235. [CrossRef]

48. Alvarez, M.P.J. Migração vertical de *Lucifer faxoni* Borradaile, 1915 (*Crustacea: Decapoda*) nas águas ao Largo de Santos, Brasil. *Bol. Zool.* **1985**, *9*, 177–193. [CrossRef]

49. Krebs, C.J. *Ecología: Estudio de la Distribución y la Abundancia*, 6th ed.; Pearson Education Limited: London, UK, 2014.

50. Molles, M.C. *Ecology Concepts Applications*, 7th ed.; McGraw Hill: New York, NY, USA, 2016; ISBN 978-0077837280.

51. Townsend, C.R.; Begon, M.; Harper, J.L. *The Flux of Energy and Matter Through Ecosystems*, 3rd ed.; Blackwell Publishing: Oxford, UK, 2008.

52. Xu, Z.L.; Li, C. Horizontal distribution and dominant species of heteropods in the East China Sea. *J. Plankton Res.* **2005**, *27*, 373–382. [CrossRef]

53. Lemus-Santana, E.; Sanvicente-Añorve, L.; Alatorre-Mendieta, M.; Flores-Coto, C. Population structure and mating encounter rates in a marine pelagic invertebrate, *Firoloida desmarestia* (*Mollusca*). *Sex. Early Dev. Aquat. Org.* **2015**, *1*, 163–173. [CrossRef]

54. Chace, F.A. The shrimps of the Smithsonian-Bredin Caribbean expeditions with a summary of the West Indian shallow-water species (*Crustacea: Decapoda: Natantia*). *Smithson. Contrib. Zool.* **1972**, 1–179. [CrossRef]

55. Hendrickx, M.; Estrada-Navarrete, F. Temperature related distribution of *Lucifer typus* (*Crustacea: Decapoda*) in the Gulf of California. *Rev. Biol. Trop.* **1994**, *42*, 579–584. [CrossRef]

56. Morgan, M.J.; Colbourne, E.B. Variation in maturity-at-age and size in three populations of American plaice. *ICES J. Mar. Sci.* **1999**, *56*, 673–688. [CrossRef]

57. Guoping, Z.; Xiaojie, D.; Liming, S.; Liuxiong, X. Size at sexual maturity of bigeye tuna *Thunnus obesus* (*Perciformes: Scombridae*) in the Tropical waters: A comparative analysis. *Turk. J. Fish. Aquat. Sci.* **2011**, *11*, 149–156. [CrossRef]

58. Arshad, A.; Amin, S.M.N.; Osman, N.; Cob, Z.C.; Saad, C.R. Population parameters of planktonic shrimp, *Lucifer intermedius* (*Decapoda: Sergestidae*) from Sungai pulai seagrass area Johor, peninsular Malaysia. *Sains Malays.* **2010**, *39*, 877–882.

Article

Spatial Distribution of *Cyclograpsus cinereus* Dana 1851 on the Rocky Shores of Antofagasta (23°27′ S, Chile)

Patricio De los Rios-Escalante [1,2,*], Carlos Esse [3], Marco Antonio Retamal [4,†], Oscar Zúñiga [5], Maritza Fajardo [5] and Farhana Ghory [6]

[1] Departamento de Ciencias Biológicas y Químicas, Facultad de Recursos Naturales, Universidad Católica de Temuco, Casilla 15-D, Temuco 4780000, Chile

[2] Núcleo de Estudios Ambientales, Universidad Católica de Temuco, Temuco 4780000, Chile

[3] Unidad de Cambio Climático y Medio Ambiente—UCCMA, Instituto Iberoamericano de Desarrollo Sostenible—IIDS, Universidad Autónoma de Chile, Temuco 4780000, Chile; carlos.esse@uautonoma.cl

[4] Facultad de Ciencias Naturales & Oceanográficas, Universidad de Concepción, Concepción 4030000, Chile

[5] Departamento de Ciencias Acuáticas y Ambientales, Facultad de Ciencias del Mar y Recursos Hidrobiológicos, Universidad de Antofagasta, Antofagasta 1240000, Chile; ozr@vrt.net (O.Z.); maritzafajardo89@gmail.com (M.F.)

[6] Marine Reference & Research Collection Center, University of Karachi, Karachi 75270, Pakistan; farhanaghory@yahoo.com

*   Correspondence: prios@uct.cl; Tel.: +56-9-5828-6881

†   Deceased.

Citation: Rios-Escalante, P.D.l.; Esse, C.; Retamal, M.A.; Zúñiga, O.; Fajardo, M.; Ghory, F. Spatial Distribution of *Cyclograpsus cinereus* Dana 1851 on the Rocky Shores of Antofagasta (23°27′ S, Chile). *Diversity* **2022**, *14*, 418. https://doi.org/10.3390/d14060418

Academic Editors: Patricia Briones-Fourzán, Michel E. Hendrickx and Michael Wink

Received: 20 January 2022
Accepted: 15 April 2022
Published: 24 May 2022

Publisher's Note: MDPI stays neutral with regard to jurisdictional claims in published maps and institutional affiliations.

Abstract: The decapod fauna in the intertidal zone of the rocky shores of Chile is highly diverse, especially along the northern and central mainland coasts, where the influence of the cold Humboldt Current results in high productivity. One of the most abundant species in these ecosystems is the decapod *Cyclograpsus cinereus* Dana, 1851. The aim of the present study, carried out in the spring and summer seasons between 2018 and 2020, was to determine the spatial distribution patterns of the decapod *C. cinereus* in different sites along the rocky shores of Antofagasta bay, northern Chile, in order to establish probabilistic models that explain its distribution at each site. Individuals were counted in random quadrants in the intertidal zone. The data thus obtained were processed by application of the variance/mean ratio to determine whether the distribution of individuals was random, aggregated or uniform, associated with Poisson, negative binomial or positive binomial distributions, respectively. The data revealed aggregated (negative binomial) distribution in 15 sampling events, and uniform (positive binomial) distribution in 4 events. The sampling sites were located on rocky shores in four sectors of an urban zone, and two in a protected zone; no significant differences were found between the densities of the sites in the two zones. The results of the interpretative probabilistic models indicated aggregated distribution patterns, agreeing with previously reported interpretative probabilistic models for the distribution of decapods on the rocky shores of central and southern Chile.

Keywords: *Cyclograpsus cinereus*; spatial distribution; intertidal; rocky shore; negative binomial distribution

1. Introduction

The intertidal decapods of the mainland coast of Chile are characterised by their high species diversity. This is especially the case in central and northern Chile (18–33° S), due to the strong influence of the cold Humboldt Current, associated with continuous upwelling processes, which supports high productivity and in consequence high species diversity [1,2]. In this scenario, there are many widespread decapod species in southern Peru and northern and central Chile [2–4].

The high environmental heterogeneity of rocky shores is due to the irregular character of the substrate. The typical creeks [2–4] and areas of rounded rocks give rise to microenvironments that can sustain high invertebrate species richness, as was observed in early studies of the rocky shores of central Chile [5,6], and in recent studies of the country's mainland coast [7]. The ecological importance of these species is their function as detritivores; they are also prey species for littoral rocky fishes, sea birds and larger crabs [5,6,8]. The decapod species reported include creek-dwelling Grapsid crabs, as well as smaller crab species that inhabit the spaces underneath rounded rocks, such as *Cyclograpsus cinereus* Dana, 1851, *Petrolisthes granulosus* (Guerin, 1835) and *P. violaceus* (Guerin, 1830) [2,5–7].

C. cinereus is widespread along the mainland Chilean and Peruvian coasts [2,5,7,9], and on Chilean oceanic islands [2,10,11]. This species lives mainly under rounded rocks, and forms an important component of the rocky shores of mainland Chile. According to evidence from central Chile, it alternates with Porcellanid crabs as the dominant species [2,5,6]. The literature indicates that this species is gregarious at low tide, but that the individuals spread out at high tide; they are detritivores, and are preyed on by littoral fishes [5,6,8]. Their gregarious behaviour at low tide would be a strategy to avoid desiccation, protect themselves against potential predators, and obtain more efficient use of food resources; however, there are no studies of the density or spatial patterns of these populations [5,6] to indicate whether their distribution pattern, explained by probabilistic models, is random, uniform or aggregated.

The aim of the present study was to determine the spatial distribution patterns of the decapod *C. cinereus* in different sites along the rocky shores of Antofagasta bay, northern Chile, in order to establish probabilistic models that explain their distribution at each site. Our hypothesis was that the spatial distribution of *C. cinereus* on the rocky shores of Antofagasta bay would follow a defined probabilistic pattern.

2. Materials and Methods

2.1. Study Sites

The sites were selected on rocky beaches with rounded rocks along the seafront of Antofagasta bay [12]. Two of the sites were in a protected zone designated for littoral fishery resource management, in accordance with Chilean law (El Lenguado 1 and El Lenguado 2), where no aquatic resource extraction or culture is permitted; these sites are located at the south end of Antofagasta bay, where there is a very narrow coastal plain with granitic and basaltic rocky shores [12] (Table 1, Figure 1). The other four sites were located within the urban zone of Antofagasta city (Playa Blanca, Las Almejas, Trocadero and Jardines del Norte), where there is a wide coastal plain with basaltic rocky shores. The sites were visited in September, December and February 2018–2020 at low tide (Table 1, Figure 1).

Table 1. Geographical locations of sites included in the present study.

Site	Geographical Location
Jardines del Norte	23°34′48.4″ S; 70°23′36.4″ W
Trocadero	23°34′57.2″ S; 70°23′40.5″ W
Las Almejas	23°40′13.8″ S; 70°24′35.5″ W
Playa Blanca	23°40′34.5″ S; 70°24′50.7″ W
El Lenguado 1	23°46′09.5″ S; 70°28′22.2″ W
El Lenguado 2	23°46′21.7″ S; 70°28′26.1″ W

2.2. Data Collection

Initial species identification was done in situ, based on literature descriptions [3,6,9]. Random quadrants (10 × 10 cm) were defined at each study site (n = 90 for each site) during low tide. This size of quadrant was chosen as being most suitable for observing this fast-moving species [5,6], and in view of the irregular topography of the site with creeks and rounded rocks [13,14]. Data were recorded manually in a field notebook.

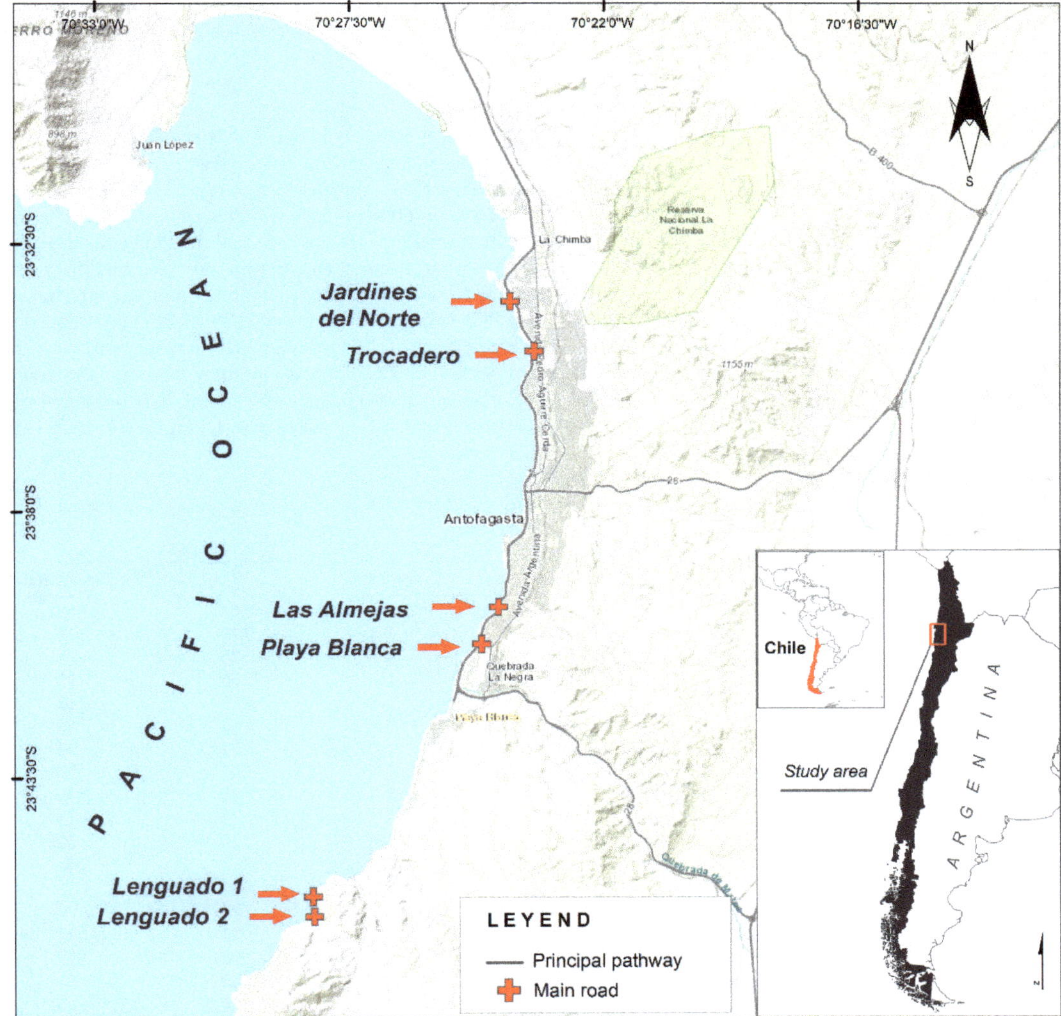

Figure 1. Map of sites (red arrows and red rectangle) included in the present study; LEGEND: Main road.

2.3. Data Analysis

In the first stage, the abundance of *C. cinereus* was compared between each sampling site/date ("event") and between the two groups of sites (protected and with human intervention). The data were tested for homoscedasticity and normality; if the results were negative, further non-parametric tests were performed, including the Kruskal–Wallis test and Dunnett's multiple comparison [15], using Python software [16]. Normal distribution and homogeneity of variance were verified; in the absence of both conditions, we carried out a Kruskal–Wallis test using the Pandas, NumPy, Matplotlib and Seaborn libraries [17–20], and Dunnett's multiple comparison test using the bioinfokit library [21].

A variance/mean ratio was obtained for each sampling event, first to determine whether the distribution was random (value equal to 1); uniform (value less than 1); or aggregated (value greater than 1) [22–24]. Once the spatial pattern had been established, we determined whether the distribution of the species was Poisson, binomial or negative binomial; this was done manually using Excel software and literature descriptions [22–24].

3. Results

The results denoted a density varying between 0.96 ind/0.01m^2 (Trocadero, February 2020), and 2.64 ind/0.01 m^2 (Las Almejas, September 2018; Table 2). The results of the Kruskal–Wallis test revealed the existence of significant differences between sampling events (T = 426.51; $p < 0.01$), while the Dunnett's multiple comparison test revealed that the events with maximum and minimum abundance showed marked differences against the remaining events (Table 3). The results of the variance/mean ratio revealed that most events presented a value greater than 1.0, implying potential association with a negative binomial distribution; the exceptions were Playa Blanca (6 February 2020, 28 December 2019), Trocadero (28 December 2019), and Playa Blanca (17 September 2019) where a variance/mean ratio value lower than 1.0 was reported, associated with positive binomial distribution (Table 1). The presence of negative binomial and positive binomial distribution patterns, associated, respectively, with variance/mean ratio greater than 1.0 and lower than 1.0, are significant for each case ($p < 0.01$; Figure 2).

Table 2. *C. cinereus* mean density (ind/grid 10 × 10 cm), variance (var) and variance/mean ratio for sampling events.

Sampling Date	Site	n	Mean	Var	Var/Mean Ratio
7 February 2020	Jardines del Norte	90	0.35	0.52	1.49
6 February 2020	Playa Blanca	90	1.02	0.79	0.77
6 February 2020	Las Almejas	90	2.17	2.35	1.08
5 February 2020	El Lenguado 2	90	1.47	1.55	1.05
5 February 2020	El Lenguado 1	90	1.62	6.19	3.82
28 December 2019	Playa Blanca	90	2.56	2.05	0.80
28 December 2019	Las Almejas	90	1.30	1.74	1.34
28 December 2019	Trocadero	90	1.78	1.03	0.58
27 December 2019	El Lenguado 1	90	1.28	1.24	0.97
27 December 2019	El Lenguado 2	90	1.13	1.31	1.15
17 September 2019	Playa Blanca	90	0.72	2.18	3.02
17 September 2019	Las Almejas	90	0.72	2.18	3.02
7 February 2019	Trocadero	90	0.96	1.30	1.36
7 February 2019	Jardines del Norte	90	1.19	2.72	2.29
7 February 2019	Las Almejas	90	1.04	1.17	1.12
7 February 2019	Playa Blanca	90	1.19	2.72	2.29
7 February 2019	El Lenguado 1	90	0.92	1.42	1.54
29 December 2018	Las Almejas	90	0.19	0.31	1.65
21 September 2018	Las Almejas	90	2.64	5.22	1.97

Table 3. *P* values for Dunnett's multiple comparison tests for *C. cinereus* densities in sampling events (p values lower than 0.05 denote significant differences, in bold).

	S	R	Q	P	O	N	M	L	K	J	I	H	G	F	E	D	C	B
A	**0.000**	1.000	**0.000**	**0.032**	0.134	**0.001**	1.000	**0.000**	1.000	**0.006**	**0.000**	**0.000**	**0.000**	**0.000**	**0.000**	1.000	**0.000**	**0.000**
B	1.000	**0.000**	**0.000**	**0.000**	**0.000**	**0.000**	**0.000**	**0.000**	**0.000**	**0.000**	1.000	1.000	**0.000**	**0.000**	**0.001**	**0.000**	**0.002**	
C	0.194	**0.000**	1.000	0.474	0.130	1.000	**0.000**	1.000	**0.000**	1.000	1.000	1.000	1.000	1.000	1.000	**0.000**		
D	**0.000**	1.000	**0.018**	1.000	1.000	0.373	0.987	0.060	1.000	1.000	**0.000**	**0.000**	**0.002**	**0.000**	**0.000**			
E	0.101	**0.000**	1.000	0.846	0.247	1.000	**0.000**	1.000	**0.000**	1.000	1.000	1.000	1.000	1.000				
F	**0.047**	**0.000**	1.000	1.000	0.491	1.000	**0.000**	1.000	**0.000**	1.000	1.000	1.000	1.000					
G	**0.009**	**0.000**	1.000	1.000	1.000	1.000	**0.000**	1.000	0.000	1.000	0.850	0.850						
H	1.000	**0.000**	0.148	**0.000**	**0.000**	**0.006**	**0.000**	**0.048**	**0.000**	**0.001**	1.000							

Table 3. *Cont.*

	S	R	Q	P	O	N	M	L	K	J	I	H	G	F	E	D	C	B
I	1.000	**0.000**	0.148	**0.000**	**0.000**	**0.006**	**0.000**	**0.048**	**0.000**	0.001								
J	**0.000**	0.061	1.000	1.000	1.000	1.000	**0.000**	1.000	0.541									
K	**0.000**	1.000	**0.004**	1.000	1.000	0.117	1.000	**0.016**										
L	**0.000**	**0.001**	1.000	1.000	1.000	1.000	**0.000**											
M	**0.000**	1.000	**0.000**	**0.000**	**0.001**	**0.000**												
N	**0.000**	**0.010**	1.000	1.000	1.000													
O	**0.000**	0.945	1.000	1.000														
P	**0.000**	0.279	1.000															
Q	**0.001**	**0.000**																
R	**0.000**																	

A = Las Almejas: 21 September 2017; B = Las Almejas: 29 December 2017; C = El Lenguado 1: 7 December 2018; D = Playa Blanca: 7 February 2019; E = Las Almejas: 7 February 2019; F = Trocadero 1: 7 February 2019; G = Trocadero 2: 7 February 2019; H = Playa Blanca: 17 September 2019; I = Las Almejas: 17 September 2019; J = El Lenguado 1: 27 December 2019; K = El Lenguado 2: 27 December 2019; L = Trocadero 2: 28 December 2019; M = Playa Blanca: 28 December 2019; N = Las Almejas: 28 December 2019; O = El Lenguado 2: 5 February 2020; P = El Lenguado 1: 5 February 2020; Q = Playa Blanca: 6 February 2020; R = Las Almejas: 6 February 2020; S = Jardines del Norte 2: 7 February 2020.

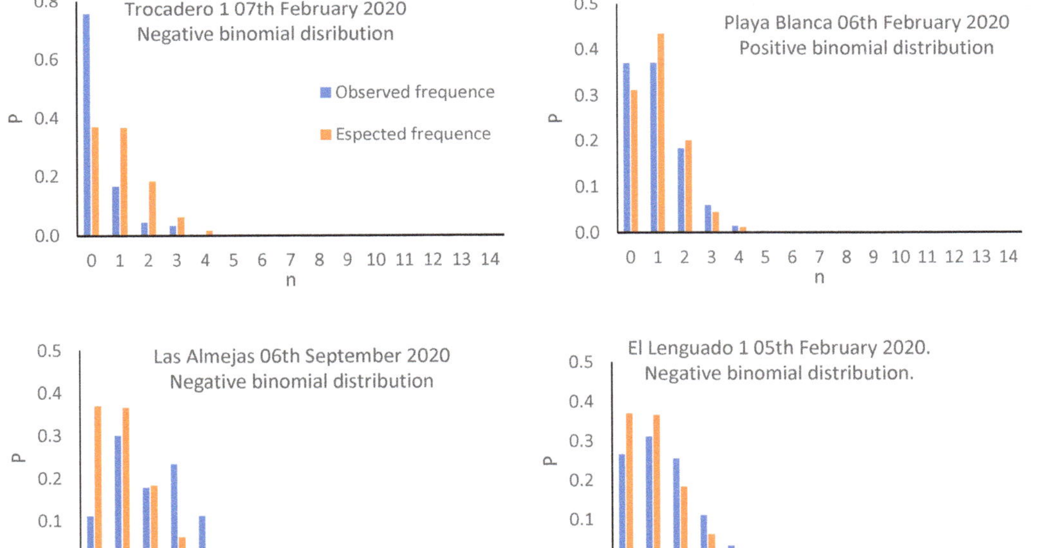

Figure 2. *Cont.*

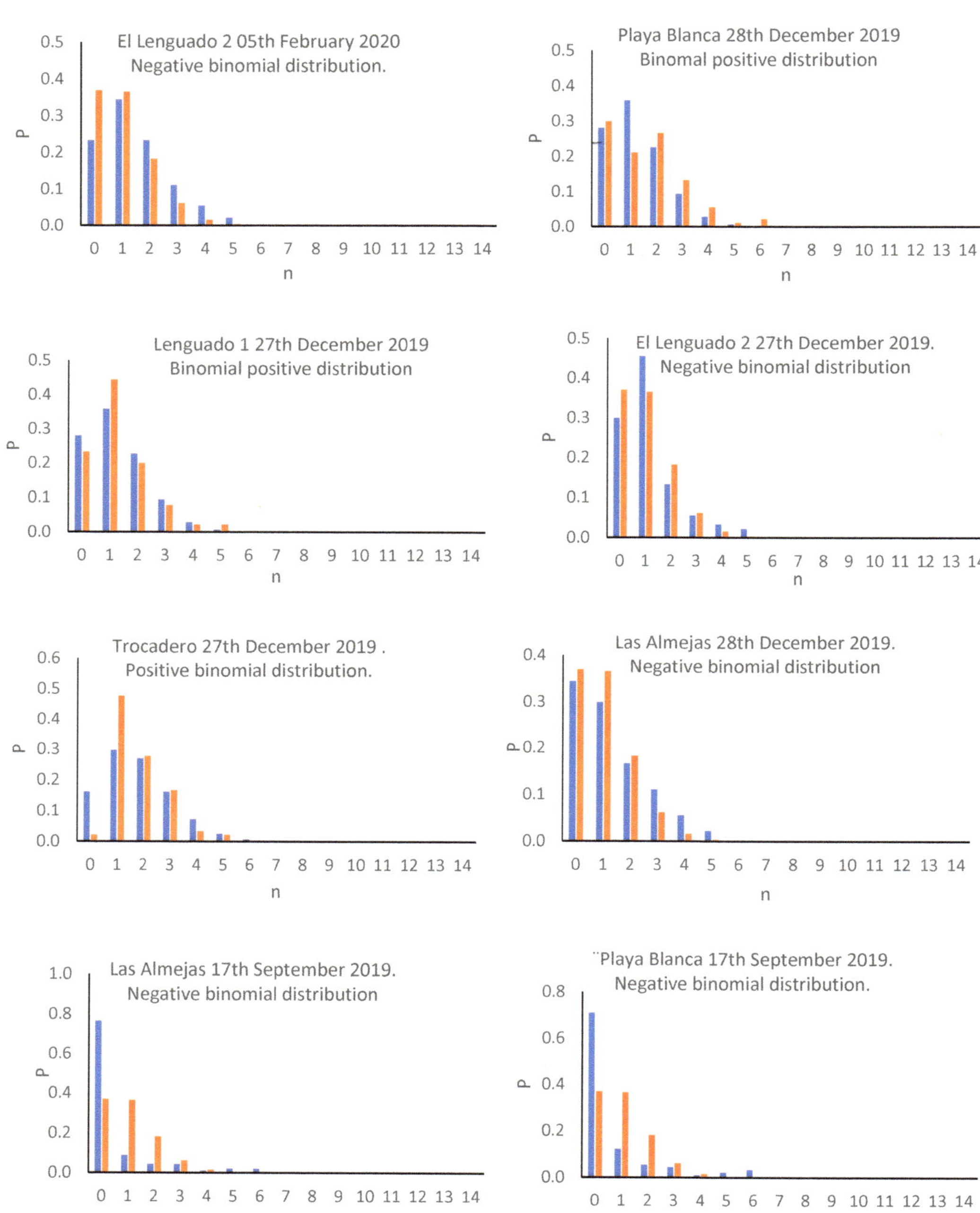

Figure 2. *Cont.*

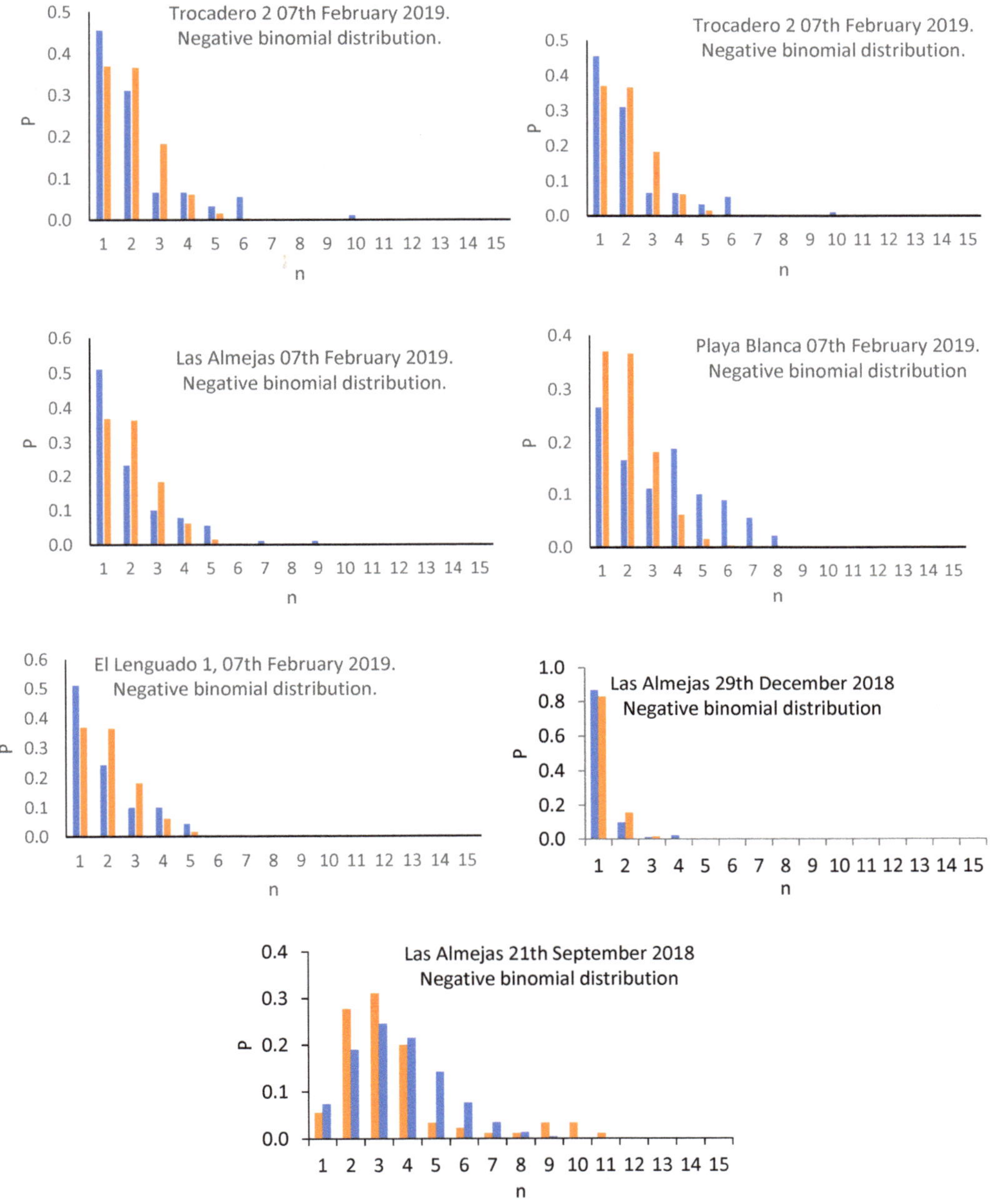

Figure 2. Distribution patterns observed for *C. cinereus* populations in sampling events.

4. Discussion

The results showed that the presence of *C. cinereus* in sites included in the present study agrees with classic literature on the biogeography of this species on the coast of mainland Chile [2,3,7]. The presence of negative binomial distribution in many of the sampling events would agree with the early literature on rocky shores in central Chile, in which this species was described as forming groups that live under rounded rocks in the intertidal zone [6]. A similar pattern was reported for *Petrolisthes granulosus* in central Chile [5], and these two species coexist along the rocky shores of this area [2,5–7]. However, these references did not report any quantitative data on population density.

The use of probabilistic models for the spatial distribution of marine crustaceans in mainland Chile was reported for the intertidal rocky shore in central Patagonia. The results show a preference for aggregated patterns, associated with negative binomial distribution [23]. This association of an aggregated pattern with negative binomial distribution has been reported by classic literature on benthic fauna [25], as well as by applied entomological studies [22]. The same pattern was reported for inland water crustaceans in Chilean Patagonia [24,26], for intertidal invertebrates on the north Patagonian coast [27], and for the gastropod *Echinolittorina peruviana* on the rocky shore of Antofagasta [12]. The existence of uniform distribution in four of the 19 sampling events is probably due to interspecific interactions, most likely competition with other species that share the same ecological niche [5–7].

The presence of negative binomial distribution associated with aggregated distribution is a survival strategy for the efficient use of trophic resources [5,6], as well as a protection against desiccation in intertidal environments [28–30]. The literature describes the existence of migration patterns influenced by the tides and day—night cycle [5,6]. At low tide, intertidal crustaceans, including *C. cinereus*, form groups under rounded rocks that create a favourable microenvironment [31], whereas at high tide the individuals are dispersed, probably to avoid predation [5,6]. In this scenario, more detailed study is needed to investigate potential migration patterns based on the tides and the day–night cycle [5,6,12,13]. In our study, the samples were collected in the middle of the day at low tide, but future studies could be extended to variations in spatial distribution at different times of day and state of the tide [5,6]. The influence of the topographical characteristics of each habitat could also be explored. In the long term, the oceanographic conditions that can affect the population structure might also be investigated.

Author Contributions: P.D.l.R.-E. carried out field work and drafted the manuscript; C.E. collaborated in data analysis and images; M.A.R., M.F., F.G. and O.Z. collaborated in drafting the manuscript and literature compilation. All authors have read and agreed to the published version of the manuscript.

Funding: This research was funded by MECESUP UCT, grant number 0804, funded by PDE, FONDECYT INCIACIÓN N°11160650 (ANID ex—CONICYTChile) funded by PDE, and the APC was funded by Universidad Católica de Temuco.

Institutional Review Board Statement: Not applicable.

Data Availability Statement: Not applicable.

Acknowledgments: The authors express their gratitude to M.I. and S.M.A. for their valuable suggestions for improving the manuscript, and Willie Barne for support in English edition.

Conflicts of Interest: The authors declare no conflict of interest. The funders had no role in the design of the study; in the collection, analysis, or interpretation of the data; in the writing of the manuscript, or in the decision to publish the results.

References

1. Santelices, B. *Algas Marinas de Chile: Distribución, Ecología, Utilización y Diversidad*; Ediciones Pontificia Universidad Católica de Chile: Santiago, Chile, 1992.
2. Retamal, M.A.; Moyano, H.I. Zoogeografia de los crustáceos decápodos chilenos marinos y dulceacuícolas. *Lat. Am. J. Aquat. Res.* **2010**, *38*, 302–328. [CrossRef]
3. Retamal, M. *Los Decápodos Chilenos. CD-Rom. ETI (Amsterdam). Universidad de Concepción*; Springer: Berlin/Heidelberg, Germany, 2000.
4. Velásquez, C.; Jaramillo, E.; Camus, P.A.; Manzano, M.; Sánchez, R. Biota del intermareal rocoso expuesto de la Isla Grande de Chiloé, Archipiélago de Chiloé, Chile: Patrones de diversidad e implicancias ecológicas y biogeográficas. *Rev. De Biol. Mar. Y Oceanogr.* **2016**, *51*, 33–50. [CrossRef]

5. Sanhueza, E.; Bahamonde, N.; Lopez, M.T. *Petrolisthes granulosus* (Guérin) en biocenosis supramareales de El Tabo (Crustacea, Decapoda, Anomura). *Bol. Del Mus. Nac. De Hist. Nat.* **1975**, *34*, 121–136.

6. Bahamonde, N.; Lopez, M.T. *Cyclograpsus cinereus* Dana, en biocenosis supramareales de Chile. *Bol. Del Mus. Nac. De Hist. Nat.* **1971**, *29*, 165–203.

7. De los Ríos-Escalante, P.; Figueroa-Muñoz, G.; Retamal, M.A.; Vega-Aguayo, R.; Esse, C. Size overlap in intertidal decapod communities along the Chilean coast. *Sci. Mar.* **2020**, *84*, 151–154. [CrossRef]

8. Medina, M.; Araya, M.; Vega, C. Alimentación y relaciones tróficas de peces costeros de la zona norte de Chile. *Investig. Mar.* **2004**, *32*, 33–47. [CrossRef]

9. Moscoso, V. Catálogo de crustáceos decápodos y estomatópodos del Perú. *Boletín Del Inst. Del Mar Del Perú* **2012**, *27*, 1–207.

10. Retamal, M.; Gorny, M. Crustáceos decápodos recolectados en las islas Robinson Crusoe, Alejandro Selkirk, San Felix y San Ambrosio, Crucero CIMAR 6–Islas Oceánicas. *Cienc. Tecnol. Del Mar.* **2004**, *27*, 71–75.

11. De los Ríos-Escalante, P.; Ibañez-Arancibia, E. A checklist of marine crustaceans known from Easter island. *Crustaceana* **2006**, *89*, 63–84. [CrossRef]

12. De los Ríos-Escalante, P.; Esse, C.; Stella, C.; Adikesavan, P.; Zúñiga, O. Spatial distribution of *Echinolittorina peruviana* (Lamarck, 1882) for intertidal rocky shore in Antofagasta, Chile. *Braz. J. Biol.* **2022**, *83*, e246889. [CrossRef] [PubMed]

13. Underwood, A.J.; Chapman, M.G. Design and analysis in benthic surveys in environmental sampling. In *Methods for the Study of Marine Benthos*. Eleftheriou, A., Mcintyre, A., Eds.; Blackwell Science: Oxford, UK, 2005; pp. 1–42.

14. Manriquez, P.H. Muestreo y análisis de comunidades intermareales de fondos duros. In *Programas de Monitoreo del Medio Marítimo Costero. Diseños Experimentales, Muestreos, Métodos de Análisis y Estadística Asociada*; Castilla, J.C., Fariña, J.M., Caamaño, A., Eds.; Ediciones Pontificia Universidad Católica de Chile: Santiago, Chile, 2021; pp. 233–256.

15. Zar, J.H. *Biostatistical Analysis*; Prentice Hall Inc.: Upper Saddle River, NJ, USA, 1999; p. 631.

16. Van Rossum, G.; Drake, F.L., Jr. Python Reference Manual. Centrum voor Wiskunde en Informatica Amsterdam, The Netherlands. Available online: https://www.python.org/ (accessed on 5 January 2022).

17. McKinney, W. Data structures for statisticala computing in Python. In Proceedings of the 9th Python in Science Conference, Austin, TX, USA, 28 June–3 July 2010; Volume 445, pp. 51–56. Available online: https://pandas.pydata.org/ (accessed on 5 January 2022).

18. Harris, C.R.; Millman, K.J.; van der Walt, S.; Gommers, R.; Virtalen, P.; Cournapeau, D.; Wieser, E.; Taylor, J.; Berg, S.; Smith, N.J.; et al. Array programming with NumPy. *Nature* **2020**, *585*, 357–362. [CrossRef] [PubMed]

19. Hunter, J.D. Matplotlib:A2D Graphics Environment. *Comput. Sci. Eng.* **2007**, *9*, 90–95. Available online: https://matplotlib.org/stable/users/project/citing.html (accessed on 5 January 2022).

20. Waskiim, M.L. Seaborn: Statistical data visualization. *J. Open Source Softw.* **2021**, *6*, 3021. Available online: https://seaborn.pydata.org/citing.html (accessed on 5 January 2022).

21. Huang, Y.; Wang, J.Y.; Wei, X.M.; Hu, B. Bioinfo-Kit: A sharing software tool for Bioinformatics. *Appl. Mech. Mater.* **2014**, *472*, 466–469. Available online: https://github.com/reneshbedre/bioinfokit (accessed on 5 January 2022).

22. Fernandes, M.G.; Busoli, A.C.; Barbosa, J.C. Distribucao espacial de *Alabama argillacea* (Hubner)(Lepidoptera: Noctuidae) em algodoeiro. *Neotrop. Entomol.* **2003**, *32*, 107–115. [CrossRef]

23. Vega, R.; Figueroa-Muñoz, G.; Retamal, M.A.; De Los Ríos, P. Spatial distribution and abundance of *Hemigrapsus crenulatus* (H. Milne-Edwards, 1837) (Decapoda, Varunidae) in the Puerto Cisnes Estuary (44° S, Aysen region Chile). *Crustaceana* **2018**, *91*, 1465–1482. [CrossRef]

24. De los Ríos-Escalante, P. Non randomness in spatial distribution in two inland water species malacostracans. *J. King Saud Univ.-Sci.* **2017**, *29*, 260–262. [CrossRef]

25. Elliot, J.M. *Some Methods for the Statistical Analysis of Benthic Invertebrates, 25*; Freshwater Biological Association of Sciences Publications: Cumbria, UK, 1983; pp. 1–157.

26. De los Ríos-Escalante, P.; Mansilla, A. Spatial patterns of *Pisidium chilense* (Mollusca Bivalvia) and *Hyalella patagonica* (Crustacea, Amphipoda) in an unpolluted stream in Navarino Island (54° S.; Cape Horn Biosphere Reserve). *J. King Saud Univ.-Sci.* **2017**, *29*, 28–31. [CrossRef]

27. De los Ríos-Escalante, P.; Carreño, E. Spatial distribution in marine invertebrates in rocky shore of Araucania region (38° S, Chile). *Braz. J. Biol.* **2020**, *80*, 362–367. [CrossRef] [PubMed]

28. Lagos, M.E.; Castillo, N.; Albarrán-Mélzer, N.; Pinochet, J.; Gebauer, P.; Urbina, M.A. 2021. Age dependent physiological tolerances explain population dynamics and distribution in the intertidal zone: A study with porcelain crabs. *Mar. Environ. Res.* **2021**, *169*, 105343. [CrossRef] [PubMed]

29. Viña, N.; Bascur, M.; Guzmán, F.; Riera, R.; Paschke, K.; Urzua, A. Interspecific variation in the physiological and reproductive parameters of porcelain crabs from the Southeastern Pacific coast: Potential adaptation in contrasting marine environments. *Comp. Biochem. Physiol. Part A* **2018**, *226*, 22–31. [CrossRef] [PubMed]

30. Yeoh, L.H. Behavioural Competition in the Intertidal Shore Crab *Petrolisthes elongatus*. Master's Thesis, University of Wellington, Wellington, New Zealand, 2020.

31. Hollowald, P.; Field, R. Can rock-rubble groynes support similar intertidal ecological communities to natural rocky shores? *Land* **2020**, *9*, 131. [CrossRef]

 *diversity*

Interesting Images

Nocturnal Predation of Christmas Tree Worms by a Batwing Coral Crab at Bonaire (Southern Caribbean)

Ellen Muller [1], Werner de Gier [2,3], Harry A. ten Hove [2], Godfried W. N. M. van Moorsel [4,5] and Bert W. Hoeksema [2,3,*]

[1] Kaya Platina 3, Nawati, Bonaire, Caribbean, The Netherlands; imagine@imaginebonaire.com
[2] Taxonomy and Systematics Group, Naturalis Biodiversity Center, P.O. Box 9517, 2300 RA Leiden, The Netherlands; werner.degier@naturalis.nl (W.d.G.); harry.tenhove@naturalis.nl (H.A.t.H.)
[3] Groningen Institute for Evolutionary Life Sciences, University of Groningen, P.O. Box 11103, 9700 CC Groningen, The Netherlands
[4] Ecosub, Berkenlaantje 2, 3956 DM Leersum, The Netherlands; vanmoorsel@ecosub.nl
[5] ANEMOON Foundation, P.O. Box 29, 2120 AA Bennekom, The Netherlands
* Correspondence: bert.hoeksema@naturalis.nl; Tel.: +31-71-7519-631

Received: 9 November 2020; Accepted: 27 November 2020; Published: 30 November 2020

Christmas tree worms (Serpulidae: *Spirobranchus*) occur in shallow parts of coral reefs, where they live as associates of a large number of stony coral species [1,2]. They dwell inside a calcareous tube, which is usually overgrown by the host coral and partly embedded deep inside the coral skeleton, except for the tube's opening and the worm's operculum [3]. Even if host corals and worm tubes become overgrown by other invertebrates, the worms continue to grow and are able to keep their tube openings free [4,5].

Despite their wide distribution, high densities, and the damage these tube-dwelling worms may cause to corals [6,7], little is known about their natural enemies. They appear well protected by the tube, which is armed by a long spine on the opening margin. Although they may live up to 40 years [3], mortality of *Spirobranchus* worms in dense populations is not uncommon, and most obvious when their vacated tubes are inhabited by small fish and crustaceans [6,8]. There are a few reports on attempted feeding of Christmas tree worms by fish and on *Spirobranchus* remnants found in fish stomachs ([9], references therein), but no information is available on other predators.

Therefore, it is surprising that a West Atlantic batwing coral crab, *Carpilius corallinus* Herbst, 1783, was observed preying on two individuals of *Spirobranchus giganteus* (Pallas, 1766) during a night dive at Playa Pabou (12°09′41.8″ N, 068°17′01.0″ W), Kralendijk, Bonaire on 18 October 2020; time 18:45–21:15 (Electronic Supplementary Material). The worms were living in a colony of the scleractinian coral *Porites astreoides* Lamarck, 1801, at a depth of 7 m. The crab was using its slender left claw to break away the thick calcareous tubes of the worms, which resulted in a deposit of limestone debris aside the coral (Figure 1a). The crab seemed to extract soft parts of the worm from the tube and manipulate them by using its second pair of pereiopods, which are the first pair of walking legs (Electronic Supplementary Material). The use of walking legs during feeding is not uncommon in other brachyuran lineages. Some spider crabs (Majidae) use walking legs to pry and wedge open gastropod and bivalve shells [10] and box crabs (Calappidae) can be seen to use the first two pairs of walking legs to rotate prey shells to find an opening for easier access and grip for their specialized right claw (W.d.G. pers. obs.).

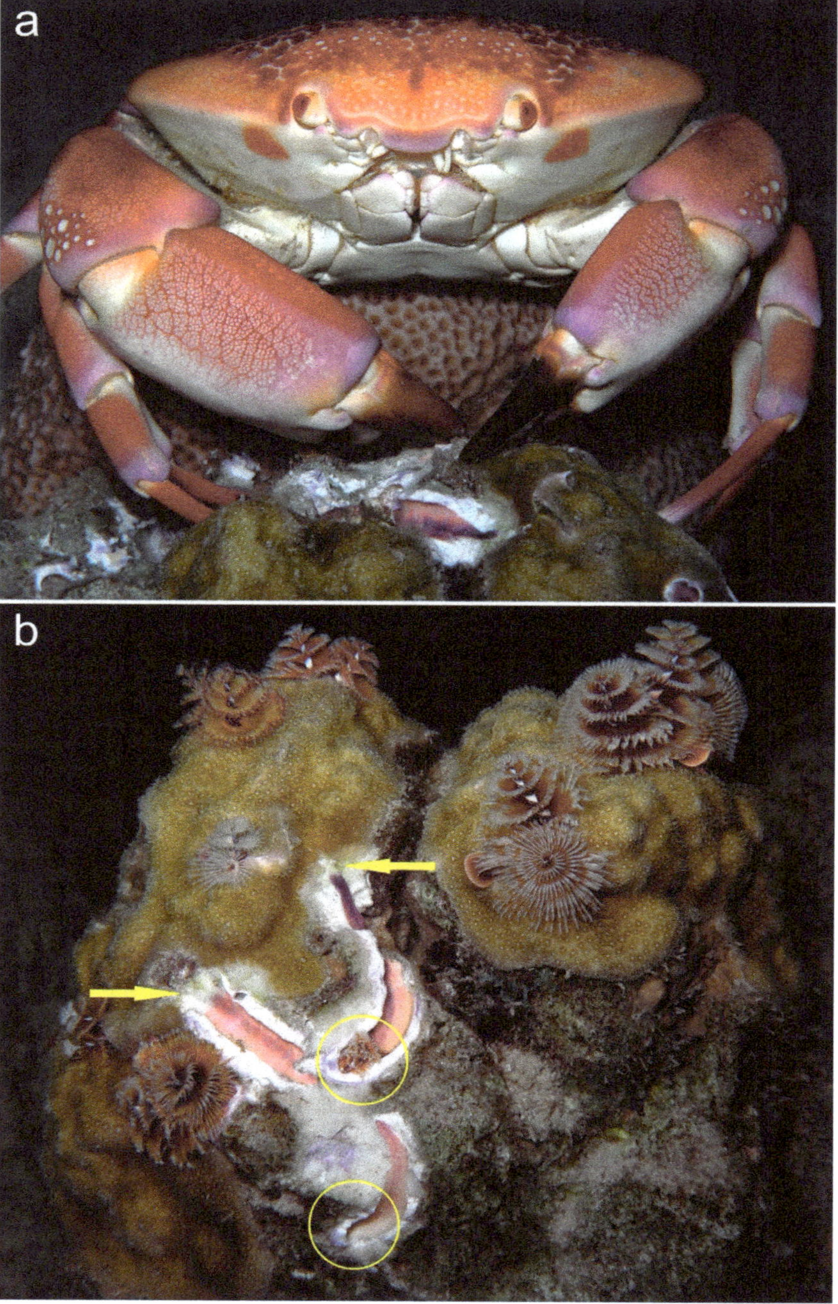

Figure 1. Predation of Christmas tree worms (*Spirobranchus giganteus*) in a coral of *Porites astreoides* by a Batwing coral crab (*Carpilius corallinus*) at Bonaire: (**a**) the crab foraging on worms; (**b**) two days later, two damaged worm tubes with one worm still alive. Arrows: position of worm tube openings. Circles: position of worm tubes exiting the coral skeleton. Photo credit: Ellen Muller.

During another night dive, two days later, the crab was no longer present but the extent of damage to the worm tubes and the coral was evident (Figure 1b). The worm in the longest of the two tubes was gone, while the other worm had survived in a part of the tube that was inside the host coral. Its operculum appeared lost (Figure 1b).

This observation is interesting because little is known about species predating on Christmas tree worms (see above), while also hardly anything has been published on the diet of the Batwing coral crab. *Carpilius corallinus* is well known for its nocturnal activities [11] and it is the only West Atlantic member of Carpiliidae, a family of three congeneric species [12]. All three species possess an enlarged right cheliped (claw-bearing leg), with a blunt molariform tooth found proximally on the cutting edge of the pollex, which is the fixed 'finger' of the claw.

In laboratory conditions in Guam, the Indo-West Pacific species *C. convexus* (Forskål, 1775) and *C. maculatus* (Linnaeus, 1758) have been observed to use their major claw to crush shells of various species of gastropods [13]. The latter crab species has also been reported as predator of a commercially important abalone, *Haliotis asinina* Linnaeus, 1758, in the Philippines [14], and was found in the field between the remains of freshly-killed gastropods on two separate occasions in Guam [15].

Individuals of the West Atlantic *C. corallinus* were also found to be feeding on gastropods in captivity, while they were also fed with sardines [16]. In another case, a female individual in an aquarium was observed to break apart shells inhabited by hermit crabs in an attempt to remove them from their homes [11]. Only one published record was found on the diet of *C. corallinus* in its natural environment, consisting of *Diadema* sea urchins [11]. There is also unpublished data concerning *C. corallinus* feeding on sea urchins, as well as on a topshell, *Calliostoma javanicum* (Lamarck, 1822), all from Bonaire (E.M., pers. obs.).

It seems that information on the diet of *Carpilius* species is rare, but considering the armor of previously reported prey species, the crushing of serpulid worm tubes seems to be within their capacity when they use their right claw. The crab at Bonaire was, however, using its slender left claw to feed from the worm tube. We do not know if the crab had initially crushed the tube using its specialized right claw and continued feeding using its left claw, or if the crab initially used its left claw to break the tube.

The extent of damage on worm tubes is striking (Figure 1). In spite of many dives on Bonaire, this kind of harm was not reported before. Because *Spirobranchus* tubes may easily become covered by coral tissue and algae [3,6,7], it is possible that damaged worm tubes may get unnoticed due to similar overgrowth. All in all, we do not expect *Spirobranchus* to be a regular part of the diet of *Carpilius corallinus*. The present observation and previously published information suggest that *Carpilius* species are not prey specific. More research on the diet and foraging behavior of these commercially important crab species will teach us more about their role in the food chains of coral reefs.

Supplementary Materials: The following is available online at http://www.mdpi.com/1424-2818/12/12/455/s1. Video footage (Batwing crab video) of the same Batwing coral crab (as in Figure 1) foraging on Christmas tree worms, destructing the worm tubes.

Author Contributions: Conceptualization and supervision, E.M. and B.W.H.; methodology, illustrations and funding acquisition, E.M. and B.W.H.; investigation, E.M., W.d.G., H.A.t.H., G.W.N.M.v.M. and B.W.H.; writing—original draft preparation, E.M., W.d.G. and B.W.H.; writing—review and editing, E.M., W.d.G., H.A.t.H., G.W.N.M.v.M. and B.W.H. All authors have read and agreed to the published version of the manuscript.

Funding: Fieldwork at Bonaire (for W.d.G., G.W.N.M.v.M., and B.W.H.) was supported by the WWF Netherlands Biodiversity Fund, the Treub Maatschappij—Society for the Advancement of Research in the Tropics, and by the Nature of the Netherlands program of Naturalis Biodiversity Center. W.d.G. received funding from the L.B. Holthuis Fund and the Jan Joost ter Pelkwijk Fund.

Acknowledgments: We thank STINAPA Bonaire (National Parks Bonaire Foundation), DCNA (Dutch Caribbean Nature Alliance) and Dive Friends (Bonaire) for logistic assistance. EM thanks in particular VIP Diving for continuous support. We are grateful to three anonymous reviewers for their constructive remarks on the manuscript.

Conflicts of Interest: The authors declare no conflict of interest. The funders had no role in the design of the study; in the collection, analyses, or interpretation of data; in the writing of the manuscript, or in the decision to publish the results.

References

1. Hoeksema, B.W.; ten Hove, H.A. The invasive sun coral *Tubastraea coccinea* hosting a native Christmas tree worm at Curaçao, Dutch Caribbean. *Mar. Biodivers.* **2017**, *47*, 59–65. [CrossRef]
2. Perry, O.; Sapir, Y.; Perry, G.; ten Hove, H.; Fine, M. Substrate selection of Christmas tree worms (*Spirobranchus* spp.) in the Gulf of Eilat, Red Sea. *J. Mar. Biol. Assoc. UK* **2018**, *98*, 791–799. [CrossRef]
3. Nishi, E.; Nishihira, M. Age-estimation of the Christmas tree worm *Spirobranchus giganteus* (Polychaeta, Serpulidae) living buried in the coral skeleton from the coral-growth band of the host coral. *Fish. Sci.* **1996**, *62*, 400–403. [CrossRef]
4. García-Hernández, J.E.; Hoeksema, B.W. Sponges as secondary hosts for Christmas tree worms at Curaçao. *Coral Reefs* **2017**, *36*, 1243. [CrossRef]
5. Hoeksema, B.W.; García-Hernández, J.E.; van Moorsel, G.W.N.M.; Olthof, G.; ten Hove, H.A. Extension of the recorded host range of Caribbean Christmas tree worms (*Spirobranchus* spp.) with two scleractinians, a zoantharian, and an ascidian. *Diversity* **2020**, *12*, 115. [CrossRef]
6. Hoeksema, B.W.; van der Schoot, R.J.; Wels, D.; Scott, C.; ten Hove, H.A. Filamentous turf algae on tube worms intensify damage in massive *Porites* corals. *Ecology* **2019**, *100*, e2668. [CrossRef] [PubMed]
7. Hoeksema, B.W.; Wels, D.; van der Schoot, R.J.; ten Hove, H.A. Coral injuries caused by *Spirobranchus* opercula with and without epibiotic turf algae at Curaçao. *Mar. Biol.* **2019**, *166*, 60. [CrossRef]
8. Böhm, T.; Hoeksema, B.W. Habitat selection of the coral-dwelling spinyhead blenny, *Acanthemblemaria spinosa*, at Curaçao, Dutch Caribbean. *Mar. Biodivers.* **2017**, *47*, 17–25. [CrossRef]
9. Hoeksema, B.W.; ten Hove, H.A. Attack on a Christmas tree worm by a Caribbean sharpnose pufferfish at St. Eustatius, Dutch Caribbean. *Bull. Mar. Sci.* **2017**, *93*, 1023–1024. [CrossRef]
10. Woods, C.M.C. Natural diet of the crab *Notomithrax ursus* (Brachyura: Majidae) at Oaro, South Island, New Zealand. *N. Z. J. Mar. Freshw. Res.* **1993**, *27*, 309–315. [CrossRef]
11. Pequegnat, L.H.; Ray, J.P. Crustaceans and other arthropods. In *Biota of the West Flower Garden Bank*; Bright, T.J., Pequegnat, L.H., Eds.; Gulf Publishing: Houston, TX, USA, 1974; pp. 231–288.
12. Wetzer, R.; Martin, J.W.; Trautwein, S.E. Phylogenetic relationships within the coral crab genus *Carpilius* (Brachyura, Xanthoidea, Carpiliidae) and of the Carpiliidae to other xanthoid crab families based on molecular sequence data. *Mol. Phylogenet. Evol.* **2003**, *27*, 410–421. [CrossRef]
13. Vermeij, G.J. Interoceanic differences in vulnerability of shelled prey to crab predation. *Nature* **1976**, *220*, 135–136. [CrossRef]
14. Aspe, N.M.; Cabales, R.G.; Sajorne, R.E.; Creencia, L.A. Survey on the predators of abalone *Haliotis asinina* from the perspective of the local fisherfolks in selected sites of Palawan, the Philippines. *J. Shellfish Res.* **2019**, *38*, 463–473. [CrossRef]
15. Zipser, E.; Vermeij, G.J. Crushing behaviour of tropical and temperate crabs. *J. Exp. Mar. Biol. Ecol.* **1978**, *31*, 155–172. [CrossRef]
16. Laughlin, R.A. Some observations on the occurrence, reproduction and mating of the coral crab *Carpilius corallinus* (Herbst, 1783) (Decapoda, Xanthidae) in the Archipielago Los Roques, Venezuela). *Crustaceana* **1982**, *43*, 219–221. [CrossRef]

*diversity*

Article

The Use of Chemical Cues by *Sargassum* Shrimps *Latreutes fucorum* and *Leander tenuicornis* in Establishing and Maintaining a Symbiosis with the Host *Sargassum* Algae

Jaime L. Frahm and William Randy Brooks *

Department of Biological Sciences, Florida Atlantic University, Boca Raton, FL 33431, USA;
jaimelfrahm@gmail.com
* Correspondence: wbrooks@fau.edu; Tel.: +1-561-297-3888

Citation: Frahm, J.L.; Brooks, W.R. The Use of Chemical Cues by *Sargassum* Shrimps *Latreutes fucorum* and *Leander tenuicornis* in Establishing and Maintaining a Symbiosis with the Host *Sargassum* Algae. *Diversity* **2021**, *13*, 305. https://doi.org/10.3390/d13070305

Academic Editors: Patricia Briones-Fourzán and Michel E. Hendrickx

Received: 15 June 2021
Accepted: 2 July 2021
Published: 6 July 2021

Publisher's Note: MDPI stays neutral with regard to jurisdictional claims in published maps and institutional affiliations.

Abstract: A mutualistic symbiosis exists between the alga *Sargassum* spp. and two shrimp species, *Latreutes fucorum* and *Leander tenuicornis*. However, little is known about how these shrimp locate and establish their host alga. Both visual and chemical cues are potentially available. A previous study has looked at both cue variables with results that are mixed. Specifically, these same shrimp species used chemical cues only when visible cues were available simultaneously. Visual cues would be presumably restricted at night, but chemical cues are potentially available continuously. This current research elaborates on the previous study to fully understand *Sargassum* shrimp chemoreception. Increases in sample sizes and both a 4-chambered and Y-maze apparatus were used to test whether the shrimp could detect *Sargassum* cues, dimethylsulfoniopropionate (DMSP) (a chemical excreted by some marine algae), and conspecific cues. Neither shrimp species showed a strong directional response to any of the chemical cues, but the *Sargassum* and DMSP cues did cause more shrimp to exhibit searching behavior. Additionally, several differences in responses between male and female shrimp were found for each cue. A lowered dilution of DMSP was also tested to determine sensitivity of *L. fucorum* shrimp to the chemical cue; although searching behavior was triggered, conclusions about quantifying the sensitivity could not be made. Overall, these results show the shrimp can detect chemical cues—in the absence of visual cues—that could affect initiating and maintaining this shrimp/algal symbiosis.

Keywords: symbiosis; *Sargassum* shrimps; chemical cues

1. Introduction

Pelagic, floating mats of primarily *Sargassum fluitans* and *Sargassum natans* algae are found in both tropical and temperate waters of the Atlantic Ocean [1,2]. These species are characterized with highly branched thalli and small air bladders (pneumatocysts) that keep the algae afloat [1]. *S. natans* differs from *S. fluitans* with thinner blades and a spine located on the pneumatocysts [3] (see Figure A1 in Appendix A).

These floating mats range greatly in size from small patches less than 0.5 m in horizontal diameter up to huge mats 50 m in diameter [1,4]. Several factors contribute to this broad range in sizes. For example, sustained calm conditions can allow *Sargassum* mats to form large aggregations, but harsher weather conditions such as high winds can break up large mats of *Sargassum* or create narrower windrows [4]. As the algae approach the shoreline, wave action can break up mats into significantly smaller sizes. Boats and other watercrafts running through the mats could also cause further disruption of *Sargassum* mat structure and size. Wind can also push *Sargassum* on to shores in the Atlantic and Caribbean, sometimes in huge mass events [5]. However, *Sargassum* can also be found in the Sargasso Sea in high quantities [1,6].

Some associated species, such as *Sargassum* shrimp, form a symbiosis with these algae and are rarely found elsewhere [1]. With so many species reliant either part of or their entire lives on the *Sargassum* habitat, it has been designated as an Essential Fish Habitat [7]. While many studies have focused on species richness and diversity within the *Sargassum* habitat, very little is known about details of interactions among the symbionts and host algae [3]. *Sargassum* shrimp are the most abundant macro invertebrates in *Sargassum* mats.

There are two species of *Sargassum* shrimp, *Latreutes fucorum* and *Leander tenuicornis*, both of which exhibit camouflage and coloration closely resembling the brown and yellow colors of *Sargassum* algae, making it challenging for the observer to locate them within the habitat [1]. These shrimp occupy positions in the algal mats based on the frond characteristics and depth of the fronds in the water column. For example, *L. tenuicornis* shows a preference for deeper floating *Sargassum* patches (10–12 cm below the surface) than shallow patches. Additionally, *L. tenuicornis* positions itself in a parallel alignment with the fronds, thus allowing the shrimp to blend with the fronds and exhibit algal morphology mimicry [8]. Such behaviors by *L. tenuicornis* and its innate camouflage maximize protection from predators such as jacks, sargassum fish (*Histrio histrio*) and the gray triggerfish (*Balistes capriscus*), the last of which is one of the most abundant fish in *Sargassum* and feeds almost exclusively on *Sargassum* shrimp [1,7]. Clearly, *Sargassum* shrimp are vital components of the *Sargassum* community and food chain. Therefore, this study focused on these two species of shrimp.

A mutualistic symbiosis is formed when two organisms live together and provide reciprocal benefits. In the symbiosis between *Sargassum* and these shrimp, the alga is a host that provides shelter and protection to the shrimp [9]. In return, shrimp and other animals provide nutrients to the algae [10]. Mutual benefits to symbionts can help explain why an association occurs, but another critical aspect is understanding how the symbiosis is initiated and maintained. Typically, the smaller, more mobile symbiont is the one to seek out and form the symbiosis with the larger symbiont or host [3].

This leads to the question about how shrimp symbionts locate the *Sargassum* patches, which are temporally and spatially highly variable in abundance [11–13]. Presumably, the shrimp must initially locate such patches of *Sargassum* algae. DNA analysis of *L. fucorum* shows the shrimp are highly dispersive during the planktonic larval stage [2]. Additionally, juveniles and those individuals separated from mats by biotic perturbations (e.g., disruptive feeding actions by large, pelagic species such as dolphinfish, *Coryphaena hippurus* [1,13–15]), and abiotic perturbations [(e.g., wind and waves that break up patches, or when mats are driven onto beaches, cf. [5]); and anthropogenic events (e.g., algal mat disruption by boats)] must locate and re-establish association with new, displaced or larger algal mats.

Visual cues are mostly available during the daytime and would be potentially available at limited distances only, based on the visual abilities of the shrimp. However, chemical cues would likely be available most of the time, and distance would be potentially less of a problem than using visual cues exclusively.

There is an abundance of chemical cues in the marine habitat, as every organism releases some type of chemical signal into the environment via metabolic activity [16]. Once a chemical is released into the water, the molecules disperse by undergoing diffusion or bulk flow, which is the movement of molecules from high to low pressure [16]. There are many marine organisms that can detect chemicals via chemoreceptors [17]. Receptors must be both highly specific and diverse for organisms to identify an array of specific chemicals within the environment [16]. These chemicals are used to locate food, avoid predators, find a mate, find a suitable habitat, homing, recognize conspecifics, and mediate social behavior (such as forming a hierarchy) [17]. Specifically, chemoreceptive organs on decapods contain sensilla, which are either hair like or rod shaped [17]. The sensilla are usually located on the antennules but can also be found on parts of the mouth and legs [17].

The American lobster (*Homarus americanus*) can detect specific chemical cues to recognize individual lobsters and form dominance hierarchies [16]. Another example of a chemoreceptive crustacean is the porcellanid crab (*Porcellana sayana*), which uses chemical cues to locate sea anemones associated with hermit crabs [18]. Yet another crustacean, the rock shrimp (*Rhynchocinetes typus*) uses chemoreception to find and select mates [19].

Dimethylsulfoniopropionate (DMSP) is a chemical cue excreted by some marine algae, e.g., dinoflagellates [20,21] and benthic *Sargassum* species [22], that is detected by a variety of marine animals [23–25]. It can potentially attract zooplankton predators to minimize grazing on the phytoplankton [26]. Potentially high DMSP concentrations are also present in areas where primary productivity and foraging activity are high, which include windrows where *Sargassum* and other potential DMSP-producing organisms aggregate to form weedlines [27]. DeBose et al. [27] showed that some fishes (e.g., jacks, *Caranx hippos* and *C. melampygus*) associated with *Sargassum* mats responded significantly to cues of DMSP. DMSP concentrations of 10^{-9} M in the Sargasso Sea were found, which indicate that this compound could be used as a cue for symbionts to locate *Sargassum* patches [28].

This current research paper is based on a prior study by Jobe and Brooks [3] on the importance of chemical and visual cues for *Sargassum* shrimp in locating and selecting habitat. The questions addressed by this previous study were whether chemical and visual cue responses to *Sargassum* algae occurred by *L. tenuicornis* or *L. fucorum*, and whether these shrimp have preferences for a different *Sargassum* species. In that study, two types of apparatus were used: (1) 4-chambered olfactometer (Figure 1) to test for chemical cues only, and (2) aquarium with, first, chemical cues blocked and visual cues available (by placing the algae in beakers thereby allowing only visual cues), then chemical cues and visual cues by placing algae directly in the water with the shrimp.

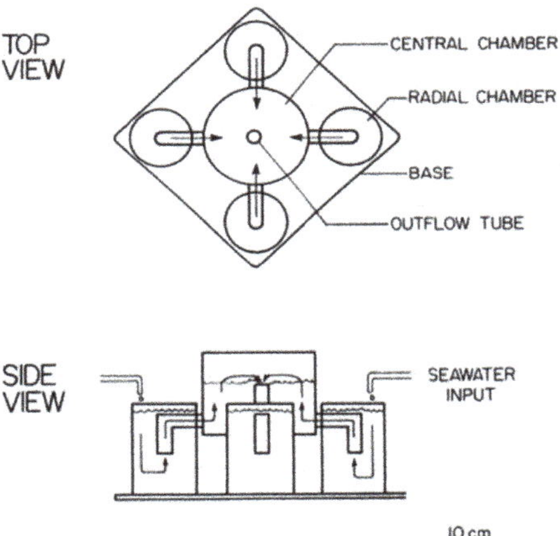

Figure 1. Four-chambered choice apparatus.

Although there were no significant differences detected in the "chemical cues only" trials, there was significance in the "visual only" and the "visual plus chemical" cue trials. Specifically, "visual cues only" trials showed several specific significant results: (1) *L. fucorum* chose live *Sargassum* over artificial *Sargassum* and *S. natans* over *S. fluitans*; (2) small-sized *L. tenuicornis* chose *S. natans* over artificial *Sargassum*, and large-sized *L. tenuicornis* chose *S. fluitans* over *S. natans* and artificial *Sargassum*. However, when both visual and chemical cues were present, there were additional significant similar and differing results: (1) neither shrimp species had a significant preference for a specific *Sargassum* species, and (2) small *L. tenuicornis* selected *S. fluitans* over artificial algae. Although chemical cues from *Sargassum* were unavailable in the "visual cues only" trials, they were still present in the latter set of trials (i.e., both visual and chemical cues were available to the shrimp). While these results do not demonstrate that *Sargassum* shrimp responded to chemical cues in the 4-chambered apparatus trials, they do not preclude the possibility that chemoreception was employed by the shrimp.

The objective of this current research was to elaborate on the previous study to fully understand *Sargassum* shrimp chemoreception. Increases in sample sizes and both a 4-

chambered and a Y-maze apparatus were used to test whether the shrimp could detect *Sargassum* cues, dimethylsulfoniopropionate (DMSP) (a chemical excreted by some marine algae), and conspecific cues.

Overall, these results showed the shrimp can detect chemical cues—in the absence of visual cues—that could affect initiating and maintaining this shrimp/algal symbiosis.

2. Materials and Methods

2.1. Collection and Maintenance of Specimens

Organisms used in this study were collected via boat off the east coast of Florida near the Boca Inlet. Using a large dip net, clumps of *Sargassum* algae were collected and put into plastic coolers. The shrimps, *Latreutes fucorum* and *Leander tenuicornis*, were separated from the algae immediately on the boat. Extra algae and any other organisms collected unintentionally were returned to the water. The algae and shrimp were then transported back to Florida Atlantic University in plastic coolers. A total of 2800 shrimp were collected for use in trials.

Organisms were placed in holding tanks with filters in the laboratory with seawater made from sea salt mix, and the salinity maintained at 32–35 ppt. (Seawater for all trials was also made using sea salt mix.) Once the temperature of the water in the coolers matched the temperature of the water in the holding tanks, the shrimp were separated by size groups (e.g., shrimp less than 5 mm, measured from rostrum to tail, were separated from larger shrimp to minimize potential aggression and predation). Shrimp were acclimated to their laboratory environment a minimum of 6 h before being used in any experimental trials. All shrimp were tested within 3 days of collection.

2.2. Four-Chambered Apparatus Trials

A 4-chambered apparatus (Figure 1) was set up with equal amounts of synthetic seawater running through each radial chamber by dripping via plastic tubing from four separate seawater/chemical sources elevated (gravity forces water flow) so that water flowed simultaneously via each of the four seawater inputs into their respective radial chambers and into the central chamber before draining out of the apparatus. Animals were subsequently placed in the central chamber and could enter any of the 4 radial chambers. Blinds were placed around the apparatus to minimize light cues and maximize the potential use of chemical cues.

For the control trials, all four radial chambers had plain (i.e., no chemical cues experimentally added) seawater running through them. This control allowed for examination of the shrimps' behavior in the absence of an added chemical cue. Groups of 30 shrimp were placed into the central chamber of the apparatus. Two size groups of *L. fucorum* were tested: Group 1 $\geq$ 10 mm, and Group 2 < 10 mm. *L. tenuicornis* was not tested using the 4-chambered apparatus due to low collection numbers. After 4 h, the location of the shrimp within the apparatus was recorded. Specifically, the number of individuals in each radial chamber after 4 h was recorded and treated as a positive response. The number of individuals that remained in the central chamber was also recorded. No shrimp individuals were used more than once in any trials in this and all other experiments. Trials were replicated with new groups of 30 shrimp a minimum of 10 times. The distribution of the animals within the apparatus (i.e., how many were in any of the radial chambers versus those still in the central chamber) after 4 h was analyzed using binomial tests.

Three types of chemical cue sources were used separately in the choice experiments, *Sargassum*, dimethylsulfoniopropionate (DMSP), and conspecifics. For the *Sargassum* chemical cue treatment, 2 kg of *Sargassum* was collected and placed in 37 L of synthetic seawater for 48 h to allow any chemical compounds to be released into the water. *Sargassum* composition can vary greatly throughout the year due to many factors such as age, amount of epibiotic growth, species present, and seasonal conditions. To minimize chemical composition variability in *Sargassum* effluent between trials, 45 mL aliquots were taken from one batch and then frozen for subsequent use throughout experimentation (see [29]). Three additional batches of *Sargassum* chemical effluent were made using *Sargassum* collected at different times of the year. This was done to minimize potential variation between chemical cues from *Sargassum* batches collected throughout three years of research. For use in trials, a 100% concentration of *Sargassum* effluent solution was used. For the DMSP chemical treatment, synthetic seawater mixture containing DMSP at a concentration of 10^{-5} was used (10^{-9} M is a concentration commonly occurring in situ; [27,28]. These *Sargassum* and DMSP solutions were the strongest concentrations used in trials and were tested first to determine whether the shrimp responded.

For *Sargassum* chemical cue trials, *Sargassum* effluent was added to one randomly selected radial chamber water source. Again, groups of *L. fucorum* were placed in the center and allowed to move between the radial chambers for a 4 h duration before the location of the shrimp was recorded. The same procedure occurred for the DMSP chemical cue trials. For conspecific chemical cues, 10 live *L. fucorum* shrimp were placed directly into a water source for the 4-chambered apparatus. A chi-square goodness of fit test was used to compare the number of shrimp that moved from the central chamber based on the two size groups, and *t*-tests were used to analyze the data. A one-way ANOVA was used to determine whether the number of shrimp that moved in the control data and the three treatments were significantly different.

2.3. Y-Maze Trials

A Y-maze apparatus (Figure 2) was set up with equal amounts of synthetic seawater running through each branch (areas A and B) and draining out of the apparatus (area C). Shrimp were tested by initially placing an individual in area C. A perforated plexiglass gate was in place to keep the shrimp in the initial area C for a 10 min acclimation period while the flow of water from both branches moved through the system. After 10 min, the perforated gate was raised using a pulley system, and the shrimp were free to explore the remaining areas of the Y-maze for an additional 20 min. To minimize observer interactions, a camera was used to monitor initial choice and subsequent movements of the shrimp. Preliminary observations showed that smaller shrimp were difficult to see in the video recordings, so only shrimp that $\geq$10 mm in length were used these trials. The sex of each shrimp was determined after use in trials by observing the presence (male) or absence (female) of the appendix masculina on the second pleopods [30]. Again, light levels were minimized by placing a blind around the apparatus.

For control trials, the two branches received synthetic seawater with no chemicals added. For treatment trials, the three chemical cues were tested in separate trials by randomly assigning the branches (either A or B) with the chemical cue using the same preparation and solution concentrations described for the 4-chambered apparatus trials. After the data were statistically analyzed for these concentrations, it was deemed appropriate to use a weaker DMSP solution in the *L. fucorum* trials to test the sensitivity of the shrimps' chemoreception to the cue. The additional concentration of 10^{-9} M (compared to the initial concentration of 10^{-5}) was used, which is a concentration commonly occurring in situ [27,28]. Approximately 100 replications each for *L. fucorum* and 50 replications each for *L. tenuicornis* were tested for the controls and three treatments.

A binomial test was used to determine whether significantly more shrimp moved out of area C in the presence of a cue. For the shrimp that moved out of area C, the binomial test was used to determine whether significantly more shrimp initially chose the cue side. Additionally, for the shrimp that moved out of area C, a paired *t*-test repeated-measures ANOVA was used to determine whether there were any differences in time spent in the arm with the cue compared to the arm without the cue between the control and treatment trials.

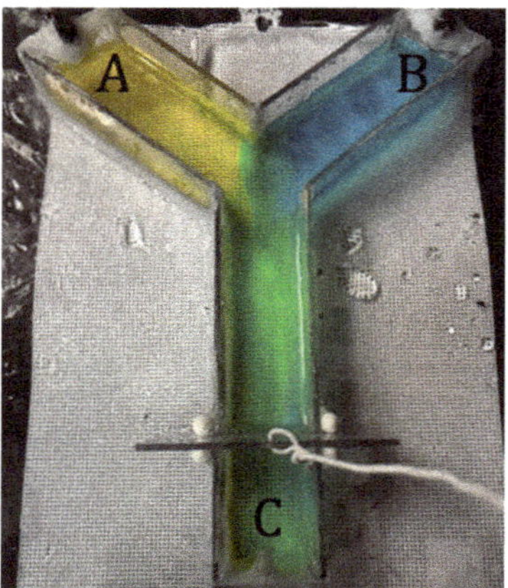

Figure 2. Y-maze apparatus. Blue and yellow dyes demonstrate symmetrical flow from each arm or branch (**A**,**B**) into the base arm (**C**) where shrimp is initially placed prior to starting a trial.

3. Results

3.1. Four-Chambered Apparatus Trials

Controls (when no chemical cues were present in any of the four chambers) showed no significant difference in choice by *Latreutes fucorum* from the null hypothesis of 25% for each of the chambers (binomial test, all $p > 0.05$), which shows the system was unbiased. *Leander tenuicornis* was not tested using the 4-chambered apparatus due to low collection numbers.

A one-way ANOVA showed the average number of *L. fucorum* that moved (regardless of their eventual chamber selection) from the center of the apparatus did not differ significantly among the control and three separate treatments ($p = 0.14$) (Figure 3). However, a specific post hoc *T*-test comparing the number of shrimp moving into radial chambers in the presence of *Sargassum* chemical cues (13.36) was slightly higher than in the control trials (10.29) ($p = 0.03$) (Figure 3). When comparing the total number of shrimp that moved and made a radial chamber selection, no significant differences were detected (chi-square goodness of fit, $p > 0.05$).

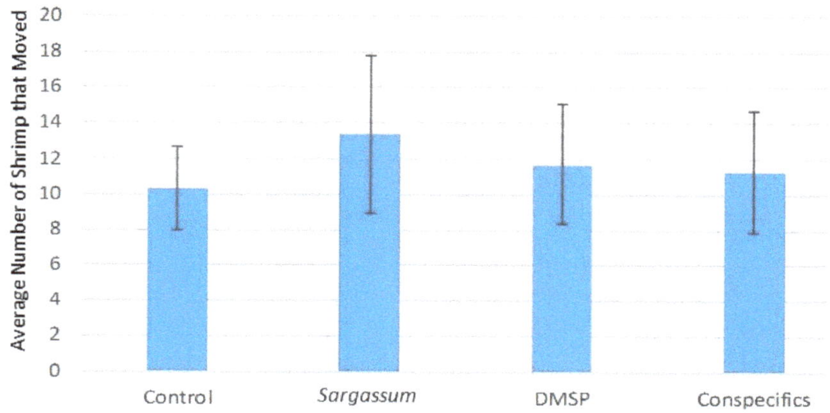

Figure 3. Average number of *Latreutes fucorum* shrimp that moved out of the central chamber of the 4-chambered apparatus for the control, *Sargassum*, DMSP and conspecifics trials (one-way ANOVA, $p = 0.14$). Post hoc *T*-test comparing the number of shrimp moving in the chemical presence of *Sargassum* (13.36) was slightly higher than in the control trials (10.29) ($p = 0.03$). Error bars indicate standard deviations.

Chi-square tests also showed no significant difference in responses to chemical cues between the two size groups of shrimp (all $p > 0.05$).

3.2. Y-Maze Trials

T-tests on the control data showed no significant difference in the amount of time spent in each arm of the apparatus for both species of shrimp; this shows the system was unbiased (*L. fucorum* $p = 0.10$, *L. tenuicornis* $p = 0.25$). Additionally, a chi-square analysis of the control data showed no significant difference between sex and the number of shrimp that moved out of holding area C (*L. fucorum* $p = 0.24$, *L. tenuicornis* $p = 0.53$).

For the treatment trials, binomial tests showed that significantly fewer shrimp for both species moved out of the area C (see Figure 2) in the presence of a *Sargassum* chemical cue (*L. fucorum* $p = 0.03$, *L. tenuicornis* $p = 0.03$) (Figures 4 and 5, respectively). Additionally, significantly fewer *L. tenuicornis* moved out of area C in the presence of conspecific chemical cues ($p = 0.01$) (Figure 5). Alternatively, significantly more *L. fucorum* moved in the presence of both DMSP chemical cue concentrations when compared with the control (DMSP 10^{-5} M $p = 0.05$, DMSP 10^{-9} M $p = 0.01$) (Figure 4).

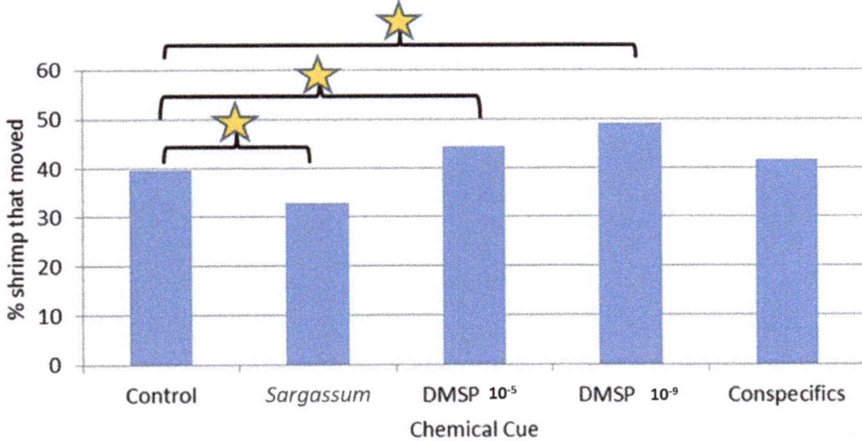

Figure 4. Number of *Latreutes fucorum* shrimp that moved out of area C in the Y-maze for the control, *Sargassum*, DMSP 10^{-5} M, DMSP 10^{-9} M, and conspecific cue trials (binomial test; *Sargassum*, $p = 0.03$; DMSP 10^{-5} M, $p = 0.05$; DMSP 10^{-9} M, $p = 0.01$; and conspecifics, $p = 0.08$). Star/bracket indicates significant differences between indicated paired groups.

The same binomial tests were also used to determine whether parsing the data by sex had any effect on significant findings. *L. fucorum* males moved less in *Sargassum* cue trials when compared to the control males ($p = 0.01$) (Figure 6), and females moved more in both DMSP concentrations compared to the control females (DMSP 10^{-5} M $p = 0.01$, DMSP 10^{-9} M $p = 0.02$) (Figure 7). *L. tenuicornis* females moved significantly less in both the *Sargassum* and the conspecifics cues (*Sargassum* $p = 0.02$, conspecifics $p = 0.01$) (Figure 8). No significance was found for *L. tenuicornis* males (all $p > 0.05$).

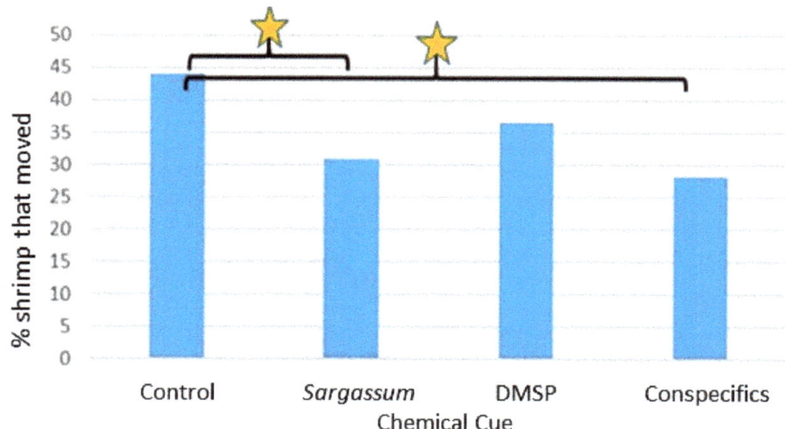

Figure 5. Number of *Leander tenuicornis* shrimp that moved out of area C in the Y-maze for the control, *Sargassum*, DMSP 10^{-5} M, and conspecific cue trials (binomial test; *Sargassum*, $p = 0.03$; DMSP, $p = 0.08$; and conspecifics, $p = 0.01$). Star/bracket indicates significant differences between indicated paired groups.

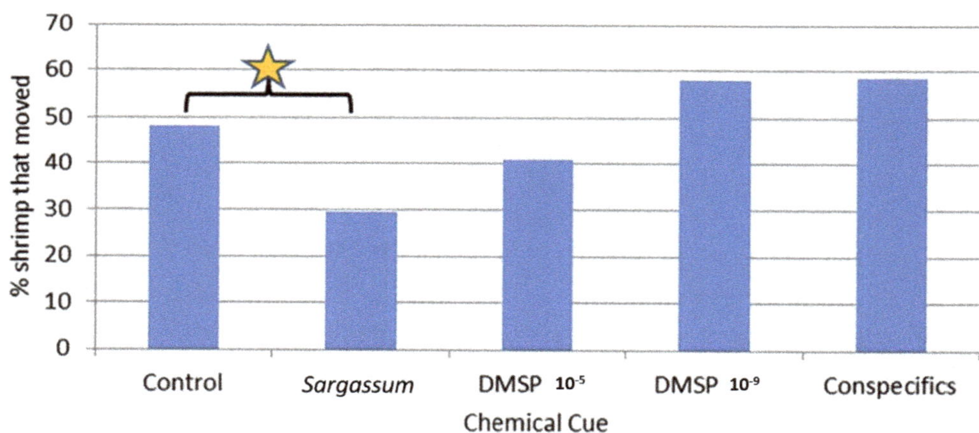

Figure 6. Number of *Latreutes fucorum* male shrimp that moved out of area C in the Y-maze for the control, *Sargassum*, DMSP 10^{-5} M, DMSP 10^{-9} M, and conspecific cue trials (binomial test; *Sargassum* $p = 0.01$; DMSP 10^{-5} M $p = 0.10$; DMSP 10^{-9} M $p = 0.08$; and conspecifics, $p = 0.08$). Star/bracket indicates significant differences between indicated paired groups.

For the shrimp that moved out of area C, a binomial test was used to determine whether significantly more shrimp initially chose the arm of the maze with a cue over the arm of the maze without a cue. *L. fucorum* males chose the cue side over the non-cue side in the conspecifics chemical cue trials ($p = 0.05$) (Figure 9). No other significance was found for *L. fucorum* in the *Sargassum* or DMSP trials, and no significance was found for *L. tenuicornis* for any of the cues (all $p > 0.05$).

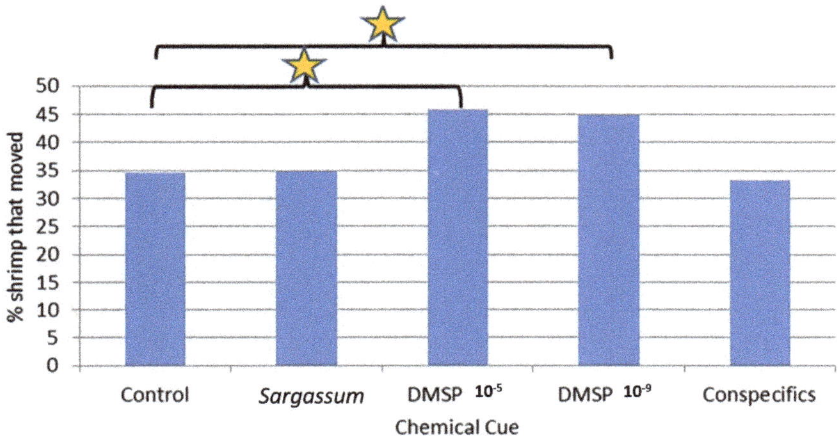

Figure 7. Number of *Latreutes fucorum* female shrimp that moved out of area C in the Y-maze for the control, *Sargassum*, DMSP 10^{-5} M, DMSP 10^{-9} M, and conspecific cue trials (binomial test; *Sargassum*, $p = 0.11$; DMSP 10^{-5}, $p = 0.01$; DMSP 10^{-9}, $p = 0.02$; and conspecifics, $p = 0.11$). Star/bracket indicates significant differences between indicated paired groups.

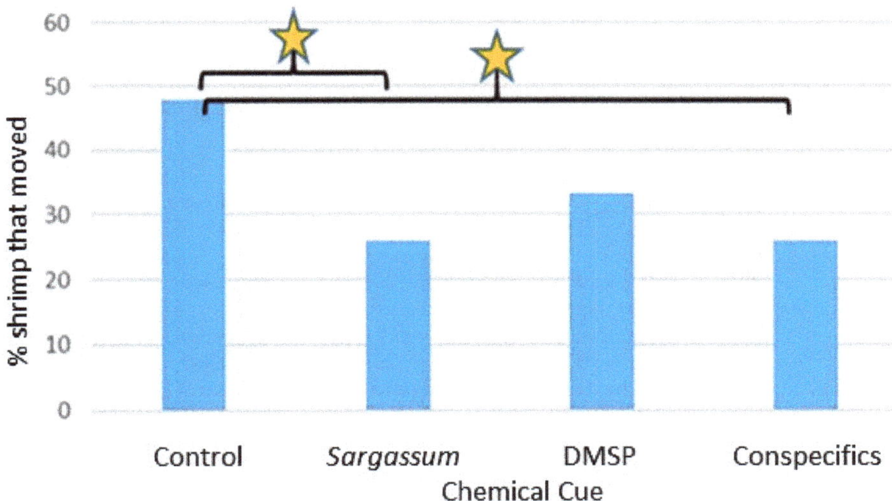

Figure 8. Number of *Leander tenuicornis* female shrimp that moved out of area C in the Y-maze for the control, *Sargassum*, DMSP, and conspecific cue trials (binomial test; *Sargassum*, $p = 0.02$; DMSP, $p = 0.05$; and conspecifics, $p = 0.01$). Star/bracket indicates significant differences between indicated paired groups.

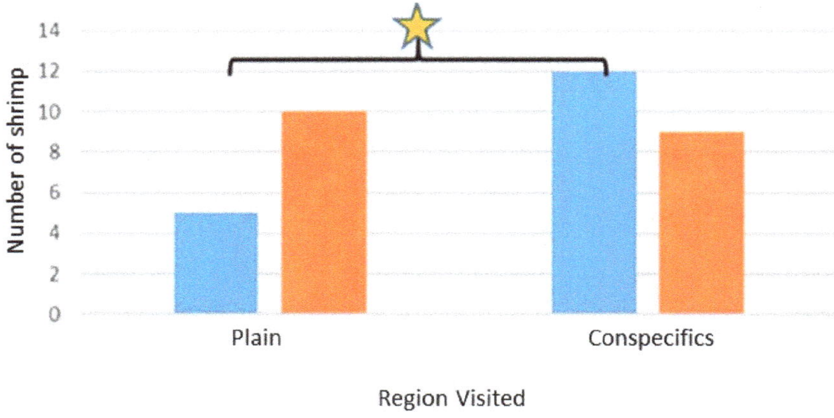

Figure 9. Initial choice in the Y-maze for the shrimp *Latreutes fucorum* between arm with the chemical cue vs. the arm with plain seawater in conspecific chemical cue trials based on sex (binomial test; males, $p = 0.05$; females, $p = 0.18$). Star/bracket indicates significant differences between indicated paired groups.

A paired *t*-test repeated-measures ANOVA for shrimp that moved in the maze showed no significance when comparing the time shrimp spent in the arm of the maze with the cue than without the cue, between the control, *Sargassum*, DMSP, and conspecific trials (all $p > 0.05$).

4. Discussion

4.1. Four-Chambered Apparatus Trials

Results of these trials for *Latreutes fucorum*, overall, were similar to those in the 2009 study by Jobe and Brooks [3], which showed minimal response by the shrimp to chemical odors when visual cues were essentially unavailable. However, in our current study a post hoc *t*-test comparing shrimp movements in the presence of *Sargassum* odors showed a significant increase. This increase in movement did not show more shrimp specifically moving into the radial chamber with the *Sargassum* chemical source but rather more shrimp moved into all of the four radial chambers in the presence of *Sargassum* compared to the other treatments. This result suggests *L. fucorum* increased its general searching behavior but was not precise in locating the directional aspects of the chemical cue source.

This slight difference between the past study and the current study, both of which used the same shrimp species, apparatus, and chemical cue type, could be related to improvements made on the preparation of the chemical cue in this study. The earlier study used a single algal frond placed in the source water at the beginning of each trial. Whereas in the current study we soaked the algal frond in seawater for 48 h prior to the start of the trials. This significantly longer algal soaking period prior to testing could potentially produce a stronger chemical cue source concentration for the shrimp to detect. Additionally, our current study increased the replications for the 4-chambered trials from 10 (in the 2009 study) to 14, which could also potentially increase the likelihood of detecting a significant response. Due to these initial, relatively low responses, serial dilutions involving potentially weaker concentrations of *Sargassum* chemical cue were not tested for either apparatus.

4.2. Y-Maze Apparatus Trials

Results of these trials for *Latreutes fucorum* and *Leander tenuicornis* were compelling in that they showed a clear ability of both shrimp species to detect and respond to chemical cues associated with *Sargassum* (including DMSP) and conspecifics Such a strong response with the Y-maze compared to outcomes with the 4-chambered apparatus (for *L. fucorum* only, due to limited availability of *L. tenuicornis*) are likely related to notable differences between the experimental procedures used with these two different types of apparatus. For example, each replicate for the 4-cambered apparatus used 30 individuals simultaneously, which means that conspecific cues from each sex were presumably readily available to each shrimp throughout those trials. Additionally, shrimp tested in groups may respond differently to shrimp tested individually. That is, movement and choices may be directly influenced by the chemical presence and direct behavior of conspecifics. Because of the high numbers of shrimp typically found in situ within *Sargassum* fronds (personal observation), it is likely that testing in groups is closer to natural conditions. However, the Y-maze has some advantages in that the number of movement choices by the shrimp is limited to two arms (versus four radial chambers) and chemical cues can be more highly controlled (e.g., conspecific, including sex, cues can be delivered individually).

In the Y-maze trials, several specific and significant responses by both shrimp species to chemical cues were observed. Although rarely was the significant difference in selection of the arm with the cue versus non-cue arm detected, there were several significant differences observed between where the shrimp were eventually located, i.e., comparing shrimp that stayed in the holding area C versus moving to either arm of the Y-maze.

Interestingly, significant responses in several instances were to actually move less than that observed in controls without chemical cues, while others involved directional movements toward the chemical odor source (but not necessarily selecting the arm with the chemical cue, i.e., increase in searching behavior but not necessarily locating the cue source, e.g., significantly more *L. fucorum* moved in the presence of both DMSP chemical cue concentrations). This dichotomy of chemical cues triggering search behavior decreases in some situations and increases in others will be discussed below for the different cues.

Significant responses to *Sargassum* for both shrimp species involved fewer individuals moving out of the holding area C. Additionally, fewer *L. tenuicornis* moved out of area C in the presence of conspecific chemical cues. *L. tenuicornis* is the larger of the two species of shrimp; therefore, interactions observed in the lab between conspecifics have occasionally been aggressive, especially larger shrimp towards smaller ones. In general, the shrimp seemed to disperse when kept together in a small container. Conspecific chemical cues for *L. tenuicornis* may have acted as a movement deterrent. For example, shrimp moving less in response to a conspecific chemical cue could be caused by conspecific shrimp releasing stress chemicals. A study on the crayfish *Procambarus clarkii* showed that individuals moved away from conspecifics because of the release of chemicals (i.e., stress cues) into the water [31]. In our study, handling of the shrimp when transferring them into the holding container (from which water dripped into an arm of the Y-maze) may have triggered the release of stress chemicals as well thereby affecting the response to conspecifics.

As pointed out previously, these shrimp are typically found in high concentrations in the fronds of *Sargassum*. Isolating a shrimp in the Y-maze may trigger anti-predator behavior. For example, the freshwater shrimp *Atya lanipes* will reduce its movement by up to 55% to avoid predation [32]. It is possible the chemical cues released from *Sargassum* effluents might have triggered a response to remain stationary. However, this would not explain why *L. fucorum* moved more in the presence of the DMSP cues. This contradiction could indicate the chemical cues released into the water from *Sargassum* are more complex than the single cue of DMSP (which presumably is released from *Sargassum*, too, along with other compounds).

This phenomenon of chemical cues triggering less search activity also occurred with *L. tenuicornis* when exposed to conspecific chemical cues. Perhaps such cues indicated to this shrimp that moving towards the source would involve interaction with additional individuals. Potential negative interactions (e.g., competition or aggression or stress chemicals released, as discussed above) among these shrimp are not well understood, even though they are found in high abundance in situ.

Alternatively, significantly more *L. fucorum* moved in the presence of both DMSP chemical cue concentrations. This demonstrates the possibility this cue could be used by shrimp in locating *Sargassum* patches, too.

Again, while an explanation of the variation of response for some chemical cues causing less shrimp movement by some and more movement for others is problematic, what is clear is that a significant difference in behaviors of the shrimp was observed for the chemical-cue treatments of *Sargassum*, DMSP and conspecifics. We were also able to further parse the significant movement differences related to the sex of the shrimp. Below, we have summarized the significant differences to the chemical cues based on shrimp species and sex.

Sargassum Cue: *L. fucorum* males and *L. tenuicornis* females moved significantly less in the Y-maze in the presence of the *Sargassum* chemical cue.

DMSP Cue: *L. fucorum* females moved statistically more in the Y-maze in the presence of both DMSP dilutions.

Conspecific Cue:

a. *L. tenuicornis* females moved in the Y-maze statistically less in the presence of conspecific cues.
b. *L. fucorum* males that moved in the Y-maze significantly chose the conspecific cue side initially.

A limitation of our study is that the shrimp were not sexed until after use in a trial. Because female shrimp are larger than males [33], when choosing shrimp from the holding aquarium, it is likely there was bias towards choosing females. This bias was especially noticeable for *L. fucorum* trials, as this species is smaller and difficult to locate. Additionally, only shrimp with a total length of 10 mm or greater were used in the Y-maze to ensure visibility in the recorded videos. This bias caused fewer males to be tested than females (*L. fucorum*, 1.88:1, female:male ratio; *L. tenuicornis*, 1.6:1, female:male ratio). Having lower replications for male shrimp decreases the power of statistical analyses, and limits conclusions that can be made about these results. Future studies may choose to incorporate all sizes of shrimp and more replicates with males. Doing so, however, may be problematic as handling smaller shrimp while preventing injury, etc., is challenging.

The two size groups of shrimp did not have any statistically different responses to any of the chemical cues tested (chi-square, all $p > 0.05$). While this may suggest no difference in response between life history stages of the shrimp, future studies should attempt to record larval shrimp responses to chemical cues, as it may be more important for the larvae to detect patches and settle into the habitat. Larvae of other species of invertebrate rely heavily on habitat chemical cues for settlement. For example, swimming larvae of the hydrozoan *Coryne uchidai* settle in response to chemical cues from *Sargassum tortile* [34]. *L. fucorum* has a planktonic larval stage that lasts 18–30 days in the lab [35]. Such a long period of time would indicate that the larvae must have a means to locate suitable habitat.

In some cases, an organism responds to certain chemical cues as a larva but may lose the capability to detect the same chemical cues as an adult. Fiddler crab (*Uca* spp.) larvae can detect cues from food it typically consumes, but adults are insensitive to those same chemical cues [36]. Those results suggest that the life history stage of an organism may influence which chemical cues they respond to based on requirements of the organism at different stages in their life history.

5. Conclusions

Sargassum habitat is part of a significant marine, pelagic holobiont (i.e., host and collective symbionts), which is a complex and diverse community, much like a floating version of a coral reef. *Sargassum* is an Essential Fish Habitat [7], and entire food chains and marine ecosystems rely on the success of this community of symbionts. Understanding how these biotic connections are established and maintained is critical. This study focused on two shrimp species that are abundant in *Sargassum*; specifically, to expand on the previous study by Jobe and Brooks [3] and clarify the role of available chemical cues for the shrimp to form and maintain their role in the *Sargassum* holobiont.

Unequivocally, we show that both *Latreutes fucorum* and *Leander tenuicornis* can detect and respond to chemical cues in the absence of visual cues from *Sargassum* patches. It is important to note that while chemoreception may be utilized, these shrimp are likely utilizing all senses available to maintain the association. As such, we envision that shrimp would use visual and chemical cues when both are available, as demonstrated by Jobe and Brooks [3], such as during the day and nights associated with full (or near) moons. The current results demonstrate responses to chemical cues exclusively could be important when visual cues are mostly limited. Additionally, both species of *Sargassum* shrimp are brooders and carry embryos until hatching [37]. Because the shrimp are hatched within the *Sargassum* patches, it is possible that planktonic larval and juvenile shrimp do not need to travel long distances to establish this symbiosis. However, natural disturbances such as storms and strong currents can disrupt and break apart *Sargassum* patches thereby potentially separating shrimp and other symbionts from their resident patch. This would require relocating patches for all ages and sizes of shrimp. Having access to both chemical and visual cues would seem advantageous.

A study on chemical feeding cues for *L. tenuicornis* found that only 5 out of the 28 single compounds used as cues were significantly stimulating the shrimp, which feed primarily on hydroids and bryozoans [38]. This shows that chemoreception can be highly specific, and thus searching for an organism's ability to use chemoreception may require testing several cues. Our study attempted to potentially isolate components of *Sargassum* by using DMSP, which did separately trigger responses. Differences in shrimp response between *Sargassum* and DMSP indicate that *Sargassum* effluents are likely highly complex, involving potentially multiple specific compounds. Future studies should conduct a chemical assay on *Sargassum* algae, potentially identifying more specific bioactive compounds that can be used to gain further insight into how this crucial mutualistic (alga supplies habitat and fish and shrimp provides nutrients via waste products); cf. [10], symbiosis is formed and maintained.

It is abundantly clear how important the shrimp are to the continuity of the *Sargassum* community. In addition to these shrimp species, several other *Sargassum* organisms have been studied. The *Sargassum* crab *Portunus sayi* is abundant in this community and was also found to detect *Sargassum* chemical cues [39]. When looking at both this current study and the previously mentioned studies on this community, it is clear that chemoreception is a major mechanism being used by several organisms in the *Sargassum* holobiont. Another study looked at habitat selection by both *L. tenuicornis* and the common fish species, *Histrio histrio*, and found that the organisms have preferred habitats based on structural complexity [8]. It is likely that all of these organisms are interconnected and play an important role in maintaining this symbiotic community, and the use of chemoreception is one vital component to the success of this symbiosis.

Finally, the *Sargassum* holobiont is in global crisis. Nutrient influx from anthropogenic sources has contributed to producing massive and sometimes harmful blooms of *Sargassum* that are not only potentially affecting the components within this ecosystem but have already profoundly affected coastal ecosystems [40] (Lapointe et al., 2021). It is more important than ever that we understand fully how *Sargassum* communities function.

Author Contributions: Conceptualization, J.L.F. and W.R.B.; methodology, J.L.F. and W.R.B.; software, J.L.F. and W.R.B.; investigation, J.L.F.; resources, J.L.F. and W.R.B.; writing—original draft preparation, J.L.F. and W.R.B.; writing—review and editing, J.L.F. and W.R.B. All authors have read and agreed to the published version of the manuscript.

Funding: This research received no external funding.

Institutional Review Board Statement: Not applicable.

Informed Consent Statement: Not applicable.

Data Availability Statement: Data are unavailable as they are from a master student's thesis that is not available at this time.

Acknowledgments: We thank Rindy Anderson and John Baldwin for their logistical and technical support. We also thank Robert Dennis, Nicky Montagna, Tracy Frahm and Gary Frahm for their financial and general support.

Conflicts of Interest: The authors declare no conflict of interest.

Appendix A

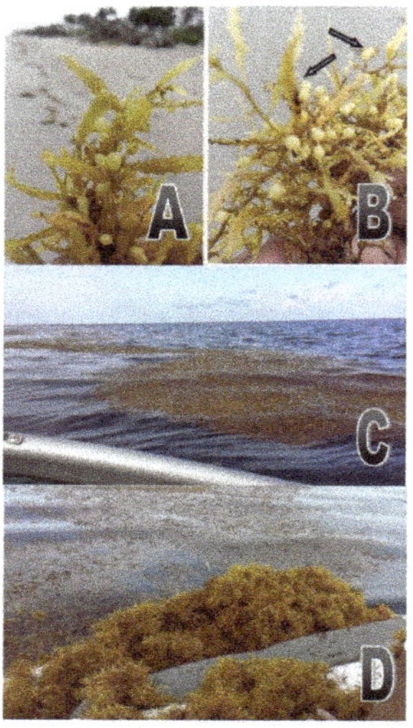

Figure A1. (**A**) indicates *Sargassum fluitans.* Notice smooth pneumatocysts (air bladders). (**B**) indicates *Sargassum natans.* Arrows indicate spines typically associated with pneumatocysts of this species. (**C**,**D**) show algal mats accumulating at sea surface and collected mats (**D**). Typically mats consist of both species shown in (**A**,**B**).

References

1. Coston-Clements, L.; Settle, L.R.; Hoss, D.E.; Cross, F.A. *Utilization of the Sargassum Habitat by Marine Invertebrates and Vertebrates—A Review*; NOAA Technical Memorandum NMFS-SEFSC-296; National Oceanic and Atmospheric Administration: Washington, DC, USA, 1991.
2. Sehein, T.; Siuda, A.N.S.; Shank, T.M.; Govindarajan, A.F. Connectivity in the slender *Sargassum* shrimp (*Latreutes fucorum*): Implications for a Sargasso Sea protected area. *J. Plankton Res.* **2014**, *36*, 1408–1412. [CrossRef]
3. Jobe, C.F.; Brooks, W.R. Habitat selection and host location by symbiotic shrimps associated with *Sargassum* communities: The role of chemical and visual cues. *Symbiosis* **2009**, *49*, 77–85. [CrossRef]
4. Marmorino, G.O.; Miller, W.D.; Smith, G.B.; Bowles, J.H. Airborne imagery of a disintegrating *Sargassum* drift line. *Deep. Sea Res. Part I Oceanogr. Res. Pap.* **2011**, *58*, 316–321. [CrossRef]
5. Hu, C.; Murch, B.; Barnes, B.B.; Wang, M.; Maréchal, J.-P.; Franks, J.; Johnson, D.; Lapointe, B.; Goodwin, D.S.; Schell, J.M.; et al. *Sargassum* Watch Warns of Incoming Seaweed. *Eos* **2016**, *97*, 10–15. [CrossRef]
6. Niermann, U. Distribution of *Sargassum natans* and some of its epibionts in the Sargasso Sea. *Helgol. Mar. Res.* **1986**, *40*, 343–353. [CrossRef]
7. Ballard, S.E.; Rakocinski, C.F. Flexible Feeding Strategies of Juvenile Gray Triggerfish (*Balistes capriscus*) and Planehead Filefish (*Stephanolepis hispidus*) Within *Sargassum* Habitat. *Gulf Caribb. Res.* **2012**, *24*, 31–40. [CrossRef]
8. Bennice, C.O.; Brooks, W.R. Habitat Selection among Fishes and Shrimp in the Pelagic Sargassum Community: The Role of Habitat Architecture. *Gulf Mex. Sci.* **2016**, *33*, 1. [CrossRef]
9. Brooks, W.R.; Hutchinson, K.A.; Tolbert, M.G. Pelagic Sargassum Mediates Predation among Symbiotic Fishes and Shrimps. *Gulf Mex. Sci.* **2007**, *25*, 5. [CrossRef]

10. Lapointe, B.E.; West, L.E.; Sutton, T.T.; Hu, C. Ryther revisited: Nutrient excretions by fishes enhance productivity of pelagic *Sargassum* in the western North Atlantic Ocean. *J. Exp. Mar. Biol. Ecol.* **2014**, *458*, 46–56. [CrossRef]

11. Stoner, A.W. Pelagic *Sargassum*: Evidence for a major decrease in biomass. *Deep. Sea Res. Part A Oceanogr. Res. Pap.* **1983**, *30*, 469–474. [CrossRef]

12. Lapointe, B.E. A comparison of nutrient-limited productivity in *Sargassum natans* from neritic vs. oceanic waters of the western North Atlantic Ocean. *Limnol. Oceanogr.* **1995**, *40*, 625–633. [CrossRef]

13. Wells, D.R.J.; Rooker, J.R. Spatial and temporal patterns of habitat use by fishes associated with *Sargassum* mats in the northwestern Gulf of Mexico. *Bull. Mar. Sci.* **2004**, *74*, 81–99.

14. Dooley, J.K. Fishes associated with the pelagic *Sargassum* complex, with a discussion of the *Sargassum* community. *Contrib. Mar. Sci.* **1972**, *16*, 1–32.

15. Rudershausen, P.J.; Buckel, J.A.; Edwards, J.; Gannon, D.P.; Butler, C.M.; Averett, T.W. Feeding Ecology of Blue Marlins, Dolphinfish, Yellowfin Tuna, and Wahoos from the North Atlantic Ocean and Comparisons with other Oceans. *Trans. Am. Fish. Soc.* **2010**, *139*, 1335–1359. [CrossRef]

16. Atema, J. Chemical signals in the marine environment: Dispersal, detection, and temporal signal analysis. *Proc. Natl. Acad. Sci. USA* **1995**, *92*, 62–66. [CrossRef] [PubMed]

17. Carr, W.E.S.; Ache, B.W.; Gleeson, R.A. Chemoreceptors of Crustaceans: Similarities to Receptors for Neuroactive Substances in Internal Tissues. *Environ. Health Perspect.* **1987**, *71*, 31–46. [CrossRef]

18. Brooks, W.R.; Rittschof, D. Chemical Detection and Host Selection by the Symbiotic Crab *Porcellana sayana*. *Invertebr. Biol.* **1995**, *114*, 180. [CrossRef]

19. Díaz, E.R.; Thiel, M. Chemical and Visual Communication During Mate Searching in Rock Shrimp. *Biol. Bull.* **2004**, *206*, 134–143. [CrossRef]

20. Dacey, J.W.H.; Wakeham, S.G. Oceanic Dimethylsulfide: Production During Zooplankton Grazing on Phytoplankton. *Science* **1986**, *233*, 1314–1316. [CrossRef]

21. Hill, R.; Dacey, J. Metabolism of dimethylsulfoniopropionate (DMSP) by juvenile Atlantic menhaden *Brevoortia tyrannus*. *Mar. Ecol. Prog. Ser.* **2006**, *322*, 239–248. [CrossRef]

22. Broadbent, A.; Jones, G.; Jones, R. DMSP in Corals and Benthic Algae from the Great Barrier Reef. *Estuar. Coast. Shelf Sci.* **2002**, *55*, 547–555. [CrossRef]

23. DeBose, J.L.; Lema, S.C.; Nevitt, G. Dimethylsulfoniopropionate as a Foraging Cue for Reef Fishes. *Science* **2008**, *319*, 1356. [CrossRef]

24. Nevitt, G.A.; Veit, R.R.; Karieva, P. Dimethyl sulfide as a foraging cue for Antarctic procelariiform seabirds. *Nature* **1995**, *376*, 680–682. [CrossRef]

25. Kowalewsky, S.; Dambach, M.; Mauck, B.; Dehnhardt, G. High olfactory sensitivity for dimethyl sulphide in harbour seals. *Biol. Lett.* **2005**, *2*, 106–109. [CrossRef] [PubMed]

26. Hay, M.E. Marine chemical ecology: What's known and what's next? *J. Exp. Mar. Biol. Ecol.* **1996**, *200*, 103–134. [CrossRef]

27. DeBose, J.L.; Nevitt, G.A.; Dittman, A.H. Rapid Communication: Experimental Evidence that Juvenile Pelagic Jacks (Carangidae) Respond Behaviorally to DMSP. *J. Chem. Ecol.* **2010**, *36*, 326–328. [CrossRef] [PubMed]

28. Vila-Costa, M.; Rinta-Kanto, J.M.; Poretsky, R.S.; Sun, S.; Kiene, R.P.; Moran, M.A. Microbial controls on DMSP degradation and DMS formation in the Sargasso Sea. *Biogeochemistry* **2014**, *120*, 295–305. [CrossRef]

29. Cox, D. The Role of Chemical Cues in Locating Pelagic *Sargassum* by the Associated Fish *Stephanolepis hispidus*. Master's Thesis, Florida Atlantic University, Boca Raton, FL, USA, 2016.

30. Penha-Lopes, G.; Torres, P.; Macia, A.; Paula, J. Population structure, fecundity and embryo loss of the sea grass shrimp Latreutes pymoeus (Decapoda: Hippolytidae) at Inhaca Island, Mozambique. *J. Mar. Biol. Assoc. UK* **2007**, *87*, 879–884. [CrossRef]

31. Schneider, R.Z.; Moore, P. Urine as a source of conspecific disturbance signals in the crayfish *Procambarus clarkii*. *J. Exp. Biol.* **2000**, *203*, 765–771. [CrossRef]

32. Crowl, T.A.; Covich, A.P. Responses of a Freshwater Shrimp to Chemical and Tactile Stimuli from a Large Decapod Predator. *J. N. Am. Benthol. Soc.* **1994**, *13*, 291–298. [CrossRef]

33. Martínez-Mayén, M.; Román-Contreras, R. Some Reproductive Aspects of *Latreutes fucorum* (Fabricius, 1798) (Decapoda, Hippolytidae) from Bahía De La Ascención, Quintana Roo, Mexico. *Crustaceana* **2011**, *84*, 1353–1365. [CrossRef]

34. Kato, T.; Kumanireng, A.S.; Ichinose, I.; Kitahara, Y.; Kakinuma, Y.; Kato, Y. Structure and synthesis of active component from a marine alga, *Sargassum tortile*, which induces the settling of swimming larvae of *Coryne uchidai*. *Chem. Lett.* **1975**, *4*, 335–338. [CrossRef]

35. Bailey, D.J. The Larval Development of the *Sargassum* Shrimps *Leander tenuicornis* (Say) (Palaemonidae) and *Latreutes fucorum* (Fabricius) (Hippolytidae) Reared in the Laboratory. Master's Thesis, Florida Atlantic University, Boca Raton, FL, USA, 1980.

36. Weissburg, M.J.; Zimmer-Faust, R.K. Ontogeny Versus Phylogeny in Determining Patterns of Chemoreception: Initial Studies with Fiddler Crabs. *Biol. Bull.* **1991**, *181*, 205–215. [CrossRef]

37. Bauer, R.T. Continuous reproduction and episodic recruitment in nine shrimp species inhabiting a tropical seagrass meadow. *J. Exp. Mar. Biol. Ecol.* **1989**, *127*, 175–187. [CrossRef]

38. Johnson, B.R.; Atema, J. Chemical stimulants for a component of feeding behavior in the common gulf-weed shrimp Leander tenuicornis (Say). *Biol. Bull.* **1986**, *170*, 1–10. [CrossRef]

39. West, L.E.; Brooks, W.R. Habitat location and selection by the symbiotic *Sargassum* crab *Portunus sayi*: The role of chemical, visual, and tactile cues. *Symbiosis* **2018**, *76*, 177–185. [CrossRef]
40. Lapointe, B.E.; Brewton, R.A.; Herren, L.W.; Wang, M.; Hu, C.; McGillicuddy, D.J.; Lindell, S.; Hernandez, F.J.; Morton, P.L. Nutrient content and stoichiometry of pelagic *Sargassum* reflects increasing nitrogen availability in the Atlantic Basin. *Nat. Commun.* **2021**, *12*, 1–10. [CrossRef] [PubMed]

Article

A Long-Term Symbiotic Relationship: Recruitment and Fidelity of the Crab *Trapezia* on Its Coral Host *Pocillopora*

H. M. Canizales-Flores [1], A. P. Rodríguez-Troncoso [1,*], F. A. Rodríguez-Zaragoza [2] and A. L. Cupul-Magaña [1]

[1] Laboratorio de Ecología Marina, Centro Universitario de la Costa, Universidad de Guadalajara, Av. Universidad No. 203, Puerto Vallarta 48280, Mexico; hazel_mariacf@hotmail.com (H.M.C.-F.); amilcar.cupul@gmail.com (A.L.C.-M.)

[2] Laboratorio de Ecología Molecular, Microbiología y Taxonomía (LEMITAX), Departamento de Ecología, Centro Universitario de Ciencias Biológicas y Agropecuarias, Universidad de Guadalajara, Camino Ramón Padilla Sánchez No. 2100 Nextipac, Zapopan 45110, Mexico; fabian.rzaragoza@academicos.udg.mx

* Correspondence: pao.rodriguezt@gmail.com or alma.rtroncoso@academicos.udg.mx; Tel.: +52-3222262319

Citation: Canizales-Flores, H.M.; Rodríguez-Troncoso, A.P.; Rodríguez-Zaragoza, F.A.; Cupul-Magaña, A.L. A Long-Term Symbiotic Relationship: Recruitment and Fidelity of the Crab *Trapezia* on Its Coral Host *Pocillopora*. *Diversity* **2021**, *13*, 450. https://doi.org/10.3390/d13090450

Academic Editors: Michael Wink, Patricia Briones-Fourzán and Michel E. Hendrickx

Received: 30 August 2021
Accepted: 15 September 2021
Published: 19 September 2021

Publisher's Note: MDPI stays neutral with regard to jurisdictional claims in published maps and institutional affiliations.

Abstract: The symbiotic relationship between the crab *Trapezia* spp. and pocilloporid corals has been characterized as obligate. Although this relationship is considered common and has been widely registered within the distribution areas of these corals, the initiation of this symbiotic relation and its potential persistence throughout the life cycle of the crustacean is still poorly described. To understand the *Trapezia–Pocillopora* symbiosis, determining the time and conditions when *Trapezia* recruits a coral colony and the factors influencing this process are key. Thus, in the present study, healthy, small and unrecruited coral fragments were attached to the substrates (using cable ties) of nearby adult *Pocillopora* colonies. All fragments were monitored for two years to measure their growth and size at the first evidence of *Trapezia* crab recruitment, as well as the abundance and permanence of the crabs on the coral fragments. Results showed a relation between the space available (coral volume) and crab recruitment as an increase in substrate complexity is required to provide protection for the crabs and hence maintain the symbiosis, while abiotic conditions such as sea temperature and the distance of the fragments from the adult coral colonies seemingly did not affect the recruitment process. In addition, crabs are able to move between colonies, thus discarding the theory that once recruited, crabs are obligate residents on this specific colony.

Keywords: ecology; symbiosis; crustacean; crab; coral

1. Introduction

Invertebrates are the most abundant and diverse group within coral reefs [1]; some of them are considered obligate associates on live corals [2] and their distribution has been fully attributed to the presence of their host [3–5]. However, their abundance and permanence not only depend on the presence of the host but also on (1) their reproduction and recruitment, both regulated by abiotic factors, and (2) the ability of the larvae to disperse, recognize and occupy suitable substrates [6]. Specifically, the benefits coral colonies provide for their commensals, in comparison to free-living species, are a permanent shelter, nursery site and in some cases nourishment [7,8].

Moreover, the obligate association invertebrate–corals can be characterized as bi-directional. In particular, different groups of crustaceans have been reported to engage in symbiotic associations with other macroinvertebrates such as sponges, sea anemones and echinoderms, among others [9]. The benefit of these symbiotic assemblages may determine their hosts' fitness and long-term survival [5]. Hermatypic corals are the base of coral communities; however, their permanence depends not only on the coral response but also on their associated organisms. The presence of hydrozoa appears to reduce the corals' susceptibility to diseases [10]. In the specific case of crustaceans, they have a relevant role in the daily maintenance of the colonies as they cleanse sediments and protect them from predators [11,12] but also have a major role as their presence can slow down the progression of white syndrome, as observed in Acroporids [13]. Therefore, the presence of a high diversity of crustaceans, not only at a species level but also at a size level, can decrease the vulnerability of the coral to different local stressors [14]. The accurate characterization of the diversity of coral-associated organisms possibly contributes not only to an increase of the knowledge on the ecology or physiology of a specific group but also to the different mechanisms controlling coral reef functioning [5].

Trapezia spp. crabs are considered both highly abundant and conspicuous brachyuran crustaceans, with a well-documented specific association as defenders of pocilloporid corals [14,15]. The success of the *Pocillopora–Trapezia* symbiosis results from the fact that the crabs share distribution with *Pocillopora*, the most abundant genus along the Eastern Tropical Pacific [16], and also from their functional role in this symbiosis. The coral harbors the crab, but in exchange, *Trapezia* crabs use *Pocillopora* colonies as a specific substrate for refuge, as a feeding ground for grazing or food capture and as a direct source of nutrition (detrital sludge, algae, etc.) [17]. In return, *Trapezia* eliminates the excess of sediment and detritus accumulated on the coral tissue, thus contributing to the fitness of their *Pocillopora* host and resulting in higher coral growth rates and an increase in their resistance to thermal stress and bleaching [18].

Trapezia crab reproduction, recruitment and even migration responses are influenced by annual changes in sea surface temperature and salinity [19]. Brachyuran reproduction includes a planktonic larval phase that can last in the water column for days or weeks, thus contributing to their local and regional dispersion [19]. Subsequently, the competent larvae will detect specific chemical cues to settle and recruit [20]. In general, in tropical areas, invertebrate larvae develop and settle during the warm season as temperature is a trigger for their reproductive activity [21] and also due to food availability for both the larvae and the early recruited organisms [1]. On the other hand, at sub-tropical coral reefs, it has been demonstrated that there are no specific temporal patterns for crustacean recruitment [22]. Therefore, recruitment depends on the organism's ability to reproduce and the optimal environmental conditions, but the availability and suitability of the substrate also represent an important factor for determining the species' abundance and composition.

Despite the relevance and life-long permanence of *Trapezia* crabs in coral communities, studies regarding their recruitment and other behavior such as migration are scarce and restricted to the presence, density, abundance at the coral genus level and the potential relationship between crabs, coral colony size and available space [23]. When recruited, crabs develop territorial behavior, protecting their host from both predators and intraspecific competitors [8]. However, *Trapezia* is not a permanent resident on a single coral colony as the crabs can migrate between coral colonies with available space to provide them shelter and food [23] or when searching for partners for reproduction [7]. Although the basis of the symbiotic relationship between corals and crabs as well as the services that are provided are well characterized, so far, we still lack understanding regarding the initiation and stability of this symbiotic relationship, information that is necessary to identify and describe the recruitment-permanence process of *Trapezia* on *Pocillopora* and to recognize the key factors involved in it. This symbiosis cannot be fully understood without the basic knowledge of the individual organisms, which will then provide insights into the role of *Trapezia* for the presence and even ability of the coral to resist and/or respond to future climate change scenarios.

To provide insights on coral–*Trapezia* symbiosis over two years, we registered the beginning of the relationship in order to identify the factors involved in the recruitment and permanence of *Trapezia* crabs in *Pocillopora* corals and thus reach a better understanding of this symbiotic relationship.

2. Materials and Methods

2.1. Study Area

Islas Marietas National Park (IMNP) is an insular natural protected area located in the Northeastern Tropical Pacific and consists of two islands (Isla Larga and Isla Redonda) and several islets (Figure 1) [24]. The region is characterized as an oceanographic transition area, with the seasonal convergence of three oceanic currents: Costa Rica Coastal Current, the California Current and the water mass of the Gulf of California [25]. Also, local conditions, such as the presence of seasonal upwelling, internal waves and runoffs influence the environmental conditions of the area [26,27].

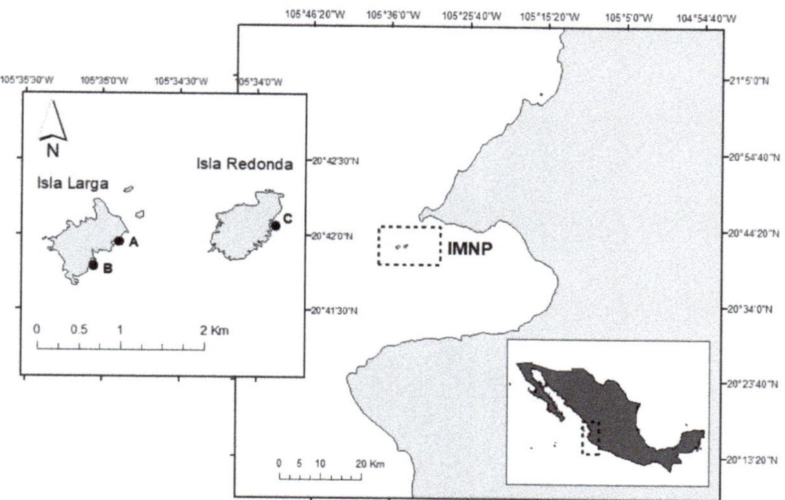

Figure 1. Study area: Islas Marietas National Park (IMNP), Nayarit, Mexico. All sampling sites harbor a high *Pocillopora* sp. coverage: (A) Restoration Zone (20.698860° N, 105.580997° W) and (B) Cueva del Muerto (20.697389° N, 105.582806° W) at Isla Larga, and (C) Pavonas Platform (20.700908° N, 105.565304° W) at Isla Redonda.

Regarding benthic fauna, IMNP harbors one of the most important coral communities of the Northeastern Tropical Pacific, composed mainly of branched corals of the genus *Pocillopora* [28] and distributed in a depth range between 0 and 20 m [24]. It is important to mention that *Pocillopora* is one of the most important reef builders in the Eastern Pacific [10], represented in the area by at least eight different morpho-species [29]. In addition to its high richness, this region is also considered an important larval dispersal and connectivity zone along the ETP [16].

For the purpose of this study and better spatial representation, three sites were selected. All sites harbor a high coral cover of healthy adult *Pocillopora* spp. colonies with a high availability of naturally fragmented corals also defined as fragments of opportunity: (1) Restoration Zone (RZ) and (2) Cueva del Muerto (CM), both located on Isla Larga and representing 11% cover of *Pocillopora* corals, and (3) Pavonas Platform (PP) on Isla Redonda, with ~9% cover, located at a depth of 6–8 m [29,30] (Figure 1). The benthic communities at the sites include massive and sub-massive corals of the genus *Porites* and *Pavona*, as well as other important morpho-functional groups such as articulated and encrusting coralline algae, fleshy macroalgae, algal turf, sponges, coral rubble, other sessile groups, rock and sand [31].

2.2. Field Experiment

To evaluate the *Trapezia* recruitment rate and the relationship with the available space, a total of 44 *Pocillopora* fragments of opportunity (~8 cm length, ~6 cm width) were hand-collected by SCUBA diving. Each fragment was visually evaluated in order to select only healthy coral fragments without evidence of bleaching, tissue damage or high abundance of external and internal bioeroders. More importantly, all fragments represented a single branch ($\approx$1 cm height) without associated *Trapezia* crab. To minimize the stress of transplantation, all fragments were carefully placed in plastic baskets and immediately transferred by divers to designated transplant areas. Each fragment was individually fixed to a rocky substrate with a plastic cable tie [32,33] and then tagged so that individual growth and overall development of each coral colony and *Trapezia* recruitment could be monitored (Figure 2).

Figure 2. Coral fragment transplantation and monitoring. (**a**) Coral fragment selection; (**b**) Fixing and labeling of *Pocillopora* fragments without *Trapezia*.

Coral growth was measured every two months during a two-year period (September 2018–October 2020). For this, maximum length, width and height of coral fragments were measured with plastic calipers (0.05 mm precision) to estimate the volume and hence determine the habitat space available for recruitment, as described by Barry [34]. It is important to emphasize the relevance of the three-dimensional structure resulting from the growth pattern of *Pocillopora*. So far, fragmentation is considered the most important reproductive strategy of *Pocillopora* in the region [35]. Since the fragments originated from asexual reproduction (fragmentation), they initially did not show a complex growth, so a 1 cm height was used in the formula to calculate the volume until the fragments presented ramification and therefore an increase in their complexity.

Trapezia recruitment (Figure 3) was assessed over the two-year period by visually recording the abundance of the crabs per colony at every visit (bi-monthly) using a thin metal rod and inserting it into the spaces between the coral branches to trigger *Trapezia* defense response. Additionally, to evaluate the relationship between crab migration and available space, the distance of each transplanted fragment (colony) to the closest adult *Pocillopora* colony was measured. Also, daily sea temperature was recorded in situ by a HOBO temperature data logger (Onset, MA, USA) and the average value of the sampling time (month in which the visits were made to record data). Furthermore, bi-monthly mean sea surface temperature was calculated to evaluate the effect of temperature on crab recruitment.

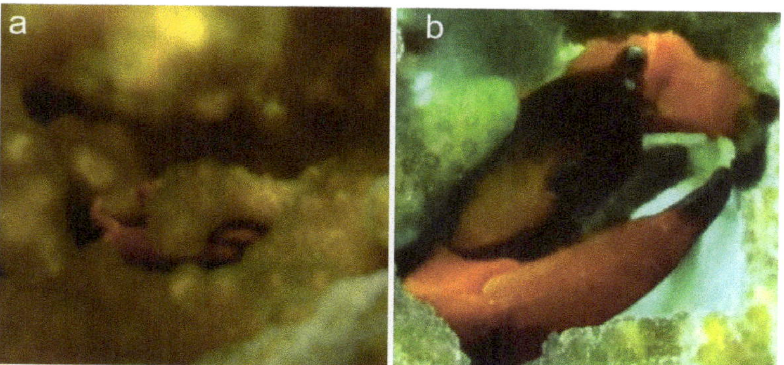

Figure 3. Recruitment of *Trapezia* in *Pocillopora* fragments. (**a**) *Trapezia* sp. hidden between coral branches; (**b**) Close up to *Trapezia* crab.

2.3. Data Analysis

Differences in *Trapezia* recruitment (density per colony) and habitat space (i.e., coral volume) through the sampling time were analyzed, each one with a one-way repeated measures analysis of variance (ANOVA) based on permutations and using the sampling time (T) as a fixed factor. These repeated measures ANOVAs based on permutations were constructed with Euclidean distance matrices following Anderson et al. [36]. The main test and pairwise comparisons were assessed using 10,000 permutations of residuals based on a reduced model and type III sum of squares. Analyses were computed using PRIMER ver. 6.1.11 + PERMANOVA v.1.0.1 software [37].

A multiple linear regression (MLR) was performed to identify the explaining variables that best explained the *Trapezia* recruitment. Thus, coral volume and the sea temperature were independent variables, while *Trapezia* crabs' abundance was the dependent variable $[Y = \beta_0 + \beta_1 X_1 + \beta_2 X_2]$. Fit to the Gaussian curve and homoscedasticity were met from model residuals. MLR was done with SigmaPlot 11.2 software. In order to explore how the natural adult *Pocillopora* colonies on the site were related to the variation in *Trapezia* recruitment from the experimental transplanted coral fragments, we plotted an exploratory graph. The data used in the graph were the total abundance of *Trapezia* of each fragment over time and the distance of each transplanted coral fragment to the nearest adult colony.

3. Results

Pocillopora fragments showed a growth of 2.52 ± 1.18 cm yr^{-1} in length and 2.12 ± 1.13 cm yr^{-1} in width during the first year, and during the second sampling year, a growth of 4.55 ± 1.43 cm yr^{-1} in length and 4.59 ± 1.91 cm yr^{-1} in width. In both, the difference in annual coral growth and the consequent development of branches resulted in an almost twofold differential increase in the colony volumes during the first year (42 to 77 cm^3), while during the second year the colonies reached +10-fold volume, with a maximum value of 1335 cm^3.

The first evidence of crab recruitment was recorded two months after transplantation (first sampling time point) in fragments with a minimum volume of 20.13 cm^3. During the study, a maximum of seven crabs per colony was observed. However, their permanence was not stable, as the number of crabs fluctuated in the colonies over time; therefore, the occupation was not permanent, since in coral colonies harboring crabs at one sampling time point, they could be absent in the next and present again in the following sampling time point, evidencing the ability of *Trapezia* to migrate between colonies (Table S1). Also, the distance between colonies was assessed with a maximum distance of 12.5 m, suggesting that the crabs were able to move among coral colonies at distances ≥10 m. Finally, the temperature data showed a sea temperature range of 19–31°C.

On the other hand, repeated measures ANOVAs based on permutations evidenced an increase of the colony volume over time (Figure 4, Table 1) and an increase of *Trapezia* recruited (Figure 5, Table 1).

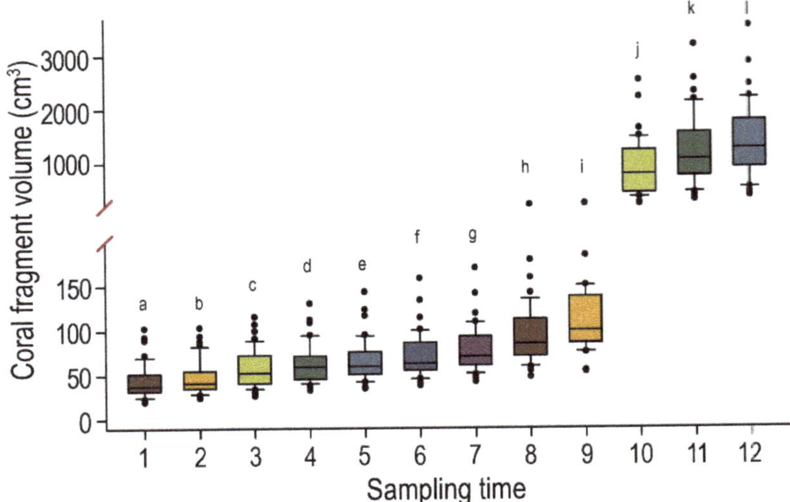

Figure 4. Boxplots of changes in colony volume over time. Corals were monitored bi-monthly for two years, with 0 representing the time when coral fragments were transplanted. The bars indicate the average (with the representation of the median) volume of fragments per time point. Y-Axis break was used to shrink a segment and enhance the readability of fragment volume change in the chart. Letters indicate significant differences between groups.

Table 1. One-way univariate ANOVA with repeated measures based on permutations evaluating differences in coral growth as the volume, and recruitment of *Trapezia*, between time (T).

Source	Habitat Space			Recruitment		
	Pseudo-F	**P (perm)**	**C.V.(%)**	**Pseudo-F**	**P (perm)**	**C.V.(%)**
Time	145.44	0.0001 *	71.7	108.1	0.0001 *	70.9
Residuals			21.9			29.1

* Results with statistical differences ($p < 0.05$).

The MLR was significant (F = 10.417, p = 0.005) with a R^2_{adj} = 63% [Y = 15.993 + (1.889 * Temperature) + (0.00183 * Volume)] where coral volume was the best predictable variable (t = 4.265, p = 0.002) for *Trapezia* recruitment and how this influence increases over time. Temperature alone was not significantly related to *Trapezia* recruitment (t = 0.823, p = 0.432), but in combination with coral volume, it did generate a significant model (F = 10.417, p = 0.005). In addition, despite not being constant, a pattern of increase in the density of crabs was found with an evidently higher abundance of *Trapezia* with the increase of habitable space (coral volume) as a result of *Pocillopora* tridimensional growth and the development of branches.

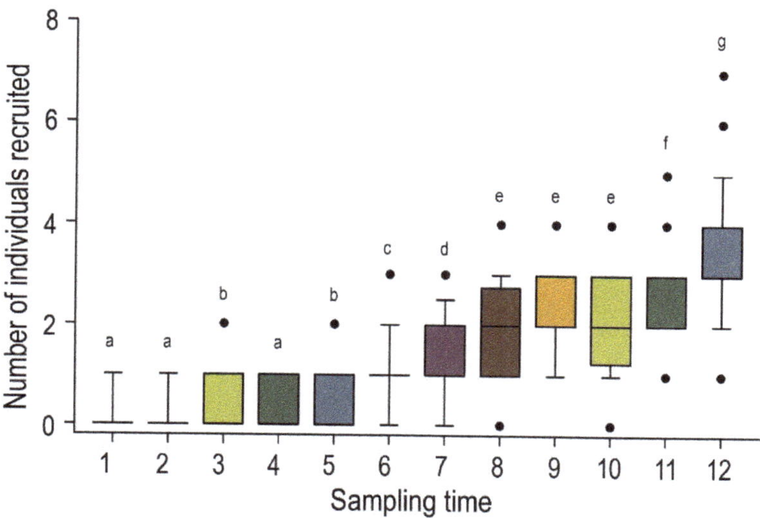

Figure 5. Boxplots of *Trapezia* crab's recruitment on *Pocillopora* fragments. A decrease or absence of *Trapezia* suggests migration between colonies. Letters indicate significant differences between groups.

4. Discussion

Crustaceans are considered the most abundant and diverse group of invertebrates associated with coral reefs [1] and *Trapezia* represents a high-richness and high-occurrence group [38]. Although the *Trapezia–Pocillopora* association and the role of both host–symbiont have been previously documented [8,14,18,38–40], no information was available regarding the crab recruitment process. In order to initiate a symbiotic relationship, the presence of both organisms is necessary; however, in the case of the coral, a recently recruited colony does not represent a habitable space for *Trapezia*, and colony volume and associated available area has been determined as both a limiting and essential factor for the commensals [34,39]. The availability of sufficient space for the presence of *Trapezia* is not attributed to their nutritional requirements or reproductive activity but to the territorial and aggressive behavior that these crabs develop [41]. In the present study, the coral colonies show differences in their volume and thus available area, whilst no differences in the recruitment of *Trapezia* related to the volume of the coral were found. In this context, it could be hypothesized that the probability that *Trapezia* recruit a colony is related to the increasing complexity of its branching morphology and not the volume per se [40]. However, our findings showed that three-dimensionality was not necessary for crab recruitment; instead, the crabs only need a healthy coral fragment without secondary branching, even though this could result in increased *Trapezia* vulnerability due to being relatively unprotected against predators.

Also, independently of the presence of branches or the lack of structural complexity of the coral colonies, crabs were not permanent residents, as indicated by their fluctuating abundance per coral colony throughout the monitoring period. Previously, the territorial behavior of the crab had led to the assumptions that (1) the presence of more than a couple of crabs per colony would be related to the available area [39], (2) that there could be a hierarchy by space within the colony, with larger crabs displacing smaller crabs to the margins of the colony or entirely outside of it [42], as the recruitment of recent recruit crabs depends on an intra-species interaction where the conspecific resident crab should accept the new resident [43], and (3) that in addition to having a high energetic cost, movements or migrations among coral hosts would increase the vulnerability of the crabs to predators [44]. Therefore, besides the area and exposure factors, the distance between hosts, predation pressure, host characteristics (size, abundance), physical stress (temperature) and sex are key factors determining the balance between costs and possible benefits of colony fidelity [7,45]. Predation pressure can be related to the distance between hosts, i.e., when predation pressure is high [46] and distances between hosts increases, the crabs reduce or cease their movements [47]. Here it was observed that the distance between tagged and adult *Pocillopora* colonies was not a limiting factor to prevent the migration of *Trapezia* between hosts (Figure 6), suggesting that at the beginning of the symbiosis, *Trapezia* presents a random distribution among *Pocillopora* corals in the reef, and as the fragment grows, habitat space and complexity and other factors such as those mentioned above can modify the distribution and permanence of crabs.

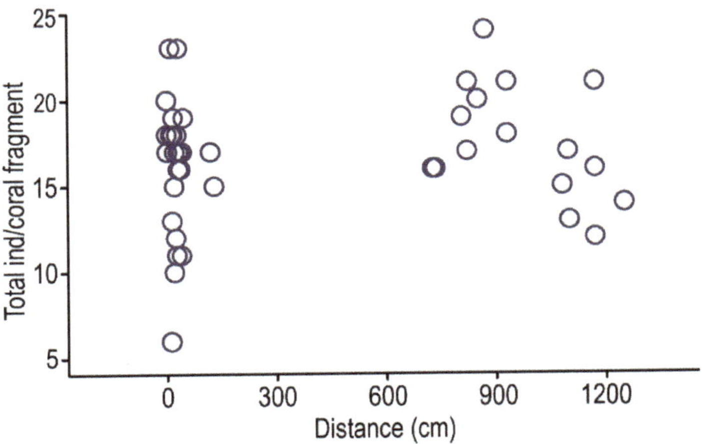

Figure 6. Distances between fragments. Distances represent migration between colonies, as there is no pattern between recruitment and distances; these do not influence the recruitment of *Pocillopora* colonies.

Temperature is one of the main environmental factors that affect physiological processes such as growth, reproduction and recruitment of crustaceans [48]. Even though *Trapezia* is able to acclimate its behavior, biochemistry and physiology in order to compensate for temperature fluctuations, after surpassing their thermo-tolerance limit, high-energy cost activities will be limited or as is the case of reproduction, can be inhibited [49]. *Trapezia* can present behavioral changes in situations of heat stress, leading these crabs to modify their role in their relationship with the coral, causing a break-down of the symbiosis and a shift from mutualistic antagonism towards an antagonistic relationship between members of coral-dwelling communities [50]. In this study, the presence of positive or negative thermal anomalies that could have caused a modification in the behavior of the crustaceans was not evidenced, though it was expected that seasonal changes in temperature would have affected crab recruitment. However, there were no significant changes in the recruitment process of *Trapezia* on *Pocillopora*, though it has been shown previously that invertebrate recruitment in sub-tropical coral communities does not present a seasonal pattern but may have intra-annual differences [22]. Therefore, this fluctuation in *Trapezia* recruitment may have been due to other biological interactions such as predation or the presence of secondary food sources (zooplankton).

The cause and, even more so, the persistence of the symbiotic relationship between the coral and the crustacean seems to be a highly complex process. Our results show that the presence, abundance and residence of *Trapezia* in *Pocillopora* were not determined by the physical characteristics of the substrate (area available) or by their reproduction patterns but were regulated by seasonal temperature fluctuations. Therefore, although the relationship between *Trapezia* and *Pocillopora* remains symbiotic [51] and the presence of the coral determines the distribution of the crab [52], there is not a permanent link at the coral–individual level. Still, understanding the process of recruitment and permanence of *Trapezia* on its host is relevant in determining their mutual subsistence. The *Trapezia–Pocillopora* long-term relationship is primarily defined at the coral community level and not so much at the individual level (colony–crab). The degradation and decline of coral reefs ecosystems also threaten their associated organisms, and the traditional view includes that the permanence of *Trapezia* is regulated by their coral host *Pocillopora*; however, the functional roles of the crustacean as cleaners and protectors from different corallivores and diseases [14,16] may contribute to the ability of *Pocillopora* to resist to multiple stressors and maintain the base of the coral community. Therefore, while healthy corals provide resources for the maintenance of *Trapezia*, at the same time, the presence of these crabs ensures the maintenance and permanence of *Pocillopora* and the biomass that directly depends on them.

Supplementary Materials: The following are available online at https://www.mdpi.com/article/10.3390/d13090450/s1, Table S1: Density of recruited *Trapezia* crabs.

Author Contributions: Conceptualization, H.M.C.-F. and A.P.R.-T.; methodology, H.M.C.-F., A.P.R.-T. and A.L.C.-M.; formal analysis, F.A.R.-Z.; resources, A.P.R.-T. and A.L.C.-M.; writing—original draft preparation, H.M.C.-F. and A.P.R.-T.; writing—review and editing, H.M.C.-F., A.P.R.-T., F.A.R.-Z. and A.L.C.-M.; funding acquisition, A.P.R.-T. and A.L.C.-M. All authors have read and agreed to the published version of the manuscript.

Funding: H.M.C.-F. was funded by a doctoral fellowship from the Consejo Nacional de Ciencia y Tecnologia (CONACYT-No. 632506). The present work was supported by the National Geographic Society (NGS-55349R-19 to APRT) and the project "Programa Integral de Fortalecimiento Institucional" of the Universidad de Guadalajara (P/PIFI-2010-14MSU0010Z-10 to ACM).

Institutional Review Board Statement: Not applicable.

Informed Consent Statement: Not applicable.

Data Availability Statement: All data is included in the manuscript, there is no other data applicable for the purpose of the present manuscript.

Acknowledgments: We thank the authorities of the Islas Marietas National Park (CONANP) for their assistance with field operations Also, the authors thank Diana Morales and Ismael Huerta for image editing, Violeta Martinez for providing photographs of *Trapezia* crabs and Adolfo Tortolero-Langarica for his help in fieldwork. A special thanks to Patricia Briones for her invitation in this Special Issue. Finally, we thank to Nadine Schubert for the English proofreading of the final version of the manuscript.

Conflicts of Interest: The authors declare no conflict of interest.

References

1. Glynn, P.W.; Enochs, I.C. Invertebrates and their roles in coral reef ecosystems. In *Coral Reefs: An Ecosystem in Transition*; Dubinsky, Z., Stambler, N., Eds.; Springer Press: Dordrecht, The Netherlands, 2011; pp. 273–325. [CrossRef]
2. Enochs, I.C.; Glynn, P.W. Corallivory in the Eastern Pacific. In *Coral Reefs of the Eastern Tropical Pacific: Persistence and Loss in a Dynamic Environment*; Glynn, P.W., Manzello, D.P., Enochs, I.C., Eds.; Springer: Berlin/Heidelberg, Germany, 2016; Volume 8. [CrossRef]
3. Garth, J.S. New brachyuran crabs from the Galapagos Islands. *Allan Hancock Pac. Exped.* **1939**, *5*, 9–49.
4. Garth, J.S. The Crustacea Decapoda (Brachyura and Anomura) of Eniwetok Atoll, Marshall Islands, with special reference to the obligate commensals of branching corals. *Micronesica* **1964**, *1*, 137–144.
5. Montano, S. The extraordinary importance of coral-associated fauna. *Divers* **2020**, *12*, 357. [CrossRef]
6. Karlson, R.H. *Dynamics of Coral Communities*, 1st ed.; Kluwer Academic Publishers: London, UK, 1999; pp. 40–55.
7. Castro, P. Movements between coral colonies in *Trapezia ferruginea* (Crustacea: Brachyura), an obligate symbiont of scleractinian corals. *Mar. Biol.* **1978**, *46*, 237–245. [CrossRef]
8. Stella, J.S.; Pratchett, M.; Hutchings, P.; Jones, G. Coral-associated invertebrates: Diversity, ecological importance and vulnerability to disturbance. *Oceanogr. Mar. Biol.* **2011**, *49*, 43–104.
9. Baeza, J.A. Crustaceans as symbionts: An overview of their diversity, host use and life styles. In *Lifestyles and Feeding Biology. The Natural History of the Crustacea*, 1st ed.; Thiel, M., Watling, L., Eds.; Oxford University Press: New York, NY, USA, 2015; Volume 2, pp. 163–189.

10. Montano, S.; Fattorini, S.; Parravicini, V.; Berumen, M.L.; Galli, P.; Maggioni, D.; Arrigoni, R.; Seveso, D.; Strona, G. Corals hosting symbiotic hydrozoans are less susceptible to predation and disease. *Proc. R. Soc. Lond. B Biol. Sci.* **2017**, *284*, 20172405. [CrossRef]

11. Pratchett, M.S. Influence of coral symbionts on feeding preferences of crown-of-thorns starfish *Acanthaster planci* in the western Pacific. *Mar. Ecol. Prog. Ser.* **2001**, *214*, 111–119. [CrossRef]

12. Rouzé, H.; Lecellier, G.; Mills, S.C.; Planes, S.; Berteaux-Lecellier, V.; Stewart, H. Juvenile *Trapezia* spp. crabs can increase juvenile host coral survival by protection from predation. *Mar. Ecol. Prog. Ser.* **2014**, *515*, 151–159. [CrossRef]

13. Pollock, F.J.; Katz, S.M.; Bourne, D.G.; Willis, B.L. Cymo melanodactylus crabs slow progression of white syndrome lesions on corals. *Coral Reefs* **2013**, *32*, 43–48. [CrossRef]

14. McKeon, C.S.; Moore, J.M. Species and size diversity in protective services offered by coral guard-crabs. *PeerJ* **2014**, *2*, 574. [CrossRef] [PubMed]

15. Garth, J.S. Decapod crustaceans inhabiting reef-building corals of Ceylon and the Maldive Islands. *J. Mar. Biol. Assoc. India.* **1974**, *15*, 195–212.

16. Glynn, P.W.; Ault, J.S. A biogeographic analysis and review of the far eastern Pacific coral reef region. *Coral Reefs* **2000**, *19*, 1–23. [CrossRef]

17. Lee, S.Y. Ecology of brachyura. In *Treatise on Zoology-Anatomy, Taxonomy, Biology. The Crustacea*, 1st ed.; Castro, P., Davie, P.J.F., Guinot, D., Schram, F.R., von Vaupel Klein, J.C., Eds.; Brill: Boston, MA, USA, 2015; Volume 9, Part C (2 vols); pp. 469–541.

18. Stewart, H.L.; Holbrook, S.J.; Schmitt, R.J.; Brooks, A.J. Symbiotic crabs maintain coral health by clearing sediments. *Coral Reefs* **2006**, *25*, 609–615. [CrossRef]

19. Mclay, C.L.; Becker, C. Reproduction in brachyura. In *Treatise on Zoology-Anatomy, Taxonomy, Biology. The Crustacea*, 1st ed.; Castro, P., Davie, P.J.F., Guinot, D., Schram, F.R., von Vaupel Klein, J.C., Eds.; Brill: Boston, MA, USA, 2015; Volume 9, Part C (2 vols); pp. 185–243.

20. Hadfield, M.G. Biofilms and marine invertebrate larvae: What bacteria produce that larvae use to choose settlement sites. *Annu. Rev. Mar. Sci.* **2011**, *3*, 453–470. [CrossRef]

21. Sanford, E.; Kelly, M.W. Local adaptation in marine invertebrates. *Annu. Rev. Mar. Sci.* **2011**, *3*, 509–535. [CrossRef] [PubMed]

22. Rodríguez-Troncoso, A.P.; Rodríguez-Zaragoza, F.A.; Mayfield, A.B.; Cupul-Magaña, A.L. Temporal variation in invertebrate recruitment on an Eastern Pacific coral reef. *J. Sea Res.* **2019**, *145*, 8–15. [CrossRef]

23. Patton, W.K. Community structure among the animals inhabiting the coral *Pocillopora damicornis* at Heron Island Australia. In *Symbiosis in the Sea*, 1st ed.; Vernberg, W., Ed.; Univ. South Carolina Press: Columbia, SC, USA, 1974; pp. 219–243.

24. CONANP (Comisión Nacional de Áreas Naturales Protegidas). *Programa de Conservación Y Manejo, Parque Nacional Islas Marietas*; Secretaria de Medio Ambiente y Recursos Naturales (SEMARENAT): México City, México, 2007; pp. 21–57.

25. Wyrtki, K. Surface currents of the eastern tropical Pacific Ocean. *Bull. Inter-Am. Trop. Tuna Commn.* **1965**, *9*, 268–305.

26. Plata, L.; Filonov, A. Marea interna en la parte noroeste de la Bahía de Banderas México. *Cienc. Mar.* **2007**, *33*, 197–215.

27. Portela, W.; Beier, E.; Barton, E.D.; Castro, R.; Godínez, V.; Palacios Hernández, E.; Fiedler, P.C.; Sánchez-Velazco, L.; Trasviña, A. Water masses and circulation in the tropical pacific off Central Mexico and surrounding areas. *J. Phys. Oceanogr.* **2016**, *46*, 3069–3081. [CrossRef]

28. Reyes-Bonilla, H.; Calderón Aguilera, L.E.; Cruz-Piñón, G.; Medina-Rosas, P.; López-Perez, R.A.; Herrero-Perezrul, M.D.; Leyte-Morales, G.E.; Cupul-Magaña, A.L.; Carriquiry-Beltran, J.D. *Atlas de Corales Pétreos (Anthozoa: Scleractinia) del Pacífico Mexicano*, 1st ed.; Sociedad Mexicana de Arrecifes Coralinos AC, CICESE, CONABIO, CONACYT, DBM/UABCS, CUC/UdeG, Umar: Mexico City, Mexico, 2005; p. 129.

29. Hernández-Zulueta, J.; Rodríguez-Zaragoza, F.A.; Araya, R.; Vargas-Ponce, O.; Rodríguez-Troncoso, A.P.; Cupul-Magaña, A.L.; Díaz-Pérez, L.; Ríos-Jara, E.; Ortiz, M. Multi-scale analysis of hermatypic coral assemblages at Mexican Central Pacific. *Sci. Mar.* **2017**, *81*, 91–102. [CrossRef]

30. Sotelo-Casas, R.C.; Rodríguez-Troncoso, A.P.; Rodríguez-Zaragoza, F.A.; Solís-Marín, F.A.; Godínez-Domínguez, E.; Cupul-Magaña, A.L. Spatial-temporal variations in echinoderm diversity within coral communities in a transitional region of the northeast of the eastern pacific. *Estuar. Coast. Shelf Sci.* **2019**, *227*, 106346. [CrossRef]

31. Cruz-García, R.; Rodríguez-Troncoso, A.P.; Rodríguez-Zaragoza, F.A.; Mayfield, A.; Cupul-Magaña, A.L. Ephemeral effects of El Niño–Southern Oscillation events on an eastern tropical Pacific coral community. *Mar. Freshw. Res.* **2020**, *71*, 1259–1268. [CrossRef]

32. Edwards, A.J. *Reef Rehabilitation Manual*; The Coral Reef Targeted Research and Capacity Building for Management Program: Queensland, Australia, 2010; pp. 64–108.

33. Tortolero-Langarica, J.J.A.; Cupul-Magaña, A.L.; Rodríguez-Troncoso, A.P. Restoration of a degraded coral reef using a natural remediation process: A case study from a Central Mexican Pacific National Park. *Ocean Coast. Manag.* **2014**, *96*, 12–19. [CrossRef]

34. Barry, C.K. Ecological Study of the Decapod Crustaceans Commensal with the Branching Coral *Pocillopora meandrina* var. Nobilis Verrill. Master's Thesis, University of Hawaii, Honolulu, HI, USA, 1965.

35. Highsmith, R.C. Reproduction by fragmentation in corals. *Mar. Ecol. Prog. Ser.* **1982**, *7*, 207–226. [CrossRef]

36. Anderson, M.J.; Gorley, R.N.; Clarke, K.R. *PERMANOVA+ for PRIMER: Guide to Software and Statistical Methods*; PRIMER-E: Plymouth, UK, 2008; p. 214.

37. Clarke, K.R.; Gorley, R.N. *Primer V6: User Manual/Tutorial*; PRIMER-E: Plymouth, UK, 2006; p. 192.

38. Castro, P. Animal symbioses in coral reef communities: A review. *Symbiosis* **1988**, *5*, 161–184.

39. Abele, L.G.; Patton, W.K. The size of coral heads and the community biology of associated decapod crustaceans. *J. Biogeogr.* **1976**, 35–47. [CrossRef]
40. Stewart, H.L.; Price, N.N.; Holbrook, S.J.; Schmitt, R.J.; Brooks, A.J. Determinants of the onset and strength of mutualistic interactions between branching corals and associate crabs. *Mar. Ecol. Prog. Ser.* **2013**, *493*, 155–163. [CrossRef]
41. Preston, E.M. A computer simulation of competition among five sympatric congeneric species of xanthid crabs. *Ecology* **1973**, *54*, 469–483. [CrossRef]
42. Huber, M.E. Aggressive behavior of *Trapezia* intermedia miers and *T. digitalis* Latreille (Brachyura: Xanthidae). *J. Crust. Biol.* **1987**, *7*, 238–248. [CrossRef]
43. Vannini, M. A shrimp that speaks crab-ese. *J. Crust. Biol.* **1985**, *5*, 160–167. [CrossRef]
44. Roughgarden, J. Evolution of marine symbiosis—A simple cost-benefit model. *Ecology* **1975**, *56*, 1201–1208. [CrossRef]
45. Thiel, M.; Zander, A.; Baeza, J.A. Movements of the symbiotic crab *Liopetrolisthes mitra* between hosts, black sea urchins *Tetrapygus niger*. *Bull. Mar. Sci.* **2003**, *72*, 89–101.
46. Knowlton, N. Sexual selection and dimorphism in two demes of a symbiotic, pair-bonding snapping shrimp. *Evolution* **1980**, *34*, 161–173. [CrossRef]
47. Bell, J.L. Changing residence: Dynamics of the symbiotic relationship between *Dissodactylus mellitae* Rathbun (Pinnotheridae) and *Mellita quinquiesperforata* (Leske) (Echinodermata). *J. Exp. Mar. Biol. Ecol.* **1984**, *82*, 101–115. [CrossRef]
48. Hartnoll, R.G.; Bryant, A.D. Growth to maturity of juveniles of the spider crabs *Hyas coarctatus* Leach and *Inachus dorsettensis* (Pennant) (Brachyura: Majidae). *J. Exp. Mar. Biol. Ecol.* **2001**, *263*, 143–158. [CrossRef]
49. Azra, M.N.; Aaqillah-Amr, M.A.; Ikhwanuddin, M.; Ma, H.; Waiho, K.; Ostrensky, A.; Prestes dos Santos Tavares, C.; Abol-Munafi, A.B. Effects of climate-induced water temperature changes on the life history of brachyuran crabs. *Rev. Aquac.* **2020**, *12*, 1211–1216. [CrossRef]
50. Stella, J.S.; Munday, P.L.; Walker, S.P.W.; Pratchett, M.S.; Jones, G.P. From cooperation to combat: Adverse effect of thermal stress in a symbiotic coral-crustacean community. *Oecologia* **2013**, *174*, 1187–1195. [CrossRef] [PubMed]
51. Tsuchiya, M.; Taira, A. Population structure of six sympatric species of *Trapezia* associated with the hermatypic coral *Pociliopora damicornis* with a hypothesis of mechanisms promoting their coexistence. *J. Jpn. Coral Reef Soc.* **1999**, 9–17. [CrossRef]
52. Knudsen, J.W. *Trapezia* and *Tetralia* (Decapoda, Brachyura, Xanthidae) as obligate ectoparasites of pocilloporid and acroporid corals. *Pac. Sci.* **1967**, *21*, 51–57.

Review

Lobster Distribution and Biodiversity on the Continental Shelf of Brazil: A Review

Raul Cruz [1,*], Marina T. Torres [1], João V. M. Santana [2] and Israel H. A. Cintra [3]

1   Fundação Cearense de Apoio ao Desenvolvimento Científico e Tecnológico, Av. Oliveira Paiva 941, Cidade dos Funcionários, Fortaleza 60822-130, Brazil; marinatorresrodriguezm@gmail.com
2   Instituto Federal de Educação, Ciências e Tecnologia do Ceará (IFCE), Campus de Acaraú. Av. Des. Armando de Sales Louzada, s/n, Acaraú 60000-000, Brazil; joao.vicente@ifce.edu.br
3   Instituto Socioambiental e dos Recurso Hídricos, Universidade Federal Amazonia (UFRA), Ave. Presidente Tancredo Neves, Belém 60000-000, Brazil; israel.cintra@ufra.edu.br
*   Correspondence: rcruzizquierdo@gmail.com

Abstract: The continental shelf of Brazil is home to a wide range of lobster species, with varying body size, color, habitat preference, and geographic and bathymetric distribution. Spiny lobsters (*Panulirus*) and slipper lobsters (Scyllaridae) are exploited for export and for the domestic market. Deep sea lobsters (Nephropidae and Polychelidae) have no commercial potential, and little is known about their biology. In this review, we identified 24 lobster species from benthic ecosystems off Brazil (Palinuridae 25%, Scyllaridae 29%, Nephropidae 25%, Polychelidae 17%, Enoplometopidae 4%). We designed a simplified theoretical scheme to understand the role of lobsters in the ecosystem, based on available evidence of distribution, biodiversity, life cycle, connectivity, and abundance. Finally, we propose a theoretical scheme of trophic top-down control, with interactions between a large decapod (spiny lobster), a demersal predator (red snapper), an apex predator (small tuna), benthic invertebrates and fishing exploitation.

Keywords: lobster; life cycle; predator-prey; food chain; Brazil

Citation: Cruz, R.; T. Torres, M.; Santana, J.V.M.; Cintra, I.H.A. Lobster Distribution and Biodiversity on the Continental Shelf of Brazil: A Review. *Diversity* **2021**, *13*, 507. https://doi.org/10.3390/d13110507

Academic Editors: Patricia Briones-Fourzán, Michel E. Hendrickx and Michael Wink

Received: 28 August 2021
Accepted: 15 October 2021
Published: 20 October 2021

Publisher's Note: MDPI stays neutral with regard to jurisdictional claims in published maps and institutional affiliations.

1. Introduction

A highly coveted marine resource for over 100 years, marine lobsters sustain one of the most profitable artisanal fishing industries in the world. In Brazil, exports of whole lobster were on the rise since 2013, with USD 92.46 million in revenues in 2019 [1].

Lobsters fall into several families: spiny lobsters (Palinuridae), slipper lobsters (Scyllaridae), flaming reef lobsters (Enoplometopidae), clawed lobsters (Nephropidae), and blind lobsters (Polychelidae) [2]. Despite their morphological differences, all the above are referred to as 'lobsters'.

Marine lobsters in the tropical western Atlantic belong to three infraorders of crustacean decapods: Achelata Scholtz & Richter, 1995, Astacidea Latreille, 1802, and Polychelida Scholtz & Richter, 1995. Achelata includes not only spiny lobsters (family Palinuridae, Latreille 1802), but also slipper lobsters (family Scyllaridae, Latreille 1825) and one species of furry lobster (currently classified in the family Palinuridae). In contrast, Astacidea (Nephropidae Dana, 1852 and Enoplometopidae Saint Laurent, 1988) have chelae on the first three pairs of pereiopods, while Polychelida (Polychelidae Wood-Mason, 1875), known as blind lobsters, have chelae on four or even all five pairs of pereiopods [3,4].

The taxonomy and diversity of marine lobsters in the tropical western Atlantic is reviewed in this paper, lobster feed into areas of ongoing research in Brazil. Worth highlighting is the analysis of lobster diversity and communities published by Silva et al. [5] for the Great Amazon Reef System (GARS). We also provide the first record in the GARS of *Scyllarus chacei* Holthuis, 1960, based on species with prehatch eggs, extending the geographic range for this species.

Silva, et al. [4] filled important gaps in our knowledge of crustacean biodiversity on the continental shelf off the Amazon River. Cruz, et al. [6] described the life cycle of spiny lobsters (*Panulirus* spp.) and connectivity between oceanic regions, but little is known about the dynamics of recruitment and migration and the influence of climate and fisheries on lobster biodiversity [7,8].

Spiny lobsters (*Panulirus* White, 1847) are heavily exploited in commercial and recreational fisheries in the Wider Caribbean and along the coast of Brazil. The slipper lobster (Scyllaridae) is occasionally caught as by-catch in spiny lobster traps or shrimp trawls and sold on the Brazilian domestic market. In the waters off southern Brazil, clawed lobsters are a common by-catch in shrimp trawls. The species *Metanephrops rubellus* (Moreira, 1903), sometimes referred to as langoustine, Uruguayan lobster, or deep-sea lobster, is a valuable export commodity [9].

Fishing, coastal infrastructure developments, and water pollution are the anthropogenic factors that most impact marine ecosystems. Pollution is most harmful to organisms in the early stages of the life cycle (eggs, larvae, nursery areas, juveniles). Urban developments modify the natural biotope (coastal area), with repercussions on migration, spawning, and nurseries. Fishing directly or indirectly alters and degrades marine ecosystems (especially in coastal regions where it tends to be more intense and combined with other anthropogenic disturbances) and the depletion of fishing stocks poses a threat to biodiversity [10]. In such scenarios, trophic and nontrophic interactions may be severely compromised [11], as observed for decapods [12].

In Brazil, with the enforcement of management measures in the period 2012–2020, fishing pressure on spiny lobster stocks decreased and a tendency for recovery was observed [13]. However, using a variety of methods and gear (for example, rectangular baited traps with one or two entrances and bottom gillnets [14]), artisanal lobster fishermen still cause significant damage to benthic habitats and the depletion of nontarget species [15]. In addition, artificial shelters ('marambaias') of varying design were since 2000 deployed on fishing grounds off northeastern Brazil [16]. As the gear is nonselective, a variety of nontargeted organisms are captured as by-catch (commercial fish, mollusks, crabs, etc.), including species that prey on lobsters. Moreover, fishermen often harvest lobsters from artificial shelters or reefs by hook or hand and using a compressor connected to the boat's motor [17], in detriment to the health of both divers and the benthic ecosystem.

In this review, we address aspects of the distribution, biodiversity, connectivity, length assessment and life cycle of lobster species occurring on the continental shelf of Brazil (CSB). For this reason, we review publications, analyze information on species, search databases, and look into field surveys stratified by depth. Although available information on slipper, clawed, and blind lobsters is scarce, limiting our knowledge of their biology, ecology, and population dynamics, we believe the material gathered and discussed in this review is highly relevant for both researchers and fishing resource managers in Brazil. We also look at time series of spiny lobster and red snapper landings to understand more about predator-prey interactions. Finally, we identify major knowledge gaps and offer suggestions for future studies on this emerging and highly promising research topic.

2. Material and Methods

2.1. Lobster Site Description

Lobsters are distributed throughout the Brazilian continental shelf (CSB) from Amapá (04°26′ N 51°32′ W) to Rio Grande do Sul (29°46′ S 53°10′ W). Lobster stocks are concentrated in four regions along the Brazilian coast: the North (Great Amazon Reef System), the Northeast (NE), the Southeast (SE), and the South (S).

In the North, Moura, et al. [18] described the Great Amazon Reef System (GARS) as a 1000 km-long mesophotic coral reef ecosystem under low-light conditions, located approximately 80 km from the mouth of the Amazon River and covering about 9,500 km², bridging the gap between the border of Amapá (5° N 1° S to 44° N 51° W) and the coast of Maranhão (0° S 46° W to 2° S 41° W), at a depth of 30 m to 90–120 m (shelf break).

The continental shelf off Northeastern Brazil (NE) (2° S 4° W to 9° S 35° W, Pernambuco) is relatively wide and shallow (3–60 m), followed by an abrupt drop-off starting at a depth of approximately 60 m. Lobster habitats are diverse, including reef structures and calcareous algae beds, especially off Ceará and Rio Grande do Norte, possibly due to the absence of large rivers and, consequently, small amounts of fluvial discharge [19].

Between Pernambuco and Espirito Santo (9° S 35° W to 21° S 40° W) in Southeastern Brazil (SE), the shelf is narrow and lobster landings are small.

The climate in Southern Brazil (S) (34° S to 22° S) is subtropical to temperate, with complex topographic and hydrodynamic regimes. Lobster habitats at varying latitudes and depths of the CSB are often covered with crustose coralline algae (rhodoliths) [20,21]. The entire shelf measures almost 270,000 km^2, with a maximum width of 70–230 km and a break at 100–180 m depth [22].

2.2. Length Assessment

The most accurate way to measure lobsters is by carapace length (CL, the distance between the ridge of the supraorbital spines and the posterior end of the carapace). In this study, we used total length (TL, the distance between the rostrum and the posterior edge of the telson) for the genus *Panulirus*. Fixed ratios between the different body parts make it possible to convert CL into TL, and vice-versa. Such equations were published by Cruz [23] for the species *P. argus*. The equations are as follows:

(a)   Males: TL = 32.8431 + 2.5197 * CL.
(b)   Females: TL = 36.3535 + 2.5402 * CL.
(c)   Total (both sexes): TL = 9.8740 + 2.9290 * CL.
(d)   Males: CL = −11.7974 + 0.3925 * TL.
(e)   Females: CL = −2.3439 + 0.3374 * TL
(f)   Total (both sexes): CL = −11.6569 + 0.3838 * TL.

Wt = 0.002065 × CL$^{2.7920}$ where Wt is total weight expressed in grams (g)

When the maximum carapace length (MCL) was reported, the above equations were used to convert MCL to MTL (maximum total length).

2.3. Lobster Data Collection and Time Series

During fishery trials in the Great Amazon Reef System between 1996 and 1998, lobsters were collected with bottom trawl shrimp nets at different times of the day and night, on different types of sediments, and stratified by depth (41–626 m), as part of the REVIZEE Program (Assessment of Sustainable Potential of Living Resources in the Exclusive Economic Zone). The species were identified at the CEPNOR Crustacean Laboratory and were deposited in the collection of crustaceans of the National Center for Research and Conservation of Northern Marine Biodiversity (CEPNOR) and the Chico Mendes Institute for Biodiversity Conservation (ICMBio). The species were classified according to the check list of marine lobsters by Chan [24] and the guidelines by Silva et al. [4,5], and grouped by family (Palinuridae, Scyllaridae, Nephropidae and Polychelidae) and depth. The sex of each specimen was determined, whereas the total length (TL) and carapace length (CL) were measured with a 300 mm Vernier caliper.

In the period of 1999–2000, monthly random field surveys were undertaken on the fishing grounds off Ceará at 10–50 m (2°53′22″ S 41°13′46″ W to 4°06′06″ S 38°08′56″ W) using traps and gillnets. The total size composition of lobsters caught with traps and gillnets in shallow waters was used to calculate the maximum carapace length of the target groups (the genus Panulirus and the family Scyllaridae).

In both regions, maximum body length was recorded for each lobster species, with habitats stratified by depth, to determine the maximum growth age. When only one specimen was available, its length was used as maximum value.

Currently, annual lobster landing data (1960–2020) are provided by major fishing companies, exporters, and producers affiliated with Sindfrio (a union of fishing industries and cold chains in the State of Ceará), representing 63% of the national lobster industry in

2018–2020. Figures from nonmember fishing companies were obtained from the Comex Stat database. Brazilian government agencies no longer keep yearly updated databases of red snapper (*Lutjanus purpureus*, Poey, 1866) landings, but a time series is available for the period 1960–2004.

To test for possible associations between red snapper and Brazilian red spiny lobster *Panulirus argus* (Latreille, 1804) production, we performed a cross-correlation analysis [25] of the annual production (t) and the lag tendency. Diggle [25] suggested the approximate formula $2/\sqrt{n}$, where n is the total number of months ($n = 34$), to assess the significance at 95% of the cross-correlation function if the calculated value exceeds the equation.

3. Results and Discussion

3.1. Connectivity of Spiny Lobsters (Panulirus spp.)

Available evidence from Cruz, et al. [6] supports the hypothesis that the *P. argus* pueruli found in the Lesser Antilles originate from larvae carried by rings of the North Brazil Current. In addition, phyllosomata of the green spiny lobster *Panulirus laevicauda* (Latreille, 1817) collected in the Caribbean Sea constitute compelling evidence of south-to-north connectivity between lobster populations in the Western Atlantic, contradicting the view that the Amazon River and the Orinoco River represent insurmountable biogeographic barriers [3]. This is supported by the fact that *P. laevicauda* is more abundant in Brazil than in the Caribbean and by the discovery by Rocha et al. [26] that populations of *Halichoeres radiatus* (Linnaeus, 1758) (Labridae) in the Caribbean and at offshore islands in the southwestern Atlantic are genetically identical.

The complex life cycle of spiny lobsters involves the occupation of habitats on continental and insular shelves, islands, and banks. Due to their year-round reproductive activity, large patches of phyllosomata of variable density are constantly moving with the currents, coinciding with the plankton distribution [6,8]. Current observations allow to infer that (i) larval recruitment is inversely associated with the Amazon-Orinoco River discharge, (ii) the range of salinity required for phyllosoma survival (33–35‰) is compatible with the salinity gradients reported for oceanic waters, and (iii) the North Brazil Current retroflexion supports self-recruitment [6]. Conceivably, patches of lobster larvae become trapped in the GARS and NE, leading to larval transport, puerulus settlement, and self-recruitment.

The biogeographic barrier adduced to explain allopatric speciation, as in the case of the newly proposed species *Panulirus meripurpuratus* [27], does not seem to prevent regular larval migration from south to north, as previously claimed by Sarver et al. [28,29]. It should be kept in mind that the GARS is an extensive mesophotic reef ecosystem at a depth between 50 and 100 m, located approximately 80 km from the mouth of the Amazon River and heavily exploited for fish and lobster [18].

The fact that the Brazilian lobster species *P. meripurpuratus* was caught in Venezuela and Florida [29], Buck Island St Croix, USVI (pers. comm. by Tom Mathews) and in the Mexican Caribbean [30] makes it difficult to sustain the segregation hypothesis of Giraldes & Smyth [27] (Figure 1). The presence of this species in the Caribbean region raises the possibility of hybridization between species or subspecies. We also looked into the possible existence in Brazil of hybrids between sympatric lobster species of the genus *Panulirus*, as suggested by differences in color patterns and morphology [6]. To help understand the relationship between size, feeding behavior, and nutrition preferences stratified by depth, more research on carotenoid pigments in the carapace, abdomen, and legs would be useful.

Most available biogeographic maps of panulirid species [27] are inexplicit and do not show in detail how biological taxa are spatially distributed, whereas the information gathered by Cruz, et al. [6] provides a fairly reliable picture of the abundance, distribution, and depth preferences of each species. The incorporation of data from fishing expeditions allows a more realistic view of the dynamics of lobster stocks in Brazil and the Wider Caribbean (Figure 1). The question marks on the maps (Figure 1A) indicate the absence of lobsters, contradicting the actual distribution of the species in the GARS (Figure 1C).

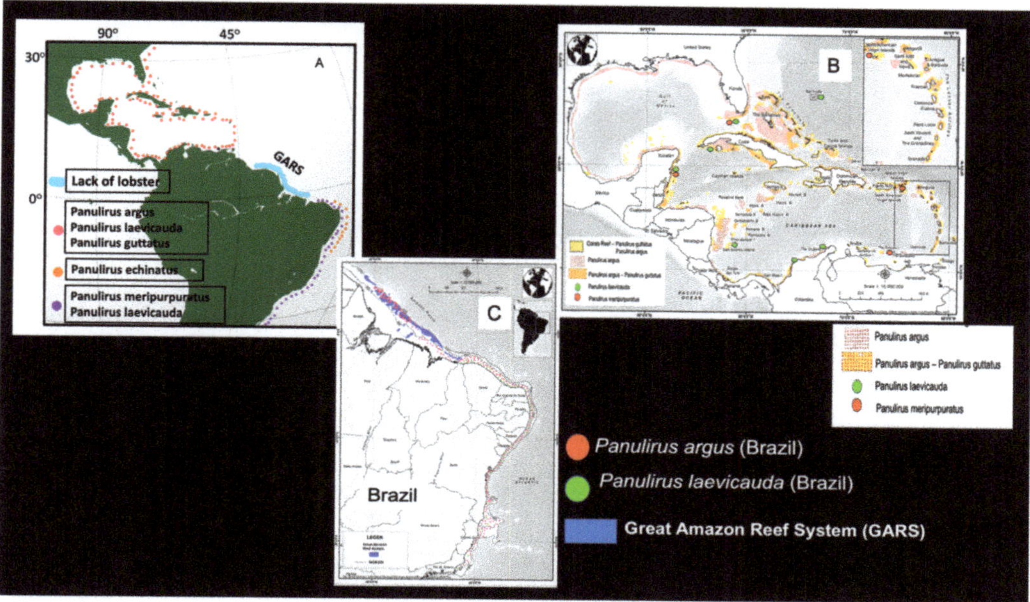

Figure 1. Spatial distribution of lobsters of genus *Panulirus* (White, 1847). (**A**) Map adapted from Giraldes & Smyth [27]. (**B**) Map adapted from Cruz et al. [6]. (**C**) Map adapted from Cruz et al. [13].

However, since the Caribbean spiny lobster and *P. meripurpuratus* are not completely segregated geographically, we believe the hypothesis of oceanic connectivity (spiny lobster larvae migrating between populations on either side of the Amazon River) is warranted and lends support to the paradigm of two subspecies. Giraldes & Smith [27] did not present compelling evidence to substantiate the notion of two different panulirid species, and genetic data alone are insufficient to redefine a species; in fact, the use and interpretation of genetic data depend on our concept of species [31]. We therefore chose to employ the name *P. argus* for both the Brazilian form of red spiny lobster and the form of Caribbean spiny lobster which occurs in the Wider Caribbean.

The co-occurrence of *P. argus* and *P. laevicauda* in the same geographic regions needs to be investigated with molecular analysis and morphological comparisons between species collected in the north and south of GARS, NE, SE, and in different regions of the Caribbean (Mexico Caribbean, Florida, the Colombian continental coast and the Lesser Antilles). Such studies will help clarify the identity of species and potential subspecies.

3.2. Lobster Distribution and Diversity

The benthic ecosystems along the Brazilian coast are inhabited by at least 24 lobster species belonging to 5 families, most importantly Palinuridae (6 species; 25%) and Scyllaridae (7 species; 29%). Nephropidae (25%) includes 6 species and 3 genera. Polychelidae (17%) (deep-sea blind lobsters) is represented by 4 species, and Enoplometopidae is represented by one (4%). These 24 species are widely distributed on the CSB.

Recently, Silva, et al. [4,5] surveyed the lobster populations of the deep waters of the GARS. The Brazilian form of *P. argus* is the most exploited panulirid in the Northern and Central GARS [5], while *Palinustus truncatus* A. Milne Edwards, 1880 is rarely captured [32] (Figure 2) and *P. laevicauda* and *Panulirus echinatus* Smith, 1869 (painted lobster) are absent in this region. Inshore habitats are unattractive for lobsters due to low salinity and pH, turbidity, and silt from the Amazon River [18]. Juveniles and adults are absent from coastal waters [6].

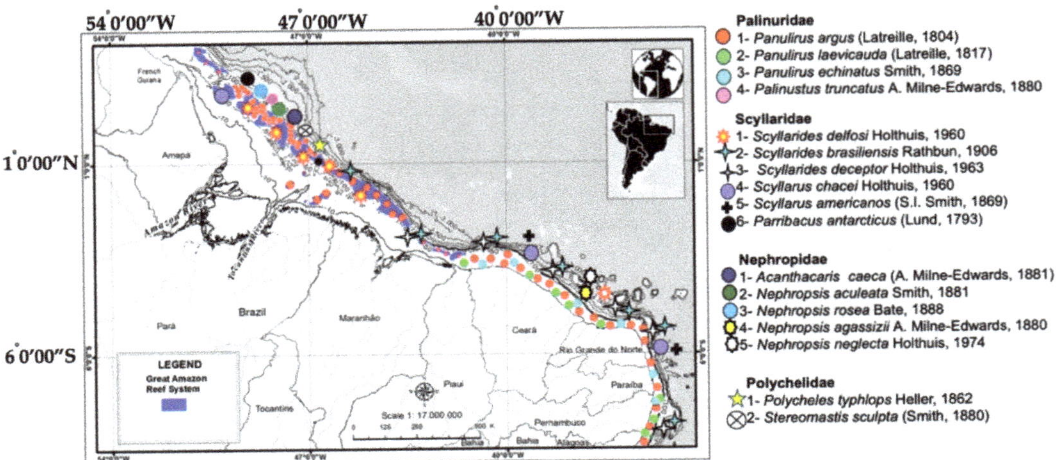

Figure 2. Spatial distribution and diversity of lobsters in Great Amazon Reef System (GARS) and in northeastern section (NE) of continental shelf of Brazil (CSB).

The genus *Scyllarides* is represented by 13 species [33] distributed worldwide. The slipper lobsters *Scyllarides delfosi* Holthuis, 1960, *Parribacus antarcticus* (Lund, 1793) (Figure 2), *Scyllarides brasiliensis* Rathum, 1906, and *Scyllarides deceptor* Holthuis, 1963 occur in NE and SE, from Maranhão to São Paulo [34,35], although Tavares et al. [36] included the coastline between Rio de Janeiro and Buenos Aires (23.5°S to 39°S) in the distribution and reported taxonomic differences between slipper lobsters of similar color patterns (Figures 2 and 3). These lobsters are caught incidentally by pink shrimp trawlers [36–38]. *Scyllarus ramosae* [39] is an invalid taxon in Chan [24]. *Scyllarides aequinoctialis* (Lund, 1793) is common in southern Brazil [3,40], but some authors believe they occur along the entire coast of Brazil [41,42]. Nevertheless, the species is not mentioned in the surveys of Coelho & Ramos Porto [35], Melo [34], Dall'Occo [43,44], Coelho et al. [45], or Serejo et al. [46]. According to the latter authors, the species is restricted to the wider Caribbean, but genetic analyses are necessary to confirm this.

The genus *Scyllarus* [47] is restricted to the Atlantic Ocean and comprises nine species, of which four (44%) were recorded from Brazilian waters: *Scyllarus chacei* Holthuis, 1960 occurs in the GARS [47], in Bahia and in Rio de Janeiro [43]. *Scyllarus americanus* (S.I. Smith, 1869) and *Scyllarus depressus* (S.I. Smith, 1881) are known from SE [39] (Figures 2 and 3).

Nephropidae (clawed lobsters) includes seven species, three of which in the GARS: *Acanthacaris caeca* (A. Milne-Edwards, 1881), *Nephrosis aculeata* Smith, 1881, and *Nephrosis rosea* Bate, 1888). The species *Nephrosis agassizii* (A. Milne-Edwards, 1880) and *Nephrosis neglecta* Holthuis, 1974 occur in NE and SE [48], while *N. aculeata* and *N. rosea* are reported from SE [33,43,46,49,50] (Figures 2 and 3).

The family Polychelidae (blind lobsters) is represented by four species on the CSB, out of a total of 38 species in the world [51]. *Stereomastis sculpta* (Smith, 1880) and *Polycheles typhlops* C. Heller, 1862 are found throughout the GARS [9], but the latter also occurs on the central Brazilian slope (11° to 22° S), along with *Pentacheles laevis* Bate, 1878 and *Pentacheles validus* A. Milne Edwards, 1880 [50] (Figures 2 and 3). *Polycheles sculptus* [50] is an invalid taxon in Chan [24].

Justitia longimanus H. Milne Edwards, 1837 (Palinuridae) is small and rather uncommon, while *Enoplometopus antillensis* Lütken, 1865 (Enoplometopidae) occurs in southern Brazil [3]. Finally, *Palinurellus gundlachi* von Martens, 1878 (Palinuridae) occurs in NE [3,43] from Paraiba to Alagoas. The last two species were first recorded at the marine reserve at Rocas Atoll [52].

The Uruguayan lobster (Nephropidae), *Metanephrops rubellus* (Moreira, 1903), is found between 22° S (Rio de Janeiro) and 26° S (Paraná) [53] (Figure 3).

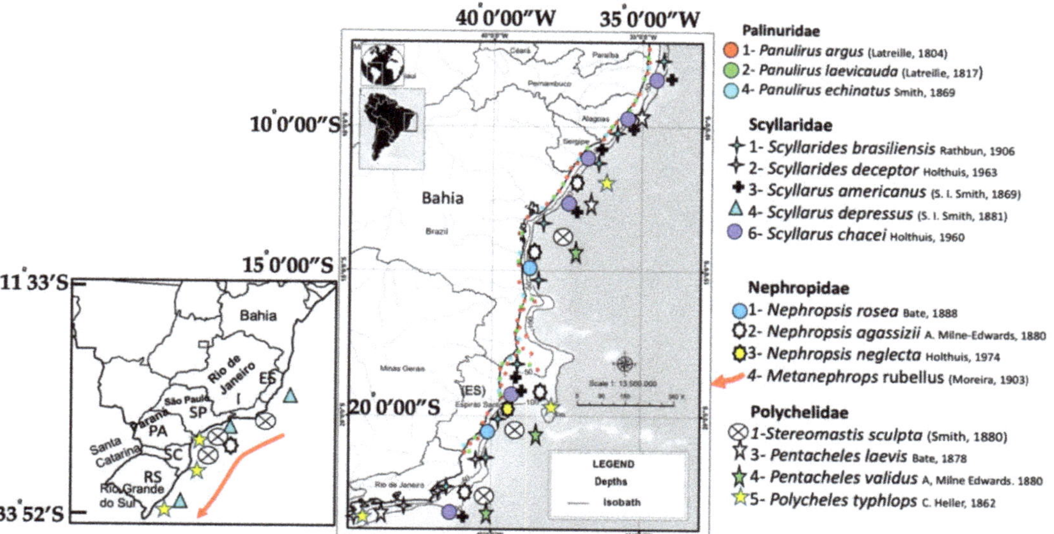

Figure 3. Spatial distribution and diversity of lobsters in southeastern (SE) and southern (S) sections of continental shelf of Brazil (CSB).

Appendix A provides a list of the species surveyed, with author and taxon date. Appendix B credits the source photographs [54–62].

3.3. Bathymetric Distribution

With approximately 7491 km of coastline, the CSB comprises some of the world's most important lobster fishing grounds (Figure 4). *S. brasiliensis* and the Brazilian form of *P. argus* are the only species found in all environments along the shelf. The main target of commercial fisheries, red spiny lobster, is captured at depths between 40 m and 110 m in the GARS [5], and between 10 and 50 m in NE and SE [6].

P. laevicauda and *P. echinatus* coexist in some regions but differ with regard to depth distribution. The former is found at 1–40 m (no exploratory survey registered it at depths greater than 50 m) [50], while the latter was only observed in shallow waters (1–25 m) and almost always in association with reef structures [34]. *P. truncatus*, a rare species restricted to the GARS, with no commercial potential [5], prefers deeper waters [50,63,64] (Figure 4).

Lobsters predominantly dwell on substrates of benthic calcareous algae [65], including red algae (Rhodophyceae) and green algae (Chlorophyceae). Slipper lobsters are found at variable depths and on a wide variety of substrates, especially on muddy sand and gravel bottoms and organogenic rock [5,44]. Species of the genera *Scyllarides* and *Scyllarus* are common on the mesopelagic slope at 200–1000 m, except *S. americanus* and *P. antarcticus* (Figure 4).

In southern Brazil, shrimp trawlers catch the species *M. rubellus* at 50–270 m [53]. The remaining species have no commercial potential and are mostly found between 100 and 1000 m depth (Figure 4).

Finally, deep-sea blind lobsters (Polychelidae) occur in the GARS and in southeastern and southern Brazil (Figure 4), but very little is known about their life cycles.

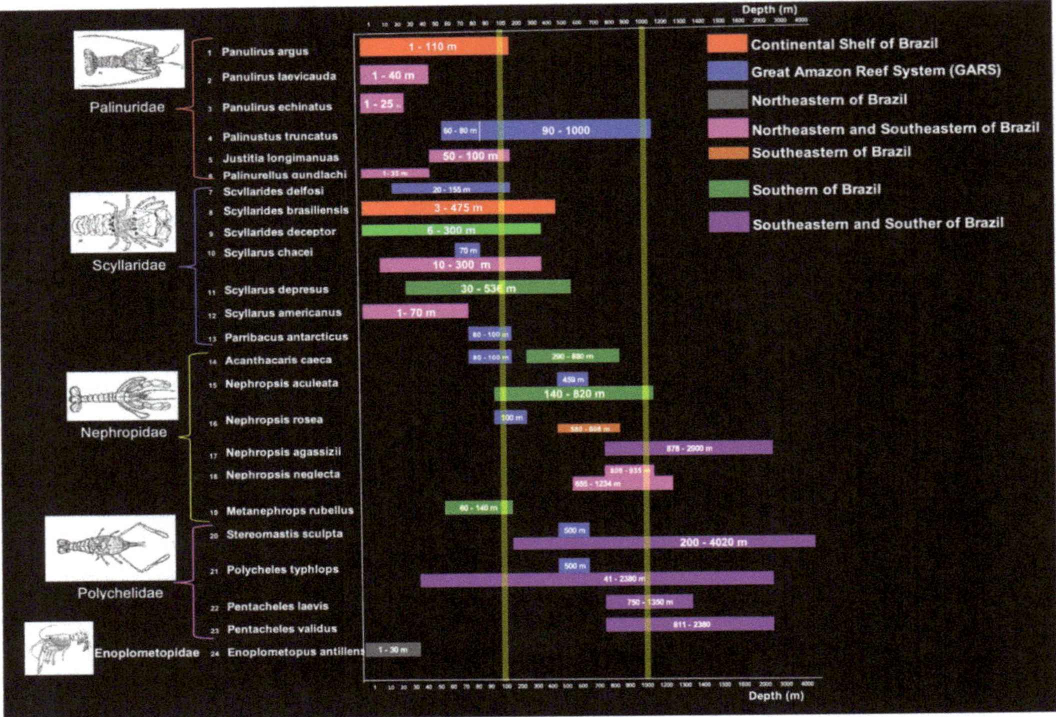

Figure 4. Bathymetric distribution of lobster species according to sector of continental shelf of Brazil (CSB).

3.4. Maximum Total Length (MTL) of Lobster Species

Knowledge of the MTL of different lobster species can help researchers to clarify important aspects of bioecology and population dynamics (longevity, clutch size, growth, reproductive potential, predator size, maximum catch size, etc.) and subsidize the implementation of regulatory measures.

The 24 lobster species occurring on the CSB display a wide range of body length (30–620 mm MTL). MTL is not associated with habitat depth (Figure 5, Table 1).

The mean MTL (MTL ± 95% confidence interval) was 313.60 ± 175.63 mm (range: 118–620) for Palinuridae, 157.00 ± 59.67 mm (range: 30–270) for Scyllaridae, 164.17 ± 57.17 mm (range: 110–300) for Nephropidae, and 80.83 ± 30.71 mm (range: 47–150) for Polychelidae. The largest mean MTL was that of the genus *Panulirus*: 433.33 ± 183.33 mm (range: 328–620) (Table 1).

Oversized ('giant') red spiny lobsters (620 mm, MTL) are the largest crustaceans and the top predators in benthic marine ecosystems on the CSB and possibly in other parts of the Western Tropical Atlantic (Figure 5). For example, animals measuring 802 mm MTL and weighing ~6 kg were observed in a marine protected area off Bermuda (Table 1). The giant Brazilian form of *P. argus* is the most abundant palinurid in Brazil and the keystone of the benthic community. Size, depth, and ecological parameters are strongly associated for red spiny lobsters [6], making it possible to infer that lobster size and prey size are positively correlated [3].

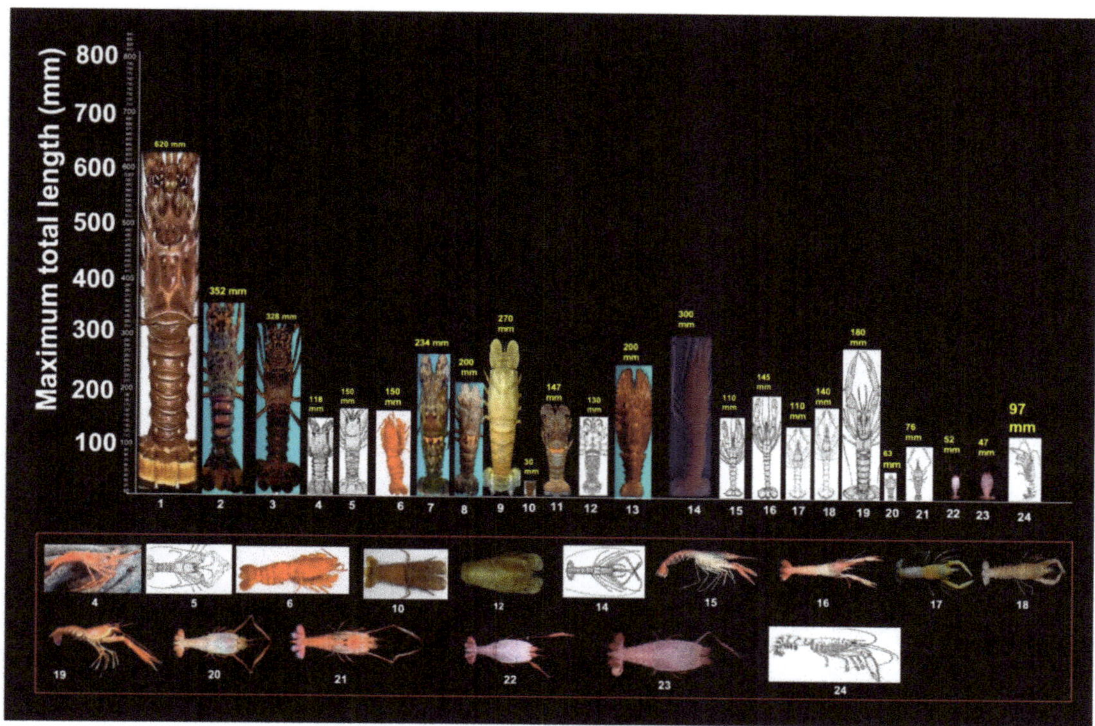

Figure 5. Maximum total length (MTL) of 24 lobster species occurring on continental shelf of Brazil (Table 1). Lower frame features enlarged pictures of small lobster species. Species (Appendix A) were distributed in following families: **Palinuridae**: 1. Brazilian form of *Panulirus argus*; 2. *Panulirus laevicauda*; 3. *Panulirus echinatus*; 4. *Palinustus truncatus*; 5. *Justitia longimanus*, 6. *Palinurellus gundlachi*. **Scyllaridae**: 7. *Scyllarides delfosi*; 8. *Scyllarides brasiliensis*; 9. *Scyllarides deceptor*; 10. *Scyllarus chacei*; 11. *Scyllarus depressus*; 12. *Scyllarus americanus*; 13. *Parribacus antarcticus*. **Nephropidae**: 14. *Acanthacaris caeca*; 15. *Nephropsis aculeata*; 16. *Nephropsis rosea*; 17. *Nephropsis agassizii*; 18. *Nephropsis neglecta*; 19. *Metanephrops rubellus*. **Polychelidae**: 20. *Stereomastis sculpta*; 21. *Polycheles typhlops*, 1862; 22. *Pentacheles laevis*; 23. *Pentacheles validus*. 24. *Enoplometopus antillensis*.

Table 1. Morphometric measurements of 24 lobster species reported from continental shelf of Brazil: maximum total length (MTL), maximum carapace length (MCL), total weight (Wt), region, depth range (m), and sex.

Species	MTL (mm)	MCL (mm)	Wt (g)	Region	Depth (m)	Sex	Reference
1- Brazilian form of *Panulirus argus*	620 (a)	233	3075 (b)	Br. GARS	1–110	M	Data base UFRA, Brazil
	802 (a)	302 (b)	6000	Ber.	-	M	Internet (*)
2- *Panulirus laevicauda*	352 (a)	127		Br.	1–40	M	Cruz et al. [13]
3- *Panulirus echinatus*	328 (a)	115		Br.	1–25	M	Cruz et al. [13]
4- *Palinustus truncatus*	118	-		Br.	60–1000	F	Silva et al. [55]
5- *Justitia longimanus*	150			Br.	50–100	M	Dall'Occo [51]
6- *Palinurellus gundlachi*	150			Cu.	1–35		Cruz et al. [15]
7- *Scyllarides delfosi*	234			Br. GARS	20–155	M	Silva et al. [55]

Table 1. *Cont.*

Species	MTL (mm)	MCL (mm)	Wt (g)	Region	Depth (m)	Sex	Reference
8- *Scyllarides brasiliensis*	200	-		Br.	3–475	-	Holthuis [3]
9- *Scyllarides deceptor*	270	101		Br.	6–300	F	Tavares et al. [16]
10- *Scyllarus chacei*	30	14	12.4	Br. GARS	10–300	F	Silva et al. [17]
11- *Scyllarus depressus*	147	27.7		Br.	30–536		Puciarelli & Rego [18]
12- *Scyllarus americanus*	130	25.3		Br.	1–70		No restrictions
13- *Parribacus antarcticus*	200			Br.	80–100		Holthuis [3]
14- *Acanthacaris caeca*	300			Br. GARS	80–880		Silva et al. [10]
15- *Nephropsis aculeata*	110			Br. GARS	140–820		Silva et al. [10], Legal & Poupin [19]
16- *Nephropsis rosea*	145			Br. GARS	100–808		Silva et al. [10]
17- *Nephropsis agassizii*	110			Br. GARS	878–2900	F	Silva et al. [10]
18- *Nephropsis neglecta*	140			Br.	655–1234	F	Alves–Júnior et al. [20]
19- *Metanephrops rubellus*	180			Br.	60–140	M	Holthuis [3]
20- *Stereomastis sculpta*	63			Br. GARS	200–4020		Silva et al. [10]
21- *Polycheles typhlops*	76			Br. GARS	41–2380		Silva et al. (10)
22- *Pentacheles laevis*	52			Br.	750–1350		Serejo et al. [53]
23- *Pentacheles validus*	47			Br.	811–2380		Serejo et al. [53]
24- *Enoplometopus antillensis*	97			Br.	1–30		Holthuis [3]

(a,b) = see Section 2 (2.4. Maximum total length). Legends: Br = Brazil; Ber = Bermuda; GARS = Great Amazon Reef System; Cu = Cuba. M = male and F = female. (*) https://www.washingtonpost.com/news/local/wp/2016/10/18/huge-14-pound-lobster-caught-in-bermuda-after-hurricane/, accessed on 15 August 2021.

Interspecific competition is assumed to have a strong influence on lobster population dynamics, although this is not easily demonstrated in the wild. In any case, the number of competitors for food and natural shelter decreases at depths below 50 m since neither the green spiny lobster nor the painted spiny lobster occurs in this range [54]. Living exclusively in mesophotic (water with low light penetration) reef ecosystems in the GARS, where mature adults make up more than 50% of the lobster stock, giant red spiny lobsters reach an average CL of 100.55 ± 0.2855 mm and have a very high reproductive potential [6]. Their versatile feeding habits include a wide range of benthic and infaunal organisms such as crabs, soft crabs, and anomurans [4].

3.5. Life Cycle

Cruz et al. [6,8] described the circulation and dispersal of larvae on the CSB during the planktonic stage, with subsequent recruitment, but due to ontogenetic vertical migration, it is still unknown how far from the parental stock the larvae travel. Briefly, by the end of the last phyllosoma stage (XI), the animals molt into pueruli. The pueruli become solitary juveniles (algal phase), then gregarious juveniles. They are recruited to the fishing grounds 27–33 months after hatching, depending on spatial and temporal factors (Figure 6).

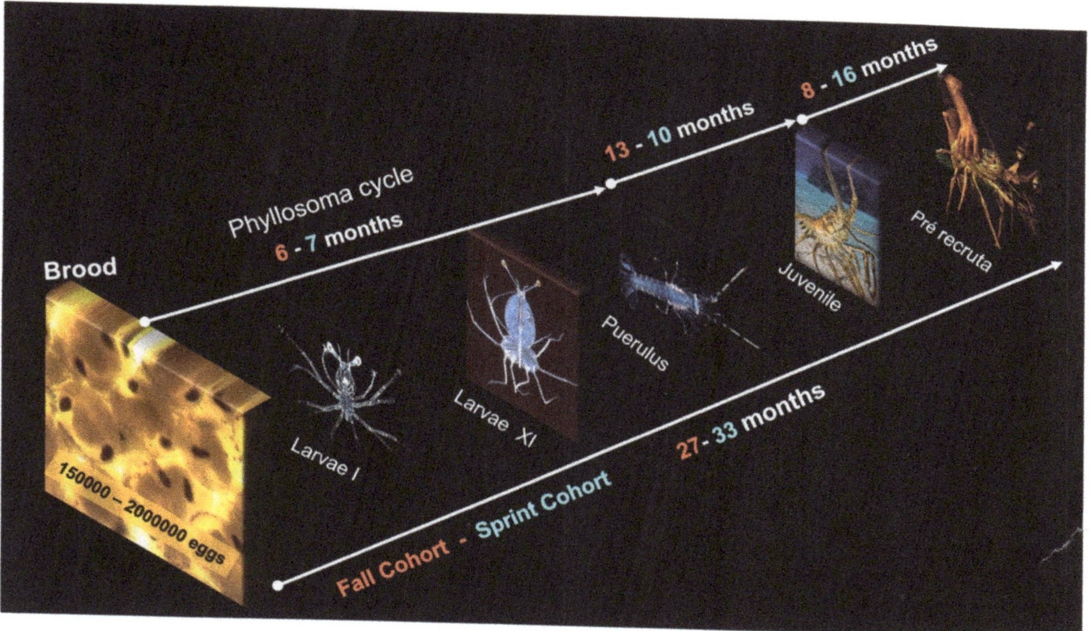

Figure 6. Summary of life cycle of Brazilian form of *Panulirus argus* (Latreille, 1804) showing evolution of a cohort from moment of hatching. Reproductive cycle is divided into four stages: spawning, larval recruitment to Brazilian coast, juvenile recruitment to nurseries, and recruitment to fishing grounds. Adapted from Cruz et al. [6].

Based on Cruz, et al. [6,8,66], the present review contributes to current knowledge of the spiny lobster's life cycle and provides evidence of similar ecological behavior for Brazilian and Caribbean populations of *P. argus*, including the duration of the planktonic phyllosoma stage in the ocean, carapace length stratification, reproductive cycle, and preferred habitats during ontogenetic development.

Brazilian studies on scyllarids have focused on taxonomy, phylogeny, and distribution, with some details on the life cycle of *P. antarcticus* inhabiting mesophotic reef ecosystems. In the GARS, three females of and two male species of *P. antarcticus* were collected at depths between 76 and 101 m [67]. The females carried hard black spermatophores and had pleopods with long setae (Figure 7A), as observed in spiny lobster species [68]. Females carrying eggs with fresh spermatophores (Figure 7B) is evidence of repeated spawning without molting; that is, females can produce two or more broods separated by three or four weeks. In the Caribbean, female *P. argus* larger than 100 mm CL spawn at least twice, separated by three or four weeks [69,70].

The metamorphosis of eggs into naupliosoma larvae was not observed for *P. antarcticus* or any other scyllarids [71]. Johnson et al. [72] estimated the length of larval life in Hawaiian waters to be approximately nine months, with 11 stages and a final phyllosoma size of 83 mm (TL). On isolated oceanic reefs over 220 km from continental Australia, Palero et al. [73] identified eleven larval stages in the plankton, measuring between 75 and 80 mm TL. These 'giant phyllosomata' are probably the largest decapod larvae reported so far (Figure 8).

Figure 7. Dorsal and ventral view of two female *Parribacus antarcticus* (Lund, 1793). (**A**) Black spermatophores and pleopods with long setae. (**B**) Eggs concomitant with fresh spermatophores. Adapted from Cintra, et al. [67].

In NE, two transparent nistos measuring 24.64 mm CL (46.51 mm TL) and 26.62 mm CL (49.57 mm TL) (nistos are equivalent to panulirid pueruli) were collected off Flecheiras (39°13.5′ W, 3°14.5′ S) in an area with extensive marine seaweed beds. Johnson, et al. [72] reported nistos in the wild to measure 20–21 mm CL. When captured, the body was transparent (except for the pigmented eyes) and depressed, with short, blade-shaped antennas. The lateral margin of the four antennas and the carapace were slightly serrated. The pleopods were highly developed and the shell was poorly calcified and devoid of spines.

Nistos look like adult *P. antarcticus* (Figure 8), but the duration of the stage is unknown. Individuals probably become solitary bottom dwellers before reaching their final habitat. No live juveniles of *P. antarcticus* were ever sampled. Nistos subsequently molt into juveniles. The spatial and temporal recruitment to the adult fishing grounds was not estimated.

Settlement patterns and larval distribution were not determined for this species, but current retroflection and stationary eddies (Figure 8) very likely cause the retention of phyllosomata near their hatching sites, ensuring larval recruitment to parental stock, as with palinurids and other scyllarids. On the other hand, larvae of *P. antarcticus* hatched in the GARS are trapped in this oceanic system by a cyclone-anticyclone system [8], suggesting this area is the main source of retention and supply of surviving larvae, thus of self-recruitment.

3.6. Biological Controls

Changes in the abundance of predators and food or shelter can strongly affect lobster population trends, and fisheries can alter ecosystem functioning and condition. The interaction between ecological factors and fisheries is complex and needs to be better understood to mitigate fluctuations in Brazilian lobster stocks.

However, further research on seasonal variations in local currents, current retroflection and ring formation is needed to clarify important aspects of larval transport, nistos settlement, and recruitment.

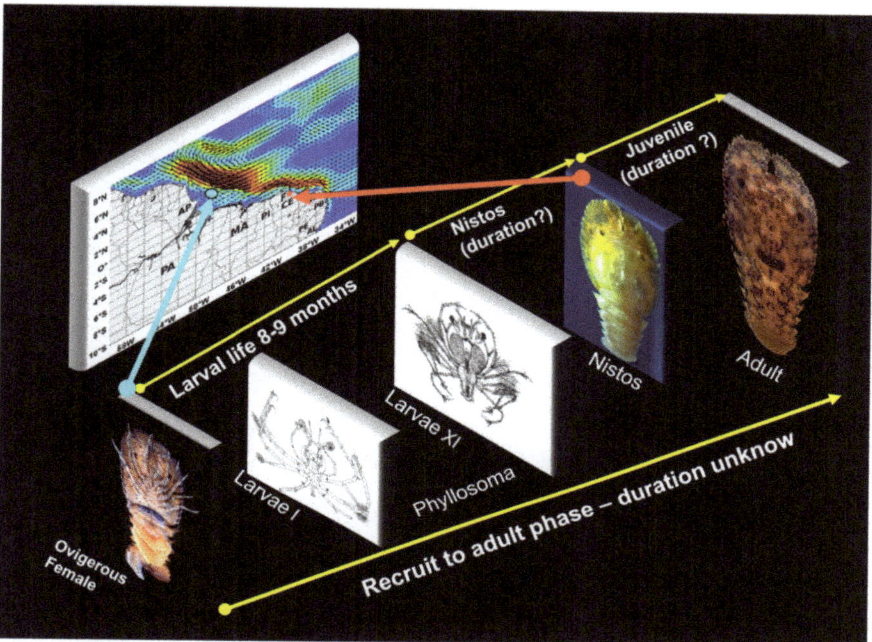

Figure 8. Theoretical scheme of life cycle of *Parribacus antarcticus* (Lund, 1793). Left: female carrying black spermatophores. Phyllosoma stages I to XI. Duration of the nistos and juvenile stages is unknown. Larvae and adults look alike. Map shows direction (arrow), intensity (color) and formation of retroflection eddies in North Brazil Current (Adapted from Cruz, et al. [8]). Light blue and red dot indicate locations in Great Amazon Reef System (GARS) and off Flecheiras (39°13.5′ W 3°14.5′ S) where ovigerous females and nistos were sampled, respectively.

3.6.1. Predators

Lobsters in the benthic phase (postlarvae, juveniles, adults) are preyed on by demersal fish, sharks, octopi, and dolphins. Our review of lobster predators in the Caribbean was based on Buesa [74], Munro [75], Aitken [76], Kanciruk [77], Cruz et al. [78] and Cruz & Phillips [79]. Juvenile lobsters were found in the stomachs of nurse sharks (*Ginglymostoma cirratum*) but many other fish species are known to prey on lobsters, including burrfishes, the Cuban snapper (*Lutjanus cyanopterus*), the grey snapper (*Lutjanus griseus*), the yellowtail snapper (*Ocyurus chrysurus*), triggerfishes (Balistidae), the red grouper (*Epinephelus morio*), and the old wife (*Enoplosus armatus*).

The lane snapper (*Lutjanus synagris*) and the spotted moray (*Gymnothorax moringa*) all eat juvenile lobsters. Occasionally, different types of sharks, turtles and octopi consume lobsters. Octopi are major lobster predators in Western Australia [80]. Their presence along the Brazilian coast suggests they prey on lobsters in this region as well, but little is known about their abundance.

As shown by Santana, et al. [15], lobster trap by-catch in NE consists primarily of snappers and grunts (60%), including species that prey on juvenile or mature lobsters (lutjanids, serranids, balistids, octopi).

The red snapper (*L. purpureus* Poey, 1866) occurs in the GARS and NE at depths between 25 and 160 m, where it was an important commercial fishing resource since 1961 [81]. Ivo & Hanson [82] hypothesized a migratory spawning pattern for the red snapper, with adults migrating to oceanic banks in NE (Ceará, Rocas Atoll, Caiçara) and juveniles recruiting to the nursery area close to the Amazon River mouth. The species feeds mostly on fish (89%), followed by crustaceans (8–10%, including lobsters) and a variety of benthic invertebrates (1–3%, including holothurians, mollusks, squids, annelids, and crustacean larvae) [83]. Other authors have included foraminifera, coelenterates, sponges, and encrusted bryozoans in the diet [84].

Dolphins (Delphinidae) are abundant on the CSB, with 21 species represented [85], all of which are carnivorous. Some prefer crustaceans (e.g., lobsters) [75], others eat a variety of shrimp, crabs, octopi, and cuttlefish. Dolphins are sighted both inshore and at high sea, including many lobster habitats. However, based on currently available information, dolphins cannot be confirmed to be key lobster predators.

According to Gupta, et al. [86], the little tunny (*Euthynnus affinis* Cantor, 1849) is an opportunistic pelagic predator of crustacean larvae, including scyllarid phyllosomata and spiny lobster pueruli (0.78%). In some cases, as observed on the coast of India (Cuddalore), they prey on larvae in the coastal zone, prior to settlement. Likewise, a preliminary analysis of the stomach contents of the skipjack tuna (*Katsuwonus pelamis* Linnaeus, 1758) revealed phyllosomata (pers. comm. by Raul Cruz).

Nine species of whales were sighted along the CSB [85]. Most feed on planktonic organisms, especially krill (*Euphausia superba* Dana, 1850), but the diet may also include small fish, copepods and some species of shrimp [87]. Available evidence supports the hypothesis that planktonic lobster larvae are ingested by whales, but the scarcity of information on the distribution of phyllosomata in Brazilian waters prevented a deeper understanding of this prey-predator relationship.

3.6.2. Feeding Behavior

Lobsters are opportunistic carnivores, preying on a wide variety of benthic invertebrates such as crustaceans, bivalves, annelids, echinoderms and gastropods. To a lesser degree, lobsters eat stomatopods, fish, polychaetes, ophiuroids, anemones, polyplacophores and asteroids, incidentally ingesting isopods, sipunculans, amphipods, and fragments of vegetation [88,89]. According to Herrera et al. [88], in response to ecological requirements spiny lobster populations migrate within the wider Caribbean in search of coarser sediments and macroinfauna (organisms >5 mm) with higher biomass density.

To minimize exposure to predation, large lobsters and scyllarids forage mainly at night and return to their protective shelter before dawn [79]. However, the feeding strategy of lobsters in deep waters is still unknown.

The abundance of oceanic phyllosomata [90,91] coincides with the relative abundance of gelatinous zooplankton, suggesting phyllosomata prey on this source [92]. The presence of jellyfish, ctenophores and salpids in the digestive glands of phyllosomata [93,94] indicates larvae are carnivorous predators at this stage.

In the last phyllosoma stage, when preparing to metamorphose into transparent nektonic pueruli, larvae stop feeding and subsist on stored energy [95]. In this study, we identified a nisto (transparent stage) of *P. antarcticus* very similar to the nisto of *Petrarctus brevicornis* (Holthuis, 1946) described by Wakabayashi, et al. [96]. In the post-puerulus stage, bottom-dwelling juveniles feed on copepods, amphipods, isopods, holothurians, and foraminifera [97].

3.6.3. Benthic Habitats and Length Assessment

Lobsters occur in benthic habitats from tide pools to the deep sea (4000 m) (see Table 1), but the large spiny lobsters exploited by the fishing industry are found at depths between 1 and 110 m [66]. Coral communities on the CSB differ very much from those in the wider Caribbean, in part due to heavy silting from the coast.

Covering 9500 km^2, the GARS features extensive reef platforms, reef walls, rhodoliths beds, sponge bottoms [98], and some 25 species of red algae [5] which serve as refuges and feeding grounds for benthic invertebrates such as lobsters and demersal fish. The great diversity of sponges (61 known species) in the GARS [99,100] increases the availability of shelters for lobsters at different stages of development and, consequently, ensures local

The scheme in Figure 13 shows how fisheries may be impacted by low abundance of predators (red snappers) and high abundance of prey (lobsters) (Figure 13A). Response to exploitation (Figure 13B) shows that the decreasing abundance of demersal predators (reduced predation) is associated with increased lobster landings.

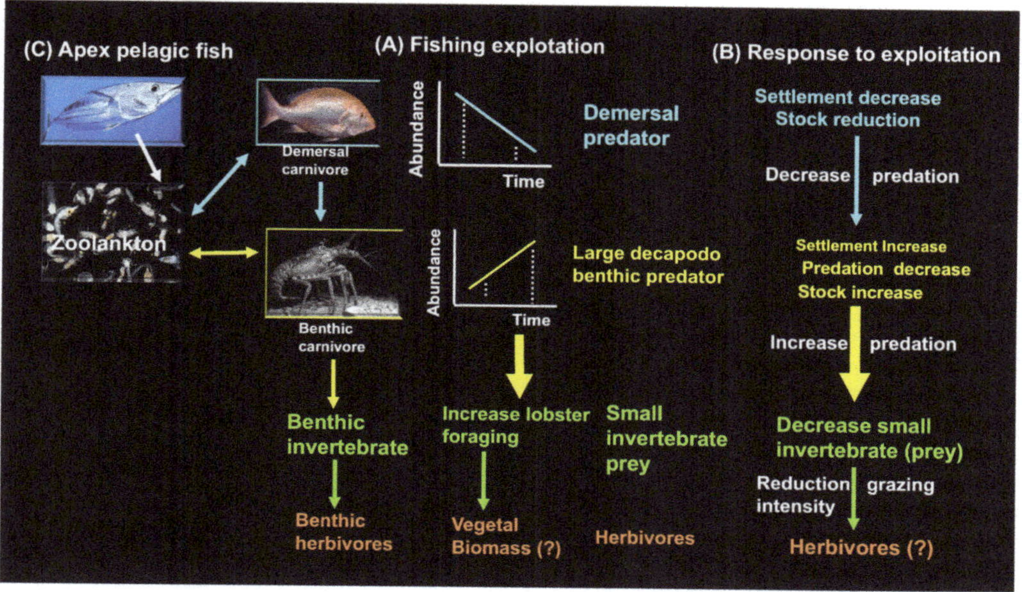

Figure 13. Theoretical model of top-down control. (**A**) Fishing exploitation of top predators (demersal carnivore) and red spiny lobster (benthic keystone) on continental shelf of Brazil (CSB). Decreasing production (abundance) of demersal predator (snapper) leads to increased prey abundance (see Figure 11). (**B**) Response to exploitation. The decreasing abundance of the demersal predator populations leads to reduce predation on the prey (red spiny lobster), which in turn leads to an increase in abundance and increased predation in small invertebrate (prey). (**C**) Apex pelagic fish. The increase of predation on zooplankton, leads to a decrease the abundance of larval recruitment. Thick arrows represent strong trophic interactions; thin arrows represent weak interactions.

As explained above, no predator-prey interactions exist between red snapper larvae and lobster phyllosomata. Animal species mostly interact through predation and competition [109], whether directly or indirectly. It is therefore reasonable to assume that increased predation on zooplankton by pelagic apex predators (Figure 13C) like skipjack (*K. pelamis*) that feed primarily upon small schooling fishes, mollusks, eggs, and crustacean larvae and pueruli in the zooplankton, leads to a decrease in zooplankton biomass, reducing the abundance of certain types of fish larvae, eggs, and phyllosomata.

Unfortunately, the variation in the biomass of benthic invertebrates and herbivores over time is unknown. But, as pointed out by Cury et al. [7], there is no need to measure all parameters to visualize the overall dynamics.

4. Conclusions

This review lists 24 lobster species distributed along the continental shelf of Brazil, from shallow to deep waters (Palinuridae 25%, Scyllaridae 29%, Nephropidae 25%, Polychelidae 17%, Enoplometopidae 4%). The maximum total size (MTL) of these species ranged from 30 to 620 mm.

The collected evidence supports the notion that the red spiny lobster (the Brazilian form of *P. argus*) is the largest and—due to its broad distribution—most important crustacean apex predator in the benthic community. The giant lobster is the most abundant palinurid in Brazil and the keystone of the benthic community. Changes in red spiny lobster abundance, for example, caused by anthropogenic impacts, can potentially compromise predator-prey interactions and the food-habitats of other species.

We know several species of slipper lobster (Scyllaridae) caught as associated fauna in spiny lobster traps are traded on the market, despite the lack of life-cycle information, stock assessments, or resource management. Deep-sea lobsters (Nephropidae and Polychelidae) have no commercial potential, and little is known about their bioecology.

Our review includes general aspects of the life cycle of the slipper lobster *P. antarcticus* and some new evidence (females carrying eggs with fresh spermatophores, indicating the ability to produce at least two broods in the same season). In addition, the capture of two nistos near the shore suggests slipper lobster larvae disperse in the ocean like *Panulirus* and settle in shallow waters.

We also provide evidence that changes in snapper abundance (predator) may have a strong effect on lobster abundance (prey). At this level, reciprocal predator-prey effects can change the abundance or productivity of a population through the food chain (top-down control).

Author Contributions: R.C. designed research, conceptualization, analyzed the data, original figures preparation, and wrote the manuscript; M.T.T. conducted the literature review, table preparation and interpretation, and review the manuscript; J.V.M.S. and I.H.A.C. collection, conservation and analyzed of biological material in different regions of Brazil (Great Amazon Reef System and the Northeastern). All authors have read and agreed to the published version of the manuscript.

Funding: This study was supported by Fundação Cearense de Apoio ao Desenvolvimento Científico e Tecnológico (FUNCAP) and Sindicato das Indústrias de Frio e Pesca do Ceará, Brazil (SINDFRIO).

Institutional Review Board Statement: Not applicable.

Informed Consent Statement: Not applicable.

Data Availability Statement: Not applicable.

Acknowledgments: The authors would like to thank FUNCAP (Fundação Cearense de Apoio ao Desenvolvimento Científico e Tecnológico, Ceará, Brazil), SECITECE (Secretaria da Ciência, Tecnologia e Educação Superior), SEDET (Secretaria de Desenvolvimento Econômico e Trabalho, Ceará, Brazil), IFCE (Instituto Federal de Educação, Ciências e Tecnologia do Ceará, Campus Acaraú), ISARH (Instituto Socioambiental e dos Recursos Hídricos), and UFRA (Universidade Federal Rural da Amazônia) for logistic support. Gratitude is extended to the exporters and producers affiliated with SINDFRIO (Sindicato das Indústrias de Frio e Pesca do Ceará, Brazil). The autors are also grateful to SAP/MAPA (Departamento de Registro e Monitoramento de Aquicultura e Pesca). We also like to extend our special thanks to anonymous referees and academy editor. Finally, many thanks to Tin–Yam Chan/Muséum national d'Histoire Naturelle, Joseph Poupin/École Navale, scientifique du Muséum de Paris, and Darryl L. Felder/University of Louisiana at Lafayette, LA, USA, for allowing us to reproduce photographs of lobster species.

Conflicts of Interest: The authors declare no conflict of interest.

Appendix A. Provides a List of the Species Surveyed

List of marine lobsters by family on the continental shelf of Brazil.
Palinuridae Latreille, 1802 (spiny lobsters)
1. Brazilian form of *Panulirus argus* (Latreille, 1804).
2. *Panulirus laevicauda* (Latreille, 1817).
3. *Panulirus echinatus* Smith, 1869.
4. *Palinustus truncatus* A. Milne-Edwards, 1880.
5. *Justitia longimanus* (H. Milne Edwards, 1837).
6. *Palinurellus gundlachi* von Martens, 1878.

Scyllaridae Latreille, 1825 (slipper lobsters)
7. *Scyllarides delfosi* Holthuis, 1960.
8. *Scyllarides brasiliensis* Rathbun, 1906.
9. *Scyllarides deceptor* Holthuis, 1963.
10. *Scyllarus chacei* Holthuis, 1960.
11. *Scyllarus depressus* (S. I. Smith, 1881).
12. *Scyllarus americanus* (S. I. Smith, 1869).
13. *Parribacus antarcticus* (Lund, 1793).

Nephropidae Dana, 1852 (clawed lobsters)
14. *Acanthacaris caeca* (A. Milne-Edwards, 1881).
15. *Nephropsis aculeata* Smith, 1881.
16. *Nephropsis rosea* Bate, 1888.
17. *Nephropsis agassizii* A. Milne-Edwards, 1880.
18. *Nephropsis neglecta* Holthuis, 1974.
19. *Metanephrops rubellus* (Moreira, 1903).

Polychelidae Wood-Mason, 1874 (blind lobsters, primitive decapods)
20. *Stereomastis sculpta* (Smith, 1880).
21. *Polycheles typhlops* C. Heller, 1862.
22. *Pentacheles laevis* Bate, 1878.
23. *Pentacheles validus* A, Milne Edwards, 1880.

Enoplometopidae de Saint Laurent, 1988.
24. *Enoplometopus antillensis* Lütken, 1865.

Appendix B.

List of the source photographs by authors.

List of Species (Appendix A)	Authors (Photo Credits)
1, 2, 3, 7, 8	Cruz et al. [54].
4,13,14,16	Silva et al. [55].
5,6	Cruz et al. [56].
9	Tavares et al. [36].
10	Silva et al. [47].
11	Puciarelly & Rego [57].
12	No restrictions
15,16	Photo (live color) Poupin J. in Legal & Poupin [58].
17,18	Photo (live color) Alves-Junior et al. [48].
19	Photo (live color) Scarabino, F. [59].
20	Photo (live color) Felder, D. [60].
21,22	Photo (live color) Tin-Yam Chan/Muséum national d'Histoire naturelle" [24]
23	Ahyong & Shane [61].
24	Holthuis [3].

List of Species (Appendix A)	Line Drawing Credits
4, 14, 17, 18, 19, 9, 25	Holthuis [3].
5	Cruz et al. [56].
15, 16, 20, 21	Silva et al. [5].
12	Adapted from Lavalli & Spanier [62].

References

1. Alencar, C.A.G.; Tavares, L.S.; Cintra, I.H.A. Current State of lobster exports in Brazil Research. *Soc. Dev.* **2020**, *9*, e312985804. [CrossRef]
2. Phillips, B.F.; Cobb, J.S.; George, C.R.W. General biology. In *The Biology and Management of Lobsters*; Cobb, J.S., Phillips, B.F., Eds.; Academic Press: New York, NY, USA, 1980; Volume 1, pp. 1–82.
3. Holthuis, L.B. Marine lobsters of the world. An annotated and illustrated catalogue of marine lobsters known to date. In *FAO Species Catalogue*; FAO: Italy, Rome, 1991; Volume 13, pp. 1–292.

4. Silva, K.C.A.; Cintra, I.H.A.; Ramos-Porto, M.; Viana, G.F.S.; Abrunhosa, F.A.; Cruz, R. Update on crustaceans known from the amazonian continental shelf and adjacent oceanic areas. *Crustaceana* **2020**, *93*, 687–701. [CrossRef]
5. Silva, K.C.A.; Cruz, R.; Cintra, I.H.A.; Abrunhosa, F.A. Structure and diversity of the lobster community on the Amazon continental shelf. *Crustaceana* **2013**, *86*, 1084–1102. [CrossRef]
6. Cruz, R.; Borda, C.A.; Santana, J.V.M.; Barreto, C.G.; Paiva, B.P.; Gaeta, J.C.; Torres, M.T.; Silva, J.L.S.; Cintra, I.H.A. Life cycle and connectivity of the spiny lobster, *Panulirus* spp.: Case studies from brazil and the wider caribbean (decapoda, achelata). *Crustaceana* **2021**, *94*, 603–645. [CrossRef]
7. Cury, P.; Shannon, L.; Shin, Y.J. The functioning of marine ecosystems: A fisheries perspective. In *Responsible Fisheries in the Marine Ecosystem*; Sinclair, M., Valdimarsson, G., Eds.; FAO: Italy, Rome; CABI Publishing: Wallingford, UK, 2003; pp. 103–123.
8. Cruz, R.; Teixeira, C.E.; Menezes, M.O.B.; Santana, J.V.M.; Neto, T.M.; Gaeta, J.C.; Freitas, P.P.; Silva, K.C.A.; Cinbtra, I.H.A. Large-scale oceanic circulation and larval recruitment of the spiny lobster *Panulirus argus* (Latreille, 1804). *Crustaceana* **2015**, *88*, 298–323. [CrossRef]
9. Rebelo-Neto, J.E. Considerações sobre a pescaria do lagostim (Metanephrops rubellus) na região sudeste/sul do Brasil. *CEPSUL—Documentos Técnicos* **1986**, *10*, 1–33.
10. Worm, B.; Barbier, E.B.; Beaumont, N.; Duffy, J.E.; Folke, C.; Halpern, B.S.; Jackson, J.B.C.; Lotze, H.K.; Micheli, F.; Palumbi, S.R.; et al. Impacts of Biodiversity Loss on Ocean Ecosystem Services. *Science* **2006**, *314*, 787–790. [CrossRef]
11. Wineland, S.M.; Kistner, E.J.; Joern, A. Non-consumptive interactions between grasshoppers (Orthoptera: Acrididae). And wolf spiders (Lycosidae) produce trophic cascades in an old-field ecosystem. *J. Orthoptera Res.* **2015**, *24*, 41–46. [CrossRef]
12. Boudreau, S.A.; Worm, B. Ecological role of large benthic decapods in marine ecosystems: A review. *Mar. Ecol. Prog. Ser.* **2012**, *469*, 195–213. [CrossRef]
13. Cruz, R.; Santana, J.V.M.; Barreto, C.G.; Borda, C.A.; Torres, M.T.; Gaeta, J.C.; Silva, J.L.S.; Saraiva, S.Z.R.; Salazar, I.O.; Cintra, I.H.A. Towards the rebuilding of spiny lobster stocks in Brazil: A review. *Crustaceana* **2020**, *93*, 957–983. [CrossRef]
14. Fonteles-Filho, A.A. The state of the lobster fishery in northeast Brazil. In *Spiny Lobster Management*; Phillips, B.F., Kittaka, J., Eds.; Blackwell: Oxford, UK, 2000; pp. 121–134.
15. Santana, J.V.M.; Neves, S.D.S.; Saraiva, S.Z.R.; Adams, C.; Cruz, R. Current Management and externalities in lobster fisheries exploitation on the continental shelf of Ceará, Brazil. *Arq. Ciên. Mar.* **2015**, *48*, 5–18.
16. Nascimento, R.C. Impactos Sócio-Ambientais de Marambaias Para a Pesca da Lagosta: O Caso de Ponta Grossa Icapuí-Ce. Tese Mestrado, Universidade Federal do Ceará, Fortaleza, Brazil, 2006; 86p.
17. Cruz, R.; Silva, K.C.A.; Neves, S.D.S.; Cintra, I.H.A. Impact of lobster size on catches and prediction of commercial spiny lobster landings in Brazil. *Crustaceana* **2013**, *86*, 1274–1290. [CrossRef]
18. Moura, R.L.; Amado-Filho, G.M.; Moraes, F.C.; Brasileiro, P.S.; Salomon, P.S.; Mahiques, M.M.; Bastos, A.C.; Almeida, M.G.; Silva, J.M.; Araujo, B.F., Jr.; et al. An extensive reef system at the Amazon River mouth. *Sci. Adv.* **2016**, *2*, e1501252. [CrossRef]
19. Coutinho, P.N.; Morais, J.O. Distribución de los sedimentos en la plataforma continental norte y nordeste del Brasil. *Arq. Ciên. Mar.* **1970**, *10*, 79–90.
20. Foster, M.S. Rhodoliths: Between rocks and soft places. *J. Phycol.* **2001**, *37*, 659–667. [CrossRef]
21. Amado-Filho, G.M.; Bahia, R.G.; Pereira-Filho, G.H.; Longo, L.L. South Atlantic rhodolith beds: Latitudinal distribution, species composition, structure and ecosystem functions, threats and conservation status. In *Rhodolith/Maërl Beds: A Global Perspective*; Riosmena-Rodríguez, R., Nelson, W., Aguirre, J., Eds.; Springer International Publishing (Coastal Research Library): London, UK, 2017; pp. 299–317.
22. Mahiques, M.M.; Siegle, E.; Francini-Filho, R.B.; Thompson, F.L.; De Rezende, C.E.; Gomes, J.D.; Asp, N.E. Insights on the evolution of the living Great Amazon Reef System, equatorial West Atlantic. *Sci. Rep.* **2019**, *9*, 13699. [CrossRef]
23. Cruz, R. *Manual de Métodos de Muestreo para la Evaluación de las Poblaciones de Langosta Espinosa*; FAO Documento Técnico de Pesca; FAO: Rome, Italy, 2002; pp. 1–43.
24. Chan, T.Y. Updated checklist of the world's marine lobsters. In *Lobsters: Biology, Fisheries and Aquaculture*; Radhakrishnan, E.V., Phillips, B.F., Achamveetil, G., Eds.; Springer: Singapore, 2019; pp. 35–64.
25. Diggle, P. *Time Series: A Bioestatistical Introduction*; Oxford University Press: Oxford, UK, 1990; 272p.
26. Rocha, L.A.; Craig, M.T.; Bowen, B.W. Phylogeography and the conservation of coral reef fishes. *Coral Reefs* **2007**, *26*, 501–512. [CrossRef]
27. Giraldes, B.W.; Smyth, D.M. Recognizing *Panulirus meripurpuratus* sp. nov. (Decapoda: Palinuridae) in Brazil—Systematic and biogeographic overview of *Panulirus* species in the Atlantic Ocean. *Zootaxa* **2016**, *4107*, 353–366. [CrossRef]
28. Sarver, S.K.; Silberman, J.D.; Walsh, P.J. Mitochondrial DNA sequence evidence supporting the recognition of two subspecies or species of the Florida spiny lobster *Panulirus argus*. *J. Crustacean Biol.* **1998**, *18*, 177–186. [CrossRef]
29. Sarver, S.K.; Freshwater, D.W.; Walsh, P.J. The occurrence of the provisional Brazilian subspecies of spiny lobster (*Panulirus argus westonii*) in Florida waters. *Fish. Bull.* **2000**, *98*, 870–873.
30. Briones-Fourzán, P.; Barradas-Ortiz, C.; Negrete-Soto, F.; Segura-García, I.; Lozano-Álvarez, E. Occurrence of *Panulirus meripurpuratus* and *Panulirus laevicauda* (Decapoda: Achelata: Palinuridae) in Bahía de la Ascensión, México. *Lat. Am. J. Aquat. Res.* **2019**, *47*, 694–698. [CrossRef]
31. Knowlton, N. Molecular genetic analyses of species boundaries in the sea. *Hydrobiologia* **2000**, *420*, 73–90. [CrossRef]

32. Fausto-Filho, J. Sobre a ocorrência *de Palinustrus truncatus* (H. Milne-Edwards, 1880), no litoral brasileiro e de *Panulirus echinatus* Smith 1860 no litoral do Estado do Ceará, Brasil (Crustácea, Decapoda, Palinuridae). *Arq. Ciên. Mar.* **1977**, *17*, 75–76.

33. Santana, W.; Pinheiro, A.P.; Oliveira, J.E.L. Additional records of three Scyllarides species (Palinura: Scyllaridae) from Brazil, with the description of the fourth larval stage of Scyllarides species (Palinura: Scyllaridae) from Brazil, with the description of the fourth larval stage of *Scyllarides aequinoctialis*. *Nauplius* **2007**, *15*, 1–6.

34. Melo, G.A.S. *Manual de Identificação dos Crustácea Decapoda do Litoral Brasileiro: Anomura: Thalassinidea*; Palinuridea e Astacidea: Cidade de São Paulo Plêiade: São Paulo, Brazil, 1999; 551p.

35. Coelho, P.A.; Ramos-Porto, M. Sinopse de crustáceos decápodos brasileiros (famílias Scyllaridae, Palinuridae, Nephropidae, Parastacidae e Axiidae). *An. Univ. Fed. Rural. Pernamb.* **1985**, *8*, 38–47.

36. Tavares, M.; Santana, W.; Pinheiro, A. On the type material of *Scyllarides Deceptor*, Holthuis, 1963 (Crustacea: Decapoda: Scyllaridae). *Papéis Avulsos Zool. (Museo Zool. Univ. São Paulo)* **2009**, *49*, 539–545. [CrossRef]

37. Oliveira, G.; Freire, A.S.; Bertuol, P.R.K. Reproductive biology of the, slipper lobster Scyllarides deceptor (Decapoda:Scyllaridae) along the southern Brazilian coast. *J. Mar. Biol. Assoc.* **2008**, *88*, 1433–1440. [CrossRef]

38. Duarte, L.F.A.; Severino-Rodrigues, E.; Gasalla, M.A. Slipper lobster (Crustacea, Decapoda, Scyllaridae) fisheries off the southeastern coast of Brazil: I. Exploitation patterns between 23°00′ and 29°65′ S. *Fish. Res.* **2010**, *102*, 141–151. [CrossRef]

39. Tavares, M. *Scyllarus ramosae*, new species, from the Brazilian continental slope, with notes on congeners occurring in the area (Decapoda: Scyllaridae). *J. Crustacean Biol.* **1997**, *17*, 716–724. [CrossRef]

40. Tavares, C.R.; Young, P.S. Nephropidae (Crustacea, Decapoda) collected by the Revizee Score-Central Program from off Bahia to Rio de Janeiro states, Brazil. *Arq. Mus. Nac.* **2002**, *60*, 77–88.

41. Schmitt, W. Crustacea Macrura and Anomura of Porto Rico and the Virgin Islands—Scientific Survey of Porto Rico and Virgin Islands. *N. Y. Acad. Sci.* **1935**, *15*, 125–262.

42. Robertson, P.B. The early larval development of the Scyllarid lobster Scyllarides aequinoctialis (Lund) in the laboratory, with a revision of the larval characters of the genus. *Deep Sea Res.* **1969**, *16*, 557–586. [CrossRef]

43. Dall'Occo, P.L.; Bento, R.T.; Melo, G.A.S. Range extensions for lobsters off the Brazilian Coast (Crustacea, Decapoda, Palinura, Astacidea). *Biociências* **2007**, *15*, 47–52.

44. Dall'Occo, P.L. Taxonomia e Distribuição das Lagostas (Crustacea: Decapoda: Achelata e Polychelida) no Oceano Atlântico. Ph.D. Thesis, Universidade Estadual Paulista, Rio Claro, Brazil, 2010; 432p.

45. Coelho, P.A.; Almeida, A.O.; Bezerra, L.E.A.; Souza-Filho, J.F. An updated checklist of decapod crustaceans (infraorders Astacidea, Thalassinidea, Polychelida, Palinura, and Anomura) from the northern and northeastern Brazilian coast. *Zootaxa* **2007**, *1519*, 1–16.

46. Serejo, C.S.; Young, P.S.; Cardoso, I.C.; Tavares, C.; Rodrigues, C.; Almeida, T.C. Abundância, diversidade e zonação dos crustáceos no talude da costa central do Brasil (11°–22° S) coletados pelo programa REVIZEE\Score Central: Prospecção pesqueira. In *Biodiversidade da Fauna Marinha Profunda na Costa Central Brasileira*; Costa, P.A.S., Olavo, G., Martins, A.S., Eds.; Rio de Janeiro Museo Nacional: Cidade de Rio de Janeiro, Brazil, 2007; pp. 133–162.

47. Silva, K.C.A.; Fransen, C.H.J.M.; Ramos-Porto, M.; Paiva, K.S.; Cintra, I.H.A.; Cruz, R. Report of Scyllarus chacei, Holthuis, 1960 (Decapoda, Scyllaridae) in Amapá state sontinental shelf of Brazil. *Crustaceana* **2012**, *85*, 1171–1177.

48. Alves-Junior, F.A.; Araujo, M.S.L.C.; Souza-Filho, J.F. Distribution of two species of *Nephropsis* Wood-Mason, 1872 (Crustacea: Decapoda: Nephropidae) from northeastern Brazil. *Zootaxa* **2016**, *4114*, 090–094. [CrossRef]

49. Tavares, M. Malacostraca-Eucarida. Nephropidae. In *Catalog of Crustacea from Brazil*; Young, P., Ed.; Museu Nacional: Rio de Janeiro, Brazil, 1999; pp. 377–378.

50. Silva, K.C.A.; Cintra, I.H.A.; Ramos-Porto, M.; Viana, G.F.S. Lagostas capturadas durante pescarias experimentais para o programa Revizee/Norte. (Crustacea, Nephropoidea, Eryonoidea, Palinuroidea). *Bol. Téc. Cient. Cepnor Belém* **2003**, *3*, 21–35.

51. Galil, B.S. Crustacea Decapoda: Review of the genera and species of the family Polychelidae Wood-Mason, 1874. In *Résultats des Campagnes Musorstom*; Crosnier, A., Ed.; Mémoires du Muséum National d'Histoire Naturelle: Paris, France, 2000; pp. 285–387.

52. Gaeta, J.C.; Silva, M.B.; Godoy, T.; Cruz, R. Update on the lobster species from Rocas Atoll Marine Reserve, Brazil. *Check List* **2015**, *11*, 1–7. [CrossRef]

53. Severino-Rodrigues, E.; Hebling, N.J.; Melo, G.A.S.; Graça-Lopes, R. Biodiversidade no produto da pesca dirigida ao lagostim Metanephrops rubellus (Moreira,1903) no litoral do Estado de São Paulo, Brasil, com ênfase à carcinofauna. *Inst. Pesca* **2007**, *33*, 171–182.

54. Cruz, R.; Conceição, R.N.L.; Marinho, R.A.; Barroso, J.C.; Holanda, J.S.; Félix, C.S.; Martins, M.E.O.; Santos, F.S.; Silva, K.C.A.; Furtado-Neto, M.A.A. Metodologias de amostragem para avaliação das populações de lagosta: Plataforma Continental do Brasil. Edição bilingüe português/espanhol). *Colecao Habitat* **2011**, *6*, 1–142.

55. Silva, K.C.A.; Cintra, I.H.A.; Ramos-Porto, M.; Viana, G.F.S. Lagostas capturadas na plataforma continental do estado do Amapá (Crustacea, Nephropoidea, Palinuroidea). *Bol. Téc. Cient. Cepnor Belém* **2007**, *7*, 173–184. [CrossRef]

56. Cruz, R.; Baisre, J.A.; Díaz, E.; Brito, R.; Blanco, W.; García, C.; Carrodeguas, C. *Atlas Biológico-Pesquero de la Langosta en el Archipiélago Cubano*; Revista Mar y Pesca: Ciudad Habana, Cuba, 1990; 125p.

57. Puciarelly, P.; Rego, A.B. The Crustacean Collection, Museum National d'Histoire Naturelle. Paris, France. 2020. No Restrictions to Copy and Redistribute the Material in Any Medium or Format. Available online: https://creativecommons.org/licenses/by/4.0/deed.en (accessed on 15 August 2021).

58. Legall, N.; Poupin, J. CRUSTA: Database of Crustacea (Decapoda and Stomatopoda), with Special Interest for Those Collected in French Overseas Territories. 2021. Available online: http://crustiesfroverseas.free.fr/ (accessed on 15 August 2021).

59. Scarabino, F. *Metanephrops rubellus* (Moreira, 1903). (scampi, cigala, Uruguayan lobster). Photography of the body in dorsal view. Annotated catalogue and bibliography of marine and estuarine shrimps, lobsters, crabs and their allies (Crustacea: Decapoda) of Argentina and Uruguay (Southwestern Atlantic Ocean). In *Frente Marítimo*; 2019; Volume 26, 179p.

60. Felder, D.L. The emergence of lobsters: Phylogenetic relationships, morphological evolution and divergence time comparisons of an ancient group (decapoda: Achelata, astacidea, glypheidea, polychelida). *Syst. Biol.* **2014**, *63*, 457–479.

61. Ahyong, S.T.; Shan, T.Y. Polychelidae from the Bohol and Sulu seas collected by Panglao 2005 (Crustacea: Decapoda: Polychelidae). *Raffles Bull. Zool.* **2008**, *19*, 63–70.

62. Lavalli, K.L.; Spanier, E. The biology and fisheries of the slipper lobster. In *Crustaceans*; Von, R., Ed.; CRC Press: Leiden, The Netherlands, 2007; 420p.

63. Fischer, W. *Lobsters: FAO Species Identification Sheets for Fishery Purposes*; FAO: Rome, Italy, 1978; 1700p.

64. Takeda, M. Crustaceans. In *Crustaceans and Mollusks Trawled off Suriname and French Guiana*; Takeda, M., Okutani, T., Eds.; Japan Marine Research Center: Tokyo, Japan, 1983; 354p.

65. Fonteles-Filho, A.A. Population dynamics of spiny lobsters (Crustacea: Palinuridae) in Norhteast Brazil. *Ciência Cultura* **1992**, *44*, 192–196.

66. Cruz, R.; Silva, K.C.A.; Gaeta, J.C.; Santana, J.V.M.; Cintra, I.H.A. Reproductive potential and stock recruitment of the Caribbean and Brazilian metapopulations of the spiny lobster, *Panulirus argus* (Latreille, 1804). *Crustaceana* **2014**, *87*, 1315–1337. [CrossRef]

67. Cintra, I.H.A.; Martins, D.E.G.; Alves-Júnior, F.A.; Santos, W.C.R.; Valente, H.M.; Klautau, A.G.C.M.; Silva, K.C.A. Report of the sculptured slipper lobster *Parribacus antarcticus* (Lund, 1793) (Decapoda: Scyllaridae) on the Great Amazon Reef System, Pará, Brazil. *Bol. Téc. Cient. Cepnor Belém* **2003**, *3*, 21–35.

68. Lipcius, R.N.; Herrnkind, W.F. Control and coordination of reproduction and molting in the spiny lobster *Panulirus argus*. *Mar. Biol.* **1987**, *96*, 207. [CrossRef]

69. Cruz, R. Fecundidad y madurez sexual en la langosta comercial Panulirus argus (Latreille, 1804) Crustácea: Palinuridae en Cuba. *Rev. Cubana Investig. Pesq.* **1980**, *5*, 2–27.

70. MacDiarmid, A.B.; Kittaka, J. Breeding. In *Spiny Lobsters: Fisheries and Culture*, 2nd ed.; Phillips, B.F., Kittaka, J., Eds.; Fishing News Books; Blackwell Science: Oxford, UK, 2000; pp. 485–507.

71. Sekiguchi, H.; Booth, J.D.; Webber, W. Early Life Histories of Slipper Lobsters. In The Biology and Fisheries of the Slipper Lobster. Lavalli, K.L., Spanier, E., Eds.; CRC Press: Leiden, The Netherlands, 2007; pp. 69–90.

72. Johnson, M.W. The phyllosoma larvae of slipper lobsters from the Hawaiian Islands and adjacent areas (Decapoda, Scyllaridae). *Crustaceana* **1971**, *20*, 77–103. [CrossRef]

73. Palero, A.F.F.; Guerao, G.B.; Hall, M.C.; Chan, T.Y.D.; Clark, P.F.E. The 'giant phyllosoma' are larval stages of *Parribacus antarcticus* (Decapoda: Scyllaridae). *Invertebr. Syst.* **2014**, *28*, 258–276. [CrossRef]

74. Buesa, R.J. *Biologia de la langosta Panulirus argus, Latreille, 1804 (Crustacea, Decapoda, Reptantia) en Cuba*; INPP/CIP: Ciudad Habana, Cuba, 1965; 228p.

75. Munro, J.L. *The Biology, Ecology, Exploitation and Management of Aribbean Reef Fishes: Crustaceans (Spiny Lobsters and Crabs)*; Zoology Department, University of the West Indies: Kingston, Jamaica, 1974; pp. 1–51.

76. Aitken, K.A. *Jamaica Spiny Lobster Investigations*; FAO Fish. Rep.; FAO: Rome, Italy, 1977; pp. 11–22.

77. Kanciruk, P. Ecology of juvenile and adult Palinuridae (spiny lobsters). In *The Biology and Management of Lobster*; Cobb, J.S., Phillips, B.F., Eds.; Academic Press: New York, NY, USA, 1980; Volume 2, Part 1; pp. 59–92.

78. Cruz, R.; Brito, R.; Díaz, R.; Lalana, R. Ecología de la langosta (*Panulirus argus*) al SE de la Isla de la Juventud. II. Patrones de movimiento. *Rev. Investig. Mar.* **1986**, *7*, 19–35.

79. Cruz, R.; Phillips, B.F. The artificial shelters (pesqueros) used for the spiny lobster (*Panulirus argus*) fisheries in Cuba. In *Spiny Lobster Management: Fisheries and Culture*, 2nd ed.; Phillips, B.F., Kittaka, J., Eds.; Blackwell: Oxford, UK, 2000; pp. 400–419.

80. Joll, L.M. *The Predation of Pot Caught Western Rock Lobster (Panulirus fongipes cygnus) by Octopus*; Department of Fisheries and Wildlife: Washington, DC, USA, 1977; pp. 1–58.

81. Ivo, C.T.C. *Estudo Sobre la Biologia da Pesca do Pargo Lutjanus Purpureus Poey, No Nordeste Brasileiro, Arquivo de Ciências do Mar*; Universidade Federal do Ceará (UFC): Fortaleza, Brazil, 1973; Part 2; pp. 113–116.

82. Ivo, T.; Hanson, A.J. Aspectos da Biologia e Dinâmica Populacional do Pargo, Lutjanus purpureus Poey, no Norte e Nordeste do Brasil. *Arq. Ciên. Mar, Fortaleza* **1982**, *22*, 1–4.

83. Pinheiro, R.; Silva, K.C.A. Dinâmica Alimentar do Pargo *Lutjanus purpureus* (Poey, 1875) Capturado na Plataforma Continental Amazônica. Bachelor's Thesis, Universidade Federal Rural da Amazônia, Belém, Brazil, 2019; 37p.

84. Silva, B.B.; Aragão, J.A.N.; Freire, J.L.; Lutz, I.A.F.; Sarmento, G.C.; Gomes, T.; Lima, W.P.G.; Santos, R.R.F. *Documento Técnico Sobre a Situação Atual das Pescarias do Pargo na Região Norte do Brasil*; Universidade Federal do Pará—Campus Bragança Instituto de Estudos Costeiros Laboratório de Bioecologia Pesqueira—LABIP: Ciudad de Bragança, Brazil, 2017; 132p.

85. Monteiro-Filho, E.L.; Oliveira, L.V.; Monteiro, K.D.K.A.; Filla, G.F.; Quito, L.; Godoy, D.F. Guia ilustrado mamíferos marinhos-Brasil. In *Projeto Boto-Cinza*; Instituto de Pesquisas de Cananéia (IPeC): Ciudad de São Paulo, Brazil, 2013; 106p.

86. Gupta, K.S.; Kizhakudan, S.J.; Kizhakudan, J.K.; Myousuf, K.S.S.; Raja, S. Preliminary observations on dominance of crustacean larvae in the diet of little tunny *Euthynnus affinis* (Cantor, 1849) caught off Chennai and Cuddalore coasts. *Indian J. Fish* **2014**, *61*, 40–44.

87. Werth, A. Feeding in marine mammals. In *Feeding. Chapter 16*; Hampden-Sydney College: Sindey, Australia, 2000; pp. 487–526.

88. Herrera, A.; Diaz-Iglesias, E.; Brito, R.; González, G.; Gotera, G.; Espinosa, J.; Ibarzábal, D. Alimentación natural de la langosta Panulirus argus en la región de los Indios (plataforma SW de Cuba) y su relación con el bentos. *Rev. Investig. Mar.* **1991**, *12*, 172–182.

89. Cox, C.; Hunt, J.H.; Lyons, W.G.; Davis, G.E. Nocturnal foraging of the Caribbean spiny lobster (*Panulirus argus*) on offshore reefs of Florida, USA. *Mar. Freshw. Res.* **1997**, *48*, 671–680. [CrossRef]

90. Lewis, J.B. The phyllosoma larvae of the spiny lobster *Panulirus argus*. *Bull. Mar. Sci.* **1951**, *1*, 78–103.

91. Goldstein, J.S.; Matsuda, H.; Takenouchi, T. A description of the complete development of larval Caribbean spiny lobster, *Panulirus argus* (Latreille, 1804) in culture. *J. Crust. Biol.* **2008**, *28*, 306–327. [CrossRef]

92. Greer, A.T.; Briseño-Avena, C.; Deary, A.L.; Cowen, R.K.; Herandez, F.J.; Graham, W.M. Associations between lobster phyllosoma and gelatinous zooplankton in relation to oceanographic properties in the northern Gulf of Mexico. *Fish. Oceanogr.* **2017**, *26*, 693–704. [CrossRef]

93. Chow, S.; Suzuki, S.; Matsunaga, T.; Lavery, S.; Jeffs, A.G. Investigation on natural diets of larval marine animals using peptide nucleic-directed Polymerase Chain Reaction clamping. *Mar. Biotechnol.* **2010**, *13*, 305–313. [CrossRef] [PubMed]

94. O'Rorke, R.; Lavery, S.; Chow, S.; Takeyama, H.; Tsai, P.; Beckley, L.E.; Thompson, P.A.; Waite, A.M.; Jeffs, A.G. Determining the diet of larvae of Western Rock Lobster (*Panulirus Cygnus*) using High-Throughput DNA Sequencing Techniques. *PLoS ONE* **2012**, *7*, e42757. [CrossRef] [PubMed]

95. Phillips, B.F.; Mcwilliam, P.S. The pelagic phase of spiny lobster development. *Can. J. Fish. Aquat. Sci.* **1986**, *43*, 2153–2163. [CrossRef]

96. Wakabayashi, K.; Yang, C.-H.; Chan, T.Y.; Phillips, B.F. The final phyllosoma, nisto, and first juvenile stages of the slipper lobster *Petrarctus brevicornis* (Holthuis, 1946) (Decapoda: Achelata: Scyllaridae). *J. Crustacean Biol.* **2020**, *40*, 237–246. [CrossRef]

97. Lalana, R.; Ortiz, M. Contenido estomacal de puérulos y post-puérulos de la langosta Panulirus argus en el Archipiélago de los Canarreos, Cuba. *Rev. Investig. Mar.* **1991**, *12*, 107–116.

98. Francini-filho, R.B.; Asp, N.E.; Siegle, E.; Hocevar, J.; Lowyck, K.; D'Avila, N.; Vasconcelos, A.S.; Baitelo, R.; Rezende, C.E.; Omachi, C.Y.; et al. Perspectives on the Great Amazon Reef: Extension, biodiversity and threats. *Front. Mar. Sci.* **2018**, *5*, 103–123. [CrossRef]

99. Muricy, G.; Moraes, F.C. Marine sponges of Pernambuco State, NE Brazil. *Rev. Bras. Oceanogr.* **1998**, *46*, 213–217. [CrossRef]

100. Muricy, G.; Lopes, D.A.; Hajdu, E.; Carvalho, M.S.; Moraes, F.C.; Carla, M.K.; Pinheiro, M.U. *Catalogue of Brazilian Porifera*; Museo Nacional Livros Serie: Recife City, Brazil, 2011; Volume 46, 300p.

101. Caddy, J.F. Why do assessments of demersal stocks largely ignore habitat? *ICES J. Mar. Sci.* **2014**, *71*, 2114–2126. [CrossRef]

102. Igarashi, M.A. Note on photographic register and preliminary observations from puerulus to juvenile of *Panulirus argus* after settlement in *Amansia* sp. macroalgae in Brazil. *Ciências Agrárias Londrina* **2010**, *31*, 767–772. [CrossRef]

103. Oliveira, G. *A Pesca da Lagosta Sapatera Scyllarides Deceptor HOLTHUIS, 1963 (CRUSTACEA: DECAPODA:SCYLLARIDAE) no sul do Brasil—Subsídios ao Conhecimiento da Biologia Reprodutiva e à Conservação da Espécie. Tese Apresentada como Requisito Parcial à Obtenção do grau de Mestre em Ciências Biológicas—Área de Concentração Zoologia*; Programa de Pós-graduação em Zoologia, Setor de Ciências Biológicas da Universidade Federal do Paraná: Curitiba, Brazil, 2008; pp. 1–62.

104. Linderman, K.; Lee, T.N.; Wilson, W.D.; Claro, R.; Jerald, A. *Transport of Larvae Originating in Southwest Cuba and the Dry Tortugas: Evidence for Partial Retention in Grunts and Snappers*; Gulf and Caribbean Fisheries Institute: Corpus Christi, TX, USA, 2001; pp. 732–747.

105. Jeffs, A. Revealing the natural diet of the phyllosoma larvae of spiny lobster Bull. *Fish. Res. Agen.* **2007**, *20*, 9–13.

106. Larrañeta, M.G. Ecología de la relación stock-reclutamiento en los peces marinos. *Oceanides* **1986**, *11*, 55–187.

107. Doi, M.; Ohno, A.; Kohno, H.; Taki, Y.; Singhagraiwan, T. development of feeding ability in red snapper *Lutjanus argentimaculatus* early larvae. *Fish. Sci.* **1997**, *63*, 845–853. [CrossRef]

108. Cruz, R.; Adriano, R. Regional and seasonal prediction of the Caribbean lobster (*Panulirus argus*) commercial catch in Cuba. *Mar. Freshwat. Res.* **2001**, *52*, 1633–1640. [CrossRef]

109. Lang, J.M.; Benbow, M. Species Interactions and Competition. *Nat. Educ. Knowl.* **2013**, *4*, 1–8.

Review

Distribution and Ecology of Decapod Crustaceans in Mediterranean Marine Caves: A Review

Carlo Nike Bianchi [1,*], Vasilis Gerovasileiou [2,3], Carla Morri [1] and Carlo Froglia [4]

[1] Department of Earth, Environmental and Life Sciences, University of Genoa, 16132 Genova, Italy; carlonike.bianchi.ge@gmail.com (C.N.B.); carla.morri.ge@gmail.com (C.M.)

[2] Department of Environment, Faculty of Environment, Ionian University, 29100 Zakynthos, Greece; vgerovas@ionio.gr or vgerovas@hcmr.gr

[3] Hellenic Centre for Marine Research (HCMR), Institute of Marine Biology, Biotechnology and Aquaculture (IMBBC), 71500 Heraklion, Crete, Greece

[4] CNR-IRBIM, National Research Council, Institute for Marine Biological Resources and Biotechnologies, 60125 Ancona, Italy; c.froglia@alice.it

[*] Correspondence: carlonike.bianchi.ge@gmail.com

Abstract: Decapod crustaceans are important components of the fauna of marine caves worldwide, yet information on their ecology is still scarce. Mediterranean marine caves are perhaps the best known of the world and may offer paradigms to the students of marine cave decapods from other geographic regions. This review summarizes and updates the existing knowledge about the decapod fauna of Mediterranean marine caves on the basis of a dataset of 76 species from 133 caves in 13 Mediterranean countries. Most species were found occasionally, while 15 species were comparatively frequent (found in at least seven caves). They comprise cryptobiotic and bathyphilic species that only secondarily colonize caves (secondary stygobiosis). Little is known about the population biology of cave decapods, and quantitative data are virtually lacking. The knowledge on Mediterranean marine cave decapods is far from being complete. Future research should focus on filling regional gaps and on the decapod ecological role: getting out at night to feed and resting in caves during daytime, decapods may import organic matter to the cave ecosystem. Some decapod species occurring in caves are protected by law. Ecological interest and the need for conservation initiatives combine to claim for intensifying research on the decapod fauna of the Mediterranean Sea caves.

Keywords: species inventory; zoogeography; ecology; species richness; depth preference; cave zonation; secondary stygobiosis; trophic depletion; protected species

Citation: Bianchi, C.N.; Gerovasileiou, V.; Morri, C.; Froglia, C. Distribution and Ecology of Decapod Crustaceans in Mediterranean Marine Caves: A Review. *Diversity* **2022**, *14*, 176. https://doi.org/10.3390/d14030176

Academic Editors: Patricia Briones-Fourzán, Michel E. Hendrickx and Michael Wink

Received: 29 January 2022
Accepted: 22 February 2022
Published: 27 February 2022

Publisher's Note: MDPI stays neutral with regard to jurisdictional claims in published maps and institutional affiliations.

1. Introduction

Members of order Decapoda are among the crustaceans most familiar to the general public and include species of commercial interest for fisheries, such as lobsters, shrimps and prawns [1]. They total over 14,000 extant species [2], which have colonized virtually all aquatic habitats, with a few species even thriving on land. Crustacean decapods are one of the most important groups—in terms of both species richness and ecological roles [3]—in the marine realm, where they occur from the upper intertidal zone to hadal depths [4], with several species adapted to extreme environments, such as hydrothermal vents and cold seeps [5].

Marine caves are another example of an extreme environment [6] that hosts a rich decapod fauna [7–9]. Rocky reefs and coral reefs around the world harbour cavities of various size at or below sea level and, therefore, filled with marine water. Several decapod taxa new to science have been described from, and occur only in, marine caves. Yet, information on the distribution and ecology of cave decapods is still largely fragmentary.

The rocky coasts of the Mediterranean Sea are particularly rich in marine caves [10], whose geology, biology and ecology have been studied with continuity for decades [11–13].

Thus, Mediterranean marine caves are perhaps the best known of the world ocean and may offer paradigms and theories to the students of marine cave decapods from other geographic regions.

This review summarizes and updates the existing knowledge about the decapod fauna of Mediterranean marine caves on the basis of a dataset assembled within the frame of a broader initiative aiming at cataloguing and assessing the biodiversity of Mediterranean marine caves [12,14]. The database incorporates records from either scientific publications or grey literature; unpublished observations by the authors in marine caves of Greece and Italy were purposely added for this review. The final decapod checklist will be mobilised to the World Register of Marine Cave Species (WoRCS, http://www.marinespecies.org/worcs/, accessed on 21 February 2022), a thematic species database of the World Register of Marine Species (WoRMS, http://www.marinespecies.org, accessed on 21 February 2022), which aims at creating a comprehensive taxonomic and ecological database of species known from marine caves worldwide [15]. When available, ecological and spatial information on the caves was also considered [14]. A thorough ecological analysis has been possible for the Grotta Marina di Bergeggi (NW Italy), which has been extensively studied since the 1970s [16–18] and where the distribution of decapod species within the cave can be compared with the available topographic information and environmental parameters [19,20]. Anchialine caves, which contain water bodies of marine origin but have reduced or no connection with the sea [21] and harbour a distinctive fauna [22], are not covered by this review: Mediterranean anchialine caves are still too little studied for decapods [23].

2. Species Inventory and Frequency

Only 68 out of the 370 literature sources examined (18%) provided information on decapod occurrence. All the decapod species found in at least one Mediterranean marine cave were considered. In total, 76 decapod species have been reported from 133 caves (or cave groups) in 13 countries (Figure 1). Most of these caves (52) are located in Italy, followed by Greece (23 caves), Montenegro (13) and France (11). Spain and Turkey provided nine caves each, Croatia seven and Albania four, while Cyprus, Israel, Lebanon, Malta and Morocco provided only one cave (or cave group) each.

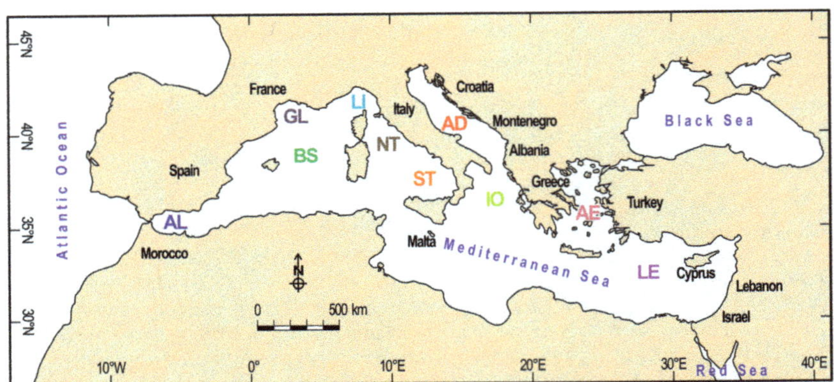

Figure 1. Map of the Mediterranean Sea with countries and marine sectors where cave decapods have been reported. AL = Alboran Sea; BS = Balearic and Sardinia seas; GL = Gulf of Lion; LI = Ligurian Sea; NT = North Tyrrhenian Sea; ST = South Tyrrhenian Sea; AD = Adriatic Sea; IO = Ionian Sea; AE = Aegean Sea; LE = Levantine Sea.

The present total of 76 species (Table 1) represents a momentous increase with respect to the earliest accounts of Riedl [7], who listed 11 decapod species from Mediterranean marine caves, and Parenzan [24], who listed seven species from Italian marine caves. The first study entirely devoted to the decapod fauna in Mediterranean marine caves was that of Gili and Macpherson [25], who collected 11 species in two caves of Mallorca (Balearic Islands, Spain). Pessani and Manconi [26] found 15 species in five marine caves of the Sorrentine Peninsula (Italy), while Pipitone and Vaccaro [27] recorded eight species in four marine caves of the Island of Ustica (Sicily, Italy). The synthesis by Manconi and Pessani [28] reported 25 species from 21 Italian marine caves. Denitto et al. [29,30] studied the decapod fauna (14 species) of a marine cave in SE Italy, focussing on *Palaemon* spp. and the west-African species *Herbstia nitida*, recorded for the first time in the Mediterranean Sea. In semi-submerged marine caves of Montenegro, Mačić et al. [31] found five species. Two species new to science were described from Mediterranean marine caves: *Odontozona addaia*, from Menorca (Balearic Islands, Spain) [32], and *Salmoneus sketi*, from Lavernaka Island (Croatia) [33].

Table 1. List of the species of decapod crustaceans found in marine caves of the Mediterranean Sea.

Infraorder	Family	Genus and Species
Stenopodidea	Stenopodidae	*Odontozona addaia* Pretus, 1990
		Stenopus spinosus Risso, 1827
Caridea	Palaemonidae	*Balssia gasti* (Balss, 1921)
		Brachycarpus biunguiculatus (Lucas, 1846)
		Gnathophyllum elegans (Risso, 1816)
		Palaemon adspersus Rathke, 1837
		Palaemon elegans Rathke, 1837
		Palaemon serratus (Pennant, 1777)
		Palaemon xiphias Risso, 1816
		Periclimenes amethysteus (Risso, 1827)
		Periclimenes scriptus (Risso, 1822)
		Urocaridella pulchella Yokes & Galil, 2006
	Alpheidae	*Alpheus dentipes* Guérin, 1832
		Athanas nitescens (Leach, 1814)
		Salmoneus sketi Fransen, 1991
		Synalpheus gambarelloides (Nardo, 1847)
	Hippolytidae	*Caridion* sp.
		Hippolyte holthuisi Zariquiey Alvarez, 1953
		Saron marmoratus (Olivier, 1811)
	Lysmatidae	*Lysmata nilita* Dohrn & Holthuis, 1950
		Lysmata seticaudata (Risso, 1816)
	Pandalidae	*Plesionika narval* (Fabricius, 1787)
	Thoridae	*Eualus cranchii* (Leach, 1817)
		Eualus occultus (Lebour, 1936)
		Eualus sollaudi (Zariquiey Cenarro, 1936)
Astacidea	Nephropidae	*Homarus gammarus* (Linnaeus, 1758)
Achelata	Palinuridae	*Palinurus elephas* (Fabricius, 1787)
	Scyllaridae	*Scyllarides latus* (Latreille, 1803)
		Scyllarus arctus (Linnaeus, 1758)
		Scyllarus pygmaeus (Bate, 1888)
Anomura	Galatheidae	*Galathea dispersa* Bate, 1859
		Galathea intermedia Lilljeborg, 1851
		Galathea nexa Embleton, 1836
		Galathea strigosa (Linnaeus, 1761)
	Munididae	*Munida rugosa* (Fabricius, 1775)
	Porcellanidae	*Pisidia bluteli* (Risso, 1816)

Table 1. *Cont.*

Infraorder	Family	Genus and Species
Brachyura	Diogenidae	*Porcellana platycheles* (Pennant, 1777)
		Calcinus tubularis (Linnaeus, 1767)
		Clibanarius erythropus (Latreille, 1818)
		Dardanus arrosor (Herbst, 1796)
		Dardanus calidus (Risso, 1827)
		Diogenes pugilator (P. Roux, 1829)
	Paguridae	*Cestopagurus timidus* (P. Roux, 1830)
		Pagurus anachoretus Risso, 1827
		Pagurus chevreuxi (Bouvier, 1896)
		Pagurus cuanensis Bell, 1845
		Pagurus prideaux Leach, 1815
	Dromiidae	*Dromia personata* (Linnaeus, 1758)
	Homolidae	*Homola barbata* (Fabricius, 1793)
	Ethusidae	*Ethusa mascarone* (Herbst, 1785)
	Eriphiidae	*Eriphia verrucosa* (Forskål, 1775)
	Leucosiidae	*Ilia nucleus* (Linnaeus, 1758)
	Progeryonidae	*Paragalene longicrura* (Nardo, 1869)
	Epialtidae	*Acanthonyx lunulatus* (Risso, 1816)
		Herbstia condyliata (Fabricius, 1787)
		Herbstia nitida Manning & Holthuis, 1981
		Lissa chiragra (Fabricius, 1775)
		Pisa armata (Latreille, 1803)
		Pisa nodipes Leach, 1815
	Inachidae	*Achaeus cranchii* Leach, 1817
		Inachus dorsettensis (Pennant, 1777)
		Macropodia czernjawskii (A.T. Brandt, 1880)
		Macropodia rostrata (Linnaeus, 1761)
	Majidae	*Eurynome aspera* (Pennant, 1777)
		Maja crispata Risso, 1827
		Maja squinado (Herbst, 1788)
	Pilumnidae	*Pilumnus hirtellus* (Linnaeus, 1761)
		Pilumnus minutus De Haan, 1835
		Pilumnus spinifer H. Milne Edwards, 1834
	Portunidae	*Carupa tenuipes* Dana, 1852
		Charybdis hellerii (A. Milne-Edwards, 1867)
		Portunus hastatus (Linnaeus, 1767)
	Xanthidae	*Xantho pilipes* A. Milne-Edwards, 1867
	Grapsidae	*Pachygrapsus marmoratus* (Fabricius, 1787)
	Percnidae	*Percnon gibbesi* (H. Milne Edwards, 1853)
	Pinnotheridae	*Nepinnotheres pinnotheres* (Linnaeus, 1758)

The species most frequently reported from Mediterranean marine caves is *Stenopus spinosus*, found in 41 caves (out of 133: 30.8%), followed by *Palaemon serratus* (34 caves: 25.6%) and *Herbstia condyliata* (32 caves: 24.1%). Six other species were reported from more than ten caves and a further six species from at least seven caves (Figure 2). All the remaining species were found in six (<5%) caves or less, with 30 species having been found in a single cave only. Thus, the majority of the 76 species reported from Mediterranean marine caves should apparently be considered as accidental in such habitats with the exception of *S. sketi*, which is hitherto known only from the cave where it has been described [33], and *O. addaia*, known only from two marine caves: one in Menorca (Spain) [32] and one in Marseilles (France) [34].

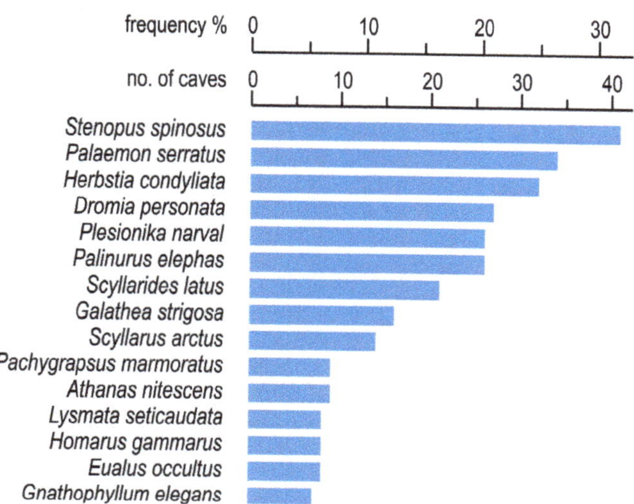

Figure 2. The 15 most common decapod species in Mediterranean Sea caves. Species found in less than seven caves (5%) are not considered.

3. Regional Variations in Species Richness

The 76 decapod species are not distributed homogenously among the 133 caves. The most species-rich cave is the Grotte de l'Île Plane (Gulf of Lion), with 19 species, followed by the Grotta Marina di Bergeggi (Ligurian Sea), with 15 species. The Grotta della Cala di Mitigliano (South Tyrrhenian Sea) and the Grotte del Ciolo (Ionian Sea) host 13 species each, while 12 species have been reported from the Grotte de la Triperie (Gulf of Lion) and the Spilja Vrbnik (Adriatic Sea), and 11 species from the Túnel Llarg (Balearic Sea) and the Fará Cave (Aegean Sea). All the remaining caves have less than ten species each, 59 of them having only one species.

Marine caves are highly fragmented habitats, whose connectivity is expectedly low [35]. In addition, the Mediterranean Sea is compartmentalized into fairly isolated sub-basins, which display a great variety of climatic and hydrologic conditions. Bianchi et al. [36] recognized a number of geographic sectors within the Mediterranean Sea, different for oceanographic characteristics and biota composition. Information on cave decapods is available for 10 sectors (Figure 1): (1) the Alboran Sea, immediately east of the Straits of Gibraltar; (2) the Balearic Sea and the Sea of Sardinia, from eastern Spain to western Sardinia; (3) the Gulf of Lion, from Cap de Creus (Catalonia) to the Giens Peninsula (France); (4) the Ligurian Sea, from the Giens Peninsula to the Piombino Promontory (Italy); (5) the North Tyrrhenian Sea, from the Piombino Promontory to the Pontine Islands (Italy); (6) the South Tyrrhenian Sea, from the Pontine Islands to north Sicily; (7) the Adriatic Sea, between Italy and the Balkan Peninsula; (8) the Ionian Sea, between Italy and western Greece; (9) the Aegean Sea, between Greece and Turkey; (10) the Levantine Sea, from south Turkey to Israel.

This choice of sectors patently suffers from the heterogeneity of research effort (in terms of number of caves explored), which has been maximum in the northern Mediterranean and minimum in southern and extreme western areas. Perhaps not coincidentally, the greatest number of decapod species has been recorded in the sectors of the Gulf of Lion and Ligurian Sea. Clearly, the number of species found is positively correlated to the number of caves explored [14], which varied from a minimum of three, in the Alboran Sea, to a maximum of 22, in the Ionian Sea, being comprised between seven and 20 in the remaining sectors (Table 2).

Table 2. Summary of the main parameters and coefficients of the species/caves curves.

Sectors	N	S_N	c	z	r_{dir}	S_∞	r_{rec}
Alboran Sea	3	5	1.7	1.00	1.000	32	0.993
Balearic and Sardinia seas	10	18	5.1	0.58	0.989	27	0.999
Gulf of Lion	7	29	7.5	0.74	0.996	45	0.999
Ligurian Sea	13	24	3.8	0.74	0.996	48	1.000
North Tyrrhenian Sea	16	19	3.1	0.69	0.991	32	0.999
South Tyrrhenian Sea	12	20	3.3	0.74	0.998	37	0.999
Adriatic Sea	20	19	2.1	0.75	0.996	32	0.999
Ionian Sea	22	22	2.3	0.64	0.997	36	0.999
Aegean Sea	18	17	3.1	0.60	0.990	23	0.999
Levantine Basin	9	15	2.8	0.78	0.997	37	1.000

N: number of caves surveyed in the different sectors; S_N: cumulative number of species in N caves; c and z: coefficients calculated from the relation $S = c \cdot N^z$; r_{dir}: correlation coefficient of the curve resulting from the direct plot of the cumulative number of species against the number of caves (see Figure 3); S_∞: theoretical maximum number of species in each sector, by extrapolation; r_{rec}: correlation coefficient between the reciprocals of both the cumulative number of species and the number of caves.

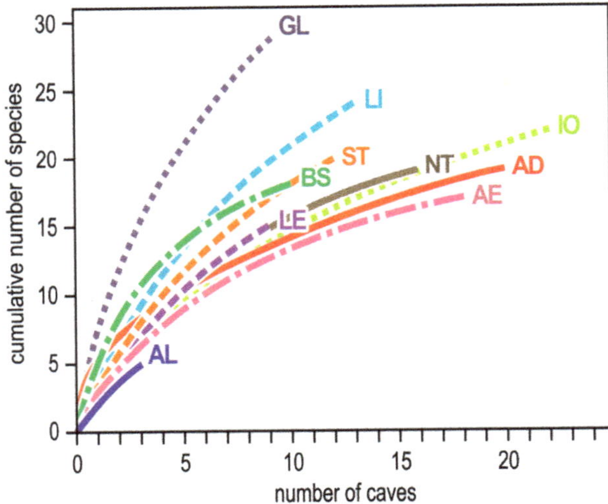

Figure 3. Species accumulation curves for cave decapods in ten Mediterranean Sea sectors. AL = Alboran Sea; BS = Balearic and Sardinia seas; GL = Gulf of Lion; LI = Ligurian Sea; NT = North Tyrrhenian Sea; ST = South Tyrrhenian Sea; AD = Adriatic Sea; IO = Ionian Sea; AE = Aegean Sea; LE = Levantine Sea (see Figure 1 for sectors location).

Thus, to make comparable the cave decapod species richness in the different sectors of the Mediterranean Sea, the cumulative number of species has been plotted against the number of caves surveyed in each sector (Figure 3). Curves were fitted to these plots according to the equation $S = c \cdot N^z$, where S is the number of species, N is the number of caves, c gives the number of species that may be expected in one cave and z is the slope of the regression line relating S and N [37,38]. Plotting the reciprocals of the cumulative number of species against the reciprocals of the number of caves was also tried. This allowed an extrapolation to show the hypothetical number of species to be found with an 'infinite' number of caves, i.e., the theoretical maximum number of cave species in each sector [39]: this number of species is useful for inter-regional comparisons but must not be taken for a real prediction.

The equation $S = c \cdot N^z$ fitted well to all ten species/caves curves (Table 2). The curve for the Gulf of Lion lays above all the remaining curves (Figure 3) due to high values of both c and z. When z exceeds 0.5, species richness may be sufficiently explained by environmental heterogeneity alone [40]: as z values were distinctly higher than 0.5 for all sectors, it can be inferred that differences among individual caves are of major importance for increasing the regional cave species pool.

Similarly, the plots of the reciprocals of species against cave numbers were well fitted by regression lines of the form $1/S = a \cdot 1/N + b$: the reciprocal of the intercept b represents the number of species S_∞ to be found in an infinite (i.e., $1/0$) number of caves. The highest value of S_∞ was found in the Gulf of Lion, followed by the Ligurian Sea. In all, the Mediterranean sectors richest in cave decapod species are the most northern ones, independent of research effort (number of caves explored, the actual number of surveys within each individual cave being not available).

4. Zoogeography

Most species have a distributional range restricted to a geographical area, and species having similar ranges can be grouped in chorological categories. The fauna of the Mediterranean Sea is composed of species belonging to several chorological categories [41]. These categories are not uniformly distributed through the whole basin but tend to occur more or less abundantly in the different geographic sectors of that sea [36,42]. The most 'typical' Mediterranean fauna, rich in endemics [43], occurs in the central sectors and especially in the western basin [44]. The northern and colder sectors harbour boreal Atlantic species (ice-ages remnants, especially of the Würm glacial), while the southern and warmer sectors are colonized by (sub)tropical species [45]. The Alboran Sea, located immediately east of Gibraltar, exhibits stronger Atlantic affinities thanks to the continued penetration of Atlantic fauna with the incoming flux of water [46]. On the contrary, the Levantine Sea is experiencing an important influx of Red Sea species since the opening of the Suez Canal [47]—a phenomenon known as 'Lessepsian immigration' in recognition of Ferdinand de Lesseps, the French diplomat who promoted the cut of the canal [48].

The present decapod dataset provides an opportunity to test whether Mediterranean marine caves exhibit the same chorological composition as the geographic sectors to which they belong or represent a filter that favours some chorological categories with respect to others. Mediterranean marine caves are considered as refuges for archaic fauna, often of Tethyan or Pliocene origin, which escaped the competitive pressure by more modern species [6,49]. Thus, marine caves would be expected to host a higher proportion of endemic and tropical species (both Tethyan and Pliocene fauna being typically tropical in character) and little or no recent immigrants in comparison to other shelf habitats.

Mediterranean cave decapod species have been assigned to six chorological categories, based on the review of d'Udekem d'Acoz [50] integrated with subsequent studies [27,51–54], to compute the chorological spectra of the Mediterranean as a whole and of the ten geographical sectors outlined above (Figure 4): (1) Mediterranean endemics; (2) Atlantic–Mediterranean species; (3) Lusitanian(–Boreal) species; (4) Mauritanian(–west African) species; (5) circum(sub)tropical species; (6) non-indigenous species (NIS), including both aliens and cryptogenics.

The endemics include six cave decapod species. Three are long known to live exclusively in the Mediterranean Sea (see Table S21 by C. Froglia in [55]): *Hippolyte holthuisi*, which has recently been recognized as the Mediterranean vicariant of the Atlantic congeneric *H. varians* Leach, 1814 [56], *Maja squinado* and *Periclimenes amethysteus*. *Odontozona addaia* and *Salmoneus sketi* are hitherto known only from the Mediterranean Sea [32–34]. Finally, *Caridion* sp. was originally identified with the boreal species *C. steveni* Lebour, 1930 by Ledoyer [57], who found it in a cave at Villefranche-sur-Mer (France), but considered as a distinct, still undescribed species by subsequent authors [50,58].

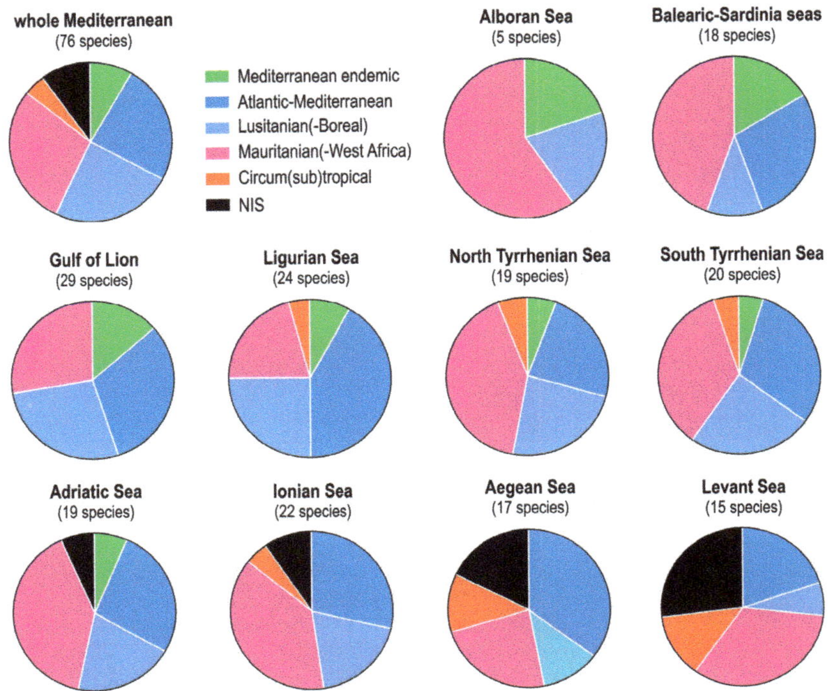

Figure 4. Chorological spectra of cave decapods in the Mediterranean Sea as a whole and in ten marine sectors. NIS = non-indigenous species.

Atlantic–Mediterranean species typically thrive in the East Atlantic from the English Channel (La Manche) to the North, to Morocco and Mauritania to the South, including Macaronesia (Azores, Madeira and Canary Islands) and the Mediterranean Sea [59,60]. Five decapod species found in Mediterranean caves strictly belong to this category: *Eriphia verrucosa*, *Pachygrapsus marmoratus*, *Palinurus elephas*, *Pisidia bluteli*—a senior synonym of *P. longimana* (Risso, 1816) and distinct from the similar *P. longicornis* (Linnaeus, 1767) according to Ferreira and Tavares [61]—and *Scyllarus arctus*. Thirteen other species in this category have a wider distribution in the East Atlantic, reaching up to Norway to the North and Cape Verde Islands or even Namibia to the South: *Achaeus cranchii*, *Athanas nitescens*, *Diogenes pugilator*, *Dromia personata*, *Eualus cranchii*, *E. occultus*, *Eurynome aspera*, *Galathea intermedia*, *Pagurus cuanensis*, *P. prideaux*, *Palaemon elegans*, *Pilumnus hirtellus* and *Xantho pilipes*.

Lusitanian species are here defined as those Mediterranean–Atlantic species that are restricted to the Mediterranean and the western European coasts from Portugal to Brittany and do not extend southward [59]. The term Lusitanian is here adopted in its classical acceptation and according to its etymology (Lusitania being the name that Romans used to indicate a region that roughly corresponds to the present-day Portugal); other authors used the term in a stricter or wider sense [62–64]. Lusitanian decapods reported from Mediterranean caves include 11 species: *Cestopagurus timidus*, *Clibanarius erythropus*, *Eualus sollaudi*, *Lissa chiragra*, *Lysmata seticaudata*, *Macropodia czernjawskii*, *Munida rugosa*, *Nepinnotheres pinnotheres*, *Pagurus chevreuxi*, *Pisa nodipes* and *Synalpheus gambarelloides*. Nine other species stretch further north (up to Norway), thus displaying a boreal character: *Galathea dispersa*, *G. nexa*, *G. strigosa*, *Homarus gammarus*, *Inachus dorsettensis*, *Macropodia rostrata*, *Palaemon adspersus*, *P. serratus* and *Porcellana platycheles*.

Similarly, Mauritanian species are those whose range, outside the Mediterranean Sea, reaches Mauritania or even Cape Verde Islands but does not extend northward [59,65]. Cave decapods belonging to this category include 13 species: *Dardanus calidus*, *Ethusa mascarone*, *Gnathophyllum elegans*, *Ilia nucleus*, *Lysmata nilita*, *Maja crispata*, *Pagurus anachoretus*, *Palaemon xiphias*, *Paragalene longicrura*, *Periclimenes scriptus*, *Pilumnus spinifer*, *Scyllarides latus* and *Scyllarus pygmaeus*. Nine other species extend their range further southward along the coasts of west Africa: *Acanthonyx lunulatus*, *Alpheus dentipes*, *Balssia gasti*, *Calcinus tubularis*, *Herbstia condyliata*, *Homola barbata*, *Pisa armata*, *Portunus hastatus* and *Stenopus spinosus*.

Three Mediterranean cave decapod species have a circum(sub)tropical range, thriving in all warm waters of the world ocean: *Brachycarpus biunguiculatus*, *Dardanus arrosor* and *Plesionika narval*.

Finally, Mediterranean caves host seven non-indigenous species (NIS) of decapods with different origin: *Percnon gibbesi* is an amphiamerican and amphiatlantic (sub)tropical species rapidly spreading in the Mediterranean Sea thanks to the passive drift of larvae with currents [66]; *Herbstia nitida* comes from tropical west Africa [30]; *Carupa tenuipes*, *Charybdis hellerii*, *Pilumnus minutus* and *Saron marmoratus* are Indo-Pacific species that possibly penetrated into the Mediterranean Sea through the Suez Canal [67]; *Urocaridella pulchella* represents an example of a species first described from the Mediterranean Sea, where it is obviously an alien of Indo-Pacific origin [68], and only later found in its native range [69]. As a whole, all these alien species share a clear tropical affinity.

Analysing the chorological spectra of the decapod fauna in Mediterranean marine caves (Figure 4), the proportion of NIS results is 9%, i.e., less than half the figure for the whole Mediterranean [55], suggesting that marine caves, and especially internal cave portions, are comparatively less receptive to newcomers [70]. The highest proportion of alien decapods is found in the eastern sectors, as expected [71]. The proportion of endemics in caves is 7.9%, i.e., slightly less than the corresponding figure of 9.9% for the total decapod fauna of the Mediterranean [55]. Endemics are comparatively more represented in the north-western sectors than in south-eastern sectors, conforming to a common pattern for the East Mediterranean marine fauna [36,72]. On the contrary, the Aegean Sea and the Levantine Basin contain more circum(sub)tropical species with respect to the northern sectors. Such a picture, however, is probably undergoing change as sea water warming is allowing warm-water species to establish also in the northern reaches of the Mediterranean Sea [73]: a still unpublished—and, therefore, not included in the present dataset—record of *B. biunguiculatus* comes from a marine cave near Marseilles (https://doris.ffessm.fr/Forum/Crevette-des-grottes-38429, accessed on 21 February 2022). Differences among geographic sectors are apparent also for Atlantic–Mediterranean, Lusitanian(–Boreal) and Mauritanian(–west African) species and partly mirror those of the total decapod fauna of the Mediterranean. These general patterns, although apparently reliable, should be taken with caution: more research is needed, especially in those geographic sectors where cave decapod fauna has not been sufficiently investigated (see, for instance, the Alboran Sea, with only five species reported from three caves).

5. Depth and Cave Zone Preferences

Information about the water depth of the seafloor at which the cave opens was available for 89 out of the 133 (66.9%) Mediterranean marine caves stored in the decapod dataset, representing nine countries (Albania, Croatia, Cyprus, France, Greece, Israel, Italy, Spain and Turkey) and ten geographic sectors (Adriatic Sea, Aegean Sea, Alboran Sea, Balearic and Sardinia seas, Gulf of Lion, Ionian Sea, Levantine Basin, Ligurian Sea, North Tyrrhenian Sea and South Tyrrhenian Sea). The water depth at the base of cave entrances ranges from 1 m to 52 m. In these 89 caves, 59 decapod species have been recorded, but only the 14 species found in at least five caves have been considered for statistical analysis.

Most of these 14 species have a wide depth range (Figure 5), but *Palaemon serratus* and *Lysmata seticaudata* clearly prefer shallow waters, never extending deeper than 20 m. *P. serratus* is commonly found outside caves in infralittoral habitats [74]. *Scyllarides latus*,

Herbstia condyliata and *Dromia personata* also prefer shallow depths but were recorded down to 30 m. *Plesionika narval* and *Homarus gammarus*, on the contrary, avoid the shallowest caves and were more frequently recorded in deep caves, down to 52 m and 45 m, respectively. Outside caves, *P. narval* is mostly found at epibathyal depths [75]. All the remaining species prefer intermediate depths and typically show a circalittoral affinity outside caves [76].

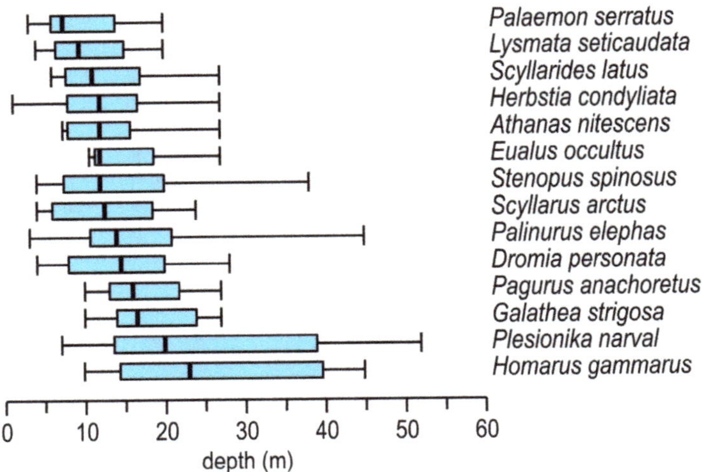

Figure 5. Depth preference of selected cave decapod species (depth is that of the seafloor at the entrance of the cavity). Boxes depict the 25–75 percent quartiles; the thick vertical line inside the box indicates the median; the whiskers represent minimal and maximal values.

Independent of the depth at which they open, marine caves are topographically and ecologically heterogeneous in their inside, exhibiting dramatic environmental gradients [77]: within a few metres, there are variations of light, water movement and trophic input, which, in the external environment, can take place within tens or even hundreds of metres [11]. These environmental gradients generate a marked zonation of cave communities [78]. Riedl [7] distinguished six biotic zones, based on species replacement across the outside–inside gradient of blind-ended caves. Bianchi and Morri [79] also distinguished six ecological zones, but, rather than species replacement, they considered change in growth forms, trophic guilds, three-dimensional structure and biotic cover of the sessile communities. Due to the great influence of the French school on Mediterranean marine ecologists, the most widely accepted and followed model of cave zonation is that of Pérès and Picard [44], who distinguished two basic situations based on the occurrence of characteristic species: the semidark cave, and the dark cave.

For the purpose of the present review, it has been possible to distinguish three cave zones where decapods have been recorded: the entrance, the semi-dark zone and the dark zone. Information about which zone decapods came from is available from 39 caves (29.3%) situated in all the ten geographical sectors (Adriatic Sea, Aegean Sea, Alboran Sea, Balearic and Sardinia seas, Gulf of Lion, Ionian Sea, Levantine Basin, Ligurian Sea, North Tyrrhenian Sea and South Tyrrhenian Sea) and belonging to nine countries (Albania, Croatia, France, Greece, Italy, Lebanon, Malta, Spain and Turkey). In these 39 caves, 60 decapod species have been found. However, only the twelve species recorded in more than five caves have been considered to make possible an estimation of their relative frequency in the three cave zones (Figure 6).

Three species, namely *Athanas nitescens*, *Lysmata seticaudata* and *Scyllarides latus*, have been more frequently reported from the entrance of the caves. On the contrary, *Plesionika narval* and *Stenopus spinosus* were more frequent in the dark zone. For the remaining seven species (*Dromia personata*, *Eualus occultus*, *Galathea strigosa*, *Herbstia condyliata*, *Palaemon serratus*, *Palinurus elephas* and *Scyllarus arctus*), no clear pattern is discernible. Pessani and Manconi [26] analysed decapod zonation in the Grotta della Cala di Mitigliano (South Tyrrhenian Sea, Italy): out of the 14 species found, only *H. condyliata* reached the innermost part of the cave, while the remaining species were localized near the entrance and in the semidark zone. Harmelin [80] ascribed *Homarus gammarus*, *L. seticaudata*, *P. serratus*, *P. elephas*, *S. latus* and *S. arctus* to the semidark zone; *G. strigosa*, *H. condyliata* and *S. spinosus* to the dark zone.

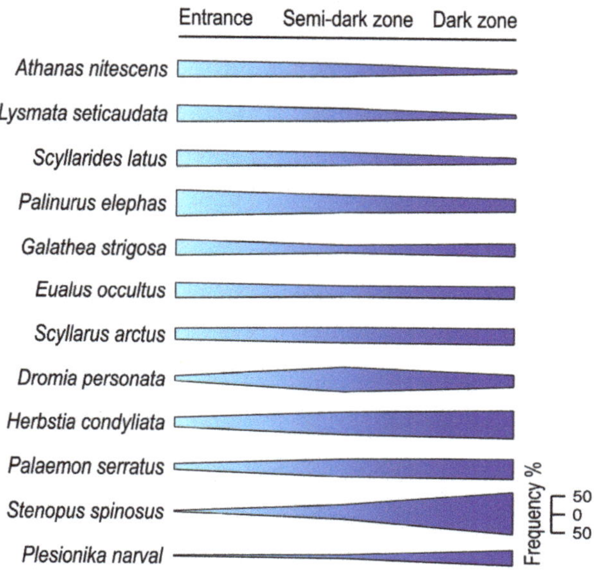

Figure 6. Kite diagrams illustrating cave zone preference for selected cave decapod species.

6. Ecology and Distribution in the Grotta Marina di Bergeggi

The distribution of decapods within Mediterranean marine caves has been described by Gili and Macpherson [25] in Spain and by Pessani and Manconi [26] in Italy. However, no study has addressed the relation between decapod species occurrence within a marine cave and environmental data. Sgorbini et al. [19] took measurements in situ and collected water and sediment samples for laboratory analysis in order to define the environmental characteristics of the Grotta Marina di Bergeggi, a comparatively small but morphologically complex marine cave in the Ligurian Sea, NW Italy (Figure 7).

Besides a wide emerged part, the Grotta Marina di Bergeggi has a submerged part that develops between the sea surface and 7 m depth, has a length of about 40 m and is articulated in five main topographic features. The Remo's Cavern is the outer cave, with an environmental situation comparable to the adjacent sea tract, but a reduced illumination. The Gulley, the two lateral chambers (First Chamber and Lights' Chamber) and the Hall are obscure and lie along the overall tunnel path of the cavity. The inner 'lakes' (Lemons' Lake and Lake through the Hole) represent blind-ended portions, obscure and subject to meteoric water infiltrations. The parameters taken into account included light ($\mu W \cdot cm^{-2}$), current velocity ($cm \cdot s^{-1}$), water temperature ($^{\circ}C$), salinity (psu), dissolved oxygen (ppm), pH, sediment mean grain size (mm), suspended organic matter (SOM, $mg \cdot L^{-1}$), chlorophyll *a* ($\mu g \cdot L^{-1}$) and nitrogen to carbon (N/C) percent ratio (Table 3). The data were taken in July 1986; the methodological details about measurement and sampling can be found in Sgorbini et al. [19].

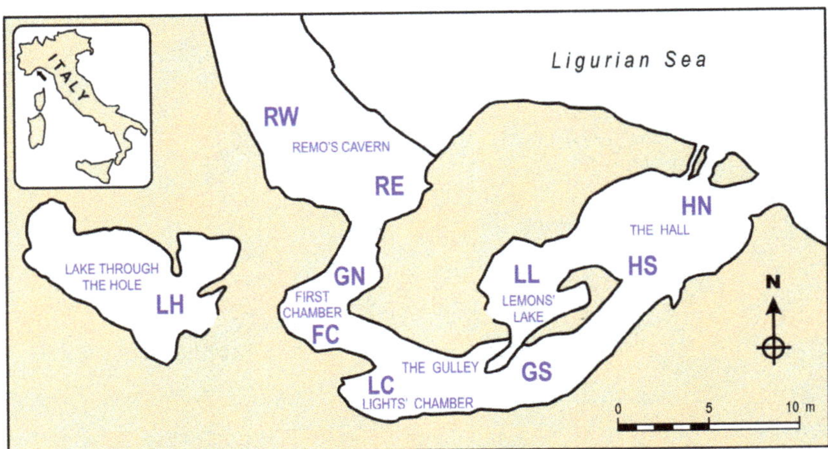

Figure 7. Simplified map of the submerged part of the Grotta Marina of Bergeggi with main toponyms and sampling points. RW = Remo's Cavern West; RE = Remo's Cavern East; GN = Gulley North; GS = Gulley South; FC = First Chamber; LC = Lights' Chamber; HN = Hall North; HS = Hall South; LL = Lemons' Lake; LH = Lake through the Hole. Inset: location of Bergeggi in Italy.

Table 3. Environmental data measured in July 1986 in different stations of the Grotta Marina di Bergeggi [19,20].

Parameters	RW	RE	GN	GS	FC	LC	HN	HS	LL	LH
Light (μW·cm^{-2})	29.0	32.5	0.4	0.1	0.2	0.0	1.3	0.2	0.1	0.0
Current Velocity (cm·s^{-1})	20.3	15.9	14.3	14.1	7.5	1.6	15.9	14.3	7.6	2.7
Temperature (°C)	19.0	19.0	18.8	18.5	19.0	18.5	18.3	18.0	17.5	17.0
Salinity (psu)	37.8	37.8	37.7	37.5	37.1	37.5	37.6	37.2	26.8	27.8
O$_2$ (ppm)	6.25	5.55	6.49	6.13	6.35	6.10	5.35	5.80	6.45	3.53
pH	8.22	8.18	8.15	8.10	8.11	8.09	8.14	8.14	7.82	7.90
Mean Grain Size (mm)	7.5	10.0	6.9	7.0	7.9	4.8	9.0	7.5	0.5	3.6
Suspended Organic Matter (mg·L^{-1})	1.05	1.05	0.79	0.39	0.79	0.25	1.74	1.96	0.73	0.39
Chlorophyll *a* (μg·L^{-1})	0.30	0.30	0.30	0.25	0.30	0.20	0.30	0.26	0.14	0.15
Nitrogen/Carbon %	11.2	11.8	10.3	10.0	9.8	9.3	8.7	8.1	9.8	9.4

RW = Remo's Cavern West; RE = Remo's Cavern East; GN = Gulley North; GS = Gulley South; FC = First Chamber; LC = Lights' Chamber; HN = Hall North; HS = Hall South; LL = Lemons' Lake; LH = Lake through the Hole.

Knowing the value of these parameters within the cave allows evaluating the relative importance of three main ecological factors: illumination, hydrological confinement and trophic depletion [13]. The decrease in light has been considered the most obvious factor influencing species distribution, through the limitation of the development of photophilic assemblages [81]. The threshold values of light intensity in the different cave zones have rarely been measured: Southward et al. [82] showed that, at the cave entrance, with light intensity equal to 17% of that of the surface, the assemblages are still photophilic; at 3%, the assemblages become sciaphilic, below 0.8%, only animals are found. According to Harmelin et al. [6], light intensity in the dark zone is lower than 0.01% of the sea-surface value. Early observations by Harmelin [83] showed that tunnel-shaped dark caves exhibited a richer biota than blind-ended dark caves. This led to the idea that, besides light, hydrological confinement was a major driver of fauna distribution in marine caves [84]. Hydrological confinement is a complex and rather abstract quantity: it is mainly a hydrodynamic notion, essentially depending on water exchange, a factor that has rarely been measured. Bianchi et al. [85] measured current velocities ranging from less than 2 cm·s^{-1}

to 6–10 cm·s^{-1}, but sea conditions may greatly alter these figures. Water circulation affects a series of hydrological parameters, such as temperature, salinity, oxygen concentration, pH and sedimentation. The absence of vegetal life, and, hence, of autochthonous primary production, makes marine cave communities completely dependent on the input of food from the external environment [77]. Therefore, animals living in the innermost portions of the cave, far from the entrance, suffer trophic depletion [86] in terms of both the quantitative decrease of the nourishment (e.g., reduced amount of suspended organic matter and chlorophyll *a* content) and its qualitative degradation (e.g., low nitrogen to carbon ratio). The amount of suspended particulate matter decreases significantly from the entrance to the innermost portions of the marine caves due to the progressive sedimentation of the suspended particles and their capture by filter-feeders [87]. Chlorophyll *a* characterizes 'fresh' vegetal organic matter, rich in live phytoplankton cells, and its concentration in the water decreases dramatically in the innermost parts of the cave [88]. The nitrogen to carbon ratio is an index of the nutritional value of the organic matter: food with high carbon and little nitrogen content (e.g., cellulose) is poorly nutritious compared with food that is proportionally richer in nitrogen (e.g., proteins). Inside caves, the ratio can get lower than 6%, a threshold value for animal consumption [89,90].

Cluster analyses on all these environmental parameters in the Grotta Marina di Bergeggi allowed recognizing five locales that correspond well to the topographic features of the cavity [20]. In two stations for each locale (Figure 7), the presence of decapod species was recorded visually by scuba diving in July 1986 during daytime. Their abundance was estimated semi-quantitatively in the field using the following codes [91]: 1 = one individual; 2 = two to five individuals; 3 = more than five individuals.

In total, 15 species were found: *Herbstia condyliata* (Figure 8a), *Stenopus spinosus* (Figure 8b), *Palaemon serratus* (Figure 8c), *Lysmata seticaudata* (Figure 8d) and *Scyllarus arctus* (Figure 8e), in the order, were the most abundant and common; *Dromia personata* (Figure 8f), *Palinurus elephas* and *Alpheus macrocheles* were scarcer, while *Clibanarius erythropus* (Figure 8g), *Eriphia spinifrons*, *Inachus dorsettensis* (Figure 8h), *Pachygrapsus marmoratus*, *Pilumnus hirtellus*, *Scyllarides latus* (Figure 8i) and *Xantho pilipes* were occasional (only one individual found). The highest number of species was found in the Hall, followed by the Gulley; the lowest number was found in the lakes (Figure 9).

A correspondence analysis was applied to the data matrix of 15 decapod species × 10 stations to explore the ecological gradients [92]. Groups of species were individuated by cluster analysis using Euclidean distances and minimum variance clustering [93]. All the statistical analyses were performed using the free software PaSt [94].

The cluster analysis distinguished four groups of species (Figure 10). The first group contains four species (*P. marmoratus*, *A. macrocheles*, *E. spinifrons* and *P. hirtellus*) that were restricted to the Remo's Cavern. The second group includes seven species (*S. latus*, *X. pilipes*, *P. elephas*, *L. seticaudata*, *C. erythropus*, *S. arctus* and *I. dorsettensis*) that preferred the parts of the cave nearest to the entrances, such as the First Chamber and the northern stations of the Gulley and the Hall. The third group is composed by three species (*H. condyliata*, *D. personata* and *S. spinosus*) that, on the contrary, preferred the innermost parts of the cave but not the lakes: the southern stations of the Hall and Gulley and the Light's Chamber. Finally, the fourth group is made by *P. serratus*, the only species that—despite being typically infralittoral—preferred the lakes.

An interesting property of correspondence analysis is the possibility of plotting both station-points and species-points on the same factorial plane, which allows for an immediate reading of gradients and affinities. Only the first axis from the correspondence analysis was significant ($p < 0.05$) according to the Lebart's test [95]. Both species-points and station-points of the Grotta Marina di Bergeggi ordered along the first axis according to the external–internal gradient (Figure 10).

To explore which of the environmental parameters measured is most likely responsible for that gradient, a correlation analysis was performed between the station-point scores of the first axis and the values of the individual parameters in such stations (Figure 11). The first axis scores were highly significantly correlated ($p < 0.01$) with light intensity, and significantly correlated ($p < 0.05$) with the nitrogen to carbon ratio. No significant correlation was found between the first axis scores and the remaining parameters illustrative of hydrological confinement (current velocity, temperature, salinity, oxygen, pH and mean grain size) or trophic depletion (suspended organic matter and chlorophyll *a*). This result suggests that light is the main factor influencing the ecological distribution of decapods in marine caves. Apart from some of the species occasionally found at the cave entrance, as in the Remo's Cavern, decapods entering caves are sciaphilic species that find shelter during the day and get out at night to feed in the external environment [76]. Motility allows decapods to avoid the negative effects of hydrological confinement and trophic depletion, which strongly influence the abundance and distribution of the sessile fauna [13].

Figure 8. Underwater photographs of decapod species in the Grotta Marina of Bergeggi: (**a**) *Herbstia condyliata*; (**b**) *Stenopus spinosus*; (**c**) *Palaemon serratus*; (**d**) *Lysmata seticaudata*; (**e**) *Scyllarus arctus*; (**f**) *Dromia personata*; (**g**) *Clibanarius erythropus* (in a shell of *Cerithium vulgatum*); (**h**) *Inachus dorsettensis*; (**i**) *Scyllarides latus*.

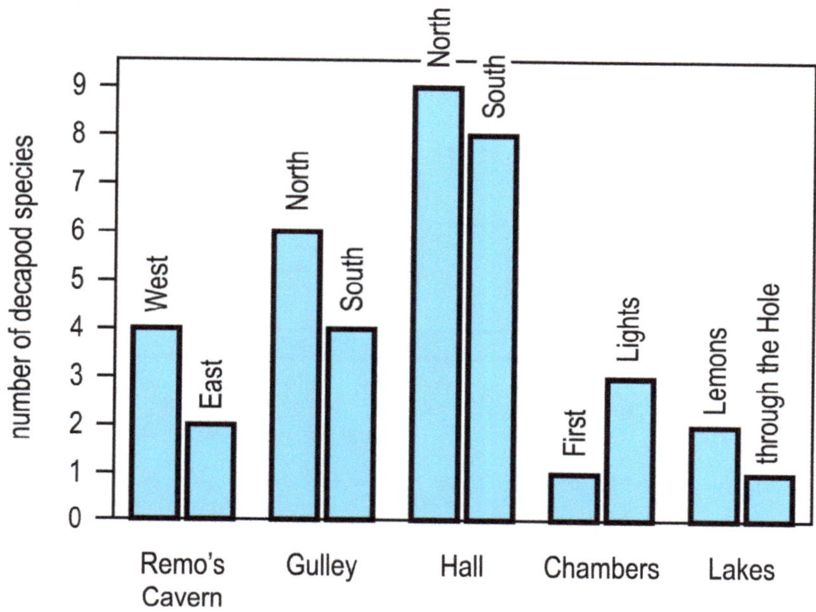

Figure 9. Number of decapod species in the different locales and stations of the Grotta Marina of Bergeggi.

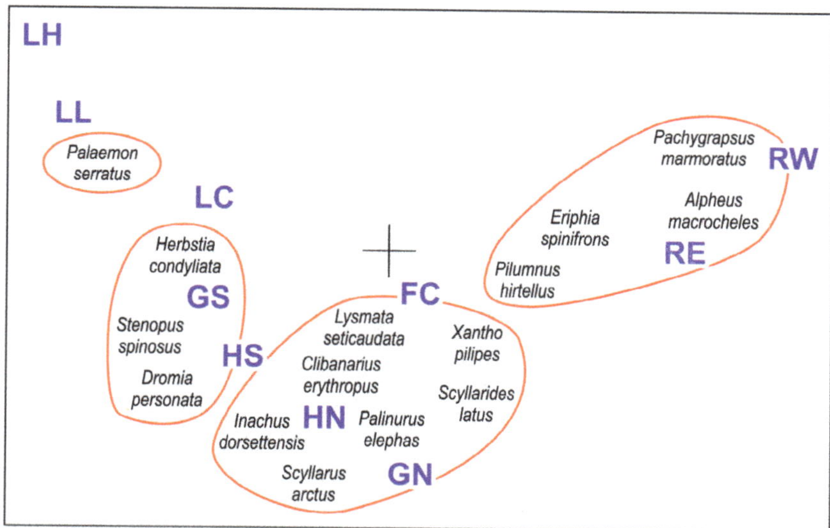

Figure 10. Bivariate plot on the plane of the first two axes from correspondence analysis of decapod species and stations in the Grotta Marina of Bergeggi. First (horizontal) axis explains 46.1% of the total variance; second (vertical) axis explains 25.4% of the total variance. Only first axis is significant ($p < 0.05$). Species are grouped according to cluster analysis (Euclidean distance, minimum variance clustering). LH = Lake through the Hole; LL = Lemons' Lake; LC = Lights' Chamber; GS = Gulley South; HS = Hall South; HN = Hall North; GN = Gulley North; FC = First Chamber; RE = Remo's Cavern East; RW = Remo's Cavern West.

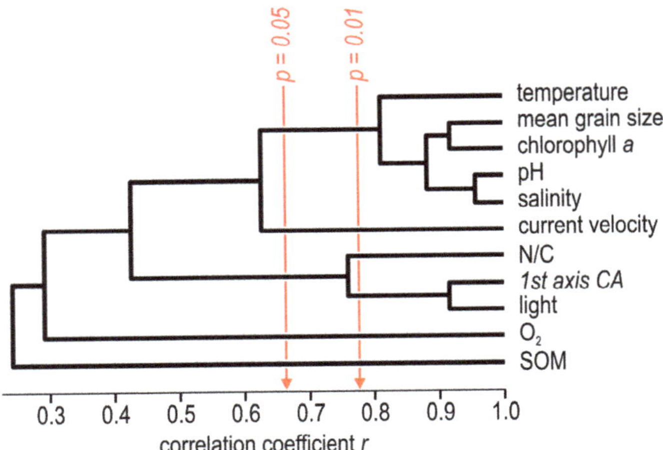

Figure 11. Dendrogram (minimum variance clustering) of the correlation coefficient *r* between the scores of the first axis from correspondence analysis, applied to the matrix of 15 decapod species × 10 stations in the Grotta Marina di Bergeggi, and a set of environmental parameters measured within the cave in the same stations where decapod species were recorded. *p* is the significance level.

7. Final Remarks

Thanks to the exploitation of a large dataset originally developed by Gerovasileiou and Voultsiadou [14], integrated with new information purposely assembled, the present review unveiled for the first time the decapod species richness of Mediterranean marine caves. The total of 76 species reported from 133 caves in 13 countries corresponds to a momentous increase with respect to previous regional or partial inventories.

The greatest number of cave decapod species has been recorded in the northernmost sectors of the Mediterranean, such as the Gulf of Lion and the Ligurian Sea, which is only in part due to the higher number of caves explored [14]. A general decrease in species richness from the north-western to south-eastern sectors is a common pattern for the whole Mediterranean Sea biota [55]. The proportion of endemic species in marine caves is comparatively low, suggesting that the role of refuge that marine caves exert for the survival of relict species known for other taxa [6,49] is not confirmed in the case of decapods. Consistently, marine caves, although less receptive than other habitats, are nonetheless subjected to the colonization by alien decapods, as anticipated for other faunal groups [70]; the presumed Lessepsian species *Urocaridella pulchella*, first reported from a marine cave in Turkey [96], has rapidly spread to marine caves of Cyprus [97] and Greece [98].

Most decapod species were found only in a few marine caves and can, therefore, be considered as 'accidental', having entered caves by chance. Many of them are simply sciaphilic species widespread in infra- or circalittoral habitats, which enter marine caves in search of shelter. The case of the Grotta Marina di Bergeggi demonstrated that decapod occurrence in marine caves is mainly correlated to the decrease in light intensity and not to other factors (hydrological confinement, trophic depletion) that characterize the marine cave environment [13].

The fact that the bulk of the decapod fauna in marine caves is made by accidental species, drawn by the regional species pool, implies that the chorological spectra of decapods found in marine caves mostly reflect that of the geographical sectors where the caves are located. Warm-water species are more represented in the south-eastern sectors, and so are non-indigenous decapods, which typically exhibit tropical affinities. Mediterranean endemics are more represented in the western sectors (Balearic and Sardinia seas, Gulf of Lion), conforming to the general pattern for the Mediterranean fauna [36].

Only 15 decapod species were comparatively frequent (more than 5% of the caves), but none of them seem exclusive to caves and have, therefore, to be defined as 'stygoxenes' (occur in caves but do not complete their life cycle within caves) or at most 'stygophiles' (can complete their life cycle within caves but also thrive in suitable habitats outside caves). Unfortunately, little is known about the population biology of cave decapods. Seasonal observations through one year allowed Denitto et al. [29] to detect large swarms of juveniles of *Palaemon* sp. (identified as *P. elegans* by the authors; however, considering the cave zones where those shrimps were collected, we do believe that, in reality, they belonged to *P. serratus*, whose juveniles have a shorter rostrum and can be misidentified with *P. elegans*) in spring and especially in summer in the innermost and dark portion of the Grotta di Ciolo (SE Italy), which suggests recruitment within the cave. In the Grotta Marina di Bergeggi (NW Italy), juveniles of *Herbstia condyliata* were observed in October (CNB, unpublished observations). All decapods produce planktonic larvae, but whether species living in caves release their larvae within or outside the cave is not known; however, juveniles have been seen in caves, and zoeae have been collected in cave plankton [99].

Other putative stygophiles include *Stenopus spinosus* and *Plesionika narval*, which combined frequent occurrence in caves and preference for the dark zone. The concept of 'secondary stygobiosis' has been conceived for those crevicular (cryptobiotic) or bathyphilic species that only secondarily colonize caves and become abundant there [100]. While commonly reported from marine caves, *S. spinosus* can be found also in rock crevices and biogenic reef anfractuosities and inside shipwrecks [101]. *P. narval* is actively fished with bottom trawls or pots in epibathyal grounds 200 to 400 m deep [75] but its swarms can be easily observed in marine caves 20 to 50 m deep [102]: consistently, the analysis of a large Mediterranean dataset for the present review highlighted that it prefers deep caves.

Morphological and behavioural adaptations to cave life typical of stygobiotic species can be observed in caridean and stenopodid shrimps as well as in galatheid and brachyuran crabs found in marine caves [103]. Two species have not been—at least yet—reported from habitats other than marine caves and thus could perhaps be considered as 'stygobionts' (i.e., cave-exclusive): the stenopodid *Odontozona addaia*, from the Balearic Sea and the Gulf of Lion [34], and the alpheid *Salmoneus sketi*, from the Adriatic Sea [33]. Future research in cryptic and deep-sea habitats might substantiate or refute their presumed stygobiosis.

The existing knowledge on Mediterranean marine cave decapods is far from being complete: only a small number of existing caves have been explored and, mostly, in a superficial and incomplete manner, while many caves (especially the deepest ones) are still unknown. Future research should focus on filling regional gaps (e.g., south-eastern Mediterranean, Alboran Sea) and is expected to lead to an increment in the number of species known and to help answer many fundamental questions in cave biology and ecology [104]. An aspect of cave decapod biology that deserves further investigation is their role in cave ecosystem functioning. As other motile cave species do, decapods likely exit the caves to feed in nearby external habitats at night and return to rest in caves during daytime. During their stay in the cave, they release faecal pellets, which increase the internal trophic load [13]. This way, they import organic matter, thus mitigating trophic depletion in caves [105]. Such a pivotal role has been thoroughly investigated in the swarm-forming mysid *Hemimysis speluncola* [106] and in the schooling cardinal fish *Apogon imberbis* [107] but not in decapods. In addition, large decapod crustaceans, such as *Dromia personata*, provide an opportunity for the transport (phoresy) of sessile filter-feeding organisms, which settle as epibionts on their carapace [86]. These epibionts are allowed feeding outside at night and represent a source of larvae capable of maintaining pseudo-populations of sessile stygoxenic taxa in caves, thus contributing to their biodiversity [13].

Quantitative data on cave decapod populations in different periods of the year are virtually lacking [29] but are badly needed to tackle population biology topics, which are of major importance to understand behavioural adaptations and for conservation purposes. Mediterranean marine caves are a priority habitat, protected by the Habitats Directive of the European Union and by the Mediterranean Action Plan of the United Nations Environment

Programme, but nonetheless threatened by many anthropogenic pressures [18,87,108,109]. The lists of Mediterranean marine invertebrates that are protected, according to Appendix III of the Bern Convention (https://www.coe.int/en/web/bern-convention, accessed on 21 February 2022), or whose exploitation is regulated, according to Annex III of the Barcelona Convention (https://ec.europa.eu/environment/marine/international-cooperation/regional-sea-conventions/barcelona-convention/index_en.htm, accessed on 21 February 2022), include six species of decapods that occur inside marine caves: *Homarus gammarus*, *Maja squinado*, *Palinurus elephas*, *Scyllarides latus*, *Scyllarus arctus* and *S. pigmaeus*. For *H. gammarus* and all Palinuridae (hence including *P. elephas*), European Council Regulation No 1967/2006 defines the minimum catchable size, prohibits capturing and selling berried females and imposes that the specimens caught accidentally be promptly released back to the sea. Basic ecological interest and the need for conservation initiatives combine to claim for intensifying the investigations on the decapod fauna of the Mediterranean Sea caves.

Author Contributions: Conceptualization, C.N.B. and C.M.; methodology, C.N.B., C.F., V.G. and C.M.; software, V.G.; validation, C.F.; formal analysis, C.N.B. and C.M.; investigation, C.N.B., V.G. and C.M.; resources, C.N.B., C.F., V.G. and C.M.; data curation, C.N.B. and V.G.; writing—original draft preparation, C.N.B., C.F., V.G. and C.M.; writing—review and editing, C.N.B., C.F., V.G. and C.M.; visualization, C.N.B. and C.M.; supervision, C.F.; project administration, C.M.; funding acquisition, C.N.B. and C.M. All authors have read and agreed to the published version of the manuscript.

Funding: This review received no external funding.

Institutional Review Board Statement: Not applicable.

Data Availability Statement: Data are available from the authors upon request: V.G. for the Mediterranean Sea marine caves database, C.N.B. for the Bergeggi cave dataset.

Acknowledgments: We thank Valerio Zupo (Ischia, Italy) for stimulating an earlier version of our review, and Bella Galil (Haifa, Israel) for advice on *Urocaridella pulchella*. Giovanni Diviacco (Genoa, Italy) led the earliest marine biology studies on the Grotta Marina di Bergeggi. C.N.B. and C.M. wish to dedicate this paper to their unforgettable colleague and friend Riccardo Cattaneo-Vietti (1949–2021), who coined the concept of secondary stygobiosis, here applied to crustacean decapods for the first time.

Conflicts of Interest: The authors declare no conflict of interest.

References

1. Anderson, S.C.; Mills Flemming, J.; Watson, R.; Lotze, H.K. Rapid global expansion of invertebrate fisheries: Trends, drivers, and ecosystem effects. *PLoS ONE* **2011**, *6*, e14735. [CrossRef] [PubMed]
2. De Grave, S.; Pentcheff, N.D.; Ahyong, S.T.; Chan, T.Y.; Crandall, K.A.; Dworschak, P.C.; Felder, D.L.; Feldmann, R.M.; Fransen, C.H.J.M.; Goulding, L.Y.D.; et al. A classification of living and fossil genera of decapod crustaceans. *Raffles Bull. Zool. Suppl.* **2009**, *21*, 1–109.
3. Abele, L.G. Species diversity of decapod crustaceans in marine habitats. *Ecology* **1974**, *55*, 156–161. [CrossRef]
4. Noël, P.; Monod, T.; Laubier, L. Crustacea in the biosphere. In *Treatise on Zoology—Anatomy, Taxonomy, Biology. The Crustacea*; von Vaupel Klein, J.C., Charmantier-Daures, M., Schram, F.R., Eds.; E.J. Brill: Leiden, The Netherlands, 2014; Volume 4, part B, revised, updated edition; pp. 3–115.
5. Martin, J.W.; Haney, T.A. Decapod crustaceans from hydrothermal vents and cold seeps: A review through 2005. *Zool. J. Linn. Soc.* **2005**, *145*, 445–522. [CrossRef]
6. Harmelin, J.G.; Vacelet, J.; Vasseur, P. Les grottes sous-marines obscures: Un milieu extrême et un remarquable biotope refuge. *Téthys* **1985**, *11*, 214–229.
7. Riedl, R. *Biologie der Meereshöhlen*; Paul Parey: Hamburg, Germany, 1966; pp. 1–636.
8. Holthuis, L.B. Decapoda. In *Stygofauna Mundi: A Faunistic, Distributional, and Ecological Synthesis of the World Fauna Inhabiting Subterranean Waters (Including the Marine Interstitial)*; Botosaneanu, L., Ed.; E.J. Brill: Leiden, The Netherlands, 1986; pp. 589–615.
9. Anker, A. A worldwide review of stygobiotic and stygophilic shrimps of the family Alpheidae (Crustacea, Decapoda, Caridea). *Subterr. Biol.* **2008**, *6*, 1–16.
10. Giakoumi, S.; Sini, M.; Gerovasileiou, V.; Mazor, T.; Beher, J.; Possingham, H.P.; Abdulla, A.; Çinar, M.E.; Dendrinos, P.; Gucu, A.C.; et al. Ecoregion-based conservation planning in the Mediterranean: Dealing with large-scale heterogeneity. *PLoS ONE* **2013**, *8*, e76449. [CrossRef]

11. Bianchi, C.N.; Cattaneo-Vietti, R.; Cinelli, F.; Morri, C.; Pansini, M. Lo studio biologico delle grotte sottomarine: Conoscenze attuali e prospettive. *Boll. Mus. Ist. Biol. Univ. Genova* **1996**, *60*, 41–69.

12. Gerovasileiou, V.; Chintiroglou, C.; Vafidis, D.; Koutsoubas, D.; Sini, M.; Dailianis, T.; Issaris, Y.; Akritopoulou, E.; Dimar-chopoulou, D.; Voultsiadou, E. Census of biodiversity in marine caves of the Eastern Mediterranean Sea. *Medit. Mar. Sci.* **2015**, *16*, 245–265. [CrossRef]

13. Gerovasileiou, V.; Bianchi, C.N. Mediterranean marine caves: A synthesis of current knowledge. *Oceanogr. Mar. Biol. Ann. Rev.* **2021**, *59*, 1–88.

14. Gerovasileiou, V.; Voultsiadou, E. Mediterranean marine caves as biodiversity reservoirs: A preliminary overview. In *Proceedings of the 1st Mediterranean Symposium on the Conservation of Dark Habitats*; Langar, H., Bouafif, C., Ouerghi, A., Eds.; UNEP/MAP–RAC/SPA: Tunis, Tunisia, 2014; pp. 45–50.

15. Gerovasileiou, V.; Martínez, A.; Álvarez, F.; Boxshall, G.; Humphreys, W.F.; Jaume, D.; Becking, L.E.; Muricy, G.; van Hengstum, P.J.; Dekeyzer, S.; et al. World Register of marine Cave Species (WoRCS): A new thematic species database for marine and anchialine cave biodiversity. *Res. Ideas Outcomes* **2016**, *2*, e10451. [CrossRef]

16. Bianchi, C.N.; Cevasco, M.G.; Diviacco, G.; Morri, C. Primi risultati di una ricerca ecologica sulla grotta marina di Bergeggi (Savona). *Boll. Mus. Ist. Biol. Univ. Genova* **1986**, *52*, 267–293.

17. Parravicini, V.; Guidetti, P.; Morri, C.; Montefalcone, M.; Donato, M.; Bianchi, C.N. Consequences of sea water temperature anomalies on a Mediterranean submarine cave ecosystem. *Estuar. Coast. Shelf Sci.* **2010**, *86*, 276–282. [CrossRef]

18. Montefalcone, M.; De Falco, G.; Nepote, E.; Canessa, M.; Bertolino, M.; Bavestrello, G.; Morri, C.; Bianchi, C.N. Thirty year ecosystem trajectories in a submerged marine cave under changing pressure regime. *Mar. Environ. Res.* **2018**, *137*, 98–110. [CrossRef] [PubMed]

19. Sgorbini, S.; Bianchi, C.N.; Degl'Innocenti, F.; Diviacco, G.; Forti, S.; Morri, C.; Niccolai, I. Méthodologie d'une étude hydrobi-ologique dans la grotte marine de Bergeggi (mer Ligure). *Rapp. Comm. Int. Mer Médit.* **1988**, *31*, 119.

20. Morri, C.; Bianchi, C.N.; Degl'Innocenti, F.; Diviacco, G.; Forti, S.; Maccarone, M.; Niccolai, I.; Sgorbini, S.; Tucci, S. Gradienti fisico-chimici e ricoprimento biologico nella Grotta Marina di Bergeggi (Mar Ligure). *Mem. Ist. Ital. Speleol. Ser. II* **1994**, *6*, 85–94.

21. Stock, J.H.; Iliffe, T.M.; Williams, D. The concept 'anchialine' reconsidered. *Stygologia* **1986**, *2*, 90–92.

22. Bianchi, C.N. Flora e fauna: Lineamenti generali e prospettive. In *Grotte Marine: Cinquant'anni di Ricerca in Italia*; Cicogna, F., Bianchi, C.N., Ferrari, G., Forti, P., Eds.; Ministero dell'Ambiente e della Tutela del Territorio: Roma, Italy, 2003; pp. 137–146.

23. Gràcia, F.; Jaume, D.; Ramis, D.; Fornós, J.J.; Bover, P.; Clamor, B.; Gual, M.À.; Vadell, M. Les coves de Cala Anguila (Manacor, Mal-lorca). II: La Cova Genovesa o Cova d'en Bessó. Espeleogènesi, geomorfologia, hidrologia, sedimentologia, fauna, paleontologia, arqueologia i concervació. *Endins Pub. Espeleol.* **2003**, *25*, 43–86.

24. Parenzan, P. *Animalia Speluncarum Italiae*; Congedo: Galatina, Italy, 1989; (published posthumous in 2002); pp. 1–224.

25. Gili, J.M.; Macpherson, E. Crustáceos decápodos capturados en cuevas submarinas del litoral Balear. *Investig. Pesq.* **1987**, *51* (Suppl. 1), 285–291.

26. Pessani, D.; Manconi, R. I Crostacei Decapodi delle grotte della Penisola Sorrentina. *Mem. Ist. Ital. Speleol. Ser. II* **1994**, *6*, 199–202.

27. Pipitone, C.; Vaccaro, A. Studio dei crostacei decapodi dell'Isola di Ustica: Censimento faunistico, distribuzione e biogeografia. *NTR-IRMA* **2003**, *70*, 1–24.

28. Manconi, R.; Pessani, D. Crostacei decapodi. In *Grotte Marine: Cinquant'anni di Ricerca in Italia*; Cicogna, F., Bianchi, C.N., Ferrari, G., Forti, P., Eds.; Ministero dell'Ambiente e della Tutela del Territorio: Roma, Italy, 2003; pp. 187–193.

29. Denitto, F.; Moscatello, S.; Belmonte, G. Occurrence and distribution pattern of *Palaemon* spp. shrimps in a shallow submarine cave environment: A study case in South-eastern Italy. *Mar. Ecol.* **2009**, *30*, 416–424. [CrossRef]

30. Denitto, F.; Pastore, M.; Belmonte, G. Occurrence of the Guinean species *Herbstia nitida* Manning & Holthuis, 1981 (Decapoda, Brachyura) in a Mediterranean submarine cave and a comparison with the congeneric *H. condyliata* (Fabricius, 1787). *Crustaceana* **2010**, *83*, 1017–1024.

31. Mačić, V.; Panou, A.; Bundone, L.; Varda, D.; Pavićević, M. First inventory of the semi-submerged marine caves in South Dinarides Karst (Adriatic coast) and preliminary list of species. *Turk. J. Fish. Aquat. Sci.* **2019**, *19*, 765–774. [CrossRef]

32. Pretus, J.L. Description of *Odontozona addaia* spec. nov. (Crustacea: Decapoda: Stenopodidae) from a marine cave in the island of Minorca, western Mediterranean. *Zool. Meded.* **1990**, *63*, 343–357.

33. Fransen, C.H.J.M. *Salmoneus sketi*, a new species of alpheid shrimp (Crustacea: Decapoda: Caridea) from a submarine cave in the Adriatic. *Zool. Meded.* **1991**, *65*, 171–179.

34. Chevaldonné, P.; Pretus, J.L. Rediscovery of the rare Mediterranean marine cave stenopodid shrimp *Odontozona addaia* Pretus, 1990, 30 years after its original description (Crustacea: Decapoda: Stenopodidea). *Zootaxa* **2021**, *4950*, 137–148. [CrossRef]

35. Rastorgueff, P.A.; Chevaldonné, P.; Arslan, D.; Verna, C.; Lejeusne, C. Cryptic habitats and cryptic diversity: Unexpected patterns of connectivity and phylogeographical breaks in a Mediterranean endemic marine cave mysid. *Mol. Ecol.* **2014**, *23*, 2825–2843. [CrossRef]

36. Bianchi, C.N.; Morri, C.; Chiantore, M.; Montefalcone, M.; Parravicini, V.; Rovere, A. Mediterranean Sea biodiversity between the legacy from the past and a future of change. In *Life in the Mediterranean Sea: A Look at Habitat Changes*; Stambler, N., Ed.; Nova Science: New York, NY, USA, 2012; pp. 1–55.

37. Dengler, J.; Oldeland, J. Effects of sampling protocol on the shapes of species richness curves. *J. Biogeogr.* **2010**, *37*, 1698–1705. [CrossRef]

38. Bianchi, C.N.; Azzola, A.; Cocito, S.; Morri, C.; Oprandi, A.; Peirano, A.; Sgorbini, S.; Montefalcone, M. Biodiversity monitoring in Mediterranean marine protected areas: Scientific and methodological challenges. *Diversity* **2022**, *14*, 43. [CrossRef]

39. Bianchi, C.N.; Dando, P.R.; Morri, C. Increased biodiversity of sessile epibenthos at subtidal hydrothermal vents: Seven hypotheses based on observations at Milos Island, Aegean Sea. *Adv. Oceanogr. Limnol.* **2011**, *2*, 1–31. [CrossRef]

40. Williamson, M. Relationship of species number to area, distance and other variables. In *Analytical Biogeography. An Integrated Approach to the Study of Animal and Plant Distributions*; Myers, A.A., Giller, P.S., Eds.; Chapman and Hall: London, UK, 1990; pp. 92–115.

41. Bianchi, C.N.; Morri, C. Marine biodiversity of the Mediterranean Sea: Situation, problems and prospects for future research. *Mar. Pollut. Bull.* **2000**, *40*, 367–376. [CrossRef]

42. Tortonese, E. How is to be interpreted a 'Mediterranean' species? *Thalassographica* **1978**, *2*, 9–17.

43. Tortonese, E. Distribution and ecology of endemic elements in the Mediterranean fauna (fishes and echinoderms). In *Mediterranean Marine Ecosystems*; Moraitou-Apostolopoulou, M., Kiortsis, V., Eds.; NATO Conference Series 8; Plenum Press: New York, NY, USA, 1985; pp. 57–83.

44. Pérès, J.M.; Picard, J. Nouveau manuel de bionomie benthique de la mer Méditerranée. *Rec. Trav. Stat. Mar. Endoume* **1964**, *31*, 1–137.

45. Fredj, G.; Bellan-Santini, D.; Menardi, M. État des connaissances sur la faune marine méditerranéenne. *Bull. Inst. Océanogr. Monaco* **1992**, *no. spécial 9*, 133–145.

46. Harmelin, J.G.; d'Hont, J.L. Transfers of bryozoan species between the Atlantic Ocean and the Mediterranean Sea via the Strait of Gibraltar. *Oceanol. Acta* **1993**, *16*, 63–72.

47. Zenetos, A.; Gofas, S.; Verlaque, M.; Çinar, M.E.; García Raso, J.E.; Bianchi, C.N.; Morri, C.; Azzurro, E.; Bilecenoglu, M.; Froglia, C.; et al. Alien species in the Mediterranean Sea by 2010. A contribution to the application of European Union's Marine Strategy Framework Directive (MSFD). Part 1. Spatial distribution. *Medit. Mar. Sci.* **2010**, *11*, 381–493. [CrossRef]

48. Galil, B.S. Lessepsian migration: New findings on the foremost anthropogenic change in the Levant basin fauna. In *Symposium Mediterranean Seas 2000*; Della Croce, N.F.R., Ed.; University of Genova, Istituto di Scienze Ambientali Marine: Santa Margherita Ligure, Italy, 1993; pp. 307–323.

49. Logan, A.; Bianchi, C.N.; Morri, C.; Zibrowius, H. The present-day Mediterranean brachiopod fauna: Diversity, life habits, biogeography and paleobiogeography. *Sci. Mar.* **2004**, *68* (Suppl. 1), 163–170. [CrossRef]

50. D'Udekem d'Acoz, C. *Inventaire et Distribution des Crustacés Décapodes de l'Atlantique Nord-Oriental, de la Méditerranée et des Eaux Continentales Adjacentes au Nord de 25° N*; Patrimoines Naturels 40; Muséum National d'Histoire Naturelle: Paris, France, 1999; pp. 1–383.

51. Bakir, K.; Katagan, T.; Aker, H.V.; Ozcan, T.; Sezgin, M.; Ates, A.S.; Koçak, C.; Kirmim, F. The marine arthropods of Turkey. *Turk. J. Zool.* **2014**, *38*, 765–831. [CrossRef]

52. Marco-Herrero, E.; Abello, P.; Drake, P.; García-Raso, J.E.; Gonzalez-Gordillo, J.I.; Guerao, G.; Palero, F.; Cuesta, J.A. Annotated checklist of brachyuran crabs (Crustacea: Decapoda) of the Iberian Peninsula (SW Europe). *Sci. Mar.* **2015**, *79*, 243–256. [CrossRef]

53. García Raso, J.E.; Cuesta, J.A.; Abello, P.; Macpherson, E. Updating changes in the Iberian decapod crustacean fauna (excluding crabs) after 50 years. *Sci. Mar.* **2018**, *82*, 207–229. [CrossRef]

54. Gonzalez, J.A. Checklists of Crustacea Decapoda from the Canary and Cape Verde Islands, with an assessment of Macaronesian and Cape Verde biogeographic marine ecoregions. *Zootaxa* **2018**, *4412*, 401–448. [CrossRef] [PubMed]

55. Coll, M.; Piroddi, C.; Steenbeek, J.; Kaschner, K.; Ben Rais Lasram, F.; Aguzzi, J.; Ballesteros, E.; Bianchi, C.N.; Corbera, J.; Dailianis, T.; et al. The biodiversity of the Mediterranean Sea: Estimates, patterns, and threats. *PLoS ONE* **2010**, *5*, e11842. [CrossRef] [PubMed]

56. Terossi, M.; De Grave, S.; Mantelatto, F.L. Global biogeography, cryptic species and systematic issues in the shrimp genus *Hippolyte* Leach, 1814 (Decapoda: Caridea: Hippolytidae) by multimarker analyses. *Sci. Rep.* **2017**, *7*, e6697. [CrossRef] [PubMed]

57. Ledoyer, M. Écologie de la faune vagile des biotopes méditerranéens accessibles en scaphandre autonome (région de Marseille principalement). IV.—Synthèse de l'étude écologique. *Rec. Trav. Stat. Mar. Endoume* **1968**, *44*, 125–295.

58. García, L.; Gràcia, F. Sobre algunes espècies de crustacis decàpodes interessants de les illes Balears (Crustacea: Decapoda). *Boll. Soc. Hist. Nat. Balears* **1996**, *39*, 177–186.

59. Tortonese, E. La fauna del Mediterraneo e i suoi rapporti con quella dei mari vicini. *Pubbl. Stn. Zool. Napoli* **1969**, *37*, 369–384.

60. Briggs, J.C. *Marine Zoogeography*; McGraw-Hill: New York, NY, USA, 1974; pp. 1–475.

61. Ferreira, L.A.A.; Tavares, M. *Pisidia longimana* (Risso, 1816), a junior synonym of *P. bluteli* (Risso, 1816) (Crustacea: Decapoda: Anomura: Porcellanidae) and a species distinct from *P. longicornis* (Linnaeus, 1767). *Pap. Avul. Zool.* **2020**, *60*, e20206036. [CrossRef]

62. Spalding, M.D.; Fox, H.E.; Allen, G.R.; Davidson, N.; Ferdaña, Z.A.; Finlayson, M.A.X.; Halpern, B.S.; Jorge, M.A.; Lombana, A.; Lourie, S.A.; et al. Marine ecoregions of the world: A bioregionalization of coastal and shelf areas. *BioScience* **2007**, *57*, 573–583. [CrossRef]

63. Briggs, J.C.; Bowen, B.W. A realignment of marine biogeographic provinces with particular reference to fish distributions. *J. Biogeogr.* **2012**, *39*, 12–30. [CrossRef]

64. Briggs, J.C.; Bowen, B.W. Marine shelf habitat: Biogeography and evolution. *J. Biogeogr.* **2013**, *40*, 1023–1035. [CrossRef]

65. Morri, C.; Cattaneo-Vietti, R.; Sartoni, G.; Bianchi, C.N. Shallow epibenthic communities of Ilha do Sal (Cape Verde Archipelago, Eastern Atlantic). *Arquipél. Life Mar. Sci.* **2000**, *2*, 157–165.

66. Sparrow, A.; Badalamenti, F.; Pipitone, C. Contribution to the knowledge of *Percnon gibbesi* (Decapoda, Grapsidae), an exotic species spreading rapidly in Sicilian waters. *Crustaceana* **2001**, *74*, 1009–1017. [CrossRef]

67. Zenetos, A.; Gofas, S.; Morri, C.; Rosso, A.; Violanti, D.; García Raso, J.E.; Çinar, M.E.; Almogi-Labin, A.; Ates, A.S.; Azzurro, E.; et al. Alien species in the Mediterranean Sea by 2010. A contribution to the application of European Union's Marine Strategy Framework Directive (MSFD). Part 2. Introduction trends and pathways. *Medit. Mar. Sci.* **2010**, *13*, 328–352. [CrossRef]

68. Yokes, B.; Galil, B.S. New records of alien decapods (Crustacea) from the Mediterranean coast of Turkey, with a description of a new palaemonid species. *Zoosystema* **2006**, *28*, 747–755.

69. Ďuriš, Z. Palaemonid shrimps (Crustacea: Decapoda) of Saudi Arabia from the 'Red Sea biodiversity survey' 2011–2013, with 11 new records for the Red Sea. *Mar. Biodiver.* **2017**, *47*, 1147–1161. [CrossRef]

70. Gerovasileiou, V.; Voultsiadou, E.; Issaris, Y.; Zenetos, A. Alien biodiversity in Mediterranean marine caves. *Mar. Ecol.* **2016**, *37*, 239–256. [CrossRef]

71. Klaoudatos, D.; Kapiris, K. Alien crabs in the Mediterranean Sea: Current status and perspectives. In *Crabs: Global Diversity, Behavior and Environmental Threats*; Ardovini, C., Ed.; Nova Science: New York, NY, USA, 2014; pp. 101–159.

72. Morri, C.; Puce, S.; Bianchi, C.N.; Bitar, G.; Zibrowius, H.; Bavestrello, G. Hydroids (Cnidaria: Hydrozoa) from the Levant Sea (mainly Lebanon), with emphasis on alien species. *J. Mar. Biol. Ass. UK* **2009**, *89*, 49–62. [CrossRef]

73. Bianchi, C.N.; Caroli, F.; Guidetti, P.; Morri, C. Seawater warming at the northern reach for southern species: Gulf of Genoa, NW Mediterranean. *J. Mar. Biol. Ass. UK* **2018**, *98*, 1–12. [CrossRef]

74. Morais, S.; Narciso, L.; Calado, R.; Nunes, M.L.; Rosa, R. Lipid dynamics during the embryonic development of *Plesionika martia martia* (Decapoda; Pandalidae), *Palaemon serratus* and *P. elegans* (Decapoda; Palaemonidae): Relation to metabolic consumption. *Mar. Ecol. Progr. Ser.* **2002**, *242*, 195–204. [CrossRef]

75. Thessalou-Legaki, M.; Frantzis, A.; Nassiokas, K.; Hatzinikolaou, S. Depth zonation in a *Parapandalus narval* (Crustacea, Decapoda, Pandalidae) population from Rhodos Island, Greece. *Estuar. Coast. Shelf Sci.* **1989**, *29*, 273–284. [CrossRef]

76. Zariquiey Alvarez, R. Crustáceos Decápodos Ibéricos. *Investig. Pesq.* **1968**, *32*, 1–510.

77. Bianchi, C.N. Biologia delle grotte marine del Mediterraneo. In *Mare ed Ecologia*; Fiorentini, A., Ed.; Provincia di Genova, Unione Regionale Province Liguri and Marevivo: Genoa, Italy, 1994; pp. 35–44.

78. Bianchi, C.N.; Morri, C. Biologia ed ecologia delle grotte sottomarine. In *Speleologia Marina*; Barbieri, F., Ed.; Gribaudo: Cavallermaggiore, Italy, 1999; pp. 113–160.

79. Bianchi, C.N.; Morri, C. Studio bionomico comparativo di alcune grotte marine sommerse; definizione di una scala di confinamento. *Mem. Ist. Ital. Speleol.* **1994**, *6* (Suppl. 2), 107–123.

80. Harmelin, J.G. Les peuplements des substrats durs circalittoraux. In *Les Biocénoses Marines et Littorales de Méditérranée. Synthèse, Menaces et Perspectives*; Bellan-Santini, D., Lacaze, J.C., Poizat, C., Eds.; Patrimoines Naturels 19; Muséum National d'Histoire Naturelle: Paris, France, 1994; pp. 118–126.

81. Cinelli, F.; Fresi, E.; Mazzella, L.; Pansini, M.; Pronzato, R.; Svoboda, A. Distribution of benthic phyto- and zoocoenoses along a light gradient in a superficial marine cave. In *Biology of Benthic Organisms*; Keegan, B.F., Ceidigh, P.O., Boaden, P.J.S., Eds.; Pergamon Press: Oxford, UK, 1977; pp. 173–183.

82. Southward, A.J.; Kennicutt, M.C.; Alcalà-Herrera, J.; Abbiati, M.; Airoldi, L.; Cinelli, F.; Bianchi, C.N.; Morri, C.; Southward, E.C. On the biology of submarine caves with sulphur springs: Appraisal of ^{13}C/^{12}C ratios as a guide to trophic relations. *J. Mar. Biol. Ass. UK* **1996**, *76*, 265–285. [CrossRef]

83. Harmelin, J.G. Bryozoaires des grottes sous-marines obscures de la région marseillaise, faunistique et écologie. *Téthys* **1969**, *1*, 793–806.

84. Morri, C. Confinamento idrologico. In *Grotte Marine: Cinquant'anni di Ricerca in Italia*; Cicogna, F., Bianchi, C.N., Ferrari, G., Forti, P., Eds.; Ministero dell'Ambiente e della Tutela del Territorio: Rome, Italy, 2003; pp. 291–296.

85. Bianchi, C.N.; Abbiati, M.; Airoldi, L.; Alvisi, M.; Benedetti-Cecchi, L.; Cappelletti, A.; Cinelli, F.; Colantoni, P.; Dando, P.R.; Morri, C.; et al. Hydrology and water budget of a submarine cave with sulphur water spring: The Grotta Azzurra of Capo Palinuro (Southern Italy). *Proc. Ital. Ass. Oceanol. Limnol.* **1998**, *12*, 285–301.

86. Bianchi, C.N.; Morri, C.; Russo, G.F. Deplezione trofica. In *Grotte Marine: Cinquant'anni di Ricerca in Italia*; Cicogna, F., Bianchi, C.N., Ferrari, G., Forti, P., Eds.; Ministero dell'Ambiente e della Tutela del Territorio: Rome, Italy, 2003; pp. 297–305.

87. Rastorgueff, P.A.; Bellan-Santini, D.; Bianchi, C.N.; Bussotti, S.; Chevaldonné, P.; Guidetti, P.; Harmelin, J.G.; Montefalcone, M.; Morri, C.; Perez, T.; et al. An ecosystem-based approach to evaluate the ecological quality of Mediterranean undersea caves. *Ecol. Indic.* **2015**, *54*, 137–152. [CrossRef]

88. Fichez, R. Les pigments chlorophylliens: Indices d'oligotrophie dans les grottes sous-marines. *Compte Rendus Acad. Sci.* **1990**, *310*, 155–161.

89. Cocito, S.; Fanucci, S.; Niccolai, I.; Morri, C.; Bianchi, C.N. Relationships between trophic organization of benthic communities and organic matter content in Tyrrhenian Sea sediments. *Hydrobiologia* **1990**, *207*, 53–60. [CrossRef]

90. Rastorgueff, P.-A.; Harmelin-Vivien, M.; Richard, P.; Chevaldonné, P. Feeding strategies and resource partitioning mitigate the effects of oligotrophy for marine cave mysids. *Mar. Ecol. Progr. Ser.* **2011**, *440*, 163–176. [CrossRef]

91. Bianchi, C.N.; Pronzato, R.; Cattaneo-Vietti, R.; Benedetti-Cecchi, L.; Morri, C.; Pansini, M.; Chemello, R.; Milazzo, M.; Fraschetti, S.; Terlizzi, A.; et al. Mediterranean marine benthos: A manual of methods for its sampling and study. Hard bottoms. *Biol. Mar. Medit.* **2004**, *11*, 185–215.
92. Hirst, C.N.; Jackson, D.A. Reconstructing community relationships: The impact of sampling error, ordination approach, and gradient length. *Divers. Distrib.* **2007**, *13*, 361–371. [CrossRef]
93. Legendre, P.; Legendre, L. *Numerical Ecology*; Elsevier: Amsterdam, The Netherlands, 2012; pp. 1–1006.
94. Hammer, Ø.; Harper, D.A.T.; Ryan, P.D. PaSt: Paleontological statistics software package for education and data analysis. *Palaeontol. Electron.* **2001**, *4*, 4.
95. Lebart, L. Validité des résultats en analyse des données. *Consommation* **1977**, *1*, 41–69.
96. Öztürk, B.; Güngör, A.; Barraud, T. Marine caves biodiversity and conservation in the Turkish part of the Mediterranean Sea: Preliminary results of East Med Cave Project. In *Marine Caves of the Eastern Mediterranean Sea: Biodiversity, Threats and Conservation*; Öztürk, B., Ed.; Publication no. 53; Turkish Marine Research Foundation (TUDAV): Istanbul, Turkey, 2020; pp. 147–158.
97. Crocetta, F.; Al Mabruk, S.A.A.; Azzurro, E.; Bakiu, R.; Bariche, M.; Batjakas, I.E.; Bejaoui, T.; Ben Souissi, J.; Cauchi, J.; Corsini-Foka, M.; et al. New alien Mediterranean biodiversity records (November 2021). *Medit. Mar. Sci.* **2021**, *22*, 724–746.
98. Digenis, M.; Ragkousis, M.; Vasileiadou, K.; Gerovasileiou, V.; Katsanevakis, S. New records of the Indo-Pacific shrimp *Urocaridella pulchella* Yokes and Galil, 2006 from the Eastern Mediterranean Sea. *BioInvasions Rec.* **2021**, *10*, 295–303. [CrossRef]
99. Moscatello, S.; Belmonte, G. The plankton of a shallow submarine cave ('Grotta di Ciolo', Salento Peninsula, SE Italy). *Mar. Ecol.* **2007**, *28* (Suppl. 1), 47–59. [CrossRef]
100. Cattaneo, R.; Pastorino, M.V. Popolamenti algali e fauna bentonica nelle cavità naturali della regione litorale mediterranea. *Rass. Speleol. Ital.* **1974**, *12*, 272–281.
101. Karlovac, O. Présence du *Stenopus spinosus* Risso dans l'Adriatique. *Bil. Inst. Oceanogr. Ribar. Split* **1953**, *5*, 1–3.
102. Ferrero, L. *Parapandalus narval* (Fabricius) in una grotta dell'Isola di Giannutri—Arcipelago Toscano. *Boll. Pesca Piscic. Idrobiol. N.S.* **1969**, *23*, 163–167.
103. Iliffe, T.M.; Bishop, R.E. Adaptations to life in marine caves. In *Fisheries and Aquaculture. Encyclopedia of Life Support Systems*; Safran, P., Ed.; UNESCO and EOLSS Publishers: Oxford, UK, 2007; pp. 1–26.
104. Mammola, S.; Amorim, I.R.; Bichuette, M.E.; Borges, P.A.V.; Cheeptham, N.; Cooper, S.J.B.; Culver, D.C.; Deharveng, L.; Eme, D.; Ferreira, R.L.; et al. Fundamental research questions in subterranean biology. *Biol. Rev.* **2020**, *95*, 1855–1872. [CrossRef]
105. Ott, J.A.; Svoboda, A. Sea caves as model systems for energy flow studies in primary hard bottom communities. *Pubbl. Stn. Zool. Napoli* **1978**, *40*, 477–485.
106. Coma, R.; Carola, M.; Riera, T.; Zabala, M. Horizontal transfer of matter by a cave-dwelling mysid. *Mar. Ecol.* **1997**, *18*, 211–226. [CrossRef]
107. Bussotti, S.; Di Franco, A.; Bianchi, C.N.; Chevaldonné, P.; Egea, L.; Fanelli, E.; Lejeusne, C.; Musco, L.; Navarro-Barranco, C.; Pey, A.; et al. Fish mitigate trophic depletion in marine cave ecosystems. *Sci. Rep.* **2018**, *8*, 9193. [CrossRef]
108. Nepote, E.; Bianchi, C.N.; Morri, C.; Ferrari, M.; Montefalcone, M. Impact of a harbour construction on the benthic community of two shallow marine caves. *Mar. Pollut. Bull.* **2017**, *114*, 35–45. [CrossRef]
109. Ouerghi, A.; Gerovasileiou, V.; Bianchi, C.N. Mediterranean marine caves: A synthesis of current knowledge and the Mediterranean Action Plan for the conservation of "dark habitats". In *Marine Caves of the Eastern Mediterranean Sea: Biodiversity, Threats and Conservation*; Öztürk, B., Ed.; Publication no. 53; Turkish Marine Research Foundation (TUDAV): Istanbul, Turkey, 2019; pp. 1–13.

MDPI
St. Alban-Anlage 66
4052 Basel
Switzerland
Tel. +41 61 683 77 34
Fax +41 61 302 89 18
www.mdpi.com

Diversity Editorial Office
E-mail: diversity@mdpi.com
www.mdpi.com/journal/diversity